Chemie der Pflanzenschutz- und Schädlings-bekämpfungsmittel

Band 8

Spezielle Chemie der Herbizide · Anwendung und Wirkungsweise

Special Chemistry of Herbicides · Applications and Mechanisms

Herausgegeben von R. Wegler

Springer-Verlag Berlin Heidelberg New York 1982

Professor Dr. Richard Wegler
Auf dem Forst 2, D-5090 Leverkusen 1

CIP-Kurztitelaufnahme der Deutschen Bibliothek
Chemie der Pflanzenschutz- und Schädlingsbekämpfungsmittel/
hrsg. von R. Wegler. – Berlin; Heidelberg; New York: Springer.
NE: Wegler, Richard [Hrsg.]
Bd. 8.→Spezielle Chemie der Herbizide, Anwendung
und Wirkungsweise
*Spezielle Chemie der Herbizide, Anwendung und
Wirkungsweise* = Special chemistry of herbicides,
applications and mechanisms/hrsg. von R. Wegler. –
Berlin; Heidelberg; New York: Springer 1981.
(Chemie der Pflanzenschutz- und Schädlingsbekämpfungsmittel; Bd. 8)
ISBN-13: 978-3-642-81643-7   e-ISBN-13: 978-3-642-81642-0
DOI: 10.1007/ 978-3-642-81642-0
NE: Wegler, Richard [Hrsg.]; PT

Herstellung: Brühlsche Universitätsdruckerei, Gießen
2152/3140-543210

Vorwort

Band 8 ist den Neuentwicklungen auf dem Gebiet der Herbizide gewidmet. Die Bedeutung der Herbizide ist von 1976 bis 1980 erneut gestiegen, und weltweit beträgt der Herbizid-Verkauf wertmäßig soviel, wie der Verkaufswert der Insektizide und Fungizide zusammen! Langsam bedienen sich auch, bei intensiverem Anbau von Nahrungsmitteln und Nutzkulturen, „unterentwickelte" Länder aller Arten von Pestiziden und Herbiziden, und so ist mit einer weiteren Steigerung der Herbizid-Anwendung zu rechnen.

Von 1976 bis 1980 wurden viele neue Versuchsprodukte, aber vergleichsweise wenig neuartige Handelsprodukte bekannt, denn der Herbizid-Standard ist schon sehr hoch, sodaß es immer schwieriger wird, Herbizide mit wesentlichen Vorteilen zu finden, zu entwickeln und zum Verkauf zu bringen. Die mengenmäßig hohe Produktion wichtiger Herbizide macht es zudem schwierig, preiswertere neue Produkte herzustellen und einzuführen. Die stark angestiegenen Forschungs- und Entwicklungskosten für ein neues Pflanzenschutzmittel – etwa 100 Mill. DM – erschweren den Fortschritt außerordentlich. Wenngleich die meisten neuen Versuchs- und Handelsprodukte Substanzgruppen entstammen, deren Bedeutung schon Ende 1976 erkannt worden war, und deren erste Versuchsprodukte bereits eine Weiterentwicklung anzeigten, so wurden doch auch spektakuläre Entdeckungen gemacht, von denen es wiederum erste Versuchsprodukte gibt. Herbizide mit Aufwandmengen von etwa 20 g/ha, also mit extrem guter Wirksamkeit zeigen, daß der Weg in unerwartetes Neuland noch lange kein Ende hat.

Schon 1976 bekannte Verbindungsklassen, wie die der Diarylether-oxyalkancarbonsäuren, nicht nur von der Entdeckerfirma Hoechst AG intensiv bearbeitet, ergaben eine Fülle neuer Herbizide mit meist selektiver Gräserwirkung. Mitarbeiter der Firma Hoechst AG haben daher für diesen Band einen entsprechenden Beitrag zur Verfügung gestellt.

Das immer noch aktuelle und schwierige Problem der Wildhaferbekämpfung wird in einem Beitrag der Shell Int. Res. behandelt, eine Firma mit speziellen Erfolgen bei Flughafer-Herbiziden. Dabei werden die Schwierigkeiten deutlich, Flughafer im Getreide selektiv zu vernichten.

Die Herbizid-Fortschritte 1976 bis 1980 wurden in Ergänzung von Band 5 vom Herausgeber unter Mitarbeit eines bewährten Biologen (und Mitarbeiter der Bände 2 und 5) bearbeitet. Der Herausgeber hat auch die besonders für Forschende aus der chemischen Industrie wichtige Patentübersicht 1976 bis 1980 übernommen. Erstmals werden neben Patenten bzw. Offenlegungsschriften der Bundesrepublik Deutschland die an Bedeutung rasch zunehmenden Europa-Patente vollständig aufgeführt. In Anbetracht der wachsenden Bedeutung Japans als Erfinder, Hersteller und Verbraucher von Pflanzenschutzmitteln, sind japanische Patente mit besonderer Sorgfalt erfaßt worden. Zusammen mit der Patentergänzung aus Band 5 liegt nun eine fast lückenlose Übersicht über die wichtigsten Patente durch Anmeldung in wenigstens einem Industrieland vor. Aus den Patentergänzungen lassen sich Hinweise auf Neuentwicklungen mit zu erwartenden Versuchsprodukten ablesen, was für eine ökonomische Forschungsplanung von großer Bedeutung ist.

Bei der Zusammenstellung der Patentergänzung wurden Wiederholungen von früheren Verbindungen vermieden, es sei denn, daß (in einigen ganz wenigen Fällen) wichtige Neuerkenntnisse bekannt wurden. Sind in der Patentergänzung sog. „Safener" oder „Antidots" als Patente erfaßt, so finden sie auch einen kurzen Niederschlag im Fortschrittsbericht. „Safener" sind für einige bedeutende Herbizide bei speziellen Anwendungen (z. B. in Maiskulturen) zur Verbesserung der Selektivität, d. h. zur Vermeidung von Herbizidschäden in der zu schützenden Kultur, von steigender Bedeutung.

Mehr Raum als früher wurde den Zusammenhängen zwischen chemischer Konstitution und herbizider Wirksamkeit gewidmet. Einige japanische und US-Firmen haben auf der IUPAC-Tagung 1978 in Zürich hierzu wesentliche Beiträge veröffentlicht.

Einige wichtige Umweltschutzprobleme werden ausführlich besprochen, um dem Leser soweit wie möglich ein sachlich begründetes Urteil über aktuelle Streitfragen, z. B. Schäden durch Anwendung von 2,4,5-T, zu ermöglichen.

Für alle Versuchs- und Handelsprodukte wird die in den Chemical Abstracts verwendete chemische Bezeichnung angegeben, und zusätzlich die Brutto-Formel. Damit dürften Irrtümer hinsichtlich der chemischen Konstitution fast unmöglich sein. Zur Vereinfachung werden die für ein Produkt infrage kommenden Patente sowie erste Veröffentlichungen direkt im Anschluß an die Verbindungen genannt. Zur raschen Auffindung der Zusammenhänge älterer Wirkstoffe aus der gleichen Verbindungsklasse werden stets die Abschnittnummern dieser Verbindungsklasse aus Band 5 nebst der entsprechenden Seitenzahl angegeben, was sich auch im Inhaltsverzeichnis wiederholt.

R. Wegler

Inhaltsübersicht

Phenoxy-phenoxypropionic Acid Derivatives and Related Compounds

H. J. Nestler

Hoechst AG, Pflanzenschutzforschung-Chemie
D-6230 Frankfurt am Main 80

Contents

1. Introduction

Phenoxy-phenoxypropionic acid derivatives were detected to be a new class of grass herbicides in 1971. The selective activity of these compounds against monocotyledonous species is a surprising fact, as this group of substances may be regarded as a mere phenoxy homologous series of the predominantly broad-leaf herbicidal phenoxy fatty acids like, for example, dichlorprop, mecoprop, or fenoprop [1]. Furthermore, it is very surprising that this class of compounds was not investigated until 1971, in spite of the enormous amount of labor invested into the systematic structural variation of the auxin-like acting phenoxy-alkane-carboxylic acids since their discovery in 1941/42 [2].

Following the first report about the herbicidal properties of the phenoxy-phenoxypropionic acid derivatives in a patent application by Hoechst in 1972 [3], many companies have been starting research in this field. Nearly 140 patents and applications have been filed since, the majority of contributions originating from Hoechst, Ciba-Geigy, Ishihara Sangyo Kaisha, and Rohm and Haas[1].

It is estimated that about five to six thousand individual compounds have been synthesized and biologically tested within seven years after the first publication, providing a relatively wide field of experience and information about this scientifically and economically interesting new class of probably systemicly acting herbicides.

2. The PPP System

The general structure of the phenoxy-phenoxypropionic acid system, which will be referred to as the "PPP" system throughout this report, is given in formula *1*, to be compared with the corresponding general structure of the hormone-type broad-

---

1   Cf. Vol. 5 of this series, pp. 190–191; patent review suppl. Vols. 5 and 8, Sect. 5.7.3c

leaf herbicides 2. In the most important products of the PPP series, which will be discussed in detail in Sect. 7, X and Y are substituents like halogen, trifluoromethyl, or nitro groups, while R stands for hydrogen, an alkali metal atom, or a lower alkyl group. These compounds, especially when applied *post emergence*, are extremely potent grass herbicides, at the same time being tolerated by broadleaf plants like sugar beets, soybeans, cotton, plantation crops, and many vegetable cultures. Some PPP derivatives even show selectivity within the monocotyledonous class with a tolerance in important cereal crops. Herbicides of this type could offer solutions for urgent agricultural problems, e.g. the control of wild oat, millets, or annual blackgrass in crops like wheat or barley; these possibilities may explain the interest rapidly gained by the new compound class. In order to elucidate the relationships between herbicidal activity and chemical structure, practically all sites of the basic PPP molecule *1* have been subjected to structural variations. The results of these studies will be presented in Sect. 3 and 4 of this report.

3. Chemical Structure Variations

For a classification of the structural variations carried out, formulae *3* and *4* are introduced in which the aromatic rings of the PPP system have been denoted as ring A and B.

System *3* comprises those structures which are relatively closely related to the original PPP system, still possessing the basic structure of a 2-(4-phenoxyphenoxy)-propionic acid derivative. Compounds of this type will be discussed in Sect. 3.1, dealing with the variations of the aromatic substituents X, Y, and Z and those of the carboxylic acid functional group Q.

The compounds obtained by more incisive alterations of the basic system *1* will be discussed in Sects. 3.2 and 3.3. These variations may comprise – as described by the general formula *4* – products in which, e.g., ring A is not a benzene ring and ring B may be – as far as the two characteristic ether linkages are concerned – also an *ortho-* or *meta*-disubstituted system. The aliphatic moiety of the molecule too may considerably differ from the original propionic acid system, and instead of the oxygen bridge atoms – as symbolized by V and W – different atoms or atom groups may be introduced. It is an open question which of the derived structures still have to be reckoned among the phenoxy-phenoxyalkanecarboxylic acid derivatives. An outline will be given in Sect. 3.3. The systematic variation of all variables of systems *3* and *4* has been – besides of its value for testing structure activity correlation hypotheses – a valuable guide for synthesis programs by which great numbers of new screening compounds have been obtained.

3.1. Systems with Invariable Structure of the Basic PPP Molecule

3.1.1. Variation of the Aromatic Ring A Substituents X and Y

The variation of the substituents X and Y of the aromatic ring A in the basic PPP molecule (X, Y, m, n in *3*) proves to be one of the most effective variations with respect to the biological properties of the system.

The unsubstituted compounds (X, Y = H) do not show any herbicidal activity. The introduction of one or two equal or different electronegative substituents (X, Y, e.g., Cl, Br, CN, CF_3, NO_2; m, n = 0, 1, 2) leads to a series of highly active grass herbicides of which particularly the 4-chloro [3, 4], 2,4-dichloro [3], 4-trifluoromethyl [5, 6], and the 4-Cl,–Br,–I-2-nitro [7–9] compounds have to be mentioned. Further types of X, Y-substituted products have been claimed: 4-F [3, 10]; 4-CN and 4-NO_2 [11]; 2-CN or 4-CN together with additional ring A substituents [12]; 3-CH_3-4-Cl [3]; 2-Cl-4-Br [13]; 2-Cl, 2-Br-4-CF_3 [5, 6]; 2-Cl-4-NO_2 [7, 14]; 6-Cl-2-NO_2 and 4-CF_3-2-NO_2 [7]; 4-F_2CH–CO [15]; NO_2 combined with further substituents like lower alkyl, alkoxy, alkylthio, a carboxylic functional group, or a sulfonamide group [7]. Many different carboxylic acid functional derivatives (Sect. 3.1.3) have been prepared from these compounds with varied ring A substituents, e.g., carboxylic acids and the corresponding salts, or lower alkyl esters; even esters from unsaturated, cycloaliphatic, or heterocyclic alcohols have been described. The kind of the Q group, however, does not determine the specific biological properties which in the basic PPP series seem to be solely dependent on the ring A substitution pattern. X, Y, and Z substituents mentioned later throughout this report will mainly comprise the groups and atoms described here.

All these products are selectively active against graminaceous plants; broad-leaf cultures are practically not affected by normal range application rates. Within the series, a definite gradation is observed: compounds with a trifluoromethyl substituent generally seem to be the most potent grass herbicides. Some of them, when applied in a moderately increased dosage, even control perennial grass-weeds. No selectivity, however, does exist for the CF_3 compounds within the monocotyledonous class.

On the contrary, products with one or two halogen substituents in ring A do show this kind of selectivity; important cereal crops are among the tolerant species. The overall herbicidal potency is weaker than with the CF_3 compounds. The introduction of a nitro group further decreases biological activity. Alkyl and alkoxy ring-A substituents, though being claimed to be biologically active, are without practical significance. The introduction of a cyano substituent may yield interesting new substitution patterns; the chemical synthesis of these compounds, however, bears some problems, especially in large-scale production. By combination of these different types of substituents, it is possible to produce grass herbicides with well-defined activity and selectivity properties, matched to the solution of special weed control problems. Most of the PPP development products to be discussed in Sect. 7 belong to this group of compounds obtained by X, Y variation. For some of these substances also plant-growth regulating [12, 14, 16] and fungicidal properties [17] have been described.

3.1.2. Variation of the Ring B Substituent Z

In the basic PPP series, ring B mostly is – except for the two appertaining ether linkages – an unsubstituted benzene ring. In order to further modify the biological properties of the system, additional substituents Z have been introduced. Halogen atoms, particularly chlorine and bromine, are the most frequently added new Z-substituents, claimed in many applications [3, 7, 12, 14, 17–21]. Further substituent groups, sometimes occurring additionally to halogen, include: alkyl with 1–4 C atoms [3, 7, 14, 17, 18], trifluoromethyl [18–20], alkenyl [3, 17], or methoxy [21]. The numerical index o ranges from 1 to 4, preferably being 1 or 2.

The resulting products generally do not show higher herbicidal activity or selectivity than the corresponding substances with an unsubstituted B ring. Chemical preparation procedures for the ring B substituted compounds, however, prove to be relatively complicated so that there seems to be no advantage in developing a PPP herbicide with Z≠H. In the *meta* PPP series, however, substances with additional ring B substituents have gained some importance (3.3.3.1).

3.1.3. Variation of the Carboxylic Acid Functional Group Q

By far the most work has been carried out for the variation of the functional group Q which, in the basic PPP series, is a carboxylic acid function. The free phenoxy-phenoxypropionic acids (*3*, Q = COOH) and their inorganic and organic salts, e.g., with the cations Na, K, Ca, ammonium, alkylammonium, are generally the primary synthesis products that are used for the biological screening and, at the same time, serve as starting materials for the preparation of further functional derivatives. Among these the alkyl esters are the most important group. The alkyl esters have been systematically varied from C-1 to C-10, including straight-chain or branched systems, cycloalkyl, cyclopropyl-methyl [4], benzyl, or, generally, aralkyl with or without additional substituents in the aromatic ring. The corresponding thiol-propionic acids and their salts [16], or thiol esters have been prepared [3, 5–9, 11–13, 15, 17, 18, 21–23].

Further variations comprise C-1 to C-12 alkyl esters and alkylthio esters substituted in the alkyl moiety by various heteroatom-containing groups, e.g., hydroxy, acetoxy, aminocarbonyloxy, cyano [24], halogen, thiocyanate, alkoxy, alkylthio, alkylamino, acetylamino [25], trialkylammonium groups [6, 12, 14, 18,

22, 23], or esters from unsaturated alcohols, particularly allylic and propargylic systems [3, 4, 6–9, 12–15, 18, 22, 23] which may also carry additional substituents. Differently substituted phenyl, naphthyl, and phenylthio esters have been described [3, 5–7, 13, 18, 22, 23].

Heterocyclic hydroxy compounds have been chosen as alcohol components to prepare PPP esters and thiol esters, e.g., 2-oxo-tetrahydro-3-furanol [4], dioxolanyl-, furyl-, tetrahydrofuryl-pyridyl-, oxiranyl-methanol or -ethanol ("glycidyl") esters [22, 23, 26].

Some special derivatives like the propionyl lactates 5 [27] or the esters of PPP acids with alcohols obtained by reduction of PPP esters 6 [19, 22, 23] or bis-phenoxy-phenoxy acid esters from diols as alcohol components [28] have been reported.

$$X_m,Y_n\text{-phenyl}-O-\text{phenyl}(Z_o)-O-CH(CH_3)-C(=O)-OCH(CH_3)-C(=O)-OR \qquad 5$$

$$X_m,Y_n\text{-phenyl}-O-\text{phenyl}(Z_o)-O-CH(CH_3)-C(=O)-OCH_2-CH(CH_3)-O-\text{phenyl}(Z_o)-O-\text{phenyl}-X_m,Y_n \qquad 6$$

Finally, the large group of nitrogen-containing carboxylic acid derivatives has to be mentioned, e.g., the amides, including alkoxyalkylamides [29], cyclic amides like piperidides or morpholides, hydrazides [17], sulfonylamides [30], O- or N-hydroxylamine derivatives, oxime esters [10] together with their various N- or O-substituted homologues, as well as the anilides with numerous ring-substituted analogues [3, 5–9, 12, 14, 18, 31, 32]. In spite of having a different oxidation state, aldehydes (Q = CHO) and their derivatives [20] may be included into this section; they probably may be oxidized within the plant to reach the carboxylic acid stage.

Products without the C=O functional group, however, formally belonging to the class of carboxylic acid derivatives, include the nitriles, (Q = CN) [17, 20, 33], the thioamides and their various N-substituted derivatives, iminoethers, amidines [20, 34], or compounds in which the carboxylic C-atom has become part of a heterocyclic ring system, e.g., oxazoline or tetrazole [14, 20, 35, 36].

The majority of the patent applications cited in this section, is claiming the whole spectrum of carboxylic functional derivatives, indicating again – as mentioned in 3.1.1 – that primarily the substitution pattern of the phenoxy-phenoxy moiety is responsible for the biological properties of the compounds. In general, most of the carboxylic functional PPP derivatives discussed – suitable ring A substituents provided – have been found to be active grass herbicides. Some of the derivatives mentioned above have been reported to be even more potent than the basic PPP acids, salts, or alkyl esters, e.g., in special formulations or for special application purposes.

As a working hypothesis for the mode of action of PPP herbicides in post-emergence applications, the PPP free acids have been assumed to be the active species, the different kinds of derivatives only serving as transport forms to allow permeation through the leaf cuticula. This mechanism would explain the superior efficacy of the acid derivatives compared to the free acids in leaf application, while in soil or pre-emergence applications no differences are observed. Quantitative investigations or model studies of these effects, however, have not been published to-date.

For practical purposes, as technical products for weed control are concerned, the phenoxy-phenoxypropionic acid salts and the lower alkyl esters have gained most importance because of their synthetic availability and their favorable cost-to-benefit ratio.

3.2. Variations Including Alteration of the Basic PPP System

As indicated by general formula 7, this section will deal with structure variations of ring A and variations of the aliphatic moiety of the PPP molecule. Literally, the resulting compounds do not belong to the PPP series any more. Their general structures, however, are not too far remote from the basic system. As biological activity is concerned, particularly as to the selective herbicidal properties, a rough classification as PPP herbicides seems to be justified. As a typical structural element, the partial structure of ring B still is a *para*-substituted phenoxy system.

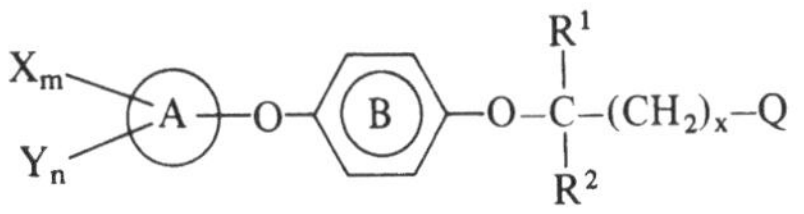

$$X_m,\ Y_n \!-\!\!\bigcirc\!\!A\!\!\!-\!O\!-\!\!\bigcirc\!\!B\!\!\!-\!O-\underset{R^2}{\overset{R^1}{C}}-(CH_2)_x-Q \qquad 7$$

The main purpose for the synthesis of these compounds – except for the preparation of new screening products – has been the aim of investigating scope and limitations of the new herbicide class.

3.2.1. Alteration of the Aromatic Ring A

The class of the PPP herbicides has been considerably extended by the introduction of different ring systems in the place of benzene ring A. Compounds of the naphthyloxy-phenoxy type 8 have been synthesized [37, 38]. They are biologically active as herbicides, particularly when derived from the β-naphthyl system. Besides their herbicidal activity against monocotyledonous weeds combined with a tolerance in certain cereal crops, some of these products are also weakly active against broad-leaf species. A modification of the herbicidal character is reached to a certain extend by introducing different substituents (X, Y) into the naphthyloxy moiety, particularly halogen atoms.

The benzene ring A has also been replaced by heterocyclic systems. The most important of these appears to be the pyridine ring, especially as a 2-pyridyloxy system 9, which may carry halogen substituents, preferably chlorine atoms in the 3- and/or 5-position [16, 21, 25, 26, 30, 34–36, 39–51].

Further types of 2-pyridyloxy-phenoxy compounds have been described, e.g., with CN groups [36], 5-bromo-6-methyl substituents [52], fluorinated 3- and/or

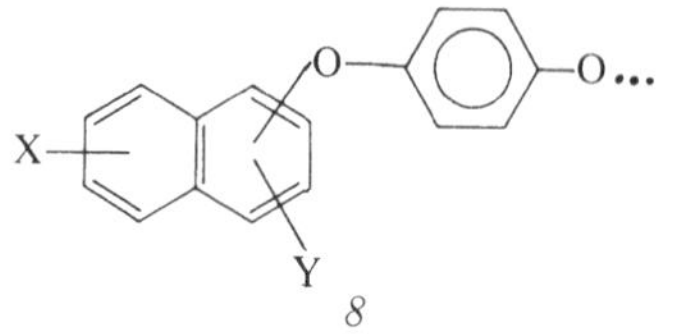

5-methyl groups like CHF_2, CF_2Cl, CF_3 [36, 53–56], or 3,5,6-trichloro products [57].

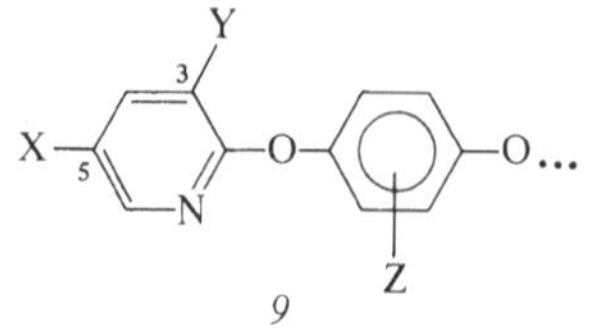

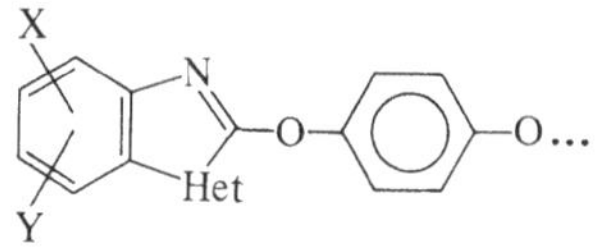

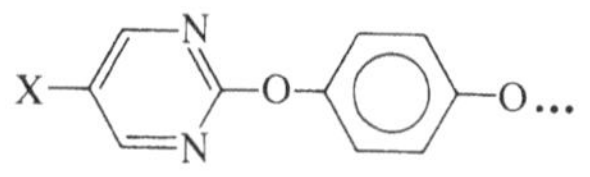

Other heteroaromatic systems reported having the general structure *10* comprise benzimidazoles (Het = NH or *N*-alkyl), benzoxazoles (Het = O), and benzothiazoles (Het = S), also including different benzene-ring substituted analogues [58]. 2-Pyrimidinyloxy compounds *11* have been described [30].

Among the heteroaromatic PPP herbicides compounds *9* are some of the most potent products (cf. Sect. 7.1). These compounds generally do not possess selective properties within the monocotyledonous species; thus, the application seems to be confined to the control of grassy weeds occurring in broad-leaf cultures, which, however, is a promising aspect, particularly regarding the very low application rates necessary with these compounds.

3.2.2. Alteration of the Propionic Acid Moiety

The structural variation of the 2-propionic acid part of the PPP molecule comprises the introduction of many different alkanecarboxylic acid systems including the preparation of corresponding alkanols. In terms of formula *7*, these changes are represented by the variation of R^1, R^2, and x together with an alteration of the functional group Q. The different carboxylic acid derivatives prepared, correspond to the changes of the functional group Q discussed in 3.1.3. It is understood that the variations of the propionic acid moiety, dealt with in this section, have been performed in combination with the various phenoxy-phenoxy systems of Sects. 3.1.1, 3.1.2, and 3.2.1.

When the 2-propionic acid moiety is replaced by the acetic acid system ($R^1 = R^2 = H$; x = 0, Q = carboxylic acid functional group) [14, 17, 18, 36, 38, 44, 45, 54, 55, 58–62], the resulting compounds lose their selective herbicidal activity against monocotyledonous plants. The herbicidal properties of the phenoxy-phenoxyacetic acid derivatives are very similar to those of commodity herbicides like 2,4-D, MCPA, 2,4,5-T, and their derivatives; so there would be no chance for a competition on an economic basis.

Alkanecarboxylic acid systems have been prepared with R^1 standing for C-1 to C-4 alkyl [3, 6, 7, 12, 14, 17, 18, 20, 54, 55], alkoxy [63], or phenyl groups [3, 17]; substituted alkyl groups have also been introduced, R^1 being aminoethyl [3, 17] or methoxymethyl [17, 39]. Products with two equal or different C-1 to C-4 substituents as R^1 and R^2 have been reported [3, 7, 12]. Synthesis and investigation of these compounds have partially been prompted by a coincident activity in the pharmaceutical field (e.g. [64, 65]). As no results of their agricultural performance have been published to-date, it is difficult to evaluate the practical significance of these substances as herbicides.

Another variation comprises the extension of the aliphatic chain of the carboxylic acid moiety by insertion of one or two additional methylene groups, i.e., the variation of x, in some cases together with a variation of R^1 from H to C_2H_5. The most important compound series of this type are those of the butyric and valeric acids (x = 2) with their different functional derivatives like esters, nitriles, or thioamides. A biochemical connection between the butyric and the acetic acid system by the β-oxidative process is supposed. R^1 again may be H or a lower alkyl group, particularly methyl [17, 38, 40, 48, 53, 58, 59, 62, 66–68]. The synthesis of 2-phenoxy-phenoxy-substituted 3-oxocarboxylic acids has also been described [47].

Recently, unsaturated systems have gained interest because of a high herbicidal potency of some of their derivatives. Derivatives have been prepared of acrylic, methacrylic, crotonic, or pentenoic acids with the phenoxy-phenoxy substituent in the 2- or 4-position [41, 69–71]. With some of these products also fungicidal activity has been observed [17]. The synthesis of phenoxy-phenoxy compounds based on dicarboxylic acids like malonic acids has been reported [42, 72].

Further variations include carbonic acid esters and amides, or even phenyl esters of acetic and sulfonic acids, in which the phenoxy-phenoxy system serves as the phenol component [17, 73]. Since these products are already far remote from the original systems 3, 4, or 7, they are generally not reckoned among the PPP compounds any more.

Alkanols where Q=OH have been obtained by the reduction of suitable derivatives of the carboxylic acids, including saturated as well as unsaturated systems. The alcohol function has been further transformed or derivatized, e.g., into halides, ethers, esters, carbamates or into the esters of phenoxy-phenoxypropionic acids [19, 43, 56, 68, 70, 74, 75]. The biological properties of the alkanols and their derivatives are similar to those of the corresponding carboxylic acids. It is assumed that – besides of a specific activity of these compounds – oxidation takes place within the biological system to form the corresponding carboxylic acids. This is the same mechanism as supposed to occur with aldehydes and their derivatives (cf. Sect. 3.1.3).

3.3. Systems Significantly Deviating from the Basic PPP Structure

Sections 3.3.1–3.3.4 will be dealing with those systems 4 that are related to the basic PPP structure, for which the designation of a *para*-phenoxy-phenoxyalkanecarboxylic acid derivate or its heterocyclic analogue, however, is no longer valid. As the general structure 4 formally also comprises systems that in no respect will belong to the PPP series even in a broader sense, an outline has to be given as to which structures should be included into a review on PPP herbicides.

Conventionally, all compounds with a phenoxy-phenoxy system or its heterocyclic analogue in combination with some sort of alkanecarboxylic acid derivative moiety – including reduction products and its derivatives – are regarded as PPP compounds, particularly when they are herbicidally active. The closely related group of phenoxy-phenol ether derivatives that have been investigated as potential insect growth regulators (IGR) will not be included in spite of a far-reaching similarity of the chemical preparation methods. The abundant material published in the IGR field, e.g., by the Stauffer Chemical Corporation, Hoffmann-La Roche or Ciba-Geigy (particularly by F. Karrer) nevertheless has initiated many preparative studies also in the PPP series.

Compounds resulting from a variation of the bridge atoms V and W in formula 4 constitute another close borderline. As will be particularly shown in the following section (3.3.1), a practicable differentiation is only to be realized with respect to the biological properties of the compounds.

3.3.1. Variation of the Bridge Atoms V and W

The oxygen-bridge atom between the aromatic rings A and B of the PPP system, $V = O$ in 4, has been replaced by the following atoms or atom groups: S [3, 20, 49], CH_2 [17, 59], OCH_2 [74, 76], CH_2S [17], CO–O– [17], SO [49]. Only with the benzyl-phenoxypropionic acid derivatives ($V = CH_2$), herbicidal activity and selectivity properties have been observed similar to those of the basic PPP products. The substances of the benzyl-phenoxy type can be used for the control of grassy weeds in many broad-leaf cultures and even in some cereal crops. Some of the other compound classes listed, while being considerably less active as herbicides, do reveal certain biological properties in different fields, e.g., as fungicides [17] or as pharmaceutics. Products with CO [64], CS [77], CH_2O, CH=N, N=CH, N=N or CO–CH=CH [78] as the bridge group V have been synthesized, but have not yet found any application as agrochemicals. The structure of these products is apparently too remote from the basic PPP system.

The same is true of compounds, in which the bridge atom W between ring B and the aliphatic part of the PPP molecule has been changed from oxygen to sulfur [3, 20] or nitrogen. The typical grass herbicidal properties of the PPP series have been observed only for the oxygen compounds; the S-products are much weaker herbicides; with $W = NR$ or NH the character of the herbicidal activity is found to be totally different. Compounds with $W = S$ are of considerable interest, however, in the *meta* series to be discussed in 3.3.3.

3.3.2. Variation of the Aromatic Ring B

Only few works have been reported on the alteration of the benzene system of ring B. The results of these variations prove to be significantly different from those of the ring A changes (3.2.1): practically all structural variations of ring B will result in products with a herbicidal activity – if any – predominantly against dicotyledonous plants, similar to that of the phenoxy-alkanecarboxylic acid derivatives 2. This activity is, however, not characteristic of the PPP system.

10

The only exception so far described, are 1,5-disubstituted naphthalene compounds of type *12* with (A) standing for the 2-pyridyl system. For these products activity against gramineous weeds has been claimed together with a surprising tolerance in rice plants [79].

12   *13*

Another variation of ring B – combined with a variation of ring A – or to be regarded as an example for a heterocyclic system, are the dibenzofuranyloxy-propionic acid derivatives *13* and their ring-substituted analogues with X, Y, and Z being predominantly halogen. With this type of compounds the limit of the structural variation of the PPP herbicides again seems to be reached. Also, these products only are selective broad-leaf herbicides, tolerant in grassy species, similar to the auxin-like acting phenoxy-alkanecarboxylic acid herbicides *2* [62].

3.3.3. *meta-Phenoxy-phenoxy-alkanecarboxylic Acid Derivatives*

Moving further away from the original *para*-PPP system, the so-called "*meta* compounds*" have to be briefly discussed. These products are formally derived from general structure *14*, in which X, Y, Z, and Q stand for the same substituents as in the basic PPP system *3*.

14   *15*

In spite of the close resemblance of the chemical structures, there are definite differences in the biological properties of the *meta*-PPP and the "normal" *para*-PPP compounds. Basically, the *meta* products are broad-leaf as well as grass herbicides with the effects against dicotyledonous plants predominating. Differences that may be observed in between single compounds of the *meta* series as far as selectivity and tolerance are concerned, are due to the substituent combinations and functional group variations. Plant growth regulating properties have been reported for some species. As a characteristic of many *meta* compounds, tolerance in rice and sometimes in maize cultures has frequently been observed.

3.3.3.1. Aromatic Ring Substituents

As an important condition for herbicidal activity – like in the *para*-PPP system – the presence of electronegative substituents in ring A, which may also be a pyridine system [80, 81], has to be mentioned. Common X and/or Y substituents are halogen, CF_3, CN, or NO_2 groups [63, 82–85].

In contrast to the *para*-PPP series, most *meta* products do have a substituent in the B ring. Substituents are preferably attached at the 2-position of ring B. Z, in many cases, is a nitro group [63, 85–90] or may also stand for halogen atoms, cyanoalkyl, and different carboxylic acid functional groups [63, 80, 85, 91–94].

3.3.3.2. Aliphatic Moieties

Besides of the propionic acid derivatives, usually prepared with Q in *14* representing many of the carboxylic acid functional groups discussed in 3.1.3, several straight-chain and branched alkanoic or alkanedicarboxylic acids (cf. 3.2.2) have been synthesized, as indicated by the general formula *15* [85].

Compounds in which the carboxylic acid functional group has been formally reduced to the alkanol (*15*, Q=OH) and with the hydroxy group transformed into ethers or esters, have also been found to be selective herbicides. Analogous compounds with halogen or amino groups in place of the alcoholic hydroxy groups have been reported to be particularly valuable because of their tolerance in paddy rice [95–104].

Special structural types within series *15* with Q=halogen, include products with a 4-alkylthio, 4-alkylsulfinyl, or 4-alkylsulfonyl substituent in ring A [105], or substances with an acetonyloxy group as the aliphatic moiety of the molecule (*15*; Q=CO–alkyl, $R^1 = R^2 = H$, x=0) [106, 107].

3.3.3.3. Bridge Atoms

The oxygen bridge V between ring A and ring B has been replaced by atom groups like $CO, OCH_2, CH_2O, S, SO, SO_2, NH, C(=CH_2), C(CH_3)OH$, and many related systems, resulting in a group of compounds of basically pharmaceutical interest, for which, however, also selective herbicidal and plant growth regulating properties have been claimed. The main field of application has been reported to be the control of gramineous weeds in cereal crops [63, 85, 108–110].

Several patents have been filed for products with a sulfur atom, a sulfinyl, or a sulfonyl group as bridge W between ring B and the aliphatic moiety in *15*. For these compounds again, the low phytotoxicity level against rice plants has been stressed [85, 111–117], making the products applicable to weed control in paddy fields. Thus, in contrast to the *para*-PPP series, the *meta* products with varied bridge atoms V and W are of considerable interest in the herbicide field.

3.3.4. ortho-Phenoxy-phenoxy-alkanecarboxylic Acid Derivatives

As the last group, the *ortho*-phenoxy-phenoxy compounds (general formula *16*) have to be mentioned. There are indications of a predominantly broad-leaf herbicidal activity and of plant-growth regulating properties with the *ortho*-PPP compounds [118].

4. Structure-Activity Correlations

4.1. Quantitative Structure-Activity Relationship (QSAR)

Basic information about qualitative structure-activity relationships has already been given in the preceding chapters dealing with the chemical structure variations and their influence on biological activity. According to our knowledge, quantitative structure-activity relationship investigations with PPP compounds have not been published to-date. QSAR treatments in the basic PPP series of compounds with different substituents in the aromatic ring A are apparently not applicable because of the substantial changes in the herbicidal spectra even with minimal changes of the substituent pattern.

With the phenoxy-phenoxypropionic acid moiety kept unchanged and only the ester part (R in *1*) of the PPP molecule systematically varied, QSAR treatments using side-chain constants according to Hansch [119] have been tried. The quantitative differences in the biological properties in these series, however, have been too weak in between the single species and not significant regarding the relatively large range of errors in greenhouse or field experiments. Thus conclusive evaluations have not been possible [120].

As a guideline for the planning of synthetic work, semi-quantitative approaches, e.g., the so-called "Topliss tree" method [121] have successfully been used [122].

4.2. Stereoisomers

In the series of the basic phenoxy-phenoxypropionic acid derivatives, two stereoisometric forms of all compounds, L-*17* and D-*17*, exist due to the presence of an asymmetric carbon atom in the 2-substituted propionic acid moiety. It has been found that – as with the auxin-type phenoxypropionic acid derivatives – the D-enantiomers are the active herbicides, with approximately twice the activity of the corresponding racemates [123]. The effect of different herbicidal properties of the D- and the L-isomers seems to be present only, however, in *post*-emergence applications. When the D-herbicides are applied before emergence, practically no

L-*17*

L-(*S*)-enantiomer

D-*17*

D-(*R*)-enantiomer

13

differences are observed between D-enantiomers and the racemates. It is assumed that racemization takes place under these conditions before the herbicide is able to reach its site of action. This would be consistent with the fact that the L-enantiomers which prove to be practically inactive in *post*-emergence applications, also do reveal some activity when applied to the soil *pre*-emergence.

The economic advantage that with the D-enantiomers only half of the amount of the corresponding racemates has to be applied in order to reach equivalent effects, has prompted several patent applications in this field. The D-isomers of the most important PPP products have been reported: phenoxy-phenoxypropionic acid derivatives with halogen, alkyl, trifluoromethyl and nitro substituents in ring A, together with many of the different functional carboxylic acid derivatives discussed in 3.1.3 [4, 10, 124] as well as those compounds in which ring A has been replaced by heterocyclic systems [124–126]. In addition to the lower application rates necessary compared with those of the corresponding racemic product, the advantages of a D-isomer application with respect to residue and toxicology questions are apparent.

5. Herbicide Combinations

PPP herbicides have been applied in mixtures with herbicidal products from different chemical classes, e.g., with thiocarbamates [112], triazines [112, 127], phenyl-cycloalkylpyrazoles [128], maleic hydrazide [129], 3,5-dihalo-4-hydroxybenzonitriles [130], or urea herbicides [131, 132]. The herbicidal spectrum can be enlarged by using these combination products, particularly broad-leaf weeds will be additionally controlled [133]. Synergistic effects that have been observed with some of the compositions mentioned above, have also been shown to be present in mixutres of PPP herbicides with synergists of the methylenedioxy-benzene type [134, 135].

Combined applications of PPP products and the closely related broad-leaf herbicidal phenoxy-substituted fatty acid herbicides have not been found to be practicable. In many reports, a definite antagonism has been demonstrated to exist between these two groups of herbicides. With the auxin-like phenoxy fatty acids present, the herbicidal activity of the PPP compounds is significantly decreased. By separate applications, however, within a few days interval, full potency of both groups will be maintained.

Selectivity of phenoxy-phenoxypropionic acid derivatives in the control of grassy weeds even in maize or rice crops, has been claimed for mixtures of PPP herbicides with the so-called antidote compound N,N-diallyl-dichloroacetamide and related products [136].

6. Chemical Synthesis

Most of the references cited in the preceding chapters contain information about the chemical procedures used in the synthesis of the PPP compounds. In this section therefore, these references will not be repeated; only those concerning the preparation of essential intermediates and starting products will be given.

In *Scheme 1*, the two main strategies are shown that have been applied to the synthesis of phenoxy-phenoxypropionic acid derivatives and analoguous compounds *1*:

Scheme 1

I. $\underset{Y}{\overset{X}{\diagup}}\!\!\diagdown$—L + HO—◯—O—CH(CH$_3$)—CO$_2$R ⟶

18   *20*

$\left[\text{or} \quad \underset{Y}{\overset{X}{\diagup}}\!\!\diagdown_{N}\text{—L} \right]$

19

$\underset{Y}{\overset{X}{\diagup}}\!\!\diagdown$—O—◯—O—CH(CH$_3$)—CO$_2$R

1 [or analoguous pyridyloxy–phenoxy compound]

II. $\underset{Y}{\overset{X}{\diagup}}\!\!\diagdown$—O—◯—OH + L—CH(CH$_3$)—CO$_2$R ⟶

21   *22*

6.1. Pathway I

This pathway will be preferred if a starting compound *18* with an appropriate leaving group L, e.g. halogen, is available. Examples are aromatic halogen compounds with additional electronegative substituents like nitro, cyano, or trifluoromethyl groups in a position activating the leaving group L. This synthetic route has become particularly important for the preparation of heteroaromatic PPP compounds, in which the aromatic ring A has been replaced by a heterocyclic system. In the synthesis of such products for example, a 2-chloropyridine (*19*) is used instead of *18* as the starting material. Pyridyloxy-phenoxypropionic acid herbicides with additional substituents in the pyridine ring like halogen, trichloromethyl, difluoromethyl, or trifluoromethyl groups are among the most potent herbicides. Halogenation reactions yielding the necessary starting compounds *19* for these products have been described [54, 55, 137–139]. Products of the *meta*-PPP series have been prepared starting from a corresponding diphenyl ether intermediate *23*.

$\underset{Y}{\overset{X}{\diagup}}\!\!\diagdown$—O—◯(Cl)—NO$_2$

23

15

Following pathway I, the PPP molecule *1* then is obtained from *18* or *19* by reaction with *20* using conventional methods, e.g., by reaction in a polar organic solvent in presence of an inorganic or organic base. The availability of the monoalkylated hydroquinone *20* is still an unsolved problem, especially in large-scale production. In order to prepare *20* in high chemical purity, the synthesis of some sort of protected intermediate *24* is necessary, (P) standing for an appropriate protecting group, e.g., benzyl.

Compound *24* is reacted with a propionic acid derivative *22* to form the intermediate *25* which finally is subjected to a selective cleavage reaction in order to obtain *20*. The direct synthesis of *20* from hydroquinone and 2-chloropropionic acid esters by selective reaction of one of the hydroxy groups has been described [140]; yields and chemical purities, however, are not satisfactory. The same problems are faced with in the preparation of the corresponding resorcinol, catechol and dihydroxynaphthalene monoethers.

6.2. Pathway II

This pathway (see *Scheme 1*) is the more convenient preparation method, if type *21* phenolic intermediates are available. The final product *1* is obtained by the reaction of *21* with *22* in a polar organic solvent in the presence of an auxiliary base, or by reaction of an alkali metal salt of *21* with the propionic acid derivative *22*. Instead of *22*, also suitable derivatives of haloacetic, -acrylic, or -malonic acids can be applied in order to prepare the products discussed in Sect. 3.2.2. The reaction of the phenol component *21* with aliphatic ketones (e.g. acetone) and chloroform yields 2,2-dialkyl-(phenoxy-phenoxy)-acetic acids [65]. The 4-substituted butyric or valeric acids have been obtained by reaction of *21* with γ-butyro- or valerolactones. For the preparation of the phenoxy-phenols *21*, many conventional synthetic procedures are available. For example, compound *18* may be reacted with the protected intermediate *24* used also for the synthesis of the monoalkylated hydroquinone. If in this case, a relatively stable protecting group (P) has been introduced, e.g. methyl, rather severe reaction conditions can be applied in a condensation reaction so that even components *18* with relatively unreactive leaving groups L will be suitable. The protecting group is then removed from the resulting intermediate *26* by selective cleavage which does not affect the aromatic ether linkage.

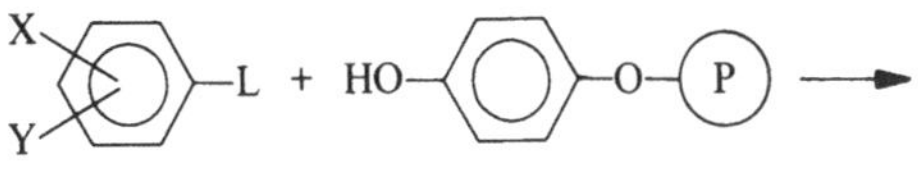

18 + *24* →

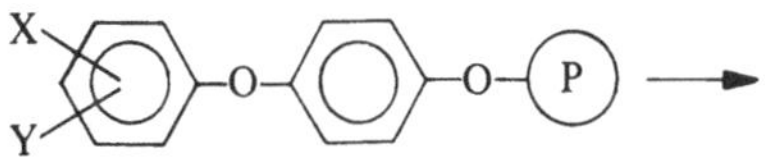

26 → *21*

Because of the rich choice of different protecting groups, together with the multifarious selective reactions for their removal, the synthesis of type *21* compounds involves no problems in the normal *para*-phenoxyphenol as well as in the corresponding *meta* series [141–144].

Hydrolysis of aromatic diazonium salts is another method of preparation of the phenoxyphenols *21* in the *para*- [145] and in the *ortho*-PPP [81] series.

In some cases, it has been possible to generate the intermediate *21* – particularly with ring A as a heterocyclic system *19* – by the reaction of stoichiometric amounts of the halo compound, e.g. 2-chloropyridine, and hydroquinone, resorcinol or catechol [45, 81, 146–150].

Another important method for the preparation of intermediates for PPP herbicides of the *meta* series, is a selective cleavage of bis-ethers *29*, available from 1,3-dihalobenzenes carrying a 4-nitro substituent (e.g. *28*) and two moles of a phenol *27*. The alkaline cleavage of *29* yields *30*, the starting product for many 3-(phenoxy)-phenoxyalkanoic acid derivatives with an additional 6-nitro substituent in the B-ring [151]. Instead of the 2,4-dichloronitrobenzene shown as an example in *28*, 3,4-dinitrochlorobenzene has been used which reacts by exchange of its 3-nitro group [152].

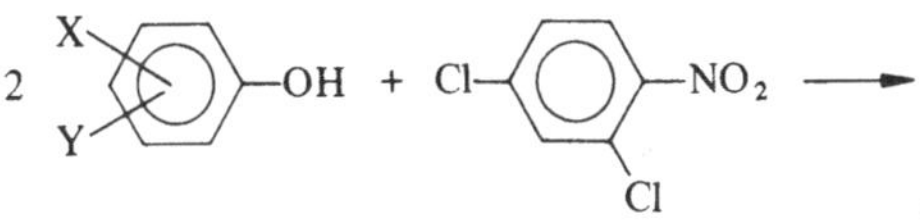

27 + *28* →

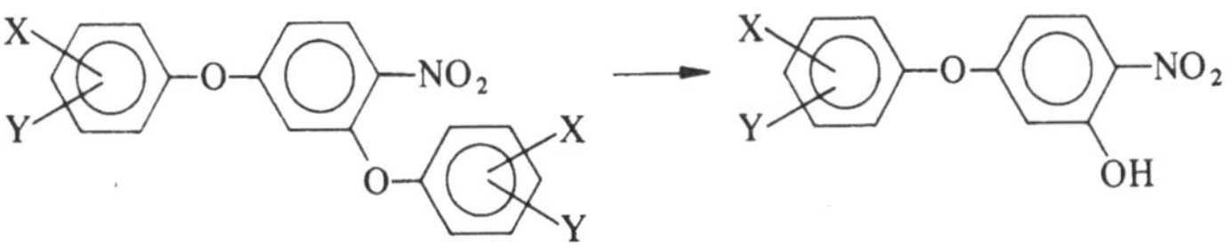

29 → *30*

If the cleavage reaction of *29* is performed in the presence of a 3-mercaptopropionic acid derivative, the corresponding 3-(6-nitro-3-phenoxyphenyl) thiopropionic acid can be directly obtained.

The alkaline cleavage of bis-aryloxybenzenes is of some importance also in the *para* and in the *ortho* series for the preparation of the phenolic starting compounds *21*. A useful variant of this reaction involves cleavage in the presence of hydroquinone or its salts; this reaction yields the phenoxy-phenols directly in good yields [153, 154]. The corresponding hetero-oxy-phenols have also been prepared by this method [150].

Several of the phenol and phenol ether intermediates have been described to possess fungicidal or herbicidal properties, particularly the 2-pyridyloxy-phenols [146] and the halogenated 3-phenoxyphenyl alkyl ethers [142, 143].

6.3. Preparation of L- and D-2-(Phenoxy-phenoxy)-propionic Acid Enantiomers

For the preparation of the enantiomeric forms of the PPP derivatives, resolution with alkaloids, e.g., cinchonine, has been applied leading to products with L-configuration [123]. Also, syntheses using chiral starting compounds have been described. The choice of the starting material determines whether D- or L-configuration of the final product will be reached. The most suitable preparation method for the optical isomers of PPP compounds is that of pathway II with *22* as a chiral component like the 2-halopropionic acid derivative L-*22* or D-*22* (the leaving group being, e.g., chlorine or bromine) or lactic acid methansulfonates or toluenesulfonates with L being $O-SO_2-CH_3$ or $O-SO_2-C_6H_4-CH_3(p)$.

$$
\begin{array}{ccc}
CO_2R & \quad & CO_2R \\
| & & | \\
L\!-\!C\!-\!H & & H\!-\!C\!-\!L \\
| & & | \\
CH_3 & & CH_3 \\
\text{L-}22 & & \text{D-}22
\end{array}
$$

6.4. Secondary Transformations and Derivatizations

The phenoxy-phenoxy-alkanecarboxylic acid system proves to be chemically very stable so that it may be subjected to a series of secondary transformations following the synthesis of the basic PPP molecule or related structures, without changing the phenoxy-phenoxy system.

The diphenyl ether moiety of the final products or the phenoxyphenol intermediates has been varied by Sandmeyer type reactions introducing, e.g., halogen atoms or cyano groups by replacement of amino groups [101, 143]. The anilines required in the first place, may be prepared from the corresponding nitro compounds.

Under certain conditions, aromatic halogen substituents can be replaced by a cyano group [12]. A nitro group – particularly in the 6-position of the *meta*-PPP system – is introduced by nitration of a 3-(phenoxy)-phenoxy-alkanoic acid derivative [88, 89] or an analogous 3-(phenoxy)-phenylthio-alkanoic acid derivative [117].

Different derivatives of phenoxy-phenoxy-alkanoic acids are prepared from the alkyl esters that most frequently will be obtained as the primary synthesis products. Suitable methods are, for example, aminolysis or transesterification. The acids themselves and their salts are prepared from the esters by hydrolysis. The metal salts have been used to obtain further derivatives, e.g., by reaction with suitable halogen compounds [26]. The most important synthetic procedure applied to the preparation of derivatives is the derivatization via reactive intermediates like the acid halides. Most of the functional derivatives discussed in Sects. 3.1.3 and 3.2.2 are available by this route utilizing common synthetic procedures. These derivatization methods also are applicable to the preparation of stereoisomeric PPP acid derivatives [124].

Access to the phenoxy-phenoxyalkanols and their derivatives is obtained by hydride reduction of the PPP esters [19, 43, 56].

7. Products

The description of products in this section will be restricted to the original PPP-type grass herbicides with a *para*-phenoxy-phenoxy system or its heterocyclic analogues. The experimental products of the *meta* series, e.g. RH-5205[2], have therefore not been considered.

7.1. Commercial Products

7.1.1. 2-[4-(2,4-Dichlorophenoxy)-phenoxy]-propionic Acid Methyl Ester (31)[2]

As the first product of the PPP series, compound *31* has been introduced into the market. Protected by German Offenlegungsschrift 2223894 [3] and having been developed under code number HOE 23,408 by Hoechst Aktiengesellschaft [155], the product is marketed under the trade names Illoxan®, Hoelon®, and Hoe-Grass®.

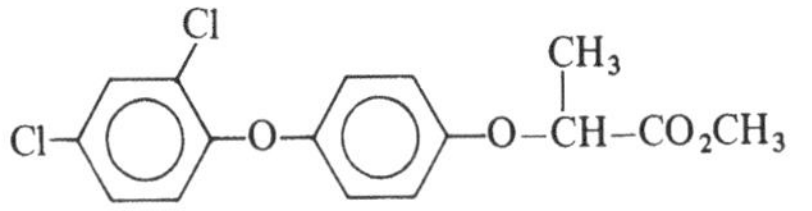

HOE 23,408 (Hoechst AG) 1975
Common name: diclofop-methyl
$C_{16}H_{14}Cl_2O_4$, mol. weight 341, mp. 39–41 °C,
bp. 175–176 °C (at 0.13 mbar)
LD_{50} acute oral (rat ♀) 563 mg a.i./kg
LD_{50} acute dermal (rat ♀) 5000 mg a.i./kg

31

The commercial product is formulated as an emulsifiable concentrate containing 19, 28, or 36% by weight diclofop-methyl. As a selective grass herbicide, it particularly controls wild oats and wild millets, lolium and phalaris species in cereal crops like wheat and barley, or in dicotyledonous crops like rapeseed, soybeans and sugar beets. Application rates are in the range of 0.7–1.0 kg a.i./ha in post-emergence application. Wild oats are most effectively controlled using diclofop-methyl in the 2 to 4 leaf stage of the weed. Pre-emergence incorporated application rates for the control of millets or bromus are about 1.5–2.0 kg a.i./ha.

After leaf application and uptake through the cuticula, the product is rapidly translocated within the plant, possibly after being hydrolyzed to the free acid. The main site of action is probably located at the vegetation point; the weed dies after having developed severe necrosis [156]. An alternative mechanism of action with separate functions of the ester and of the free acid has also been proposed [157]. The mode of action of diclofop-methyl is apparently systemic.

---

2   Cf. Vol. 5, p. 191

In the soil, diclofop-methyl is rapidly transformed by hydrolysis into 2-[4-(2,4-dichlorophenoxy)-phenoxy]-propionic acid. The acid is degraded to 4-(2,4-dichlorophenoxy)-phenol by ether cleavage. Up to 40% of the diclofop-methyl have been shown to be mineralized by degradation yielding carbon dioxide assayed as $^{14}CO_2$ [158–160].

In wheat plants, hydroxylated diclofop-methyl metabolites have been identified bearing the hydroxy group mainly in the 5-position of ring A, occurring as conjugates [161].

7.1.2. 2-[4-(3,5-Dichloropyridyl-2-oxy)-phenoxy]-propionic Acid Sodium Salt (32)

This hetero-analogous PPP compound, covered by German Offenlegungsschrift 2,546,251 [44], has been found to be a potent systemic grass herbicide. It does not show, however, selective properties within the monocotyledonous species, being tolerant only in broad-leaf cultures. The product tested under code number SL 501, has been marketed in Argentine under the trade name Hache Uno 35% (H-1) for the post-emergence control of Johnson grass, Bermuda grass, and many annual grassy weed species.

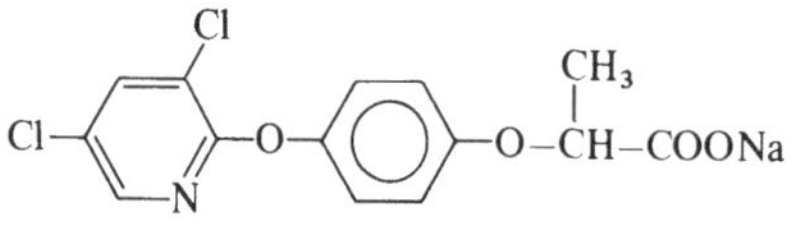

SL 501 (Ishihara Sangyo Kaisha, Ltd.) 1976
Common name: pyrifenop
$C_{14}H_{10}Cl_2NNaO_4$, mol. weight 350,
mp. 104–110 °C

32

Hache Uno 35% as a liquid concentrate formulation, has an acute oral toxicity (LD_{50}) in rats of 900 mg/kg. Application rates are 3.5–4.0 l for annual and 4.0–5.5 l for perennial grasses.

7.2. Experimental Products

7.2.1. 2-[4-(4-Chlorophenoxy)-phenoxy]-propionic Acid Isobutyl Ester (33)[3]

This grass herbicide, coded HOE 22,870, German Offenlegungsschrift 2,223,894 [3], is selectively active against millets and particularly against annual blackgrass in broad-leaf cultures and cereal crops including oats. Post-emergence application rates are 0.6 kg a.i./ha.

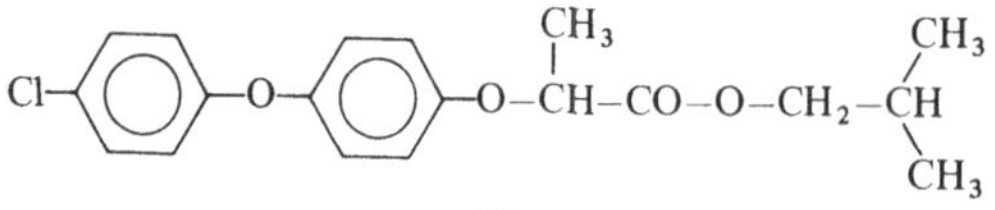

HOE 22,870 (Hoechst AG) 1975
Alopex®
Common name (proposed): clofop isobutyl

33

---

3 Cf. Vol. 5, p. 191

7.2.2. 2-[4-(4-Bromo-2-nitrophenoxy)-phenoxy]-propionic Acid Ethyl Ester (34)

Compound 34, coded B-806 – German Offenlegungsschrift 2,715,319 [9] – is a grass herbicide controlling a wide spectrum of graminaceous weeds including wild oat. Sufficient selectivity properties allow applications in wheat and barley; application rates are reported to be about 1 kg a.i./ha [162].

$$Br-\underset{NO_2}{\bigcirc}-O-\bigcirc-O-\underset{CH_3}{CH}-COOC_2H_5 \qquad \text{B-806 (Ishihara Sangyo Kaisha, Ltd.) 1979}$$

34

7.2.3. 2-[4-(4-Trifluoromethylphenoxy)-phenoxy]-propionic Acid Methyl Ester (35)

The trifluoromethyl compound HOE 29,152, reported under German patent 2,433,067 [5], also tested under code number SL 502 – German Offenlegungsschrift 2,531,643 [6] – seems to be the most active grass herbicide of the PPP series discovered up to now. Most annual grass weeds are controlled by application rates of 0.4 kg a.i./ha. Doses of 1.0 kg a.i./ha even control perennial weeds like quack-grass and Johnson grass. There is, however, – in contrast to many of the other PPP compounds – no tolerance in gramineous crops so that the product can only be applied in broad-leaf cultures.

$$F_3C-\bigcirc-O-\bigcirc-O-\underset{CH_3}{CH}-COOCH_3 \qquad \begin{array}{l} \text{HOE 29,152 (Hoechst AG) 1977} \\ \text{SL 502 (Ishihara Sangyo Kaisha, Ltd.)} \\ \text{Common name (proposed): trifop-methyl} \end{array}$$

35

7.2.4. 4-Methyl-4-[4-(4-trifluoromethylphenoxy)-phenoxy]-crotonic Acid Ethyl Ester (36)

The experimental compound KK 80 – German Offenlegungsschrift 28,29,130 [69] – is a more recent development containing as a vinylogue of the normal PPP series a phenoxy-phenoxy compound with an unsaturated carboxylic acid moiety. The first results published seem to indicate herbicidal properties similar to those of compound 35. Good results in combating grassy weeds in soybeans with post-emergence application have been reported; no damage to dicotyledonous crops has been observed [163]. Application rates for the control of annual grasses are 0.25–0.80 kg a.i./ha; Johnson grass is reported to be controlled by 0.5–1.0 kg a.i./ha.

$$F_3C-\bigcirc-O-\bigcirc-O-\underset{CH_3}{CH}-CH=CH-COOC_2H_5 \qquad \begin{array}{l} \text{KK 80 (Kumiai Chem. Ind. Co., Ltd.)} \\ \text{1979} \end{array}$$

36

8. References

Europ. Pat. is the abbreviation for "European Patent Application". The date following the patent number (given in brackets) denotes the date of application which is in most cases identical with the union priority date. The first two digits in Japanese unexamined patent numbers (Jap. Kokai Tokkyo Koho) represent the year of publication.

1. Wegler, R., Eue, L.: Chemie der Pflanzenschutz- und Schädlingsbekämpfungsmittel. Wegler, R. (ed.), Vol. 5, p. 188. Berlin, Heidelberg, New York: Springer 1977
2. Zimmermann, P.W., Hitchcock, A.E.: Contrib. Boyce Thompson Inst. Pl. Res. *12*, 321 (1942)
3. Hoechst AG: German Offen. 2,223,894 (17.5.72); C.A. *80*, 70, 543e (1974)
4. Hoffmann-La Roche AG: German Offen. 2,804,074 (Austrian Prior. 31.1.77); C.A. *89*, 215,059z (1978)
5. Hoechst AG: German Pat. 2,433,067 (10.7.74); C.A. *84*, 116,948d (1976)
6. Ishihara Sangyo Kaisha: German Offen. 2,531,643 (Jap. Prior. 17.7.74); C.A. *84*, 164,453k (1976)
7. Ciba-Geigy AG: German Offen. 2,652,384 (Swiss Prior. 20.11.75); C.A. *87*, 102,072p (1977)
8. Ishihara Mining and Chem. Co.; Jap. (unexam.) 78-121,729 (31.3.77); C.A. *90*, 87,091k (1979)
9. Ishihara Sangyo Kaisha: German Offen. 2,715,319 (Jap. Prior. 13.4.76); C.A. *88*, 70,490a (1978)
10. Hoffmann-La Roche AG: Europ. Pat. 2,246 (Swiss Prior. 15.9.78)
11. Ishihara Mining and Chem. Co.: Jap. (unexam.) 76-142,534 (30.5.75); C.A. *87*, 97,399u (1977)
12. Ciba-Geigy AG: German Offen. 2,730,591 (Swiss Prior. 9.7.76); C.A. *88*, 120,827p (1978)
13. Hoechst AG: German Offen. 2,601,548 (16.1.76); C.A. *88*, 17,322w (1978)
14. Bayer AG: German Offen. 2,754,462 (8.10.77)
15. Ishihara Mining and Chem. Co.: Jap. (unexam.) 79-55,534 (5.10.77); C.A. *91*, 140,580t (1979)
16. Ciba-Geigy AG: Europ. Pat. 2,204 (Swiss Prior. 24.11.77)
17. Hoechst AG: German Offen. 2,646,124 (13.10.76); C.A. *89*, 101,869h (1978)
18. Hoechst AG: German Offen. 2,609,461 (8.3.76); C.A. *88*, 22,357r (1978)
19. Hoechst AG: German Offen. 2,611,695 (19.3.76); C.A. *88*, 22,363q (1978)
20. Hoechst AG: German Offen. 2,613,697 (31.3.76); C.A. *88*, 1,620f (1978)
21. Ishihara Sangyo Kaisha: Jap. (unexam.) 77-131,542 (28.4.76); C.A. *88*, 89,394u (1978)
22. Hoechst AG: German Offen. 2,617,804 (23.4.76); C.A. *88*, 62,149f (1978)
23. Hoechst AG: German Offen. 2,623,558 (26.5.76); C.A. *88*, 104,932p (1978)
24. Ciba-Geigy AG: Europ. Pat. 3,517 (Swiss Prior. 3.2.78)
25. Nippon Soda Co.: Jap. (unexam.) 79-55,539 (11.10.77)
26. Ishihara Mining and Chem. Co.: Jap. (unexam.) 79-39,031 (30.8.77); C.A. *91*, 39,296e (1979)
27. Hoechst AG: German Offen. 2,628,384 (24.6.76); C.A. *89*, 23,989f (1978)
28. Ciba-Geigy AG: German Offen. 2,909,816 (Swiss Prior. 16.3.78)
29. Ciba-Geigy AG: Europ. Pat. 3,059 (Swiss Prior. 10.1.78)
30. Imperial Chem. Ind.: Europ. Pat. 3,648 (Brit. Prior. 15.2.78)
31. Ishihara Mining and Chem. Co.: Jap. (unexam.) 77-15,825 (24.7.75); C.A. *87*, 34,403v (1977)
32. Ishihara Sangyo Kaisha: Jap. (unexam.) 77-130,912 (22.4.76); C.A. *88*, 100,352b (1978)
33. Ishihara Sangyo Kaisha: Jap. (unexam.) 78-5,130 (6.7.76); C.A. *89*, 6,101t (1978)
34. Ishihara Sangyo Kaisha: Jap. (unexam.) 77-128,220 (20.4.76); C.A. *88*, 84,577w (1978)
35. Nippon Soda Co.: Jap. (unexam.) 78-111,063 (8.3.77); C.A. *90*, 72,168s (1979)
36. Ciba-Geigy AG: Europ. Pat. 2,812 (Swiss Prior. 2.1.78)
37. Ishihara Mining and Chem. Co.: Jap. (unexam.) 79-46,757 (17.9.77); C.A. *91*, 91,411c (1979)
38. Hoechst AG: German Offen. 2,748,658 (29.10.77); C.A. *91*, 193,054f (1979)
39. Ishihara Mining and Chem. Co.: Jap. (unexam.) 76-142,536 (30.5.75); C.A. *87*, 97,397s (1977)
40. Ishihara Sangyo Kaisha: German Offen. 2,715,284 (Jap. Prior. 14.4.76); C.A. *88*, 70,489g (1978)
41. Ishihara Mining and Chem. Co.: Jap. (unexam.) 76-39,622 (28.5.75); C.A. *86*, 134,891a (1977)
42. Ishihara Sangyo Kaisha: German Offen. 2,714,662 (Jap. Prior. 8.4.76); C.A. *88*, 70,488f (1978)
43. Ishihara Mining and Chem. Co.: German Offen. 2,649,706 (Jap. Prior. 29.10.75); C.A. *87*, 64,069h (1977)
44. Ishihara Mining and Chem. Co.: German Offen. 2,546,251 (Jap. Prior. 17.10.74); C.A. *85*, 78,015h (1976)
45. Ishihara Mining and Chem. Co.: Jap. (unexam.) 77-87,173 (17.1.76); C.A. *88*, 22,641d (1978)

46. Ishihara Sangyo Kaisha: Jap. (unexam.) 77-125,626 (14.4.76); C.A. *88*, 190,595k (1978)
47. Ishihara Industry Co.: Jap. (unexam.) 77-128,377 (20.4.76); C.A. *88*, 50,666j (1978)
48. Ishihara Sangyo Kaisha: Jap. (unexam.) 78-130,672 (18.4.77); C.A. *90*, 121,429p (1979)
49. Ishihara Mining and Chem. Co.: Jap. (unexam.) 79-55,575 (5.10.77)
50. Ciba-Geigy AG: Europ. Pat. 3,114 (Swiss Prior. 18.1.78)
51. Imperial Chem. Ind.: Europ. Pat. 1,473 (Brit. Prior. 12.8.77)
52. Ishihara Mining and Chem. Co.: Jap. (unexam.) 76-139,627 (28.5.75); C.A. *86*, 134,890z (1977)
53. Ishihara Sangyo Kaisha: Belg. Pat. 868,875; German Offen. 2,812,571 (Jap. Prior. 21.7.77); C.A. *90*, 152,017g (1979)
54. Imperial Chem. Ind.: Brit. Pat. 2,002,368 (27.7.78)
55. Imperial Chem. Ind.: PTC 79-094 (Brit. Prior. 9.2.78)
56. Imperial Chem. Ind.: Europ. Pat. 3,877 (Brit. Prior. 15.2.78)
57. Ishihara Sangyo Kaisha: Jap. (unexam.) 78-98,971 (10.2.77); C.A. *90*, 22,825h (1979)
58. Hoechst AG: German Offen. 2,640,730 (10.9.76); C.A. *88*, 190,816h (1978)
59. Hoechst AG: German Offen. 2,417,487 (10.4.74); C.A. *84*, 30,693e (1976)
60. Ishihara Sangyo Kaisha: Jap. (unexam.) 77-87,129 (16.1.76); C.A. *88*, 22,367u (1978)
61. Siegfried AG: German Offen. 2,750,527 (Swiss Prior. 15.11.76); C.A. *89*, 210,422c (1978)
62. Hoechst AG: German Offen. 2,732,924 (21.7.77); C.A. *90*, 163,340h (1979)
63. Ciba-Geigy AG: Europ. Pat. 4,317 (Swiss Prior. 17.3.78)
64. Devinter, S.A.: Europ. Pat. 2,151 (French Prior. 14.11.77)
65. Thiele, K., et al.: Arzneim.-Forsch. (Drug Research) *29*, 711 (1979)
66. Ishihara Mining and Chem. Co.: Jap. (unexam.) 79-5,935 (17.6.77); C.A. *90*, 186,600g (1979)
67. Ishihara Sangyo Kaisha: German Offen. 2,741,582 (Jap. Prior. 20.9.76); C.A. *89*, 101,895p (1978)
68. Hoechst AG: Belg. Pat. 858,104 (Swiss Prior. 25.3.76); C.A. *89*, 42,809p (1978)
69. Kumiai Chem. Ind. Co.: German Offen. 2,829,130 (Jap. Prior. 31.10.77); C.A. *91*, 91,343g (1979)
70. Kumiai Chem. Ind. Co.: German Offen. 2,905,458 (Jap. Prior. 18.2.78)
71. Kumiai Chem. Ind. Co.: Jap. (unexam.) 79-109,935 (14.2.78)
72. Imperial Chem. Ind.: Europ. Pat. 4,433 (Brit. Prior. 17.3.78)
73. Mitsubishi Petrochem. Co.: Jap. (unexam.) 79-2,323 (7.6.77); C.A. *90*, 181,589a (1979)
74. Celamerck GmbH u. Co.: German Offen. 2,731,214 (11.7.77); C.A. *90*, 186,622r (1979)
75. Celamerck GmbH u. Co.: German Offen. 2,729,529 (30.6.77); C.A. *90*, 116,442h (1979)
76. Hoechst AG: German Offen. 2,709,032 (2.3.77); C.A. *89*, 192,509e (1978)
77. Alfa Farmaceutici S.p.A.: German Offen. 2,918,748 (Ital. Prior. 9.5.78)
78. G. D. Searle & Co.: US Pat. 4,163,859 (18.7.77); C.A. *91*, 157,461c (1979)
79. Ishihara Mining and Chem. Co.: Jap. (unexam.) 79-32,477 (17.8.77); C.A. *91*, 91,510j (1979)
80. Ciba-Geigy AG: Europ. Pat. 176 (Swiss Prior. 29.6.77); Jap. (unexam.) 79-12,379; C.A. *90*, 181,588z (1979)
81. Ishihara Sangyo Kaisha: Jap. (unexam.) 78-50,174 (16.10.76); C.A. *89*, 109,130f (1978)
82. Ciba-Geigy AG: German Offen. 2,834,744 (Swiss Prior. 11.8.77); C.A. *90*, 168,320m (1979)
83. Bayer AG: German Offen. 2,805,981 (13.2.78); C.A. *91*, 157,473h (1979)
84. Bayer AG: German Offen. 2,805,982 (13.2.78); C.A. *91*, 157,460b (1979)
85. Ciba-Geigy AG: Europ. Pat. 5,709 (Swiss Prior. 15.3.78)
86. Mitsui Toatsu Chem.: Jap. (unexam.) 76-136,823 (16.5.75); C.A. *86*, 134,874x (1977)
87. Rohm and Haas Co.: US Pat. 4,059,435 (31.8.76); C.A. *88*, 74,222n (1978)
88. Ciba-Geigy AG: German Offen. 2,844,989 (Swiss Prior. 19.10.77); C.A. *91*, 74,359z (1979)
89. Ciba-Geigy AG: Europ. Pat. 1,641 (Swiss Prior. 25.10.77)
90. Rohm and Haas Co.: US Pat. 4,093,446 (14.3.72); C.A. *89*, 163,256q (1978)
91. Ishihara Mining and Chem. Co.: Jap. (unexam.) 79-89,022 (26.12.77); C.A. *91*, 187,954h (1979)
92. Ishihara Mining and Chem. Co.: German Offen. 2,643,438 (Jap. Prior. 27.9.75); C.A. *87*, 17,314z (1977)
93. Ciba-Geigy AG: German Offen. 2,809,541 (Swiss Prior. 8.3.77); C.A. *90*, 38,691x (1979)
94. Ishihara Sangyo Kaisha: Jap. (unexam.) 78-23,941 (13.8.76); C.A. *89*, 42,773x (1978)
95. Mitsui Toatsu Chem.: Jap. (unexam.) 76-98,326 (25.2.75); C.A. *86*, 51,583a (1977)
96. Mitsui Toatsu Chem.: Jap. (unexam.) 76-141,826 (3.6.75); C.A. *86*, 189,484a (1977)
97. Ishihara Mining and Chem. Co.: Jap. (unexam.) 77-15,822 (29.7.75); C.A. *87*, 17,310v (1977)
98. Ishihara Mining and Chem. Co.: Jap. (unexam.) 77-41,228 (27.9.75); C.A. *87*, 147,067g (1977)
99. Ishihara Mining and Chem. Co.: Jap. (unexam.) 77-47,917 (8.10.75); C.A. *87*, 113, 121e (1977)

100. Ishihara Mining and Chem. Co.: German Offen. 2,533,172 (24.7.74); C.A. *84*, 146,138g (1976)
101. Mitsui Toatsu Chem.: Jap. (unexam.) 77-72,817 (16.12.75); C.A. *88*, 46,410d (1978)
102. Nihon Nohyaku Co.: Jap. (unexam.) 77-72,816 (16.12.75); C.A. *88*, 33,191h (1978)
103. Ishihara Sangyo Kaisha: Jap. (unexam.) 77-105,214 (19.2.76); C.A. *88*, 46,395c (1978)
104. Mitsui Toatsu Chem.: Jap. (unexam.) 76-104,030 (5.3.75); C.A. *86*, 12,703v (1977)
105. Mitsui Toatsu Chem.: Jap. (unexam.) 77-28,934 (27.8.75); C.A. *87*, 48,914z (1977)
106. Mitsui Toatsu Chem.: Jap. (unexam.) 77-1,020 (24.6.75); C.A. *86*, 166,391d (1977)
107. Mitsui Toatsu Chem.: Jap. (unexam.) 79-32,427 (18.8.75); C.A. *91*, 91,346k (1979)
108. Soc. Rech. Ind. S. A.: German Offen. 2,637,098 (Brit. Prior. 20.8.75); C.A. *86*, 189,537v (1977)
109. Soc. Rech. Ind. S. A.: German Offen. 2,716,189 (Brit. Prior. 15.4.76); C.A. *88*, 50,520g (1978)
110. Unicler S.A.: German Offen. 2,910,942 (French Prior. 20.3.78)
111. Mitsui Toatsu Chem.: Jap. (unexam.) 76-73,124 (23.12.74); C.A. *86*, 66,848q (1977)
112. Mitsui Toatsu Chem.: Jap. (unexam.) 77-21,320 (8.8.75); C.A. *87*, 117,670f (1977)
113. Mitsui Toatsu Chem.: Jap. (unexam.) 77-72,815 (12.12.75); C.A. *88*, 46,392z (1978)
114. Mitsui Toatsu Chem.: Jap. (unexam.) 77-72,814 (12.12.75); C.A. *88*, 46,393a (1978)
115. Ciba-Geigy AG: Europ. Pat. 351 (Swiss Prior. 7.7.77); Jap. (unexam.) 79-16,438; C.A. *90*, 181,592w (1979)
116. Ciba-Geigy AG: Europ. Pat. 359 (Swiss Prior. 7.7.77); Jap. (unexam.) 79-16,438; C.A. *90*, 181,592w (1979)
117. Ciba-Geigy AG: Europ. Pat. 2,757 (Swiss Prior. 28.12.77); C.A. *91*, 211,108h (1979)
118. Ciba-Geigy AG: German Offen. 2,832,435 (Swiss Prior. 27.7.77); C.A. *90*, 186,607q (1979)
119. Hansch, C., Leo, A.: Substituent constants for correlation analysis in chemistry and biology. New York, Chichester, Brisbane, Toronto: John Wiley & Sons 1979
120. Nestler, H.J.: IV th Internat. IUPAC Symp. Pesticide Chemistry, Zürich 1978
121. Topliss, J.G.: J. Med. Chem. *15*, 1006 (1972)
122. Nestler, H.J. et al.: In Adv. pesticide sci., Geissbühler, H. (ed.), Part 2, p. 248. Oxford, New York: Permanon Press 1979
123. Nestler, H.J., Bieringer, H.: Z. Naturforsch. *35*b, 366 (1980)
124. Hoechst AG: Belg. Pat. 873,844 (German Prior. 24.12.77); C.A. *91*, 140,579z (1979)
125. Imperial Chem. Ind.: Europ. Pat. 2,925 (Brit. Prior. 23.12.77)
126. Imperial Chem. Ind.: Europ. Pat. 3,890 (Brit. Prior. 1.3.78)
127. Bayer AG: German Offen. 2,537,290 (21.8.75); C.A. *87*, 1,179y (1977)
128. E.I. du Pont: German Offen. 2,646,628 (US Prior. 15.10.75); C.A. *87*, 68,350r (1977)
129. Ishihara Sangyo Kaisha: Jap. (unexam.) 77-76,433 (22.12.75); C.A. *88*, 84,576v (1978)
130. Hoechst AG: German Offen. 2,755, 617 (14.12.77); C.A. *91*, 187,961h (1979)
131. Sumitomo Chem. Co.: Jap. (unexam.) 79-98,327 (20.1.78)
132. Sumitomo Chem. Co.: Jap. (unexam.) 79-98,328 (20.1.78)
133. Imperial Chem. Ind.: Europ. Pat. 4,414 (Brit. Prior. 1.3.78)
134. Hoechst AG: German Offen. 2,745,482 (10.10.77); C.A. *91*, 1,370m (1979)
135. Hoechst AG: German Offen. 2,745,869 (12.10.77); C.A. *91*, 1,371n (1979)
136. Celamerck GmbH: German Offen. 2,802,818 (23.1.78); C.A. *91*, 135,614c (1979)
137. Imperial Chem. Ind.: Jap. (unexam.) 78-53,664 (Brit. Prior. 26.10.76); C.A. *89*, 146,771m (1978)
138. Ishihara Sangyo Kaisha: Belg. Pat. 865,137 (Jap. Prior. 21.10.77); C.A. *90*, 72,070d (1979)
139. Dow Chem. Comp.: Europ. Pat. 5,064 (US Prior. 24.4.78)
140. Imperial Chem. Ind.: German Offen. 2,824,828 (Brit. Prior. 11.7.77); C.A. *90*, 168,270v (1979)
141. Ishihara Mining and Chem. Co.: Jap. Pat. 78-41,206 (16.10.71); C.A. *79*, 133,666z (1973)
142. Kumiai Chem. Ind. Co.: Jap. Pat. 78-41,207 (15.10.71); C.A. *79*, 133,672y (1973)
143. Farbwerke Hoechst AG: German Pat. 1,668,896 (11.1.68); C.A. *72*, 66,648k (1970)
144. Hoechst AG: German Offen. 2,433,066 (10.7.74); C.A. *84*, 121,453t (1976)
145. Hoechst AG: German Pat. 2,648,644 (27.10.76); C.A. *89*, 42,779d (1978)
146. Ishihara Mining and Chem. Co.: Jap. (unexam.) 76-121,516 (16.4.75); C.A. *86*, 134,876z (1977)
147. Ishihara Sangyo Kaisha: Belg. Pat. 865,136 (Jap. Prior. 20.10.77); C.A. *90*, 72,071e (1979)
148. Ishihara Sangyo Kaisha: German Offen. 2,812,649 (Jap. Prior. 20.10.77); Belg. Pat. 865,136; C.A. *90*, 72,071e (1979)
149. Nippon Soda Co.: Jap. (unexam.) 79-46,732 (21.9.77); C.A. *91*, 123,549x (1979)
150. Hoechst AG: Europ. Pat. 924 (German Prior. 30.8.77); German Offen. 2,738,963; C.A. *90*, 186,933t (1979)

151. Mitsui Toatsu Chem.: Jap. (unexam.) 77-142,031 (19.5.76); C.A. *88*, 152,226h (1978)
152. Mitsui Toatsu Chem.: Jap. (unexam.) 76-88,931 (4.2.75); C.A. *85*, 159,665x (1976)
153. Hoechst AG: German Offen. 2,658,496 (23.12.76); C.A. *89*, 108,654t (1978)
154. Ciba-Geigy AG: German Offen. 2,847,662 (US Prior. 4.11.77); C.A. *91*, 107,823b (1979)
155. Langelüddeke, P. et al.: Mitt. Biol. Bundesanst. Land.-Forstwirtsch. *165*, 169 (1975)
156. Köcher, H., Lötzsch, K.: Proc. Europ. Weed Res. Soc. Symp. Status and Control of Grassweeds in Europe *1975*, 430
157. Shimabukuro, M.A. et al.: Pesticide Biochem. Physiol. *8*, 199 (1978)
158. Smith, A.E.: J. Agr. Food Chem. *25*, 893 (1977)
159. Aßhauer, J.: IX th Internat. Congr. Plant Protection, Washington, D.C. 1979
160. Martens, R.: Pestic. Sci. *1978*, 127
161. Gorbach, S.G., Künzler, K., Aßhauer, J.: J. Agr. Food Chem. *25*, 567 (1977)
162. Nishiyama, R. et al.: Proc. 7 th Asian-Pacific Weed Sci. Soc. Conf. *1979*, 25
163. Gealy, D.R., Slife, F.W.: Proc. N. Cent. Weed Control Conf. *33*, 46 (1978)

Wild Oat Herbicides

E. Haddock, R. G. Turner

Shell Research Ltd., Shell Biosciences Laboratory, Sittingbourne Research Centre
Sittingbourne, Kent, ME9 8AG, England

Wild Oats (*Avena* Spp.) are troublesome weeds in world agriculture. Many reviews have examined this weed complex from the viewpoints of its distribution, biology and methods of control, including several recent publications [1–3]. In 1977 the Shell International Chemical Company conducted a limited survey covering the countries listed in Table 1 to examine the economic importance of this weed complex and found that approximately 2.6 million hectares of cultivated land in those areas were treated with herbicides to control wild-oats in the crops mentioned in the table.

The survey showed that 31 different crops were treated with wild-oats herbicides and 39 different chemicals were used (either singly or in combination) in programmes aimed at controlling this weed. Some of the herbicides are listed in Table 2 which includes their usage in different crops. These herbicides represent 5000 tonnes of chemicals as active ingredient valued at 40 million US dollars in 1977. Unfortunately the survey excluded N. America (USA and Canada).

This review concentrates upon the herbicides used to control wild-oats and is restricted to those herbicides of commercial importance. It examines them on the basis of their chemistry, mode of action and method of use and each of the subsequent sections covers the main groups of herbicides from these viewpoints.

Aminopropionic acid derivatives $(I)^1$ are specific herbicides for the control of all species of wild-oats (*Avena fatua, Avena ludoviciana, Avena sterilis* and *Avena barbata*) in cereal crops.

$$\text{Y} \quad \text{C=O} \\ \text{X} - \text{N} - \text{CHCO}_2\text{R} \\ \text{CH}_3$$

I

Their mode of action has been studied extensively [4–6] and is postulated to be metabolism to the free acid [5] $(I, R = H)$, the phytotoxic entity, which prevents cell elongation causing stunting of the wild-oat. Under these circumstances crop competition for light and nutrients leads to suppression and often eventual death of the wild-oat, indicating that these compounds (I) require good crop competition for maximum effect. Selectivity in the cereal crops arises owing to the different

---

1   Cf. Vol. 5 of this series, p. 194

E. Haddock, R. G. Turner

Table 1. Wild oat herbicides listed by crop usage

Crop	Country	Products
Cereals – Wheat – Barley – Rye – Oat	Austria, Australia, Benelux, France, Germany, Italy, Portugal, Morocco, Spain, Switzerland, Tunisia, U.K.	Barban, benzoylprop-ethyl, chlorfenprop-methyl, dichlofop-methyl, difenzoquat, chlortoluron, flamprop-methyl, flamprop-isopropyl, isoproturon, metoxuron, neburon, nitrofen, paraquat, triallate, trifluralin
Maize (forage)	Germany, Greece, France	Atrazine, chlorfenprop-methyl, simazine, propham, chlorpropham, EPTC
Sugar beet	Australia, Austria, Benelux, Germany, U.K.	Chlorfenprop-methyl, cycloate, diallate, flamprop-methyl, paraquat, phenmedipham, propham (and mixtures with fenuron, chlorpropham, endothal, medinoterb) pyrazon, triallate, TCA
Fodder beet	Benelux, France, Germany	Diallate, propham, TCA, triallate
Oil seed crops – Colza (Rape) – Linseed – Safflower	Australia, Chile, Benelux, Mexico, Switzerland	Asulam, carbetamide, dalapon, diallate, methachlor, trifluralin, TCA, triallate
Vegetables – Potato – Carrot – Onion – Sprouts – Pea – Cabbage – Swede – Beetroot – Parsnip Brocolli – Cauliflower	Australia, Benelux, France, Germany, U.K. Switzerland	Barban, chlorpropham, diallate diquat, dinoseb, EPTC, metobromuron, metoxuron, monolinuron, nitrofen, paraquat, propachlor, propazine, prometryne, TCA, trifluralin, triallate
Fodder crops – Alfalfa (Lucerne) – Lupins – Brassicas	Australia, Italy, Mexico, U.K.	Diallate, EPTC, nitrofen, paraquat, propachlor, trifluralin, triallate
Citrus – Orange – Lemon – Mandarins	Australia, Italy	Bromacil, paraquat
Pome and stone fruit – Apples – Olives – Peach	Italy, U.K.	Chlorthiamid, paraquat, simazine
Vine	France, Italy, Portugal	Aminotriazole, chlorthiamid, paraquat, simazine

Table 2. Summary of the types of compound available for the control of wild-oat

Common name or code	Date of introduction	Dose rate (kg/ha)	Type and time of application	Crops
1. Aminopropionate derivatives				
Benzoylprop-ethyl	1969	1.0–1.6	Post-emergence Early tillering	Wheat
Flamprop-methyl	1977	0.4–0.6	Post-emergence Early tillering	Wheat
Flamprop-isopropyl	1972	1.0	Post-emergence Early tillering	Barley
L-flamprop-iso-propyl	1977	0.6	Post-emergence Early tillering	Wheat Barley
2. Carbamic acid derivatives				
Barban	1958	0.28–0.56	Post-emergence 1–2½-leaf stage	Wheat Barley Sugarbeet
Propham	1951	4.0–6.0	Pre-emergence	Sugarbeet Maize Soyabean
Chlorpropham	1951	4.0–6.0	Pre-emergence	Sugarbeet Maize Soyabean
Asulam	1968	0.5–1.1	Pre-emergence 5 leaf-stage	Flax
3. Cyclohexenone derivatives				
Alloxydim-sodium "NP 48 Na"	1976	1.0–1.5	Post-emergence 2-leaf-early tillering	Sugarbeet Peas Beans
"NP 55"		0.2–0.6	Post-emergence 2-leaf-early tillering	Sugarbeet Peas Beans
4. Pyrazolium salt derivatives				
Difenzoquat	1973	0.75 – 1.4	Post-emergence 3–6-leaf stage	Wheat Barley
5. Haloalkanoic acid derivatives				
Chlorfenprop-methyl	1968	3.0–4.0	Post-emergence 1½–3-leaf stage	Wheat Barley
TCA	1947	1.0–5.0	Pre-emergence	Sugarbeet Peas Potatoes
Dalapon	1951	1.0–5.0	Pre-emergence	Sugarbeet Peas Potatoes
6. Nitrophenylether derivatives				
Nitrofen	1964	2.0–6.0	Pre- and early post-emergence	Winter wheat Spring cereals
Bifenox	1970	1.0–2.0	Pre- and directed post-emergence	Wheat Barley Maize

Table 2 (continued)

Common name or code	Date of introduction	Dose rate (kg/ha)	Type and time of application	Crops
M0-338		2.0–5.0	Pre-emergence	Cabbage
Flourodifen	1968	2.0–5.0	Pre-emergence	Alfalfa
HE-306		3.0–4.0	Pre- and early post-emergence	Wheat Barley Rice
7. Phenoxyphenoxyalkanoic acid derivatives				
Diclofop-methyl	1975	0.75–1.25	Post-emergence 2–3-leaf stage	Wheat Barley Broad leaf crops
Clofop-isobutyl	1975	1.0–2.0	Post-emergence	Cereals Broad leaf crops
Trifop-methyl	1979		Post-emergence	Cereals Broad leaf crops
8. Thiolcarbamate derivatives				
Triallate	1961	1.0–3.0	Pre- and early post-emergence	Wheat Barley Pea Potato
Diallate	1960	1.0–3.0	Pre- and early post-emergence	Barley Alfalfa Clover, Pea Potato
EPTC	1954	2.0–6.0	Pre-emergence (incorporation)	Sugarbeet Maize Potato Bean
Cycloate	1963	3.0–4.0	Pre-plant (incorporation)	Sugarbeet Spinach
Pebulate	1954	2.0–6.0	Pre-emergence (incorporation)	Sugarbeet Potato
Vernolate	1954	1.5–3.0	Pre-emergence (incorporation)	Soyabean Tobacco Peanut
9. Triazine derivatives				
Simazine	1956	2.0–4.0	Pre-emergence or pre-sowing	Maize Beans
Atrazine	1958	1.0–2.2	Pre-emergence or pre-sowing	Maize Beans
10. Phenylurea derivatives				
Chlortoluron	1969	2.4–3.6	Pre-emergence or early post-emergence	Winter Wheat
Isoproturon	1973	2.0–2.5	Pre- or early post-emergence	Winter Wheat

Table 2 (continued)

Common name or code	Date of introduction	Dose rate (kg/ha)	Type and time of application	Crops
Metoxuron	1968	2.4–4.0	Pre- or early post-emergence	Winter Wheat Winter Barley
Linuron	1960	0.5 – 2.0	Pre- or early post-emergence	Winter Wheat Maize Bean, Pea
Monolinuron	1958	1.0–3.0	Pre- or early post-emergence	Potato Vines ornamentals
Benzthiazuron	1966	4.0–8.0	Pre-emergence	Sugarbeet Fodderbeet

relative rates of metabolism to the phytotoxic acid in the different cereal crops [6, 7].

Benzoylprop-ethyl [1] (I, X = Y = Cl, R = CH$_2$CH$_3$), introduced in 1969, is applied post-emergence at or just before tillering of the wild-oat. Application rates vary with geographical location, probably related to crop density, but in general a dose of 1.0–1.6 kg/ha gives reliable and effective control (85–95%) of wild-oat in wheat. Benzoylprop-ethyl is selective in all wheat varieties tested to date but is not recommended for use in barley crops.

Flamprop-methyl [8] (I, X = F, Y = Cl, R = CH$_3$), introduced in 1977, is similar to benzoylprop-ethyl but the phytotoxic acid flamprop (I, X = F, Y = Cl, R = H) is approximately three times as active as benzoylprop (I, X = Y = Cl, R = H) leading to an effective dose of 0.4–0.6 kg/ha of flamprop-methyl for the control of wild-oat in wheat. Flamprop-methyl also exhibits useful herbicidal effects against other grass species [9] (*Agropyron repens* and *Alopecurus myosuroides*), but like the benzoylprop-ethyl is not recommended for use in barley. Flamprop-methyl provides reliable wild-oat control in short season cereal production (Australia and Canada) under dry conditions [8].

Flamprop-isopropyl [10] [I, X = F, Y = Cl, R = CH(CH$_3$)$_2$], introduced in 1972, is applied post emergence, at or just before tillering of the wild-oat, giving effective control (80–95%) of wild-oat at 1.0 kg/ha in barley crops.

L-flamprop-isopropyl [10] [Isopropyl R(−)N-benzoyl-N-(3-chloro-4-fluorophenyl)-L-alaninate], introduced in 1977, is the optically active (−) form of flamprop-isopropyl and can be used post-emergence in both wheat and barley at early tillering of the wild-oat giving effective control (80–95%) of wild-oat at 0.6 kg/ha with improved selectivity in the crop. Some barley varieties (i.e. Bettina) have shown some sensitivity to both flamprop-isopropyl and L-flamprop-isopropyl but selectivity in wheat appears excellent [10].

The aminopropionic acid derivatives group suffer from the disadvantage of restrictions on their use in mixtures [5, 8–10]. In common with several wild-oat herbicides this chemical group can interact biologically with a variety of different agrochemical products [17]. Nevertheless the amino-propionates are successfully used in mixtures in a variety of countries under a range of conditions in commercial practice (Table 1).

Carbamic acid derivatives (II)[2] are broad-spectrum herbicides for annual grasses, including wild-oat, and broad-leaved weeds.

$$X\text{—}C_6H_4\text{—}NHCO_2R$$

II

They exert their phytotoxic effect by inhibition of photosynthesis and blocking of cell division [11–14], resulting in growth retardation and eventual death from crop competition for light and nutrients.

Barban (II, X = Cl, R = $CH_2C \equiv C\text{–}CH_2Cl$), introduced in 1958, is applied post-emergence at the one to two and a half leaf stage of the wild-oat giving effective control at 0.28–0.56 kg/ha in crops such as wheat, barley and sugar beet [15]. Barban is phytotoxic to some varieties of wheat and barley [16] and its narrow window of application leads to some variability of performance against wild-oat. Its good performance is closely related to crop competition. Lack of competition may be overcome by increased dose rates but this may lead to crop phytotoxicity [15]. The earliness of barban application in short-season cereal production (Australia and Canada) provides useful yield increases.

Barban can be mixed with a variety of different agrochemicals for use in several different crops [17]. In common with several specific wild-oat herbicides barban can react to produce an antagonistic response. The type of interaction depends upon a variety of factors including dose, time of application, chemical mixed and its formulation [17].

Chemically related to barban are propham [II, X = H, R = $CH(CH_3)_2$] and chlorpropham [II, X = Cl, R = $CH(CH_3)_2$], introduced in 1951, which are used pre- or early post-emergence for the control of wild-oat in non-cereal crops such as sugar beet, corn and soyabean [17]. Both compounds require incorporation for maximum effect at relatively high doses (4.0–6.0 kg/ha). Even then the extent of weed control is often variable [17].

Asulam (III), a phenylsulphonylcarbamate[3], introduced in 1968, is used for the control of wild-oats in flax [19, 20]. It acts by inhibition of cell division resulting in barban-like activity [22].

$$H_2N\text{—}C_6H_4\text{—}SO_2NHCO_2CH_3$$

III

The use of a wetter in the formulation appears essential for good performance [20], wild-oats being effectively controlled at 0.5–1.1 kg/ha by a formulation

---

2   Cf. Vol. 5, p. 104
3   Cf. Vol. 5, p. 111

containing 0.125% wetter [22]. Selectivity appears highly dependent upon differential responses of the plant leaf waxes [17].

Cyclohexenone derivatives (*IV*)[4] are broad-spectrum pre- and post-emergence herbicides, first intoduced in 1976 for the control of grass weeds in broad-leaved crops [23].

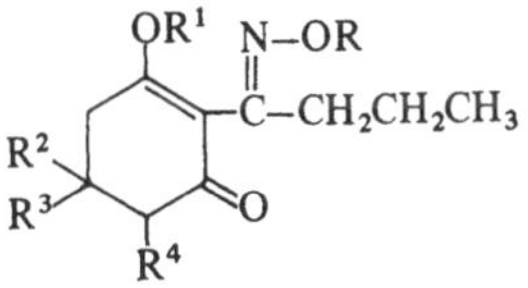

IV

Alloxydim-sodium, "NP 48 Na" (*IV*, $R=CH_2–CH=CH_2$, $R^1=Na$, $R^2=R^3=CH_3$, $R^4=CH_2CH_3$) is used for the control of wild-oat, and other grass weeds in crops such as sugar beet, peas and beans. It is absorbed through the foliage or roots of grass species and translocated to the meristem, resulting in termination of growth after two to three days and eventual death after ten to fourteen days [24]. It is applied mainly post-emergence at the two leaf to early tillering stage of the wild-oat giving good to excellent control (85–100%) at 1.0–1.5 kg/ha with no injury to sugar beet, peas or beans [25–28].

"NP 55" (*IV*, $R=CH_2CH_3$, $R^1=R^2=R^4=H$, $R^3=CH_2CH_3CH \cdot S \cdot CH_2CH_3$) has a similar spectrum of activity and mode of action to alloxydim-sodium but is four times as active being effective against wild-oat at the two leaf to early tillering stages at 0.2–0.6 kg/ha [29, 30].

Chemically similar to *IV* are compounds *V–VII* claimed to be broad-spectrum grass herbicides for use in broad-leaved crops [31].

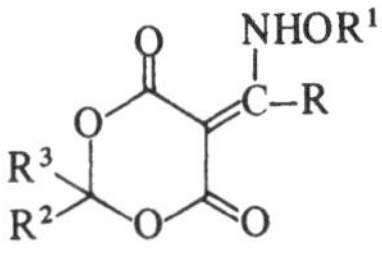

V 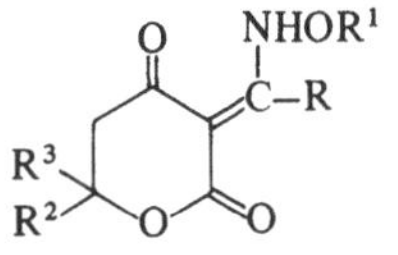　　　　　　VI 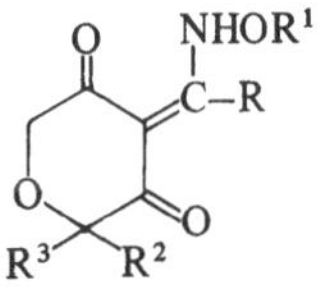　　　　　　VII

Difenzoquat[5] (*VIII*) is a specific wild-oat herbicide, introduced in 1973, which acts by disrupting cell division and terminating elongation and growth.

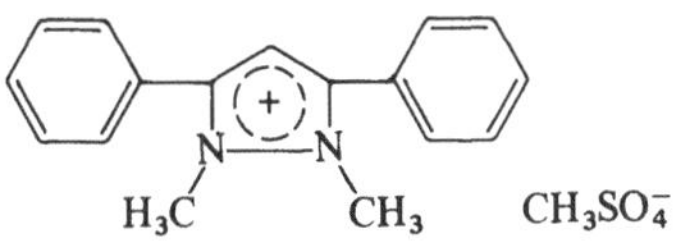

VIII

Visible symptoms do not develop for several days, although this is temperature-dependent. Inhibition of the leaves and shoots is then observed before yellowing, chlorosis, necrosis and eventual death occur [32]. Some yellowing and

---

4　Idem p. 168
5　Cf. Vol. 5, p. 282

scorch of the cereal crop may occur but this disappears within two weeks [33]. However minor height reductions do occur on certain varieties [34]. Difenzoquat appears not to be greatly dependent upon crop competition, although its performance is better in a vigourous crop [35], and applied post-emergence at the three to six leaf stage of the wild-oat [32–36] gives effective control (80–95%) at 0.75–1.4 kg/ha in both wheat and barley [34].

An ionic surfactant is essential in the application of difenzoquat, the phytotoxic effect increasing with increased surfactant concentration [32]. The use of surfactants renders difenzoquat sensitive to rainfall after application [37] and this can often have deleterious effects on its performance [38–41]. Studies on crop tolerance [26, 27, 30] have indicated that barley is less susceptible than wheat, with winter wheat being less susceptible than spring wheat, although there are exceptions in certain varieties [26–29].

Difenzoquat interacts with other herbicides [17], the nature and degree of the interaction being dependent upon a variety of circumstances. The antagonism between difenzoquat and selected broad-leaved herbicides can often be overcome in commercial practice.

Haloalkanoic acid derivatives (*IX*) are general annual and perennial grassweed herbicides whose phytotoxic effect appears to be associated with interaction with protein molecules [42]

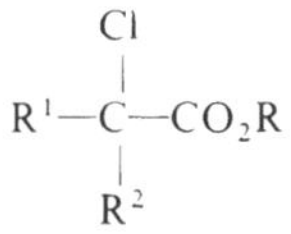

$$R^1 - \overset{\overset{\displaystyle Cl}{|}}{\underset{\underset{\displaystyle R^2}{|}}{C}} - CO_2R$$

IX

Chlorofenprop-methyl (*IX*, R=CH$_3$, R^1=H, R^2=CH$_2$-⟨◯⟩-Cl), introduced in 1968, is a specific wild-oat herbicide whose mode of action is uncertain but may involve interactions with protein molecules [43]. It is selective in cereals [44], including cultivated oat, at application rates of 3–4 kg/ha at the one-and-a-half-to three-leaf stage of wild-oat.

Certain varieties of wild-oat however appear resistant at rates suitable for crop application [45].

Chlorofenprop-methyl provides good control of wild-oat but requires a restricted set of conditions for it to exert its maximum effect. It has been reported [46] that it is sensitive to application conditions, droplet size and weed growth. The uptake criteria published [46] provide a good explanation of the variability of performance associated with its usage [47].

Chemically related to chlorofenprop-methyl are TCA (*IX*, R=H, R^1=R^2=Cl) and dalapon[7] (*IX*, R=H, R^1=CH$_3$, R^2=Cl), introduced in 1947 and 1951 respectively. They are general grass-weed, including wild-oat, herbicides used pre-emergence at high application rates (5–20 kg/ha) in non-crop situations [46] and lower rates (1–5 kg/ha) in non-cereal crops [48] such as sugar beet, peas,

---

6   Cf. Vol. 5, p. 146
7   Idem p 163

potatoes, and brassicas [17]. Despite the relative age of these products they are both still used (Table 1). Opinions differ on their mode of action [17, 42].

Nitrophenyl ethers[8] (X) are general grass-weed herbicides having some activity against broad-leaved weeds. They are postulated to act by interference with ATP production through inhibition of the Hill Reaction and mitochondrial electron transport [49].

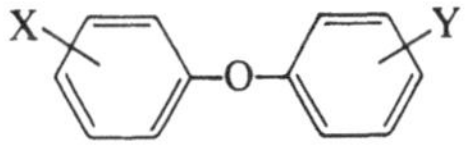

X

Nitrofen (X, X = 2, 4-Cl$_2$, Y = H) introduced in 1964, controls susceptible weeds by contact with the young shoot during emergence. It is effective pre- or early post-emergence for the control of wild-oat at 2.0–6.0 kg/ha in crops such as legumes, cabbage and cereals. Work with nitrofen at 2–4 kg/ha in winter wheat [50, 51] and spring cereals [52] has shown control of wild-oats (60–90%) with no adverse crop effects. However nitrofen is known to be variable in the extent of wild-oat control that follows pre-emergence application [52] (and it is also irregular on *Alopecurus myosuroides* [54]). This irregular performance is reflected by the prevailing usage of nitrofen in mixture with other products (including dinoterb, linuron, neburon [53]).

Chemically closely related to nitrofen are bifenox (X, X = 2,4-Cl$_2$, Y = 3-CO$_2$CH$_3$); MO-338 (X, X = 2,4,6-Cl$_3$, Y = H); fluorodifen (X, X = 2-NO$_2$-4-CF$_3$, Y = H) and HE-306 [X, X = 3,5-(CH$_3$)$_2$, Y = H], all of which have a similar spectrum of activity and mode of action.

Phenoxyphenoxyalkanoic acid derivatives[9] (XI) are broad spectrum grass herbicides used for the control of annual grasses, including wild-oat in a wide range of broad-leaved crops as well as wheat and barley.

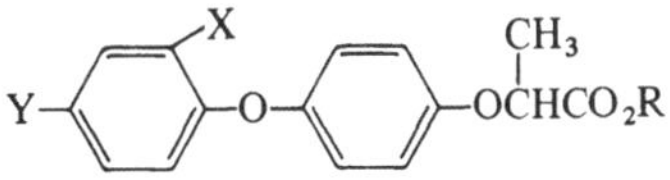

XI

Diclofop-methyl (XI, R = CH$_3$, X = Y = Cl), introduced in 1975, is thought to cause loss of menbrane integrity [54], resulting in degradation of the chloroplast, plugging of phloem cells, inhibition of transport of photosynthates which leads to root die-back, and a reduction in ATP synthesis and physiological activity [55]. It is applied post-emergence at the two-to-three-leaf stage of the wild-oat at 0.75–1.25 kg/ha in a wide range of broad-leaved crops as well as wheat and barley [56].

Its selective action in wheat and barley compared with that in wild-oat is due to the tolerant species, wheat and barley, being able to hydroxylate the "active" acid, diclofop (XI, X = Y = Cl) and remove it by further conjugation more quickly than wild-oat. [57].

---

8   Idem p. 73

9   Cf. Vol. 5, p. 191

The use of diclofop-methyl gives good control of wild-oats in a variety of crops. There are some drawbacks to its practical usage including a narrow window of application and known interactions with certain other herbicides [58]. Occasional instances of cereal phytotoxicity have been reported [39, 40, 58].

Chemically related to diclofop-methyl are clofop-isobutyl [XI, $R=CH_2CH(CH_3)_2$, $X=H$, $Y=Cl$] and trifop-methyl (XI, $R=CH_3$, $X=H$, $Y=CF_3$), introduced in 1975, which have similar activity and selectivity. Some reports claim clofop-isobutyl as being more selective than diclofop-methyl and with improved grass weed control [14, 15].

Numerous patents exist on chemically related structures (XII) involving heterocyclic systems such as pyridine [62–64], pyrimidine [65], benzothiazole [66] and other substituted aryl compounds [67–69], all of which appear to have a similar spectrum of activity.

$$Ar-X-\!\!\!\left\langle\!\!\!\bigcirc\!\!\!\right\rangle\!\!\!-OCH(CH_3)-(Y)n-CO_2R$$

XII

Where Ar is pyrid-2-yl, pyrimidin-2-yl, benzothiazol-2-yl, or substituted phenyl

X   is O, S, NR^1, CH_2
Y   is alkyl or alkenyl
n   is O–2
R   is hydroxy, acetoxy, thioalkyl, substituted amino.

Thiolcarbamates (XIII)[10] are general grass herbicides, with some activity against broad-leaved weed species, which are applied pre- or early post-emergence and require soil incorporation because of their high volatility.

$$R-S-\overset{\displaystyle O}{\overset{\|}{C}}-N\overset{R^1}{\underset{R^2}{\diagdown}}$$

XIII

The thiolcarbamates (XIII) are soil acting and exert their effect in the gaseous and/or aqueous phase. They are postulated to act by inhibition of wax formation and interference with mitosis and cell extension, however, the actual cause of death of the susceptible species is open to speculation [11].

Triallate[11] $\left[XIII,\ R=CH_2\overset{\displaystyle Cl}{\overset{|}{C}}=CCl_2,\ R^1=R^2=-CH(CH_3)_2\right]$, introduced in 1961, is a highly effective herbicide for the control of annual broad-leaved and grass weeds, particularly wild-oat, in crops such as wheat, barley, peas and potato. It is used pre- or early post-emergence at 1.0–3.0 kg/ha either as an emulsion [75], requiring incorporation, or a granule [76–80], requiring no incorporation, giving

---

10 Cf. Vol. 5, p. 92
11 Idem p. 94

effective control of wild-oat (80–95%). In general granular treatments require a slightly higher dose (1.4–2.2 kg/ha) than EC applications (1.4–1.7 kg/ha).

Diallate[12] [XIII, R=CH$_2$Ċ=CHCl (Cl), R^1=R^2=CH(CH$_3$)$_2$], introduced in 1960, has similar activity but is slightly less selective in cereal crops [81, 82], and has been largely superceded by triallate.

The volatility of the thiolcarbamates is the main disadvantage of their practical usage. However, provided the materials are thoroughly incorporated into the soil, diallate and triallate provide good wild-oat control under a variety of soil and climatic conditions [17]. They are widely used in a range of different crops.

Chemically very similar to triallate and diallate are the following compounds used for the control of wild-oat in non-cereal crops such as sugar beet, potato, soybean.

EPTC (XIII, R=CH$_2$CH$_3$, R^1=R^2=CH$_2$CH$_2$CH$_3$); Cycloate (XIII, R=R^1=CH$_2$CH$_3$, R^2=⟨⟩); Pebulate (XIII, R=CH$_2$CH$_2$CH$_3$, R^2=CH$_2$CH$_2$CH$_2$CH$_3$) and Vernolate (XIII, R=R^1=R^2=CH$_2$CH$_2$CH$_3$).

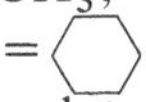

Triazine derivatives[13] (XIV) are broad-spectrum herbicides which are selective in a few crops such as maize, field beans and broad beans.

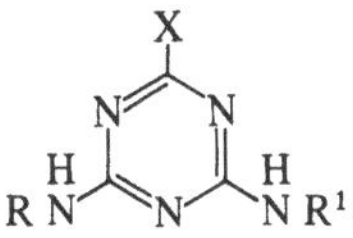

$$\underset{\text{XIV}}{\chemfig{R-\overset{H}{N}-C_6N_3X-\overset{H}{N}-R^1}}$$

XIV

They exert their phytotoxic effect by interfering with photosynthesis, probably owing to inhibition of the Hill reaction following root uptake [83].

Simazine[14] (XIV), R=R^1=CH$_2$CH$_3$, X=Cl), introduced in 1956, is applied pre-emergence or pre-sowing and at a dose rate of 2.0–4.0 kg/ha and gives effective control of wild-oat [84, 85]. Incorporation considerably increases the efficiency of simazine, a dose of 1 kg/ha giving excellent control of wild-oat [86], however selectivity in marginally resistant crops is reduced.

Atrazine [XIV, R=CH$_2$CH$_3$, R^1=CH(CH$_3$)$_2$], introduced in 1958, is similar to simazine but more predictable being active at 1.0–2.2 kg/ha when applied post-emergence or pre-sowing [87, 90], incorporation increasing its activity [88, 89]. Post-emergence applications are less effective and a dose of 2.0–2.5 kg/ha is required for control of wild-oats at the 2-4-leaf stage [90].

Phenylureas[15] (XV) are phytotoxic to mono- and dicotyledonous species, having greatest effect on the seedlings of small seeded weeds and annual grass weeds in particular. They exert their phytotoxic effect via inhibition of photosynthesis by interfering with electron transfer during the Hill reaction [7].

---

12 Idem p. 93
13 Cf. Vol. 5, p. 336
14 Idem p. 34
15 Cf. Vol. 5, p. 122

$$XV$$

Several phenylurea-type compounds (XV) have been developed primarily for the control of annual grass weeds such as blackgrass (*Alopecurus myosuroides*) in winter cereals but their activity on wild-oat, although unpredictable, is often high.

Chlortoluron[16] (XV, X = CH$_3$, Y = Cl, R = R^1 = CH$_3$), introduced in 1969, is effective pre-emergence and early post-emergence, up to the three-leaf stage [91, 95], for the control of wild-oat at 2.4–3.6 kg/ha in winter wheat [95–97] with resultant increased yields [92, 93] but wild-oat control is not always reliable. Some winter wheat varieties are susceptible [92–95] and heavy rainfall after application may increase the susceptibility of resistant varieties [95].

Isoproturon [XV, X = CH(CH$_3$)$_2$, Y = H, R = R^1 = CH$_3$], introduced in 1973, is similar to chlortoluron but appears to be better tolerated by winter wheat varieties [96, 97] and is active at lower doses both pre-emergence (2.5 kg/ha) and early post-emergence (2.0 kg/ha) [91]. Stage of weed growth is critical for good wild-oat control [17].

Chemically related to the above are the following compounds which have a similar spectrum of activity and selectivity but are less effective in their performance against wild-oat.

metoxuron   (XV, X = OCH$_3$, Y = Cl, R = R^1 = CH$_2$)

linuron   (XV, X = Y = Cl, R = CH$_3$, R^1 = OCH$_3$)

monolinuron   (XV, X = Cl, Y = H, R = CH$_3$, R^1 = OCH$_3$)

together with benzthiazuron[17] (XVI, X = H) which is used for the control of wild-oat primarily in sugar beet

$$XVI$$

and methabenzthiazuron (XVI, X = Me) which can have effects on wild-oats but is not recommended because of unreliability in practical usage [17].

There are many other herbicides including experimental materials which have been examined for their selective toxicity towards wild-oats. As these have not been developed commercially it is not the purpose of this paper to review their properties.

---

16   Idem p. 127
17   Cf. Vol. 5, p. 145

References

1. Bachthaler, G.: Symp. EWRS Status, Biology and Control of grassweeds in Europe, EWRS 2, 7 (1975)
2. Baum, B.R.: Oats: Wild and Cultivated, Biosystematics Res. Inst. Agric. Canada, Ottawa, Monograph No. 14, 1977
3. Wild Oats in World Agriculture (ed. D. Price Jones). ARC, London 1976
4. Chapman, T. et al.: Symp. New Herbicides, Versailles 2, 40 (1969)
5. Jordan, D., Bowler, D.J., Moberly, M.C.: Span. 15(1), 26 (1972)
6. Jeffcoat, B., Harries, W.N.: Pest. Sci. 4, 891 (1973)
7. Jeffcoat, B., Sampson, A.J.: 25st Symp. Fytopharm. Fytiatre. Meded. Fak. Land. Gent 38(3), 941 (1973)
8. Haddock, E. et al.: Proc. 12th Brit. Weed Conf. 1, 9 (1974)
9. Haddock, E., Jordan, D., Sampson, A.J.: Pest. Sci. 6, 273 (1974)
10. Scott, R.M., Sampson, A.J., Jordan, D.: Proc. Brit. Crop. Prot. Conf. 2, 723 (1976)
11. Corbett, J.R.: Biochemical mode of action of pesticides, p. 60. London: Academic Press 1974
12. Canvin, D.T., Friesen, G.: Weeds 7, 153 (1959)
13. Chesalin, G.A., Thiofeeva, A.A.: Dokl. Vses. Akad. Sel-Khoz. Nauk. 6, 10 (1965)
14. Dubrovin, K.P.: Proc. Intern. 16th N. Cent. and 10th West Can. Weed Cont. Conf. 1959, p. 15
15. Pfeiffer, R.K., Baker, C., Holmes, H.M.: Proc. 5th Brit. Weed Cont. Conf. 2, 441 (1960)
16. Fiddian, W.E.H.: Proc. 6th Brit. Weed Cont. Conf. 1, 203 (1962)
17. Holroyd, J. et al.: In: Wild Oats in World Agriculture (ed. D. Price Jones). ARC, London 1976
18. Butler, A.J.: Proc. 4th Weed Cont. Conf. 1, 167 (1958)
19. Molberg, E.S.: Res. Rep., West. Sect. Nat. Weed Comm. Canada, 1963, p. 90
20. Hardisty, J.A.: Res. Rep., West Sect. Nat. Weed Comm. Canada, 1971, p. 332
21. Cottrell, H.J., Heywood, B.J.: Nature 207, 655 (1965)
22. Hibbit, C.J., Foden, P.C., Savory, B.M.: Proc. 12th Brit. Weed Cont. Conf. 1, 185 (1974)
23. Hirow, Y. et al.: 1st Meet. Pest. Sci. Soc. Japan, Tokyo, 1976
24. Murayama, T., Kusagard: Japan Pesticide, Inform. No. 35
25. Ruiz, Jean, A., Santamatilde, A., Vincenzetti, P.G.: Proc. Medit. Herb. Symp., Madrid 2, 134 (1978)
26. Derycke, S.: 30th Int. Symp. Crop Protection. Mededelingen Fakulteit Landbouwwetenschappen, Gent 43(2), 1153 (1978)
27. Ingram, G.H. et al.: Weeds 761 (1978)
28. Knott, C.M.: Weeds 261 (1978)
29. Drosthin, G., Hubl, H.: Proc. EWRS (Mainz) 397 (1979)
30. Formigoni, A., Iwataki, I., Ishihara, H.: Proc. EWRS (Mainz) 403 (1979)
31. Nippon Soda: DAS 2, 524, 577 (1979)
32. Shafer, N.E.: Proc. 12th Brit. Weed Cont. Conf. 3, 831 (1974)
33. Winfield, R.J.: ibid. p. 875
34. Winfield, R.J., Caldicott, J.J.B.: Pest. Sci. 6(3), 297 (1975)
35. Feeny, R.W., Tafino, A.J.: Abst. Meet. Weed Sci. Soc. Am., 1975, p. 86
36. Richardson, W.G., Deans, M.L.: Tech. Rep. Agr. Res. Comm. Weed Res. Org. 32, 74 (1974)
37. Coupland, D., Caseley, J.C., Simmons, R.C.: Proc. 1976 Brit. Crop Protect. Conf. Weeds 1, 47 (1976)
38. Skorda, E.A. (1973): Personal communication
39. Baldwin, J.H., Finch, R.J.: Proc. Brit. Crop Protect. Conf. Weeds 1, 31 (1976)
40. Smith, J., Finch, R.J.: Proc. Brit. Crop Protect. Conf. Weeds 3, 841 (1978)
41. Baldwin, J.H., Livingstone, D.B.: Proc. Brit. Crop. Protect. Conf. Weeds 1, 25 (1976)
42. Van Overbeek, J.: Physiology and biochemistry of herbicides, pp. 387–400. London, New York: Academic Press 1964
43. Fedtke, C.: Weeds Res. 12(4), 325 (1972)
44. Fryer, J.D., Makepeace, R.J.: Weed Control Handbook 1972, Vol. II, Recommendations, 7th Ed., Blackwell, Oxford
45. Stryckers, J., van Himme, M.: Meded. Rijksuniversiteit Gent, 1972 (M), 32–35
46. Martin, T.J., Morris, D.B., Reiley, C.E.: Proc. 11th Brit. Weed Control Conf. 1, 209 (1972)
47. Eue, L.: Z. Pfl.-Krankh. Pfl.-Path. Pfl.-Schutz 1968, 211
48. Kampe, W.: 21st Symp. Fytopharm. Fytiatre. Meded. Rijksfac. Land.-Wet., Gent 34, 973 (1969)

49. Moreland, D.E. et al.: Weed Sci. *18*, 636 (1970)
50. Bouchet, F.: Defence Veg. *136*, 107 (1969)
51. Bouchet, F., Faivre-Dupaigre, R.: 9th Brit. Weed Cont. Conf., 1968, p. 40
52. Bartlett, D.H. et al.: 9th Brit. Weed Cont. Conf., 1968, p. 30
53. Index des Produits Phytosanitaires, A.C.T.A., Paris, 1980
54. Crowley, J., Prendeville, G.N.: Can. J. Plant. Sci. *59*, 275 (1979)
55. P.A.N.S. *25*, 86 (1979)
56. Langelüdecke, P. et al.: Mitt. Biol. Bund.-Anst. Land- u. Forstwirtsch., Berlin-Dahlem *165*, 169 (1975)
57. Todd, B.G.: 9th Int. Congr. Plant. Prot. and 71st Ann. Meet. Am. Phyto.-Path. Soc., 1979, 492
58. Lutman, P.J.W., Thornton, M.E.: Proc. Brit. Crop Protect. Conf. Weeds *1*, 15 (1978)
59. Hewson, R.T.: Brit. Crop. Protect. Conf. Weeds *1*, 55 (1976)
60. Richardson, W.G., Parker, C.: Techn. Rep. ARC Weed Res. Org. *39*, 56 (1976)
61. Richardson, W.G., Deans, M.L., Parker, C.: Tech. Rep. ARC Weed Res. Ord. *38*, 55 (1976)
62. ICI Ltd.: Europ. Pat. Appl. 0,002,925 (1979)
63. Dow Chem Co.: Europ. Pat. Appl. 0,000,483 (1977)
64. Ishihara Sangyo: Brit. Pat. Appl. 2,015, 995 (1979)
65. ICI (Australia) Ltd.: OLS 2,820,032 (1977)
66. Hoechst: Belg. Pat. 875,431 (1979)
67. Nippon Soda: Belg. Pat. 876,077 (1979)
68. Kumiai Chem. Ind.: OLS 2,905,458 (1979)
69. Ishihara Ind.: J5 4,055,534 (1980)
70. Morrison, J.W.: Can. J. Sci. *42*, 78 (1962)
71. Banting, J.D.: Weed Res. *7*, 302 (1967)
72. Banting, J.D.: Weed Sci. *18*, (1970)
73. Wilkinson, R.E., Smith, A.E.: Proc. 26th Ann. Meet. Weed. Sci. Soc., 1973, p. 415
74. McKercher, R.B., Ashford, R., Morgan, R.E.: Weed Sci. *23*(4), 283 (1975)
75. May, M.J.: 8th Rep. Arthur Rickwood Exp. Husb. Farm, *1973*, 23
76. Proctor, J.M., Livingston, D.B.: Proc. 11th Brit. Weed Cont. Conf. *1*, 288 (1972)
77. Hodkinson, H.D.: Proc. 11th Brit. Weed Cont. Conf., 1972, 263
78. Gummesson, G.: Proc. 14th Swed. Weed Cont. Conf. C*28*(2), 9 (1973)
79. Bouchet, F.: Notiz. Mal. Piante. (*86*), 115 (1972)
80. Klefeld, J., Weiss, Y.: Proc. 4th Israel Weed Cont. Conf., 1970, pp. 55
81. Selleck, G.W., Hannah, L.H.: Proc. 6th Brit. Weed Cont. Conf., 1962, p. 361
82. Banting, J.D.: Res. Rep. West. Sect. Nat. Weed Comm. Can., 1963, p. 57
83. Gysin, H., Knushi, E.: Proc. 4th Brit. Weed Cont. Conf., 1958, p. 225
84. Friesen, G.: Can. J. Plant Sci. *38*, 300 (1958)
85. Furtick, W.R.: Proc. 16th West Weed Cont. Conf., 1958, p. 75
86. Sexsmith, J.J.: Res. Rep. Internat. 16th N. Cent and 10th West Weed Cont. Conf. Can., 1959, p. 78
87. Chesman, J.C., Laborde, A.: 1er Conf. Com. Franc. Mauv. Herbes (COLUMA) *1961*, p. 326
88. Kampe, W.: Meded. Rijksfac. Land.-Wet. Gent *32*(3/4), 948 (1967)
89. Kampe, W.: Gesunde Pfl. *19*(2), 30 (1967)
90. Hiebel, K.: ibid. *20*(4), 89 (1968)
91. Proctor, J.M., Armsby, W.A.: Proc. 12th Brit. Weed Cont. Conf. *1*, 33 (1970)
92. Degez, L., Dencause, G., Groyenvalle, C.: 6ᵉ Conf. Com. Franc. Mauv. Herbes (COLUMA) *1971*, p. 811
93. Maynadier, M.H. et al.: 7ᵉ Conf. Franc. Mauv. Herbes (COLUMA) *1973*, p. 349
94. L'Hermite, Y., Ebner, L., Green, O.: 5ᵉ Conf. Com. Franc. Mauv. Herbes (COLUMA) *1969*. p. 349
95. Proctor, J.M., Livingston, D.B.: Proc. 11th Brit. Weed Cont. Conf. *1*, 288 (1972)
96. Smith, J.M., Tyson, D.: Proc. 10th Brit. Weed Cont. Conf. *1*, 72 (1970)
97. Hewson, R.T.: Proc. 12th Brit. Weed Cont. Conf. *1*, 75 (1974)
98. Hubbard, K.R., Livingston, D.B.: ibid. p. 67

Neue Herbizide

Fortsetzung der Patentübersicht von Band 5

R. Wegler

Auf dem Forst 2, D-5090 Leverkusen 1

L. Eue

Bayer AG, D-5090 Leverkusen

Inhalt

44

5. Organische Herbizide

5.1 Kohlenwasserstoffe, Sulfone sowie einige Ketone

(Ketone soweit Kohlenwasserstoff-Patente auch Ketone umfassen) (Bd. 5, S. 52)

Nach wie vor entstammen aus diesen Verbindungsklassen nur wenige Herbizide.

1-(Ethenoxysulfenyl)-octan

$C_{10}H_{20}O_2S$

$CH_2 = CH - SO_2 - (CH_2)_7 - CH_3$

Diese Verbindung wird als Herbizid unter dem Namen „Alvisone" beschrieben.
Die Sulfon-Gruppe spielt für die Wirksamkeit eine wesentliche Rolle, ebenso die
Länge der Alkyl-Gruppe (C_8–C_{10}). Offensichtlich ist für die Absorption durch die
Pflanze ein gewisses Gleichgewicht zwischen Hydrophilie und Lipophilie wichtig.

Lit.: E. N. Prilezhaera et al.: Dokl. Acad. SSSR *194*, 3, 7, 27 (1970); Chem. Abstr. *74*, Ref. 12068 (1971)

1-Difluormethylsulfonyl-2-nitrobenzol

$C_7H_5F_2NO_4S$

nur wissenschaftliches Präparat

Lit.: Y. u. V. Karabanor et al., s. Chem. Abstr. *87*, Ref. 1046 (1977)

5.2 Alkohole und Ether

5.2.2 *Aromatisch-aliphatische Alkohole und ihre Ether*

a) *Benzylalkohole und Derivate* (s. Bd. 5, S. 56)

Schon 1974/75 wurden von den Firmen Shell und FMC Benzylether Derivate cyclischer Acetale dreiwertiger Alkohole (z.B. Glycerin) als erfolgversprechende Versuchsprodukte, z.T. mit spezifischer Gräserwirkung eingehender praktisch untersucht. Anscheinend treten bei feldmäßigen Prüfungen unter ungünstigen Bedingungen Schwierigkeiten auf, welche durch eine unvorhergesehene Flüchtigkeit bedingt sind. Weitere Versuchsprodukte sind in der Folgezeit nach 1975 bekannt geworden.

5-[(2-Chlorphenyl)-methoxyl]-5-methyl-2-(1-methylethyl)-cis-1, 3-dioxan

$C_{15}H_{21}ClO_3$

Versuchsprod. FMC 3032

Wirksam gegen Unkräuter und Gräser. Anwendung selektiv in Sojabohnen.
Kombination mit Metribuzin

Patente: DOS 2243685, 6.9.72/29.3.73; DOS 2247030, 26.9.72/19.4.73 FMC Corp.; US Prior. 12.10.71, 21.9.71 und 9.2.72.
Lit.: T. R. Warfield, M. T. Hillson, Proc. North Cent. Weed Control Conf. *30*, 77–79 (1975); Chem. Abstr. *85*, Ref. 138463 (1976)

$C_{16}H_{24}O_3$

Versuchsprod. FMC 3071

Formel wie oben aber CH_3 anstelle von Cl.

Patente und Literatur wie zuvor.

$C_{16}H_{23}O_3$

Versuchsprod. FMC 39896

LD_{50}: >10000 mg/kg Ratte p.o. akut. Spezifisch gegen Gräser, selektiv in Baumwolle.

Patente: DOS 2739067, 30.8.77/2.3.78, FMC Corp.; US Prior. 31.8.76 und 30.6.77; US 4035178; FR 2363666; BE 858252; NE 77/953 u.a.
Lit.: Research Rep. SWSS, 32nd Ann. Meet. 1979

R. Wegler und L. Eue

$C_{14}H_{19}ClO_3$ bzw. $C_{15}H_{21}ClO_3$

Versuchsprod. FMC 39821 bzw. 39871

LD_{50}: >5000 mg/kg Ratte p.o. akut für technische Ware.
Pre-emergence Wirkung gegen Gräser und Unkräuter (FMC 39871 hierbei etwas selektiver).
Begrenzte Selektivität in Baumwolle.

Patente und Literatur wie zuvor.

Allgemeine Darstellung z.B.:

Die Darstellung wird durch Isomerenbildung wesentlich kompliziert, weshalb auch andere Darstellungswege patentiert worden sind z. B.

5.3 Aldehyde und Ketone

5.3.1 *Aliphatische und cycloaliphatische Aldehyde und Ketone*
(Bd. 5, S. 61; s.a. „Cycloaliphatische Carbonsäuren" 5.7.2, S. 168)

Diese nachfolgend beschriebene Verbindungsklasse wird zusammenfassend bei den Ketonen aufgeführt, weil Keton-Derivate mit und ohne Carboxylgruppe bekannt sind, und nachfolgend über wirksame Variationen ohne COOH berichtet wird (in Bd. 5 noch unter den Carbonsäuren registriert, S. 168 aber auch S. 62.)

Natriumsalz des 2,1-(N-Alloxyaminobutyliden-4-methoxycarbonyl)-5,5-dimethyl-cyclohexan-1, 3-dion,

$C_{17}H_{25}NO_5$ bzw. $C_{17}H_{24}NO_5Na$

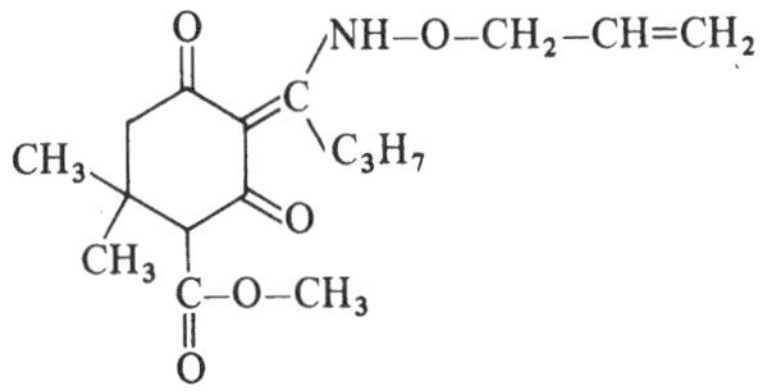

Common name: Carbodimedon und Alloxydimedon bzw. Alloxydim-sodium
Versuchsprod. der Nippon: NP 48, Kusagard®
Versuchsprod. der Schering AG: SN 75101, Fervin®
LD_{50}: 2322 mg/kg p.o. männl. Ratte, 2260 mg/kg weibl. Ratte; 1700 bzw. 1689 mg/kg intraperitonal; 1630 bzw. 1385 mg/kg männliche bzw. weibliche Hunde p.o.; 3700 bzw. 3850 mg/kg Kaninchen p.o.; 3200 bzw. 4600 und 3000 bzw. 3660 mg/kg für verschiedene Arten von Mäusen p.o.. Keine chronische Toxizität bei Ratten. Fischgiftigkeit: 48 Stdn. > 250 ppm für *Cyprimus carpio*. Andere Fischarten zeigen ähnliche Werte. In den üblichen Anwendungskonzentrationen keine Bienengiftigkeit.
Anwendung: 75 % wasserlösliche Puder des Na-Salzes 1–2 kg/ha. Gut löslich in Wasser, Methanol, Dimethylformamid. Schlecht löslich (Na-Salz) in Aceton, Xylol und Cyclohexan.
Wirkungsweise: post-emergence gegen Ungräser in dikotylen Kulturen. Aufnahme über die Blätter und z.T. über die Wurzeln. Selektiv in Zuckerrüben, Sojabohnen, Sonnenblumen, Raps, Baumwolle und Gemüse. 0,5–1 kg/ha. Als Kombination mit Basagran® (BASF) selektiv in Sojabohnen ebenso als Kombination mit Phenmedipham in Zuckerrüben.

Darstellung:

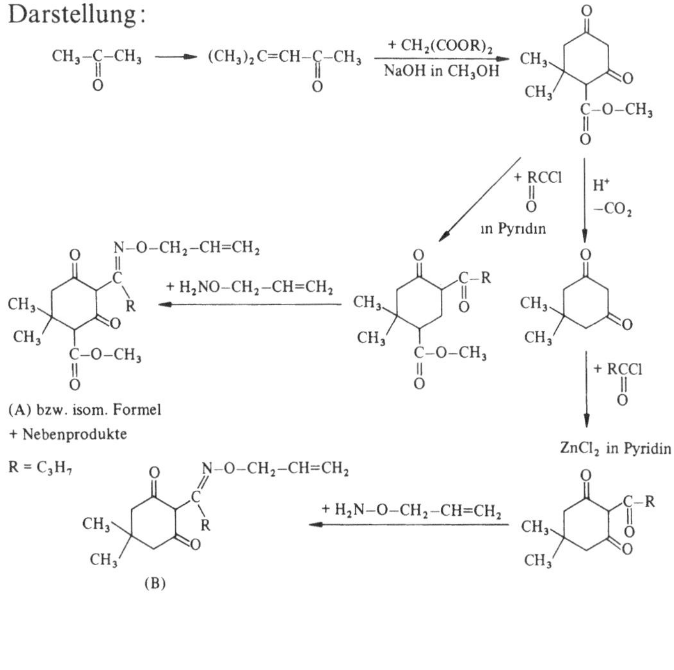

(A) bzw. isom. Formel
+ Nebenprodukte

$R = C_3H_7$

(B)

R. Wegler und L. Eue

Patente: DOS 2439104 v. 15.8.73/13.3.75; Jap. Prior. 18.8.73 und 7.3.74, Nippon Soda; Be. P. 818849.
Lit.: 41. Deutsche Pflanzenschutz-Tagung in Münster 1977, H. Hübl, E. Aster u. H. Laufersweiler: –
T. Murayama: Jap. Pestic. Inform. [35] 24–26 (1978); Weed Abstr. *27* [28], 293, Ref. 2711 (1978); *29*
[1], 40, Ref. 2711 (1980); – T. Murayama: Adv. Pest. Sci., JUPAC, Zürich 1978, Part 2 (Ed.:
U. Geissbühler p. 235, Pergamon Press, London 1979)

2-[1-Ethoxyimino-butyl]-5-(2-ethylthiopropyl-3-hydroxy-2-cyclohexen-1-on

$C_{17}H_{29}NO_3S$

Versuchsprod. NP 55 der Nippon Soda 1979 u. der BASF.

LD_{50}: 1000 mg/kg Ratte p.o., TLm_{48} Karpfen 10 ppm

oder als isomere Diketon-Form

Postemergence hochwirksam gegen Gräser in niederen Konzentrationen (0,2–0,4 kg/ha). In höheren Konzentrationen auch gegen Dauergräser. Auch dikotyle Unkräuter werden z.T. erfaßt. Wirksamer als Alloxydim-Natrium. Postemergence Anwendung, wasserlöslich wirksamer als NP 48. Selektiv in Baumwolle, Zuckerrüben, Erbsen, Raps, Tomaten, Sonnenblumen, Bohnen, Möhren, Wassermelonen, Salat und Zwiebeln.

Darstellung:

usw. siehe zuvor Carbodimedon.

Patente: DOS 2822304, 22.5.78/30.11.78, Nippon Soda; BE 867093.
Lit.: A. Formigoni et al.: Abstr. IX. Intern. Congr. Plant Protektion and 71. Ann. Meet. Amer. Phytopath. Soc. Washington/USA. August 5–11, 1979, Ref. 477 (1979)

Über Zusammenhänge zwischen chemischer Konstitution und herbizider Wirksamkeit wurde ausführlich berichtet auf der IUPAC Tagung 1978 in Zürich. Die Grundformeln

50

erlauben eine Fülle von Variationen: Eine Wasserstoff-Brückenbindung der verschiedenen isomeren Formen scheint eine günstige Voraussetzung für herbizide Wirksamkeit zu sein.

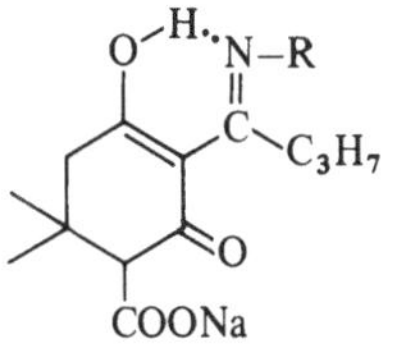

und weitere Formen.

X, Y und Z können Heteroatome sein und ergeben dann hohe pre-emergence Aktivität.

$R^1 = H$, ergibt Wirkungslosigkeit; R^2 ergibt ein Maximum bei niedrigen Alkylenen (Allyl);

R^1 u. R^4 zeigen größten Einfluß auf die Wirksamkeit;

R^4 kann $2 \times CH_3$, iso. Propyl, Cyclohexyl, Phenyl, Thienyl sein;

R^5 am wirksamsten als freie Säure oder Methylester;

$R^5 = H$, hochwirksam wenn R^1 u. $R^2 = C_2H_5$ und $R^3 = H$, $R^4 = H$. Die umfangreichen Tabellen über Variationen der Substituenten und des Grundmoleküls (X, Y u. Z = O oder N) müssen im Original des IUPAC Berichtes 1979 nachgelesen werden.

5.3.2 *Aromatische Ketone und aromatisch-aliphatische Aldehyde und Ketone* (Bd. 5, S. 63)

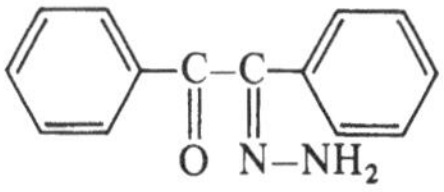

BMH, Rohm u. Haas

Anwendung als Safener gegen Herbizidschäden.

Patent: US 4009022/30.10.75/22.2.77. Rohm und Haas.
Antidot gegen Herbizidschäden durch Triazin-Herbizide

5.4 Phenol-Derivate

5.4.2 *Phenolether* (Bd. 5, S. 73–80)

Daß selbst einfache aromatisch-aliphatische Phenolether, also ohne die im folgenden ausführlich zu besprechende Diphenylether-Struktur, schon herbizid wirksam sind, wurde jüngst durch ein Versuchsprodukt belegt. N, N-Diethyl-2-[4-(phenylethinyl)-phenoxy]-ethanamin.

$C_{20}H_{23}NO$

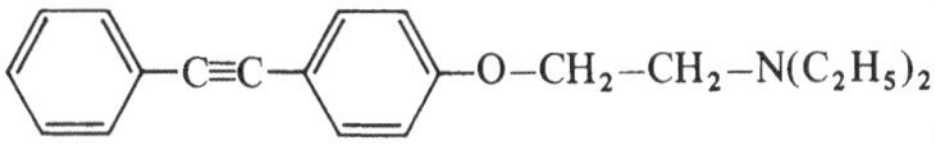

Versuchsprodukt der Fisons Ltd.

Wirkungsweise: Verhindert die Carotenoid-Biosynthese.

Lit.: B.J.Wright et al.: Phytochemistry *19*, 61–64 (1980)

Das nachfolgende Präparat zeigt, daß die aliphatische Amino-Gruppe für eine herbizide Wirkung entbehrlich ist.

Ethoxy-4-(2, 4-dichlorphenoxy)-2-buten

$C_{12}H_{14}Cl_2O_2$

Cl—⟨benzene ring, Cl⟩—O—CH_2—CH=CH—CH_2—O—C_2H_5

Pflanzenwuchsregulator und Herbizid

Die Wirksamkeit wird erklärt durch den Abbau zu 2, 4-Dichlorphenoxyessigsäure.

Lit.: C. Rius-Alonso, R. L. Wain: Appl. Biol. *88* [2], 299–308 (1978)

Cl—⟨benzene ring, Cl, Cl⟩—O—CH_2—CH_2—OH

Klorinol® der Budapest Chemical Works (1976) in Mischung mit Atrazin® als Buvinol®

LD_{50}: 1490 mg/kg männl. Ratte p.o. akut. Anwendung pre-emergence in Mais, ferner im Weinbau, Obstbau und Forst (Eichen und Buchen).

Klorinol® wird in der Pflanze z.T. zu 2, 4, 5-Trichlorphenoxyessigsäure sowie zu 2, 5-Dichlor-dioxybenzol, 4, 5-Dichlorbrenzcatechin und 2, 4-Dichlor-dioxybenzol abgebaut.

Lit.: The Development of Pesticide as a Complex Scientific Task, Budapest Chemic. Works 1976; Weed Abstr. *26* [9], Ref. 2784 (1977)

Diphenylether

Einige neue Produkte entstammen der Gruppe der Diphenylether; schon in Bd. 5, S. 73–80 wurde eine größere Anzahl beschrieben. Im Nachfolgenden werden auch Diphenylether mit einer Carboxyl-Gruppe an einem Phenyl-Rest miterfaßt, weil die Wirkungsweise sicher in erster Linie dem Diphenylether-Anteil zuzuschreiben ist. Die Carbonsäuren sind aber im Gegensatz zu den Diphenylethern besonders gegen Unkräuter wirksam. Dagegen werden die Diphenylether-oxy-essig(propion)-säuren bei den Penoxy-alkancarbonsäuren behandelt, obwohl auch hier die Diphenylether-Gruppierung für die überraschende herbizide Wirkung gegen Ungräser verantwortlich erscheint. Die intensive Bearbeitung der verschiedensten Diphenylether-Derivate hat zu einer sehr großen Zahl von Patenten geführt (siehe Patentergänzung), zahlreiche Überschneidungen sind erfolgt, eine Priorität kann nicht festgelegt werden.

Nur ein Diphenylether ohne Carboxyl-Gruppe ist seit 1976 bekannt geworden:

2,4-Dichlor-1-[3-[2-(2-ethoxyethoxy)-ethoxy]-4-nitrophenoxy]-benzol

$C_{18}H_{19}Cl_2NO_6$

Cl—⟨benzene ring, Cl⟩—O—⟨benzene ring⟩—NO_2
O—C_2H_4—O—C_2H_4—O—C_2H_5

Versuchsprod.: X 150 (Ishihara Sangyo)

5-(2,4-Dichlorphenoxy)-2-nitrobenzoesäure-K-salz

$C_{13}H_7Cl_2NO_5K$

Versuchsprod.: MC 7783 d. Mobil Chem. Co.

Der entsprechende Methylester (irrtümlich Ethyl-ester) ist im Bd. 5, S. 76 beschrieben. Postemergence Anwendung gegen Unkräuter in Sojabohnen und Erdnüssen. Im frühen Entwicklungsstadium werden auch Ungräser angegriffen.

Patente: DOS 2408432, 21.2.74/5.9.74, Mobil Chem.; Zusatz zu DOS 2019821; US 3812184 u. 3983168; GB 1232369 u. 1430496; FR 2326865 u.a.
Lit.: Techn. Information der Mobil Chem. 1978

Darstellung:

5-(2-Chlor-4-trifluormethylphenoxy)-2-nitro-benzoesaures Natrium

$C_{14}H_7ClF_3NO_5Na$

Common name Vorschlag: Acifluorfen
Versuchsprod.: RH 6201, Blazer®, Rohm u. Haas
MC 10978 Mobil Res.

Wasserlöslichkeit: > 25 g/100 ml. LD_{50}: 1540 mg/kg männl. Ratte p.o. akut; techn. Ware: 3330 mg/kg. Anwendung: Pre-emergence und besonders post-emergence gegen Unkräuter und Ungräser. Selektiv in Sojabohnen, Erdnüssen, Erbsen, Reis und Baumwolle.

Patente: DOS 2311638, 9.3.73/27.9.73; US Prior. 14.3.72 Rohm u. Haas, US 3798276; FR 2175997; s.a. DOS 2304006, 27.1.73/9.8.73, Ishihara Sangyo
Lit.: W.O.Johnson et al.: J. Agric. Food Chem. *26*, 295 (1978);
Proc. Northeastern Weed Sci. Soc. New York 4.–6.1.1978, *325*, 28–29 (1978)

Darstellung:

$C_{15}H_9ClF_3NO_3$
Methylester des Acifluorfen

Versuchsprod. MC 10108 Mobil Chem. Co.

Anwendung: Vorsaateinarbeitung und preemergence gegen Unkräuter in Sojabohnen. Auch gegen Gräser wirksam.

Patent: DOS 2311638, Rohm u. Haas (s.a. Acifluorfen)

$C_{16}H_{11}ClF_3NO_5$

ist der entspr. Ethylester. Common name Vorschlag: Acifluorfen-ethylester Versuchsprod. RH 8817 v. Rohm u. Haas, MC 10982 Mobil Res. – LD_{50}: 1770 mg/kg Ratte p.o. akut. Bodenherbizid gegen Unkräuter und Gräser. Vorsaat-Anwendung. Selektiv in Sojabohnen, Erdnüssen und Baumwolle.

Patente: Siehe oben sowie Lit.: Techn. Information d. Rohm u. Haas Co.

5-(2-Chlor-4-trifluormethyl-phenoxy)-N-methylsulfonyl-2-nitrobenzamid

$C_{15}H_{10}ClF_3N_2O_6S$

Versuchsprod. ICI (1980) PP 0221
Common name-Vorschlag: Fomesafen.

Patente: EP 3416, 27.12.78/8.8.79, ICI; GB Prior, 19.1.78, GB 2014565
Lit.: (Irving L. Adler, Brent Marco Irnes, J. Wargo jr. (Rohm u. Haas Lab.): J. Agric. Food Chem. *25*, [6] 1339 (1977)

Über den *Metabolismus* einiger Diphenylether gibt es ausführliche Untersuchungen z.B. über Oxyfluorfen und RH 2915 (s. Bd. 5, S. 75). Demnach tritt eine Abspaltung der Ethoxy-Gruppe zum Phenol ein, Reduktion des NO_2 zu NH_2 und anschließende Acetylierung der NH_2-Gruppe.
Interessant ist eine neuere Beobachtung, daß wenigstens zwei Diphenylether (CNP®, Bd. 15, 74 u. 78) sowie Nitrofen® (S. 74 u. 78) eine deutliche Wirksamkeit gegen Mosquito-Larven aufweisen (Makiko Ikeuchi et al., IUPAC Zürich 1978; Adv., Pesticide Sci. Part 3, S. 430, Pergamon Press Oxford 1979).

5.5 Kohlensäure und Thiokohlensäure-Derivate

5.5.1 *Kohlensäure und Thiokohlensäurediester* (Bd. 5, S. 87)

[Bis-(trifluormethyl)-(4-brom-3,5-dinitrophenyl)]-methylcarbonat

$C_{11}H_5BrF_6N_2O_7$

Versuchsprodukt Herbizid und Wuchsregulator

Lit.: Chem. Abstr. *93* [13], Ref. 108792 (1980), Originallit. russisch

5.5.2b *Mono und Di-thiocarbamidsäureester sekundärer aliphatischer Amine* (Bd. 5, S. 92–96)

Zusammenhänge zwischen chemischer Konstitution und herbizider Aktivität wurden in der Reihe der Benzylthiocarbamate untersucht (Bd. 5, S. 95). Verbindungen der Formel

sind in ihrer preemergencen Wirkung gegen Ungräser optimal bei $R = C_2H_5$. Eine Substitution im Phenyl-Rest durch $-CH = CH_2$ senkt die Wirksamkeit auf 1/3 bis 1/5. Die Dithio-Verbindungen sind stets weniger wirksam.

Lit.: Th.J.Giacobbe et al.: J. Agric. Food Chem. *25* [2], 320–323 (1977)

Interessant erscheint der Versuch, die herbizid wirksame Dithiocarbamidester-Gruppe an ein Polymeres anzuhängen

$(C_7H_{13}NS_2)_n, n = 138$

Wirksam gegen Echinochloa crus galli preemergence. Keine fungizide Wirksamkeit.

Lit.: H.Naruse, K.Maekawa: Chem. Abstr. *87* [19], Ref. 146932 (1977)

Auch aus Polyethyleniminen lassen sich durch Umsetzung mit Schwefelkohlenstoff und anschließende Reaktion mit Alkylchlorid höhermolekulare Dithioester gewinnen, welche fungizide und herbizide Eigenschaften aufweisen.

Lit.: H.Naruse, K.Makawa: J. Fac. Agric. Kyushu Univ.: *21* [2–3], 107–116 (1977); Chem. Abstr. *87* [9], Ref. 63924 (1977)

R. Wegler und L. Eue

In den letzten Jahren wurden Bedenken gegen einige Herbizide und Fungizide aus der Reihe der Dialkylcarbamidsäurethiolester (Bd. 5, S. 91 u. 92) hinsichtlich einer möglichen krebsverursachenden Wirkung bekannt. Da Cancerogenität und Mutagenität meist parallel verlaufen, wurden 20 entsprechende Carbamate einem Mutagenitäts-Test unterzogen (Ames-Test). Nur bei 3 Vertretern konnte Mutagenität festgestellt werden. Da diese drei Verbindungen alle eine 2-Chlorallyl-Gruppe im Ester-Teil enthielten, darf angenommen werden, daß diese Gruppierung für die Mutagenität verantwortlich ist.

Lit.: F. Lorenzo et al.: Cancer Res. *38*, 13–15 (1978)

In weiteren Untersuchungen wurde Mutagenität durch die gleichen Herbizide Diallate, Triallate und Sulfallate mit ähnlichen Testen festgestellt.

Lit.: H. Sikka, K. P. Floczy: J. Agric. Food Chem. *26* [1] 146–198 (1978)

Die Mutagenität von Diallate, Triallate und Sulfallate darf nach jüngsten Untersuchungen von J. Schuphan und Mitarbeitern als chemisch weitgehend verständlich betrachtet werden. Sulfoxyde welche in vitro und in vivo in der Leber durch Oxidation am S der Thiolcarbamate entstehen, sind nur mässig beständig, wenn in der Thiol-Hälfte in β-Stellung ein Wasserstoff steht.

Der S-Benzylester ist dementsprechend stabil.
Die Oxidationsprodukte zerfallen schon bei 25 °C in

Der S-Ethylester zerfällt erst bei 50 °C. Besteht der Thiol-Teil aus einem ein- oder mehrfach chlorierten Allylthiol, so lagern sich die Sulfoxide quantitativ um und erst die neuen Verbindungen spalten sich.
Im DOS 2518544 (Bd. 5, S. 419) wird schon diese Formel angegeben, allerdings ohne nähere Angaben der Umlagerung, weshalb eine irrtümliche Formulierung angenommen wurde.

Aus dem Trichlorsulfallat-Derivat entsteht ein Säurechlorid

56

$$\underset{R'}{\overset{R}{>}}N-\overset{\overset{\displaystyle\|}{S}}{C}-S-\text{Alkyl} \xrightarrow{\text{Oxidation}} \underset{R'}{\overset{R}{>}}N-\overset{\overset{\displaystyle\|}{\underset{\displaystyle\downarrow}{S}}}{C}-S-\text{Alkyl}$$

Sie ergeben keine den Monothio-Verbindungen entsprechende Umlagerung.
Die zuerst erwähnten Thiolcarbamate werden noch in anderer Weise metabolisiert, indem unter anderem an der α-CH-Gruppe oxidativ hydroxyliert wird. Die Metaboliten zerfallen dann weiter wie folgt:

$$>N-\overset{\overset{\displaystyle\|}{O}}{C}-S-CH_2-R \longrightarrow \left[>N-\overset{\overset{\displaystyle\|}{O}}{C}-S-\overset{\underset{\displaystyle OH}{|}}{C}H-R \right] \longrightarrow >NH + COS + \overset{\overset{\displaystyle H}{|}}{\underset{\underset{\displaystyle O}{\|}}{C}}-R$$

Auch hierbei entsteht z.B. aus Chlorallylthiol 2-Chloracrolein, aus Sulfallate aber ein Säurechlorid. Chloracroleine und Dichloracrylsäurechlorid sind Mutagene. Eine weitere Reaktionsmöglichkeit der Sulfoxide besteht in einer Umcarbamylierung, wobei diese schneller von statten geht als bei den Ausgangsverbindungen. Sehr wahrscheinlich ist diese Reaktion für die herbizide Wirksamkeit von Bedeutung.

Lit.: Ing. Schuphan, Y. Segall, J. D. Rosen: ACS-Meet. Las Vegas 1980; Privatmitteilung an den Autor.
Teilveröffentlichung: J. Schuphan, J. E. Casida: J. Agric. Food Chem. *27* [5], 1060 (1979)

5.5.2c *Carbamidsäureester aromatischer Amine mit aliphatischen Alkoholen* (Bd. 5, S. 102–106)

Aus dieser Verbindungsklasse einfacher Carbamidsäureester ist nur ein Versuchsprodukt bekannt geworden:

1-Isopropylcarbamoylethyl-3-chlorphenylcarbamat

$C_{13}H_{17}ClN_2O_3$

Common name-Vorschlag: Famoxacarb
Versuchsprod. RP 11755 Phamoxamide® (Rhone Poulenc)

Patent: GB 985362, 11.7.63/10.3.65, Rhone Poulenc.
Lit.: R. G. Harrey, R. C. Baker: Weed Res. *14*, 57–63 (1974)

Carbamidsäureester mit zwei Carbonamidsäureester-Gruppierungen oder einer Carbamidsäureester und einer Harnstoff-Gruppierung. (Bd. 5, S. 107)

Diese vielseitig abwandelbare Verbindungsklasse hat schon zu mehreren z.T. wichtigen Handelsprodukten geführt, wird aber immer noch weiter bearbeitet, wie zahlreiche Patente und Versuchsprodukte zeigen. Man beachte die *m*-Stellung der Substituenten.

o-[3-(Methylaminocarbonyloxy)-phenylaminocarbonyl]-2-propan-oxim.

$C_{12}H_{15}N_3O_4$

Versuchsprod. der VEB Fahlberg-List.

Patent: DDR 89754, 6.1.71/5.5.72, Fahlberg-List.
Lit.: S.Lang, P.Held: Wiss. Beiträge der Martin Luther-Univers. Halle-Wittenberg [10] 130–136 (1973)

3-(Ethoxycarbonylamino)-(phenyl-N-(4-fluorphenyl)carbamat)

$C_{16}H_{15}FN_2O_4$

Versuchsprod. SN-40321 Schering AG, Berlin

Anwendung: Selektiv in Bohnen und Sojabohnen; pre emergence wirksam gegen *Ambrosia artemisifolia, Chenopodium album, Amaranthus retroflexus* und *Setaria-viridis.*

Patente: DOS 1593523, 15.10.66/16.7.70, Schering AG Berlin; US 3901936; GB 1205786; FR 1547627; DDR 60302; BE 705071 u.a.
Lit.: G.H.Bayer: Proc. Northeast Weed Sci. Soc. 31–34 (1977); Chem. Abstr. *87* [17], Ref. 128645 (1977)

Darstellung z.B.:

Ein ähnliches Versuchsprodukt ist:

$C_{19}H_{22}N_2O_4$

Phenisopham, SN 58132 (Schering)

58

Über den Abbau von Carbamaten mit zwei Wirkungsgruppen ist einiges bekannt: Tandex® (Bd. 5, S. 135) erleidet an der Carbamidsäurephenylester-Gruppierung im Boden und z.B. in Gräsern sowie im Wasser eine Spaltung zum freien Phenol.

Lit.: Sami Selim, Ronald F. Cook, Bruce E. Leppert: J. Agric. Food *25* [3], 567 (1977)

5.5.3 *Harnstoffe*

5.5.3e *Aromatisch-aliphatische Harnstoffe und Thioharnstoffe* (Bd. 5, S. 122)

Die schon in Bd. 5 aufgeführten 4-Isoalkylphenyl-harnstoffe sind um ein weiteres Versuchsprodukt vermehrt worden. Ältere bekannte Harnstoffherbizide wurden von mehreren Firmen auf den Markt gebracht.

CRD der Pepro
(siehe Bd. 5, S. 124 oben)

N^1-[3-Chlor-4-(1-methyl-ethyl)-phenyl]-N, N-dimethylharnstoff

$C_{12}H_{17}ClN_2O$

Versuchsherbizid C 24680 (Ciba-Geigy)
Herbizid und Pflanzenwuchsregulator

Patente: DOS 2107774, 18.2.71/16.9.71, Ciba-Geigy; Schw. Prior. 27.2.70, 537148; GB 1293500; FR 2079082; DDR 92352 u.a.

$C_{11}H_{10}Cl_2N_2O$

Diamond Shamrock Corp.

Eine ähnliche Verbindung mit Isobutin aber nur einem Cl ist als Buturon (Bd. 5, S. 128) ausführlich beschrieben. Die neue Verbindung zeigt zwar gute herbizide Aktivität aber mangelnde Selektivität.
Lit.: W.J. Pyne et al.: J. Agric. Food Chem. *27*, 537–543 (1979)

Ein Namensnachtrag ist zu der folgenden Verbindung erforderlich, die zuvor im Bd. 5, S. 124 beschrieben worden ist: Neuer Name: HOE 16410 (Hoechst)
CGA 18731 (Ciba-Geigy)

Neuer Name: HOE 16410 (Hoechst)

Common name: Ipuron
Handelsnamen: Graminex®, (Ciba-Geigy), Melsan® (Du Pont), Arelon (Hoechst), Tolkan®, IP 50, Hytane®, Graminon®
Die Namen entsprechen verschiedenen Formulierungen.
Wasserlöslichkeit: 170 ppm bei 25 °C. Wirkungsweise: Photosynthesehemmer (Einzelheiten s. Bd. 5, S. 124)

R. Wegler und L. Eue

Das einfachste Beispiel eines *Thioharnstoffes* aus dieser Verbindungsklasse soll als Bodenherbizid Verwendung finden.

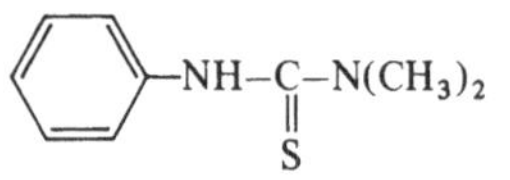

$$\text{(Phenyl)}-NH-\underset{\underset{S}{\|}}{C}-N(CH_3)_2$$

CKB 1129 Fenthioron®, VEB Chemiekombinat Bitterfeld.

Patente: DDR 110152, 110155, 110156, 110159. Anwendung bes. in Kombinationspräparaten.

Aromatisch-aliphatische Harnstoffe mit einer weiteren herbizid wirksamen Gruppierung am Phenyl-Ring
z.B. einer Carbamidsäureester-Gruppierung (s.a. Bd. 5, S. 135)

Ähnlich wie bei den Carbamidsäureestern mit einer zusätzlichen herbiziden Gruppierung am Phenyl hat Fahlberg-List ein entsprechendes Versuchsprodukt bekanntgegeben.

N-[(3-methylaminocarbonylamino)-phenyl]-isopropylcarbonat

$C_{12}H_{17}N_3O_3$

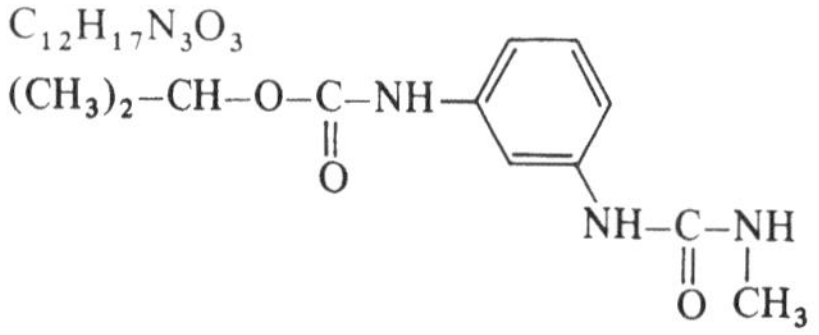

$$(CH_3)_2-CH-O-\underset{\underset{O}{\|}}{C}-NH-\text{(Phenyl)}$$

Versuchsprod. VEB Fahlberg-List

$$NH-\underset{\underset{O}{\|}}{C}-\underset{\underset{CH_3}{|}}{N}H$$

Patente: DDR 89754, 6.1.71/5.5.72, VEB Fahlberg-List.
Lit.: S.Lang, P.Held: Wiss. Beiträge der Martin Luther-Univers. Halle Wittenberg [10] 130–136 (1973)

Harnstoffe mit Phenyl am N^2 und N^1 Glied einer Cyclo-Verbindung (Bd. 5, S. 130)

Es ist ein weiteres Versuchsprodukt bekannt geworden:

N-(3,4-Difluorphenyl)-2,5-dimethyl-1-pyrrolidincarboxamid

$C_{13}H_{26}F_2N_2O$

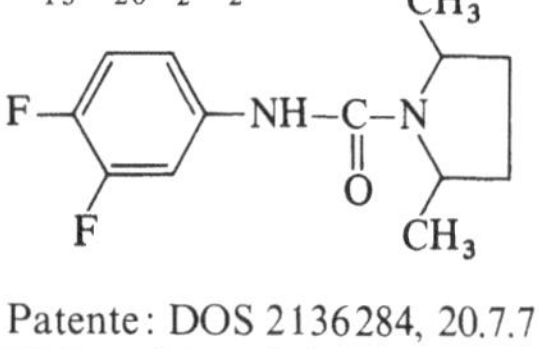

Diamond Shamrock Corp.

Patente: DOS 2136284, 20.7.71/31.8.72; US 3778247, GB 1332638.
Weitere Lit. s. Bd. 5, S. 309

Zusammenhänge zwischen chem. Konstitution und herbizider Wirkung wurden am nachfolgenden Verbindungstyp untersucht:

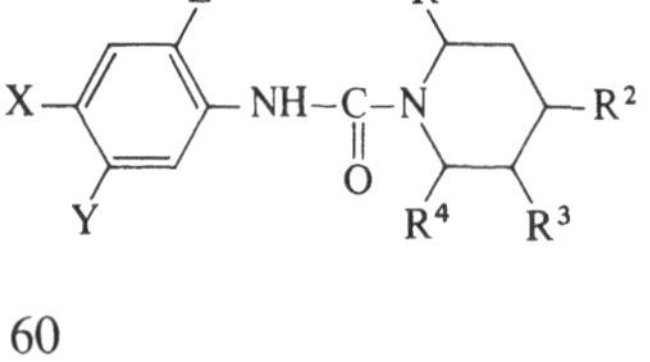

60

Bei post-emergencer Wirkung: X, Y, Z: 4 Cl > 2 Cl u. 4 CH$_3$ > 3 CH$_3$; umgekehrte Reihenfolge bei pre-emergencer Anwendung. Di- und trisubstituierte Derivate zeigen durchweg geringere Wirksamkeit.

Bestes Versuchsprodukt ist bei pre-emergencer Anwendung:

Auch Harnstoffe mit Morpholin anstelle des Piperidin wurden untersucht

Lit.: N. Haeberle et al.: Qual. Saf. Suppl. *3* (Pesticides) 696–706 (1975); Chem. Abstr. *85* [19], Ref. 138438 (1976)

Am N^1 acylierte Harnstoffe (Bd. 5, S. 136)

Einige am N^1 acylierte Harnstoffe (in Bd. 5 schon durch Patente bekannt) sind jetzt Versuchsprodukte geworden.

N^2-(4-Chlorphenyl)-N^1-methyl-N^1-thio-phenyl-harnstoff

C$_{14}$H$_{13}$ClN$_2$OS   bzw.   C$_{14}$H$_{12}$ClFN$_2$OS

Ortho 11413 bzw. Ortho 12535 (Chevron Res. Co. 1976)

Unter RE 13749 ist die 4-Fluor-Verbindung bekannt.

Darstellung:

Patente: US 3771993, 15.5.72/13.11.73, Chevron Res. Co.
Lit.: W. Prescott: Chem. and Ind. *2*, 56–63 (1978)

N-(3,4-Dichlorphenylaminocarbonyl)-N-methyl-2-propan-sulfenamid.

C$_{11}$H$_{14}$Cl$_2$N$_2$OS

Versuchsprod. RU 19331 (Roussel Uclaf)

Ohne besondere Vorteile

Patente: DOS 2256275, 16.11.72/24.5.73, Roussel Uclaf; FR 2208604, 17.11.71/28.6.74, Roussel Uclaf.
Lit.: M. Trunkenboltz, M. Prin: C. R. J. d'Études sur les Herbicides, 8. Conf. du Culuma *3*, 652–663, 753–767 (1975)

N^1-Cyclohexyldithio-N^2-(2-fluorphenyl)-N^1-methylharnstoff

C$_{14}$H$_{19}$FN$_2$OS$_2$

Versuchsprod. RE 17111, Ortho 17111 (Chevron Chem. Co.)

I

R. Wegler und L. Eue

Darstellung:

Anwendung: In Kombination mit Metribuzin in Kartoffeln.

Patente: GB 1416123, 13.3.74/3.12.75, Chevron.
Lit.: H.J.Murphy, M.J.Goven: Proc. Northeast Weed Sci. Soc. *30*, 235 (1976)

Am N^2 acylierte Harnstoffe (Bd. 5, S. 136)

S. 137: Formelfehler-Berichtigung

Methoxymarc

N-(4-Chlorphenyl)-N-[[(4-chlorphenyl)-imino]methyl]-N^1,N^1-dimethylharnstoff und das entsprechende 3-Chloranaloge.

$C_{16}H_{15}Cl_2N_3O$

Versuchsprod. SAN-224 bzw. SAN 225 (Sandoz AG)

Patente: DOS 2308943, 23.2.73/13.9.73, Sandoz AG; CH 576432 (29.2.72), GB 1418999, FR 2174103, US 3966805 u.a.

Darstellung:

Ethyl[(3,4-dichlorphenyl)-(dimethylaminocarbonyl)aminooxacetat]

$C_{13}H_{24}Cl_2N_2O_4$

Versuchsprod. R 3322 (Stauffer Chemical)

wenig aussichtsreiches Herbizid

Patente: DOS 2610804, 15.3.76/30.9.76, Stauffer. US 4018813, 4068079; GB 1484116; FR 2304602; BE 839691 u.a.
Lit.: C.J.Noll: Proc. Northeast Weed Sci. Soc. *31*, 244–246 (1977)

N^1 Methoxyharnstoffe (Ergänzung Bd. 5, S. 138)

Linuron, Afalon® (Hoechst), Lorox® (Du Pont)

In Kombination mit Trifluoralin (Treflan®) (Bd. 5, S. 236) als Mudekan® und Chandor®, Hoechst, Eli Lilly und Celamerk. In Kombination mit Nitralin (Planavin®) (Bd. 5, S. 239) als Argold® (Deutsche Shell Chemie).

N-Methoxy-N-methyl-N^1-4[2-(4-methylphenyl)-ethoxyl]-phenyl-harnstoff

$C_{18}H_{22}N_2O_3$

Versuchsprodukt S 3552 Sumitomo

Selektiv wirksam in Sojabohnen, post-emergence gegen zahlreiche Unkräuter. Schwächer wirksam gegen Gräser.

Patente: DOS 2809035, 2.3.78/7.9.78; JA Prior. 3.3.77; FR 2382434; US 4129436; GB 1555652; BE 864029 u.a.

N^2-Oxyharnstoffe (Harnstoffe des Phenylhydroxylamins) (Bd. 5, S. 140)

N-Methoxy-N^1-(1-methyl-1-phenylethyl)-N-phenyl-harnstoff

$C_{17}H_{20}N_2O_2$

Versuchsprod. der Showa Denko K.K. 1979

Dieses neue Versuchsprodukt unterscheidet sich bemerkenswert von anderen Harnstoff-Herbiziden; es ist methoxyliert am N^2, enthält aber am N^1 ein H und ist Derivat eines Benzylamins. Die Verbindung kann pre-emergence und bei Vorsaateinarbeitung angewendet werden; sie ist wirksam gegen Ungräser und Unkräuter und soll selektiv in Sojabohnen, Erdnüssen, Tomaten, Reis und Baumwolle sein.

Patente: DOS 2451418, 29.10.74/7.5.75, Showa Denko; Jap. Prior. 30.10.73; GB 1422514; US 3972909; FR 2249076; CH 604487.

Lit.: H. Kubo, N. Seki: Abstr. ACS/CSI Chem. Congr. Honolulu, Hawai 1.–6.4.1979, Part II, Ref. Pest 74

5.5.3i *Heterocyclisch-aliphatische Harnstoffe*

5.5.3i 3. *5-Ring-Verbindungen mit* $1 \times N + 1 \times O$ *und* $2 \times N + 1 \times S$, *überbrückte Harnstoff-Verbindungen* (Bd. 5, S. 147–150)

Von verschiedenen Heterocyclen sind in jüngster Zeit (1980) Versuchsherbizide bekannt geworden:

N-[5-(1,1-Dimethylethyl)-3-isoxazolyl]-N'-methylharnstoff

$C_9H_{15}N_3O_2$

Versuchsprodukt Shionogi u. Co.

Selektiv in Mais und Sojabohnen, pre-emergence Anwendung.

Patente: JA 78/86033, 4.9.77/29.7.78
Lit.: Shionogi: Chem. Abstr. *90*, Ref. 49639 (1981)

Mehr Bedeutung scheint dem N-Dimethyl-Produkt zuzukommen.

$C_{10}H_{17}N_3O_2$

Isouron®
Versuchsprod. SSH 43
Shionogi Seiyaku, 1980

Das Herbizid ist wenig toxisch. Anwendung pre-emergence gegen Unkräuter und Ungräser bis zum 1–3-Blattstadium des Unkrautes. Selektiv in Mais und Zucker-rohr.

Patente: DOS 2544367, 3.10.75/22.4.76, Shionogi; JA Prior. 4.10.74, 76/41431; US 4028376; GB 1463439; FR 2286820; CH 595762; NL 75/11694.
Lit.: Proc. Asian Pacific Weed Sci. Soc. Conf. Sydney/Australia 31–40 (1979); Weed Abstr. *29* [17], Ref. 3823 (1980)

Die schon sehr zahlreichen heterocyclisch-aliphatischen Harnstoffe des 1,3,4-Thiadiazol (siehe bes. Bd. 5, Tab. 9, 148–149) wurden um ein weiteres Versuchs-produkt vermehrt:

N-[5-(2-Chlor-1,1-dimethyl-ethyl)-1,3,4-thiadiazol-2-yl)-N,N^1-dimethyl-harnstoff.

$C_9H_{15}ClN_4OS$

Versuchsprod. EL 112
(Eli Lilly, 1979)

LD_{50}: 1000 mg/kg Ratte p.o. akut.
Herbizid gegen Unkräuter und Gräser in Zitrus-plantagen.

Patente: DOS 2325332, 6.12.73/18.5.73, US 3972706, Eli Lilly Co.; DOS 2423469, Gulf Res. a. Dev; GB 1439501, Air Prod. and Chem. Inc.
Lit.: R.D. Hicks et al.: Proc. South Weed Sci. *31*, 174–179 (1978)

Darstellung:

Schon in Bd. 5 wurde auf zahlreiche Patente (s. bes. die Patentübersicht) hingewiesen, in denen Harnstoff-Verbindungen beschrieben sind, welche durch weitere Kondensation zu einer Überbrückung der Harnstoff-Gruppe führen (Bd. 5, 156). Aus dieser vielfältig modifizierbaren Gruppe sind inzwischen weitere Versuchsprodukte bekannt geworden.

3-(5-tert.-Butyl-1,3,4-thiadiazol-2-yl)-4-hydroxy-1-methyl-2-imidazolidinon

$C_{10}H_{16}N_4O_2S$

Common name Vorschlag: Buthidazole
Ravage® VEL 5026
(Velsicol Chem. Co. 1976)

Anwendung im Obstbau

Patente: DOS 2430467, 25.6.74/23.1.75, Velsicol; BE 814958; US 3964895, 29.6.73/27.7.73/13.8.73.
Lit.: Brit. Weed Control Conf. 1976, Proc. *2*, 731 (1976)

Das Acetat von VEL 5026:
[3-[5-1,1-Dimethyl)-1,3-thiadiazol-2-yl]-1-methyl-2-oxo-4-imidazolidinyl]-
acetat

$C_{12}H_{18}N_4O_3S$ Wirkungsweise ähnlich Buthidazole

Patente: DOS 2548847, 31.10.75/28.10.76, Velsicol Chem.; US Prior. 14.4.75; US 3997321;
BE 835155; FR 2307811; GB 1545827; JA 77/100472 u. weitere Patente
Lit.: Siehe nachfolgende Verbindung

Das entsprechende Benzoat hat die Versuchs Nr. VEL 3878

$C_{17}H_{20}N_4O_3S$ Wirksam gegen Unkräuter in Mais u. Sorghum
                                                   (0,2–0,7 kg/ha)

Lit.: H.Holtmann, H.G.Friedländer: Int. Velsicol Symp. Proc. 1978, Paper Nr. 1, 7 pp; Chem. Abstr.
92 [15], Ref. 123263, 123264, 123265, 123266 (1980)

Der entsprechende Octylester ist ebenfalls ein Versuchsprodukt der Velsicol
Chem. Co.:

[3[5-(1,1-Dimethyl)-1,3,4-thiadiazol-2-yl]1-methyl-2-oxo-4-imidazolidinyl]-
octanoat

$C_{18}H_{30}N_4O_3S$

VEL 4083

Selektiv in Mais u. Sorghum gegen Gräser.

Patente: US 4052193, 18.3.76/4.10.77, Velsicol Chem.
Lit.: J.F.Kunkel, E.P.Lira: Int. Velsicol Symp. Proc. 1977, Paper 7, 12 pp; Chem. Abstr. *92* [13], Ref.
105745 (1980)

Ein Dehydratisierungsprodukt der vorangehenden Verbindungen ist ebenfalls ein
Versuchsprodukt:

1-[5-(1,1-Dimethyl)-1,3,4-thiadiazol-2-yl]-1,3-dihydro-3-methyl-2 H-imidazol-2-
on

$C_{10}H_{14}N_4OS$

Common name: Buthidazole-anhydrid
Versuchsprod. VEL 4332

Pre-emergence gegen verschiedene Unkräuter
wirksam (0,2–0,4 kg/ha)

Patente: DOS 2510513, 11.3.75/18.9.75, Velsicol Chem.; US Prior. 11.3.74; US 3933839; NL 75/2848.
Lit.: H.Holtmann, H.G.L.Friedländer: Vels. Symp. Proc. 1978, Paper Nr. 1, 7 pp; Chem. Abstr. *92*
[15], Ref. 123263/64 (1980)

R. Wegler und L. Eue

Aromatisch-heterocyclische Harnstoffe

Eine neue Kombination von aromatischen und heterocyclischen Aminen als Harnstoffe führte zu einem ersten Vertreter:

N-Phenyl-N^1-(1,2,3-thiadiazol-5-yl)-harnstoff

$C_9H_8N_4OS$

Versuchsprod. der
Schering AG Berlin 1976
SN 49537
Common name: Thiadiazuron
Handelsname: Dropp®

Defoliant für Baumwolle

Da in den letzten Jahren keine weiteren Ergebnisse bekannt wurden, dürfte der Verbindung keine größere Bedeutung zukommen.

Patente: DOS 2214632, 23.3.72/4.10.73, Schering AG; US 3883547; NL 73/04073.

5.5.3i 2. *6-Ring Verbindungen* (Bd. 5, S. 150)

Als weitestgehend neuartig darf das folgende Herbizid angesehen werden; aufgrund seiner Harnstoff-Struktur wird es hier und nicht bei den Benzolsulfonamiden besprochen:

2-Chlor-N-[[(4-methoxy-6-methylpyrimidin-2-yl)-amino]carbonyl]-benzolsulfonamid

$C_{13}H_{13}ClN_4O_4S$

Versuchsprod. Du Pont

Die Verbindung scheint bevorzugt der Ergänzung anderer Herbizide (z.B. Chlortoluron) in Getreide-Kulturen zu dienen.

Patente: DOS 2715786, 7.4.77/13.10.77, Du Pont; US Prior. 7.4.76; US 4120691; BE 853374; DDR 131519.
Lit.: Chem. Abstr. *91* [1], Ref. 1223 (1979)

Darstellung:

R = Heterocyclus

Es wurde inzwischen ein zweites ähnlich gebautes hochwirksames Herbizid (ebenfalls von Du Pont) bekannt.

66

Mit diesem Herbizid wurde eine neue Verbindungsklasse mit hoher Wirkung eingeführt:

DPX 4189 Du Pont

LD_{50}: 5545 mg/kg weibl. Ratte, 6293 mg/kg männl. Ratte, p.o. akut.

Hochaktives Herbizid gegen Unkräuter in Getreide, auch gegen einige Ungräser wirksam (Ackerfuchsschwanz). Pre- oder frühe post-emergence Anwendung. (20–140 g/ha!), z.T. schon bei 2–2,5 g/ha wirksam. Es werden auch sonst schwer bekämpfbare Unkräuter wie *Viola spp, Euphorbis spp, Veronica spp.* und *Chenopodium album* erfaßt; Alternativherbizid zu 2,4-D usw. Rasche Ausscheidung aus dem Tierkörper (innerhalb 72 Stdn. zu 99,9%); keine Mutagenität im Ames Bakterien-Test. 3 Monat-Verfütterung mit 1–500 ppm an Ratten ergab keine negativen Befunde. Halbwertzeit im Boden: 1 bis 3 Monate. Die herbizide Wirkung beruht auf der Verhinderung der Zellteilung. Kulturen wie Weizen, Hafer, Gerste und Reis ergaben in der Pflanze verträgliche Metaboliten.

Patente: DOS 2715786, 7.4.77/13.10.77, Du Pont; US Prior. 7.4.76 u. 13.10.77; US 4120691; BE 853374. Weitere Patente siehe Patentergänzung.
Lit.: North Central Weed Control Conf., Milwaukee, Dec. 3–6, 1979; P.G.Jensen: 21st Swedish Weed Conf. 1980

5.7 Carbonsäuren, Nitrile und Aldehyde (Bd. 5, S. 161)

5.7.1 *Aliphatische Carbonsäuren, ihre Ester und Amide*

a) *Monocarbonsäuren* (Bd. 5, S. 161–167)
Das auf Seite 162 erwähnte Versuchsprodukt der BASF hat den

$C_9H_{18}NO_3S$

Common name: Sulfglycapin

Erstmalig wurde in Bd. 5, S. 165 über ein herbizides aliphatisches Carbonsäure-amid ohne Chlor im Acyl-Anteil berichtet. Neue Daten über diese Verbindung liegen nun vor:

N-Benzyl-N-methylethyl-2,2-dimethylpropionsäureamid

Common name: Butam (Gulf Oil Chem.)
GPC 5544 u. S-15544
LD_{50}: 6210 mg/kg Albinoratten
2025 mg/kg Meerschweinchen p.o. akut.
Dermal: Albino Kaninchen > 2000 mg/kg
In Wasser unlöslich. Sehr gut löslich in Ethanol, Benzol, Toluol. Pre-emergence Anwendung gegen Gräser und einige wenige Unkräuter, *Amaranthus retroflexus* und *Chenopodium album* in Sojabohnen, Baumwolle, Kartoffeln, Raps, Erdnüssen und Kohl. Keine post-emergence Wirkung.

$C_{15}H_{23}NO$

Patente: DOS 2104857, 2.2.71/18.9.71, Gulf Oil Corp.; GB Prior. 6.2.70; GB 1301812; US 3690865 u. 3707366.
Lit.: 13. Brit. Weed Control Conf. Brighton, Nov. 1976

Das Gebiet der Safener oder Antidots als Zusatz zu Herbiziden zwecks Verhinderung von Herbizidschäden in Kulturen bzw. zur Verbesserung der Selektivität ist nach wie vor hoch aktuell (Bd. 5, S. 162). Es gibt über Safener für Thiolcarbamate zahlreiche Patente aus verschiedenen Verbindungsklassen. Von den vielen geprüften Säureamiden erwies sich speziell für das Herbizid EPTC das Dichlor-acetyl-diallylamid (R 25788) als das überlegenste Produkt, aber auch das Morpholid ist gut wirksam. Die Monochlor-Verbindung ist als Herbizid Randox® der Monsanto lange bekannt.

Lit.: G. R. Stephenson et al.: J. Acric. Food Chem. *26*, 137–140 (1978); *27* [3], 543–547 (1979)

Das für EPTC Gesagte gilt allgemein für Thiolcarbamate und bis zu einem gewissen Grad auch für Acetanilide mit mangelnder Selektivität in Mais. Über Zusammenhänge zwischen chemischer Konstitution von Säureamiden und ihrer Wirksamkeit als Antidots gegen Herbizidschäden durch EPTC wurden kaum Untersuchungen bekannt. Eine ungarische Publikation nennt einige sehr wahrscheinliche Gesetzmässigkeiten. Die Acylamide sollen (irgendwie) durch Umacylierung in den Herbizid-Mechanismus eingreifen. Polare und sterische Faktoren der Acylamide beeinflussen die Wirksamkeit des Antidots. Faktoren welche die Elektrophilität der Carbonyl-Gruppe beeinflussen, d.h. die Reaktivität erhöhen, bewirken eine bessere Antidot-Wirkung. Dies ist z.B. der Fall bei Cl-Substitution in α-Stellung zur Carbonyl-Gruppe und gibt bis zu den Trichloracetamiden eine Steigerung spez. bei den N-acetyl-2,2-dimethyl-1,3-oxazolidinen. Elektronen-abziehende Gruppen bei Benzoylamiden erhöhen ebenfalls die Wirksamkeit der Antidots. Amide schwächer basischer Amine führen innerhalb einer analogen Reihe zu einer Steigerung der Antidotwirkung. Längerkettige Amide alphatischer sec. Amine vermindern die Wirkung.

Lit.: F. Dutka et al.: Proc. 19th Hungarian Ann. Meet. Biochemistry, Budapest 1979; Weed Abstr. *29* [7], Ref. 2220 (1980)

Waren bis in die jüngste Zeit Dichloracetamide die wirksamsten Safener zur Verhinderung von Herbizidschäden durch EPTC in Getreide, so wird nun eine Trichloracetamid-Verbindung als besonders wirksam herausgestellt:

$C_7H_{10}ClNO_2$

$$CCl_3-\underset{\underset{O}{\parallel}}{C}-N\underset{\underset{CH_3}{CH_3}}{\diagdown}\!\!\!\diagup O$$

Patente: US 4021224, 17.12.75/3.5.77, Stauffer Chem. Co.; DOS 2757813, 23.12.77/29.6.78, E. Gy. T. Gyogyse.
Lit.: Proc. Hung. Ann. Meet. Biochem. 1979, *19*, 253; Chem. Abstr. *92* [7], Ref. 53266 (1980)

Ungesättigte aliphatische Carbonsäuren (Bd. 5, S. 166)

Neu ist ein ungesättigtes Carbonsäureamid, eine Sulfinyl-Gruppe enthaltend. Letztere kann auch durch eine Sulfenyl- oder Sulfon-Gruppe ersetzt werden.

N-Propyl-cis-β-n-butylsulfinyl-acrylamid

Darstellung:

$C_{10}H_{19}NO_2S$

$$C_4H_9-SO\diagdown\quad\overset{\displaystyle O}{\overset{\|}{C}}-NH$$
$$\underset{H}{\overset{}{}}C=C\underset{H}{\diagup}\quad C_3H_7 \qquad I$$

$$\diagup H_2O_2$$

$$C_4H_9SH \;+\; CH\equiv C-\underset{\underset{O}{\|}}{C}-\underset{C_3H_7}{NH}$$

Spezifisch wirksam gegen Gräser, z.B. *Digitaria sanguinalis Amaranthus retroflexus.*

Patente: DOS 2729672, 30.6.77/12.1.78, Kao Soap Co.; JA Prior. 30.6.76.
Lit.: T.Imasaki et al.: J. Agric. Food Chem. *25* [6], 1413 (1977)

Aminoalkylen-carbonsäuren (Bd. 5, S. 166)

Aminoalkylen-carbonsäuren, aus Naturstoffen isolierbar, zeigen wachstumshemmende Eigenschaften und sind im Grenzfall selbst herbizid wirksam. Dabei ergaben die Ester höherer Alkohole eine gesteigerte Wirksamkeit gegen Gräser.
Lauryl-DL-Valinat

$C_{17}H_{35}NO_2$

$$\underset{CH_3\diagup}{\overset{CH_3\diagdown}{}}CH-\underset{\underset{NH_2}{|}}{CH}-\underset{\overset{\|}{O}}{C}-O-C_{12}H_{25}$$

Patente: JA 69225/75, 19.10.73/10.6.75, Ajinomoto Co.; GB 1455963.
Lit.: T.Kida et al.: Agric. Biol. Chem. *40* [8], 1551–1557 (1976); *41* [6], 931–937 (1977)

5.7.2 *Cycloaliphatische Carbonsäuren* (Bd. 5, S. 168)

Siehe unter 5.3.1, Bd. 5, S. 168, cycloaliphatische Aldehyde und Ketone. Dort entspr. Verbindungen mit und ohne Carboxyl-Gruppe.

5.7.3 *Araliphatische Carbonsäuren*

b) *Phenylessigsäure-Derivate* (Bd. 5, S. 173)

Aus diesen Verbindungsklassen ist wenig Neues zu berichten. Ein Dinitril ist als Antidot für Herbizide bekannt geworden.

α-(Cyanomethoxyimino)-benzoacetonitril

$C_{10}H_7N_3O$

CGA 43089 (Ciba-Geigy)

R. Wegler und L. Eue

Darstellung:

LD$_{50}$: 2277 m/g Ratte p.o. akut.; keine herbizide Wirksamkeit. Die Verbindung wirkt als Antidot (Safener) bei Anwendung von Dual® oder seiner Kombinationen mit Herbiziden aus der Reihe der sym. Triazine. Eine Vorbehandlung mit CGA 43089 schützt Sorghum-Sämlinge vor Schäden. Auch in Reis ist CGA 43089 anwendbar.

Patente: DOS 2639405, 1.9.75/17.3.77, Ciba-Geigy; CH Prior. 4.9.78; US 4070389; FR 2322861; BE 845827; JA 77/33643.
Lit.: Techn. Information der Ciba-Geigy AG v. 15.12.1977

In enger Verwandtschaft zu dem Flughafer Herbizid Bidisin® (Bayer, Bd. 5, S. 176) wurden von anderer Seite Polychlorphenyl-propionsäurenitrile untersucht. Offensichtlich wurde aber nur die Wuchsstoffaktivität festgestellt.

Lit.: M.S.Kemp, R.L.Wain: Ann. Appl. Biol. *82* [3], 597–600 (1976)

c) *Phenoxy-essig bzw. Propion-säure-Derivate* (Bd. 5, S. 180)

Diese Verbindungsklasse wird nach wie vor stark bearbeitet, nicht zuletzt weil hier fließende Übergänge von Wuchsregulatoren zu Herbiziden bestehen.

(4-Chlor-2-formyl-phenoxy)-essigsäure

$C_9H_7ClO_4$

Versuchsprod. RP 7846
(Rhone Poulenc)
Wuchshemmer für *Poa annua* und *Poa pratensis*

Patente: FR 1186520, 26.8.59, Rhone Poulenc.
Lit.: Chem. Abstr. *84*, Ref. 145901 (1976)

2,4,5-Trifluorphenoxyessigsäure $C_8H_5F_3O_3$

wurde als herbizid hochwirksam gefunden.

Lit.: Chem. Abstr. *92* [7], Ref. 53358 (1980)

N-N-Bis-(1-methylpropyl)-2-(1-naphthalenoxy)-acetamid
$C_{20}H_{27}NO_2$

70

Von den zahlreichen geprüften Verbindungen zeigte das sec.-Butylamid die höchste Wirksamkeit bei größtmöglicher Selektivität. Die Ausnahmestellung von Amiden mit zwei sec.-Butyl-Gruppen wurde mehrfach bestätigt (Lit. siehe Aminobenzoesäureamide).

Patente: US 3718455, 20.7.70/27.2.73, Stauffer Chem. Co.

Lit.: C. Pagani et al.: Il Farmaco Ed. Sc. *33*, 332–349 (1978).

Entgegen der herrschenden Lehrmeinung, daß die Methylen-Gruppe zum Erhalt der herbiziden Wirkung nur 1 mal durch CH_3 substituiert sein dürfe, wurde jetzt ein neues Versuchsherbizid mit tert. Methylen-Gruppe bekannt. Überwiegend scheinen derartige Produkte aber Wuchsregulatoren zu sein.

$C_{20}H_{24}O_3$ bzw. $C_{10}H_{11}ClO_3$

Patente: JA 74/31819, 1.8.72/22.3.74, Sankyo Co. (In DOS 1518625 v. 10.11.64/10.4.69 wird die Synthese derartiger Verbindungen aber nur eine Verwendung als Anti-Hyperlipamid beschrieben.)

$C_{10}H_{10}Cl_2O_3$ (als Cyclohexylaminsalz)

Versuchsprodukt Antiauxin II. Wird nur als Wachstumshemmer beschrieben. Anwendung in Getreide

Lit.: Chem. Abstr. *88* [1], Ref. 1439 (1978), Original: Russisch

Höhere Phenoxyalkancarbonsäureester haben nur begrenztes Interesse gefunden

$C_{14}H_{16}Cl_2O_3$ usw.

Efiran 17–22, 17–23, 17–24

Lit.: Chem. Abstr. *90* [19], Ref. 146881 (1979)

Wird in Phenoxyessigsäure-Derivaten die Carboxyl-Gruppe durch eine Aldehyd-Gruppe bzw. ihr Oxim ersetzt, erhält man vornehmlich Pflanzenwuchsregulatoren, obwohl man einen oxydativen Abbau zu den herbizid wirksamen Carbonsäuren erwarten könnte. Über eine spezifische herbizide Prüfung ist nichts bekannt.

$C_8H_7Cl_2NO_2$

Lit.: C. R. Alonso, R. L. Wain: Appl. Biol. *84*, 409–413 (1976); Chem. Abstr. *86* [13], Ref. 84606 (1977)

Umweltgefährdung durch 2,4,5-*Trichlorphenoxyessigsäurederivate* (Bd. 5, S. 185)

Diesem Problem wurde schon in Bd. 5 ein größerer Beitrag gewidmet. Weitere Diskussionen und neue Erkenntnisse über mögliche Gefahren rechtfertigen eine Fortsetzung und Erweiterung an dieser Stelle.

Die Frage, ob 2,4,5-T, niedergeschlagen auf Gras oder Papier, bei einer Verbrennung das hochgiftige 2,3,7,8-Tetrachlordibenzdioxin ergeben, wurde jetzt quantitativ untersucht. Nur 1 bis $3 \cdot 10^{-5}\%$ des verbrannten 2,4,5-T-Materials ergeben tatsächlich „Dioxin" (oft werden die entstehenden Mengen fälschlich auf das Gewicht des Grases oder Papiers bezogen).

Lit.: R.H.Stehl, L.L.Lamparski: Science *197* [403], 1008–1009 (1977) (Dow Chem. Co.)

So persistent TCDD im Boden ist, so rasch wird es in formuliertem Zustand als Film auf der Oberfläche des Bodens oder auf Blättern durch Sonnenlicht zerstört.

Lit.: D.G.Crosby, A.S.Wong: Science *195*, 1337 u. 4284 (1977)

Die Möglichkeit einer Anreicherung von TCDD im Wasser, in Fischen und in Frauenmilch in Gebieten, in denen 2,4,5-T umfangreich verwendet wurde, ist besonders eingehend untersucht worden. Es konnte kein TCDD nachgewiesen werden (Nachweisgrenze <10 ppt!).

Lit.: L.A.Shadoff, L.Hummel et al.: Bull. Environm. Contam. Toxicol. *18* [4], 478 (1977) (Dow Chem. Co.)

Bedenken werden trotz dieser negativen Befunde vereinzelt geäußert. Doch begibt sich der Kritiker hier in Grenzbereiche der gesicherten analytischen Nachweisbarkeit und damit werden Hochrechnungen über eine mögliche Kontamination spekulativ. Eine Erwiderung und Zusammenfassung vieler Ergebnisse aus zahlreichen Versuchen findet sich bei A.H.Westing: Science *206*, 1135 (1979); E.K.Blair: *loc. cit.*, p. 1136. TCDD nach einer Daueranwendung von 2,4,5-T nachzuweisen, gelang nicht.

Über ähnliche negative Versuche siehe Bull. Environm. Contam. Toxicol. *18* [2], 123–130 (1977). Nachweisgrenze 1 ppt.

Weitere Diskussionen über die Anwendung von 2,4,5-T, 0,05 bis 0,1 ppm TCDD enthaltend, und eine mögliche TCDD-Anreicherung in Kühen auf Viehweiden, die mit 2,4,5-T behandelt wurden (was sicher nicht der Praxis entspricht), ergab Werte von 60×10^{-9} g/kg Fettgewebe der Kühe. In Reisfeldern, während zwei Jahren mit 2,4,5-T behandelt, konnte kein TCDD nachgewiesen werden (Nachweisgrenze 10×10^{-12} g). Nachprüfung der TCDD-Entstehung beim Verbrennen von 2,4,5-T ergab, daß das 10^{-6}fache der 2,4,5-T-Menge in TCDD umgewandelt wird.

Lit.: R.H.Stahl: Science *197* [4307], 1008–1009 (1977)

2,4,5-T selbst wird im Boden zu $>90\%$ innerhalb 70 Tagen und zu 99% innerhalb 1 Jahr abgebaut. 1% war nicht nachweisbar.

Lit.: D.K.Stewart, S.V.Gaul: Bull. Environm. Contam. Toxicol. *18* [2], 210–218 (1977)

Die Fütterung von Regenbogenforellen mit einer Diät, enthaltend 2,3 ppm bis 2,3 ppt 2,4,5-T, ergab keine negativen Befunde hinsichtlich Überlebensrate, Wachstum usw. bei Konzentrationen $<2,3$ ppm. (*No effect level* zwischen 2,3 ppm und 2,3 ppb.)

Lit.: L.L.Hawkes, L.A.Norris: Weed Abstr. *27* [10], Ref. 3464 (1978)

Die Diskussion, ob regelmäßige Anwendung von Herbiziden wie 2,4-D zu einer „Vergiftung" des Bodens durch eine Anreicherung führt, wurde durch Untersuchungen dahin interpretiert, daß einige Herbizide bei wiederholter Anwendung sogar rascher abgebaut werden, da sich entsprechende Abbauorganismen laufend vermehren.

Lit.: Proc. EWRS Symp. Different Meth. Weed Control, Upsala *1*, 113–120 (1977); Weed Abstr. *27* [1], Ref. 567 (1978)

Da die wiederholte Anwendung auch von reinem 2,4,5-T, z.B. zur Vernichtung von Laubunterholz in Wäldern, als gefährlich für Beerensucher betrachtet wird (s. Bd. 5, S. 185), interessieren hier einige Mitteilungen, welche diese Gefahr auch beim Verzehr von Waldbeeren kurz nach der Besprühung mit 2,4,5-T gepflückt, verneinen.

Lit.: G.Eichler, D.Leber: Allgem. Forstzeitschr. *30* [45], 994–996 (1975); Nbl. Dtsch. Pflanzenschutz-dienstes *27* [8], 118–119 (1975); Erklärung der Biologischen Bundesanstalt Braunschweig; Zusammen-fassender Expertenbericht D.J.Turner: The safety of the herbicides 2,4-D und 2,4,5-T; Forestry Commission Bull. 57, Agric. Res. Council Weed Organis. London.

Über eine weiterhin empfohlene Anwendung von 2,4-D u. 2,4,5-T im Forstbau, s. Weeds to day *10* [1], 22–25 (1979) (Oregon State Univers.).

Lit.: Weed Abstr. *29* [4], 137, Ref. 1206 (1980)

2,4,5-T bewirkt auch in höheren Dosen (über 2 Jahre), an Ratten verfüttert, keine Krebsentstehung, wohl aber eine Schädigung der Nieren und etwas Gewichtsver-lust bei 30 mg/kg täglich.

Die weitere Anwendung von 2,4,5-T scheint in den USA fast zu einer Weltan-schauungsfrage geworden zu sein. Die EPA möchte die Verwendung verbieten mit der Begründung, daß ein Zusammenhang bestehe zwischen Zeiten breiter Anwen-dung von 2,4,5-T und nachfolgenden Schwangerschaften mit leichter Steigerung der Frühgeburten. Von anderer Seite wird der EPA vorgeworfen, daß aus einem nur einmaligen zufälligen Zusammentreffen einer geringen Frühgeburtsteigerung mit einer 2,4,5-T-Anwendung eine nicht aussagekräftige Statistik über einen Zeitraum von 6 Jahren gemacht worden sei (Farm. Chemicals *142* [7], 70 (1979)). In Farm. Chemical *142* [12], 54 (1979) wird über eine eingehende Konferenz berichtet, auf der die zahlreichen behaupteten Zusammenhänge zwischen 2,4,5-T-Anwendungsschäden irgendwelcher Art an Menschen weltweit als unrichtig zurückgewiesen werden. TCDD soll bis Ende 1979 in Seveso keine teratogenen Schäden verursacht haben. Auch soll 2,4,5-T keineswegs Fehlgeburten bewirkt haben; denn auch aus Gebieten, die ursprünglich für eine Waldbesprühung vorgesehen waren, und die dann schließlich abberaumt wurde, sollen vermehrte Fehlgeburten gemeldet worden sein (Farm. Chem. p. 56, Juli 1980).

Besonders interessant erscheint die Beobachtung, daß 2,3,7,8-Tetrachlor-dibenzdioxin anticancerogene Wirksamkeit zeigt. 2,4,5-T mit Spuren von TCDD verhindert bei Mäusen Hautkrebs, verursacht z.B. durch Benzpyren. TCDD erzeugt in der Haut Monooxydase-enzyme, welche eine rasche Umwandlung des Benzpyrens in hydroxylierte Produkte bewirken sollen.

Lit.: D.Giovanni et al.: Biochem. Biophysical Res. Comm. *86* [3], 577 (1979); Weed Abstr. *29* [4], 136, Ref. 1201 (1980)

R. Wegler und L. Eue

Weitere Beiträge zur Diskussion über 2,4,5-T finden sich in Farm. Chem. p. 150, März 1979 und p. 26, Dezember 1979. Über Vermutungen und Fakten liegen umfangreichere Beiträge der Dow Chemical (Herstellerfirma von 2,4,5-T) vor (Farm. Chem. p. 23, p. 74, April 1979 und p. 20, März 1979). Eine letzte umfassende sachliche Zusammenfassung über 2,4,5-T findet sich in Rodney, W. Bovey, Alvin L. Young: The Science of 2,4,5-T and Phenoxy Herbicides, John Wiley and Sons Inc., Sommerset, NJ 08873 (1980).
In den USA scheint man zur Zeit (August 1981) von einem Verbot der 2,4,5-T Anwendung auf Grund fehlender eindeutiger negativer toxikologischer Ergebnisse, die ein Verbot rechtfertigen, abzusehen. In der Bundesrepublik Deutschland ist die Diskussion über 2,4,5-T so emotionell geworden, daß Politiker zu einem Verbot neigen, obwohl keine anderen Tatsachen als in den USA vorzuliegen scheinen.

Diphenylether-oxy-propionsäuren und ähnliche Verbindungen. (Bd. 5, S. 191)

Derartige Verbindungen wurden schon in Bd. 5 kurz erwähnt, sind inzwischen aber bei zahlreichen Firmen intensiv weiter bearbeitet worden. Wegen der großen Bedeutung speziell als Gräserherbizide mit selektiver Wirkung in Getreide, wird ihnen ein eigener Beitrag gewidmet (H. Nestler, Hoechst AG). Entsprechende Verbindungen und Produkte werden deshalb hier nur tabellarisch angeführt.
Die neueren 5-Phenoxy-2-nitropropionsäurederivate zeigen ein verändertes Wirkungsspektrum. Sie sind starke Bodenherbizide mit spezifischer Wirkung gegen dikotyle Unkräuter und z.T. Ungräser, wobei einige Kulturen wie Baumwolle und Sojabohnen gegenüber dem Herbizid tolerabel sind. Anwendung auch im Nachauflaufverfahren möglich. In Getreide sind die Herbizide im Gegensatz zu den 4,4′-Derivaten nicht einsetzbar.

Darstellung nach: H. Schönowsky, H. Bieringer: Z. Naturforschg. *85B*, 902–908 (1980)

R *nicht* = CF_3, sondern z. B. Cl

74

$$\text{Ar (oder Het)-(O, S, CH}_2\text{)-Ar}-\text{O}\left[\underset{\underset{\text{CH}_3}{|}}{\text{CH}}-\text{COOR oder }\underset{\underset{\text{CH}_3}{|}}{\text{CH}}-\text{CH}=\text{CH}-\text{COOR, }\underset{\underset{\text{CH}_3}{|}}{\text{CH}}-\text{CH}_2-\text{OH usw.}\right]$$

Chemische Konstitution	Common name	Handelsname u. Firma	spez. Anwendung	Patente u. Literatur
d-(R) Form bes. aktiv	Diclofop-methyl	HOE 23408 Illoxan° (Hoechst) Hoelon¨ Hoe-Grass°	näheres siehe Vol. 5, S. 191	D. Breidert, H. Schumacher, F. Schwertle: Sonderh. 8, Z. Pflanzenkrankh. u. Pflanzensch. 441 (1977) H. J. Nestler, H. Bieringer: Spaltung in opt. Isomere. Z. Naturforsch. B. *35* [3], 366–371 (1980)
	Clofop-isobutyl	HOE 22870 (Hoechst)		D. Breidert, H. Schumacher, F. Schwertle: Sonderh. 8, Z. Pflanzenkrankh. u. Pflanzensch. 441 (1977)
		RH 5205 (Rohm u. Haas)		D. Breidert, H. Schumacher, F. Schwertle: Sonderh. 8, Z. Pflanzenkrankh. u. Pflanzensch. 441 (1977)

Verbindungen und Produkte (Fortsetzung)

Chemische Konstitution	Common name	Handelsname u. Firma	spez. Anwendung	Patente u. Literatur
CF_3—〔〕—O—〔〕—O—CH—COO—CH_3 (CH_3) R = Isomeres	Trifop-methyl	HOE 29152		DOS 2 433 067, 10. 7. 74/29. 1. 76 Hoechst; BE 831 218, NL 7508016; DOS 2 531 643, 17. 7. 74/29. 1. 76; Ishihara Sangyo
Cl,Cl—〔〕—CH_2—〔〕—O—CH—COOCH$_3$ (CH_3)	Vorschlag: Dichlorbenz- methyl	HOE 29160	Darstellung	DOS 2 417 487, 10. 4. 74/30. 10. 75, Hoechst; DDR 128 368; FR 2 267 302; BE 827 794; JA 75/140632

Darstellung:

$$\text{(2,4-Dichlorbenzol)} + CH_2O + HCl \longrightarrow \text{(2,4-Dichlorbenzylchlorid, } CH_2Cl)$$

$$\text{oder } \text{(4-Chlorbenzylchlorid, } CH_2Cl) + Cl_2 \longrightarrow$$

$$+ \text{(Phenol)}\!-\!OH \; ZnCl_2 \longrightarrow$$

$$\text{Cl, Cl}\!-\!\text{(Ring)}\!-\!CH_2\!-\!\text{(Ring)}\!-\!OH$$

$$\xrightarrow{\text{NaOH}} \quad + \quad ClCH-\underset{CH_3}{\overset{O}{\underset{|}{\overset{\parallel}{C}}}}-OR$$

$$\longrightarrow \text{HOE } 29160$$

B 806 Ishihara Sangyo	Spez. gegen Flughafer in Weizen u. Gerste, pre- u. postemergence Anwend. Gräserherbizid	Proc. 7th Asian Pacific Weed Sci. Soc. Conf., Sydney 1979; Weed Abstr. *29* [10], Ref. 3433 (1980)
Versuchsprod. d. Hoffmann-La Roche 1980: Ro 138895	Gräserherbizid, pre emergence Anwend. in dikotylen Kulturen	Chem. Abstr. *93* [5], Ref. 39353 (1980) US 4 200 587, 20. 11. 78/29. 4. 80; EP 2 246, 28. 11. 77/ 13. 6. 79 Hoffmann-La Roche; CH Prior. 15. 9. 78
Piriphenop Hache-Uno (Ishihara Sangyo Kaisha) SL 501	Gräserherbizid. post-emerg. Mischungen mit Sencor® in Prüfung. In breitblätt. Kulturen anwendbar.	DOS 2 546 251, 15. 10. 75/29. 4. 76; Ishihara Sangyo; Jap. Prior. 17. 10. 74/ 13. 3. 75; JA 48432/ 76; BE 834 495; FR 2 288 089; DDR 123 425 u. a.

D(+) Isomeres

Chemische Konstitution	Common name	Handelsname u. Firma	spez. Anwendung	Patente u. Literatur
CF_3 — pyridine — O — phenyl — O—CH(CH_3)—COO—H	Vorschlag: Fluazifop	ICI u. Ishihara Sangyo PP 009, TF 1169, SL 236	post-emerg. gegen Gräser in breitblättr. Kulturen 0,125–0,5 kg/ha	LD_{50}: Maus 1 500; Ratte 3 000 mg/kg GB 2 002 368, 27. 9. 78; EP 1 473; DDR 139 988; DOS 2 755 536 (Ciba-Geigy) uws.
Butylester	Fluazifopbutyl			Proc. British Corp. Protect. Conf.-Weeds *1*, 29–38 (1980)
CF_3 — phenyl — O — phenyl — O—CH(CH_3)—CH=CH—COOC$_2$H$_5$ R	Vorschlag: Difenopenten	Ortho-K K 80, Bi 883, XE 733 (Kumiai Chem. 1979) Chevron 1980	post-emerg. gegen Gräser. Selekt. in Sojabohnen	DOS 2 829 130, 3. 7. 78/3. 5. 79; Kumiai Chem.; US 4 163 661; BE 871 523; JA Prior. 3. 10. 77 u. 14. 2. 78
R–COOH				
Cl — phenyl(Cl) — O — phenyl(NO$_2$) — O—CH(CH_3)—C(O)—NH—C$_2$H$_5$		Etnipromid	HOE 39106	DOS 2 632 581, 20. 7. 76/26. 1. 78; BE 857 001, FR 2 359 113; GB 1 570 153 u. a.

Die neueren 5-Phenoxy-2-nitropropionsäurederivate zeigen ein verändertes Wirkungsspektrum. Sie sind starke Bodenherbizide mit spezifischer Wirkung gegen dikotyle Unkräuter und z. T. Ungräser, wobei einige Kulturen wie Baumwolle und Sojabohnen gegenüber dem Herbizid tolerabel sind. Anwendung auch im Nachauflaufverfahren möglich. In Getreide sind die Herbizide im Gegensatz zu den 4,4'-Derivaten nicht einsetzbar.

Darstellung nach: H. Schönowsky, H. Bieringer; Z. Naturforsch. **85**B, 902–908 (1980)

K_2CO_3 in $CH_3–SO_2–CH_3$ bei 150°C

R *nicht* = CF_3, sondern z.B. Cl

CH_3COOH HBr

K_2CO_3 + $Cl–CH–COOR$ | CH_3

40% NaOH 120–130°C

+ $NH_2–C_2H_5$ in $C_2H_5–OH$

Über den Abbau der Di-Phenylether-oxypropionsäureester im Weizen ist einiges bekannt. Die Estergruppe wird leicht verseift, in 5-Stellung wird eine Hydroxyl-Gruppe eingeführt, z.T. auch in 3'- und 6'-Stellung.

Lit.: G.Grobach, K.Kuenzler, J.Asshauer: J. Agr. Food Chem. *25* [3], 507 (1977)

Für die allgemeine Gruppierung

$Ar–O(S, CH_2)$ ... (1) (4) (2) (3) $O–CH_2–C–R$ ‖ O

scheint für die 4-Stellung der aliphatischen Ether-Gruppe R = OH oder -O-Alkyl unverzichtbar zu sein für eine gute herbizide Wirksamkeit. Steht die aliphatische Ether-Gruppe in 3-Stellung, so zeigen auch Verbindungen mit R = CH_3 eine gute herbizide Wirkung. Wie bei den entsprechenden Carbonsäure-Derivaten ist aber auch hier die Wirkung nicht auf Gräser beschränkt, sondern auch dikotyle Unkräuter werden erfaßt.

1-(2-Chlor-5-[2-chlor-4-(trifluormethyl)-phenoxy]-phenoxy)-2-propanon

$C_{16}H_{11}Cl_2F_3O_3$

Hersteller:
Mitsui Toatsu Chem., Inc. (1981)
Versuchsprodukt gegen Gräser und breitblättrige Unkräuter im Reis.

Patente: JA 80/20712, 1.8.78/14.2.80.
Lit.: Chem. Abstr. *93* [17], Ref. 162719 (1980)

Aus der Reihe der schon besprochenen Diphenylether-4-oxymethylencarbonsäuren sind in jüngster Zeit (Januar 1981) spezielle Ester als Herbizide bekannt geworden.

(S-2-Propinyl)-2-(4-[(3,5-dichlor-2-pyridinyl)-oxy]-phenoxy)-propanthioat,   und der entsprechende Propansäureester

$C_{17}H_{13}Cl_2NO_3S$ und $C_{17}H_{13}Cl_2NO_4$

Wirksam gegen Ungräser.
Verwendung als Lynergisten für andere Herbizide.
Lit.: Chem. Abstr. *93* [17], Ref. 162594 (1980)

Im DOS 2837863 der Ciba-Geigy AG sind Oximderivate (s. Patentergänzung bei Sulfonen) als Antidots für diese Verbindung erwähnt.

e) *Arylaminoessigsäuren*

Die in Bd. 5, S. 194 schon ziemlich ausführlich besprochenen Herbizide Suffix® und Barnon® der Shell können durch einige Angaben ergänzt werden.

R	X	Name	Common name
C_2H_5	Cl	Suffix	Benzoylpropethyl
$\overset{CH_3}{\underset{CH_3}{CH}}$	F	Barnon	Flampropisopropyl
C_2H_5	F		Flampropethyl

Die racemische Form von Suffix® ist nur halb so wirksam wie die l-Form.
Lit.: R.M.Scott et al.: Proc. 13th Brit. Weed Contr. Conf. 723–730 (1976)
Suffix® und Barnon® wirken über die freie Säure. Beide Herbizide sind systemisch, verhindern eine Zellvermehrung und unterbinden das Streckungswachstum vorhandener Zellen. Sie sind besonders wirksam gegen Flughafer in Getreide.
Lit.: R.Ott, W.Wimschneider: Sonderh. 8 Z. Pflanzenkrankh. u. Pflanzensch. 44 (1977)

LD_{50}: mg/kg p.o. akut für techn. Wirkstoff		Barnon	Suffix
	Ratte	>4000 mg	>5000 mg
	Maus	>2500 mg	>5000 mg
	Geflügel	>1000 mg	
LD_{50}: mg/kg p.o. für die Handelsform	Ratte	>3900 mg	1472–2029 mg

f) *Heterocyclische-aliphatische Carbonsäuren* (Bd. 5, S. 196)

Nachdem schon früher verschiedene Benz-1,3-thiazolyl-2-essigsäuren als Wuchshemmstoffe und z.T. als Herbizide erkannt worden waren, wurden auch 1,2-Benzisothiazol-3-yl-essigsäuren und höhere Derivate untersucht. Es wurden weitestgehend nur Wuchsregulatoreigenschaften festgestellt. Eine alte, breit gültige

Gesetzmäßigkeit, daß nur Carbonsäuren mit 1,3 oder 5 CH_2 Gruppen gute herbizide Wirksamkeit aufweisen, wurde wieder bestätigt.

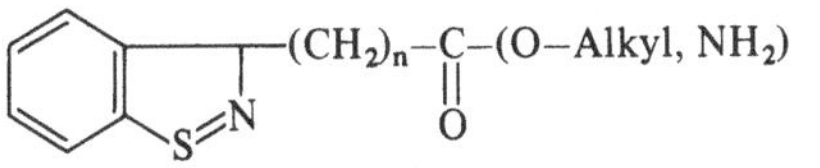

Statt der Carboxyl-Gruppe kann auch die Nitril-Gruppe stehen

Lit.: C. Branca et al.: Phytochem. *14* [12], 2545–2550 (1975)

Die Wirkungsweise soll auf einer Steigerung der Ethylen-Produktion in der Pflanze beruhen.

In der weiteren Verfolgung des schon angeführten Versuchsproduktes U 29722 (von Upjohn) wurden aus der Pyrazol-Klasse nur pre-emergence Herbizide gefunden. Dabei erwiesen sich besonders halogenierte Verbindungen als wirksam.

Lit.: G. Kornis et al.: J. Agric. Food Chem. *25*, 4884–4888 (1977)

3,4-Dibrom (bzw. Dichlor)-N,N,α-4-tetramethyl-1H-pyrazol-acetamid

$C_9H_{13}Br_2(Cl)_2N_3O$

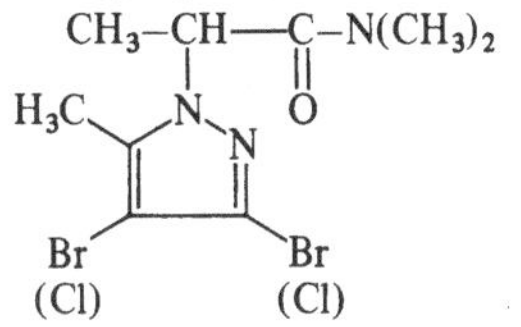

Upjohn Co.

Patente: US 3957480, 23.12.74/18.5.76, Upjohn Co.
Lit.: G. Lornis et al.: J. Agric. Food Chem. *25*, 884–888 (1977)

An anderer Stelle wird über Zusammenhänge zwischen chemischer Konstitution und herbizider Wirksamkeit berichtet.

Patente: DOS 2550566, 11.11.75/20.5.76, Upjohn Co.
Lit.: M. W. Moon et al.: J. Agric. Food Chem. *25*, 1039 (1977)

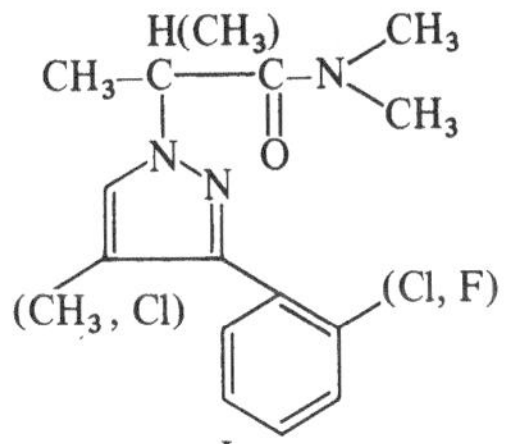

Hochwirksame pre-emergence Herbizide gegen Ungräser und Unkräuter sind die Verbindungen nach I. Die Stellung des Phenyls in 3-Stellung ist wichtig!

Darstellung:

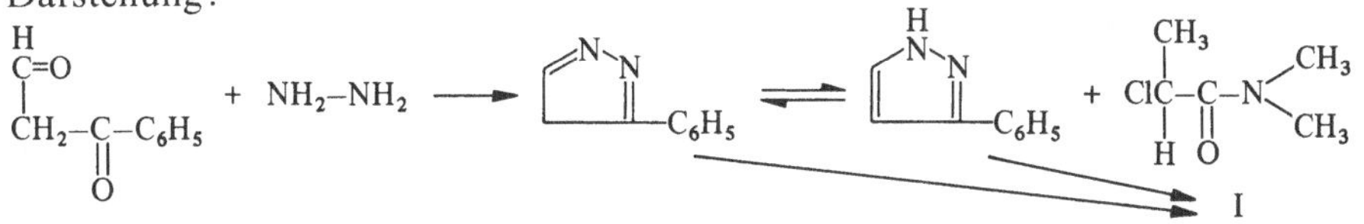

5.7.4 *Aromatische Carbonsäuren* (Bd. 5, S. 209)

Die schon in Bd. 5 erwähnten Chlorbenzoesäure-amide (S. 210) ergaben als wirksamste Verbindung einer größeren Versuchsreihe:

$C_{17}H_{26}ClNO$

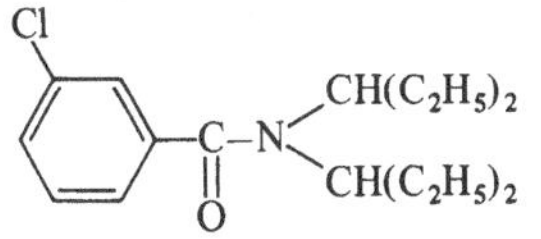

Versuchsprod. Pal 1138
Monteedison (1979)

Patente: DOS 2527113, 18.6.75/8.1.76, Monteedison.
Lit.: A. Barufini et al.: Farmaco Ed. Sci. *33*, 705–712 (1978)

R. Wegler und L. Eue

Die Verbindung

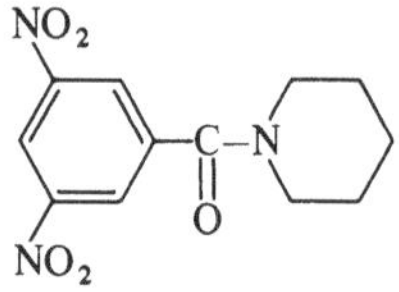

NO$_2$—(Struktur: 3,5-Dinitrobenzoesäure-piperidid)

wurde als Safener für EPTC in Getreide bekannt.

Lit.: Chem. Abstr. *92* [7], Ref. 53266 (1980)

2-Amino-6-methyl-benzoesäure

$C_8H_9NO_2$

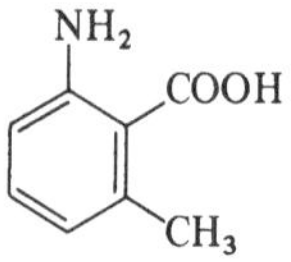

(Struktur: 2-Amino-6-methyl-benzoesäure)

Versuchsprod. der Hoffmann-La Roche.

Vorwiegend Wachstum von Gräser verzögernd aber auch herbizid wirksam gegen Graskeimlinge.

Patente: DOS 2605609, 12.2.76/26.8.76, Hoffmann-La Roche; BE 838477; GB 1499432; FR 2300503; DDR 124577.
Lit.: W.H.De Silva et al.: Z. Pflanzenkrankh. Pflanzensch. *86* [9/10] (1979)

Lediglich zu Zwecken der Untersuchung über herbizide Wirksamkeit wurde hergestellt:

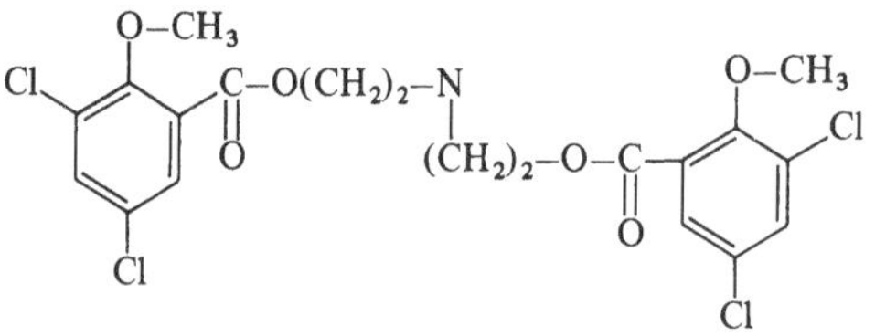

(Struktur: Bis-ester)

Lit.: T.R.Harger et al.: 1976 Meet. Weed Sci. Soc. USA II, 1976

In Anlehnung an das Handelsprodukt Chloramben (Amiben®) Bd. 5/214 wurden weitere Untersuchungen über chemische Konstitution und herbizide Wirksamkeit bekannt. Gut wirksam gegen Hirsearten waren

(Strukturen: drei substituierte Dichlorbenzoesäure-Derivate) und … und …

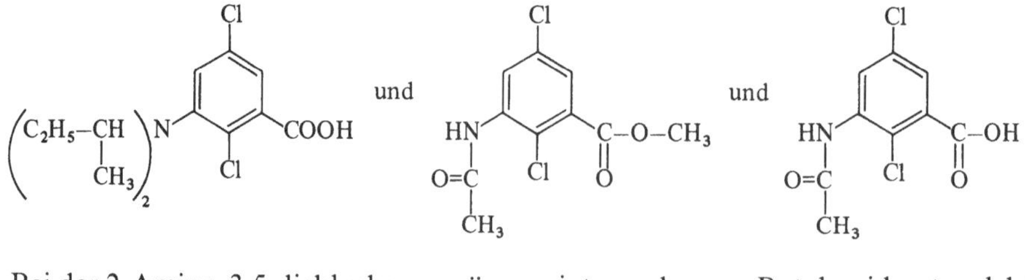

Bei der 2-Amino-3,5-dichlorbenzoesäure zeigt nur das sec.-Butylamid gute selektive Wirksamkeit.
Lit.: A.Baruffini, P.Borgna: Farmaco Ed. Sci. *33* [4], 281–287 (1978)

d) *Aromatische Carbonsäuren mit heterocyclischen Substituenten*

2-(-5-Phenyl-2-oxazolyl)-benzoesäure und ($C_{16}H_{11}NO_3$ u. $C_{16}H_{11}NO_2S$)
2-(-5-Phenyl-1,3,4-oxadiazol-2-yl)-benzoesäure ($C_{15}H_{10}N_2O_3$)

Maßgeblich für die geringe Wirkung dieser Verbindungen scheint eine Substitution in O-Stellung der Carboxyl-Gruppe durch ein konjugiertes System zu sein, das seinerseits wieder durch einen Phenyl-Rest substituiert ist.

Lit.: G.F.Katekar: Phytochemistry *15* [10], 1421–1424 (1976)

Darstellung:

5.7.5 Heterocyclische Carbonsäuren

Carbonsäuren von heterocyclischen 5-Ringen; 5-Ringe mit 1 N (Bd. 5, S. 218)

Die folgenden Carbamidsäure-amid-Verbindungen des Indolins fallen durch eine Ähnlichkeit mit den aromatisch-aliphatischen Harnstoffen auf, und die Substitutionsgesetzmäßigkeiten entsprechen auch denen der araliphatischen Harnstoffe. Gemeinsam mit heterocyclisch-aliphatischen Harnstoffen ist die herbizide Wirksamkeit auch bei Substitution mit nur einem CH_3 am N^1 (s. I und III).

pre- und post-emergence Wirkung

Lit.: M.Mazza et al.: Farmaco Ed. Sci. *31*, 46–54 (1976); *32*, 54–66 (1977)

Die folgenden beiden Verbindungen können als Harnstoffe eines sek. Anilins aufgefaßt werden oder als Carbamidsäure-Derivate einer teilhydrierten heterocyclischen Verbindung. Aus chemisch-systematischen Gründen werden sie hier als Carbamidsäureamide registriert.

2,3-Dihydro-N-methyl-1 H-indol-1-carboxamid

$C_{10}H_{12}N_2O$ und $C_{11}H_{14}N_2O$

Pre- und post-emergence Wirkung. Aufnahme durch Wurzel und Blätter

Lit.: M. Mazza et al.: Farmaco Ed. Sci. *31* [10], 746–754 (1976)

5-Ring-Verbindungen mit $1 \times N + 1 \times O$ (Bd. 5, S. 219)

1-(Chloracetyl)-2,2-dimethyl-3-oxa-1-azaspiro-[4,5]decan

$C_{12}H_{20}ClNO_2$

Versuchsprod. RE 19 790
(FMC Corp.)

Antidot für Acetanilid-Herbizide spez. Alachlor und Butachlor

Wesentlich für die Wirksamkeit ist die Chloracetylgruppe.

Lit.: G. L. Jordan, R. G. Harvey: Weed Sci. *26* [4], 313 (1978)

Darstellungs-Beispiel

Über eine andere (aber ebenfalls unbefriedigende) Darstellung von (A) siehe J. Am. Chem. Soc. *76*, 1840 (1954), *73*, 4199 (1951), *71*, 378 (1949).

Patente: DOS 2558083, 19.2.75/24.6.76, FMC Corp.; US 4069036; GB 1512540; FR 2295034.

5-Ringe $1 \times N + 1 \times S$

Kalium-3,4-dichlorisothiazol-5-carboxylat

$C_4Cl_2NO_2SK$

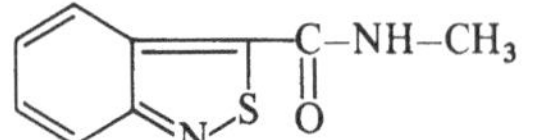

Versuchsprod. TD 1123
(Pennsalt Chem. Corp.)
Pflanzenwuchsregulator und Desiccant
verstärkt Defoliants bei Baumwolle.

LD_{50}: 325 mg/kg Ratte p.o. akut.

Anwendung 1–4 Wochen vor dem Defoliant (z.B. Def® (Mobay))

Patente: US 3393992, 5.4.67/23.7.68, Pennsalt Chem. Corp.
Lit.: G.W. Cathey, Corp. Sci. *18* [2], 301–304 (1978); Chem. Abstr. *89*, Ref. 101697 (1979). T.N.Sharer
et al.: J. Agric. Food Chem. *27* [2], 325–328 (1979)

N-Methyl-2,1-benzisothiazol-3-carboxamid

$C_9H_8N_2OS$ u. (iso C_5H_{11} statt CH_3)

Versuchsprodukt

Ob die Gruppierung —C—NH R wirklich essentiell ist, muß bezweifelt werden,
da sie ersetzt werden kann durch —NH—C—R.

Lit.: E.Coghi et al.: Ateneo Pharmense, Acta Nat. *15* [3], 211 (1979); Chem. Abstr. *92* [19], Ref.
158857 (1979)

5-Ringe $3 \times N$ (Bd. 5, S. 219)

1,2,4-Triazol-carbonsäure-Derivate (früher 1,3,4-Triazole benannt)
N-Ethyl-N-propyl-3-(propylsulfonyl)-1 H-1,2,4-triazol-1-carboxamid

$C_{11}H_{20}N_4O_3S$

Common name-Vorschlag: Epronaz
Versuchsprod. SN 533
(Schering AG)
Gegen Gräser, selektiv in dikotylen Kulturen

Patente: DOS 2132618, 30.6.71/5.1.72, Boots Pure Drug Co.
Lit.: Chem. Abstr. *87*, Ref. 112859 (1977)

Carbonsäuren von 5-Ringverbindungen $2 \times N + 1 \times S$

Eine Thiadiazol-Verbindung ist als Carboxamid mit herbizider Wirksamkeit
bekannt geworden.

5-Butoxy-N-methyl-1,2,4-thiadiazol-2(3 H)-carboxamid

$C_8H_{15}N_3O_2S$

Versuchsprod. RU 21731
Roussel Uclaf
Wirkungsweise: Photosynthesehemmer

Patente: DOS 2617339, 21.4.76/4.11.76; FR Prior. 24.4.75; FR 2308629; GB 1503256; US 4067720.
Lit.: D.E.Moreland: Ann. Rev. Plant Physiol. *31*, 597 (1980)

Carbonsäuren eines heterocyclischen 6-Ringes

In bezug auf nähere Einzelheiten eines Herbizids (Citrinin) aus einem Antibiotikum sei verwiesen auf den Beitrag in Bd. 6: Pflanzenschutzpräparate mikrobieller Herkunft.

5.8 Aromatische Aldehyde (Bd. 5, S. 226)

5.8.1 *Schiffsche Basen und Oxime des 2,6-Dichlorbenzaldehydes*

Derivate des 2,6-Dichlor-benzaldehyds, welche in der Lage sind in das hochwirksame 2,6-Dichlorbenzonitril überzugehen, haben schon längere Zeit Interesse gefunden. Sie bieten bei der Anwendung den Vorteil einer längeren Wirkung und besonders einer geringerer Flüchtigkeit. Sulfinamide des 2,6-Dichlorbenzaldehyds scheinen herbizid besonders wirksam zu sein. Vielleicht erhöht die entstehende Sulfinsäure die herbizide Wirkung.

Lit.: A.J.Friedmann, P.L.Orwick: Abstr. 179 ACS Nat. Meet., Washington 9.–14.9.1979, Pest. 43)

5.10 Aromatische Sulfonsäuren (Bd. 5, S. 232)

N-(4-Nitrophenylsulfonyl)-harnstoff

$C_7H_9N_3O_5S$

Versuchsprod. CP 49814, Monsanto Chem. Co.

Herbizid und Pflanzenwuchsregulator zur Erhöhung der Zuckerausbeute bei Zuckerrüben.

Darstellung:

$$NO_2-\langle C_6H_4\rangle-SO_2-NH_2 + KNCO \longrightarrow NO_2-\langle C_6H_4\rangle-SO_2-NH-\underset{O}{\overset{\parallel}{C}}-NH_2$$

Patente: DOS 2013848/49, 26.3.70/8.10.70, Monsanto Chem. Co.

5.11 Amine

5.11.2 *Aromatische Amine; tert.-Dinitro-aniline* (Bd. 5, S. 234)

Tert. Aniline mit Elektronen abziehenden Gruppen in 2,6- und in 2,4-Stellung sind als Herbizide in verschiedenen z.T. wichtigen Handelsprodukten enthalten. Kleinere Abwandlungen bringen immer wieder spezielle Vorteile wie die folgenden Beispiele zeigen.

α,α,α-Trifluor-N-methylallyl-2,6-dinitro-N-propyl-p-toluidin

Darstellung:

$C_{14}H_{16}F_3N_3O_4$

Common name Vorschlag:
Methalpropalin (Eli Lilly)

Patente: DOS 2511897, 19.3.75/25.9.75, Eli Lilly; GB Prior. 23.3.74; FR 2264486; BE 827038; DDR 116129; JA 129729/77.

N-[2,6-Dinitro-4-(trifluormethyl)-phenyl]-N-ethyl-benzenmethanamin

$C_{16}H_{14}F_3N_3O_4$

Versuchsprod. der Esso Research

Pflanzenwuchsregulator und Herbizid.
Anwendung im Tabakbau gegen
unerwünschte Austriebe (Suckers)

Patente: US 3780046, 24.2.70/19.12.73, Esso Res.; US 3686230.
Lit.: M. Wilcox et al.: Plant Growth Reg. Group. *1979*, 6, 96–102; Chem. Abstr. *92* [13], Ref. 105756 (1980)

Das in Bd. 5, S. 209 schon besprochene Nitralin [Planavin® (Shell)] wird mit dem altbekannten Harnstoff Monolinuron (S. 138) kombiniert und ist als Argol® (Shell) im Handel. Es ist wirksam gegen relativ resistente Unkräuter wie Klettenlabkraut, Ackerstiefmütterchen und Ehrenpreis.

Lit.: R.Ott et al.: Sonderh. 8 (1977) Z. Pflanzenkrankh. u. Pflanzenschutz

Erstmalig ist auch ein Phenylhydrazinderivat als Versuchsprodukt bekannt geworden:

3-(2,2-Diethylhydrazino)-2,4-dinitro-6-trifluormethyl-phenylamin

$C_{11}H_{14}F_3N_5O_3$

CGA 28938 Ciba-Geigy

Patente: DOS 2349228, 1.10.73/18.4.74, Ciba-Geigy; CH Prior. 2.10.72; CH 582654; FR 2201285; GB 1443749; US 3891706 u.a.

N-nitrosierbare Herbizide und Herbizide welche N-Nitroso-Verbindungen enthalten können

Im Zusammenhang mit dem Auffinden von cancerogenen Nitrosaminen in einigen Herbiziden ist es besonders in den USA zu einer heftigen Diskussion gekommen über die Anwendung von Pflanzenschutzmitteln mit nitrosierbaren NH-Gruppen oder Produkten, welche als Zwischenstufen Nitrosamine enthalten könnten. Es scheint, daß herbizide Säuren in Form von Dimethylaminsalzen stark wechselnde aber geringe Nitrosamin-Gehalte aufweisen können, wobei der Weg zur Entstehung der Nitrosamine noch nicht klar ist. Zahlreiche Behauptungen über Herbizide, welche im Boden oder auf der Pflanze nitrosiert werden könnten, hielten einer kritischen Prüfung nicht stand. Ein besonderer Angriffspunkt waren tert. Aniline, z.B. Trifluoralin (Bd. 5, S. 216). Auch hier waren die Aussagen über einen Nitrosamin-Gehalt in Feldversuchen anscheinend nicht immer gesichert und EPA lehnte eine Suspendierung der Registrierung ab. Eine geforderte strengere Überwachung bei der Herstellung von Trifluoralin bzw. eine spezielle Reinigung von Nitrosaminen als Begleitsubstanzen können ohne Zweifel kritische Beanstandungen und ernste Gefahren weitestgehend ausschalten. Eine umfassende kritische Übersicht mit zahlreichen Lit.-Stellen findet sich in Chemtech. Juni 1979, S. 366–370.

5.12 Carbonsäureamide (Bd. 5, S. 247)

b) *Acyl-Verbindungen aromatischer Amine*

Propanil = Propionsäure-3,4-dichloranilid ist einige Zeit vereinzelt in Misskredit geraten, weil die (meist rohe) Verbindungen geringe Mengen einer Azo- und Azoxy-Verbindung enthielt.

Ergänzung zu Propanil (Surcopur® Bayer): Fischtoxizität (Cyprinus carpio).

LD_{50}: 13 ppm 148 Std.

Lit.: Nachr. Chem. u. Techn. 24, 545 (1976); Y. Nishiushi: Agriculture 6 (1969)

Es wurde gefunden, daß die beiden Begleitsubstanzen potente Induktoren der Aryl-hydrocarbon-hydroxylase sind. Diese Hydroxylase ist z.B. verantwortlich für die hohe Giftwirkung des Tetrachlordibenzdioxims, das bei der Darstellung von 2,4,5-Trichlorphenol als Nebenprodukt entsteht (Bd. 5, S. 185).

Lit.: A.Poland, E.Glorer: Science *194* [4265], 627 (1976)

Wenngleich wenigstens bei der Maus, nicht aber bei Ratten, erheblich weniger giftig (Leberschäden), haben die Azoxy- und die Azo-Verbindungen doch noch beachtenswerte Giftigkeit. Einige Unfälle mit Auftreten einer Akne weisen auf die Gefahr hin. Das 3,4,3′,4′-Tetrachlorazo(xy)benzol bildet sich bei der Darstellung des 3,4-Dichloranilins durch Reduktion des 3,4-Di-chlor-nitrobenzols. Bessere Reinigung des 3,4-Dichloranilins und des Endproduktes (Propanil) vermeidet die gefährlichen Beimengungen, doch konnten in Proben von Propanil in der USA bis zu 2-2,9 mg/g (?) der Azo-Verbindung festgestellt werden.

Lit.: Chemosphere *8* [5], 283 (1979); Weed Abstr. *29* [3], 103, Ref. 883 (1980)

Ob die in Bd. 5, S. 247 mitgeteilte Entstehung eines ebenfalls giftigen Azo-Abbauproduktes aus Propanil, bei gleichzeitiger Anwendung insektizider Phosphorester Gültigkeit hat oder möglicherweise durch vorher vorhandene Nebenprodukte mitbedingt wurde, kann erst nach erneuter Überprüfung entschieden werden.

Von anderer Seite wurde gezeigt, daß Propanil im Boden in Gegenwart von $NaNO_2$ nach Hydrolyse umgewandelt wird zu

$$Cl-\text{C}_6\text{H}_3(Cl)-N{=}N-NH-\text{C}_6\text{H}_3(Cl)-Cl$$

welches toxisch ist.

Lit.: J.R.Plimmer, P.Kearney et al.: J. Agric. Food Chem. *18*, 859 (1970)

Modellversuche mit $Cl-\text{C}_6\text{H}_4-NH_2$ im Boden unter anaeroben Bedingungen und in Gegenwart bestimmter Bakterien zeigen nach Reduktion von NO_3^- zu NO_2^- bei pH 6 (nicht bei pH 7!) eine ähnliche Reaktion.

Lit.: R.Minnard et al.: J. Agric. Food Chem. *25* [4], 841–844 (1977)

Die sonst beobachtete Bildung von Azobenzolen wird auf photochemische Reaktionen sowie auf die Gegenwart von Peroxidasen zurückgeführt. Die Schlußfolgerung, daß eine Toxizitätsprüfung an Ratten nicht aussagekräftig sei, weil die Diazoamino- und die Azo-Verbindungen nur im Boden entstehen, kann man aber wohl kaum ziehen.

Einige α-Cyanalkancarbonsäure-anilide haben sich als kräftige Herbizide erwiesen.

Lit.: Proc. Asian Pacific Weed Sci. Conf., Sydney 1979; Weed Abstr. *29* [11], Ref. 3820 (1980)

R. Wegler und L. Eue

2,6-Dialkyl-acylanilide (Bd. 5, S. 255–258)

Die schon 1976 zu großer Bedeutung gelangte Verbindungsklasse der Chloracet-anilide mit ausgezeichneter Gräserwirksamkeit ist durch eine Vielzahl weiterer Patente gekennzeichnet. Es ist schwierig, klare Prioritäten bei den offensichtlich nicht seltenen Überschneidungen der Patente festzustellen. Weitere analoge Produkte aus dieser Verbindungsklasse sind bekannt geworden.

Ergänzung zum Dimethachlor (Ciba Geigy; Bd. 5, S. 257): Teridox®, LD_{50}: 1600 mg/kg Ratte. Selektiv in Kohl, Raps und Rüben gegen Ungräser und Unkräuter.

2-Chlor-N-methoxyethyl-N-phenyl-acetamid

$C_{11}H_{14}ClNO_2$

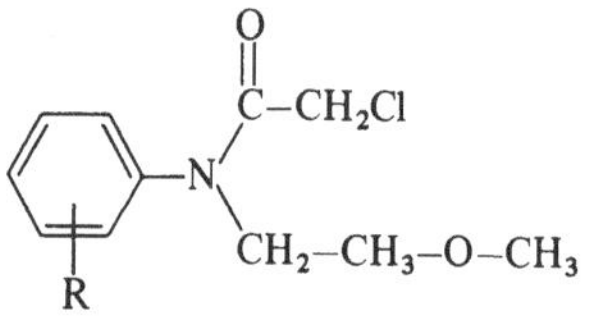

Russ. Name: Toluin
Hersteller unbekannt
R = H, in 4 $-CH_3$, in 2 $-CH_3$
über ähnliche Produkte s. Bd. 5, S. 257.

Patente: DOS 2305495, 5.2.73/23.8.73, Ciba-Geigy.
Lit.: Weed Abstr. *20* [3], 197, Ref. 1360 (1970); Chem. Abstr. *84* [19], Ref. 131335 (1976); *85* [15], Ref. 105264 (1976)

Dem Alachlor (Lasso®) und dem Butachlor (Machete®) der Monsanto sowie dem Dual® der Ciba-Geigy sowie weiteren Produkten sind nachfolgende Versuchspro-dukte ähnlich:

2-Chlor-N-(2,6-diethylphenyl)-N-(2-methoxy-1-methylethyl)-acetamid

$C_{16}H_{24}ClNO_2$

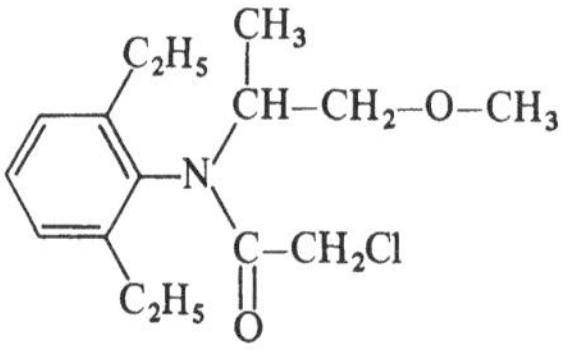

Versuchsprod. CG 119
(Ciba-Geigy)

Gräser-Herbizid.
Wirkung ähnlich Alachlor® und Butachlor®.

Patente: DOS 2328340, 4.6.73/20.12.73, Ciba-Geigy; CH Prior. 6.6.72 u. 30.3.73; CH 581426 u. 581607; US 1438311, 1438312, 3937730, 4022611; FR 2187225; GB 1438311 u. 1438312; BE 800471 u.a.

α-Chlor-2′,6′-diethyl-N(2-propoxethyl)-acetanilid

$C_{17}H_{26}ClNO_2$

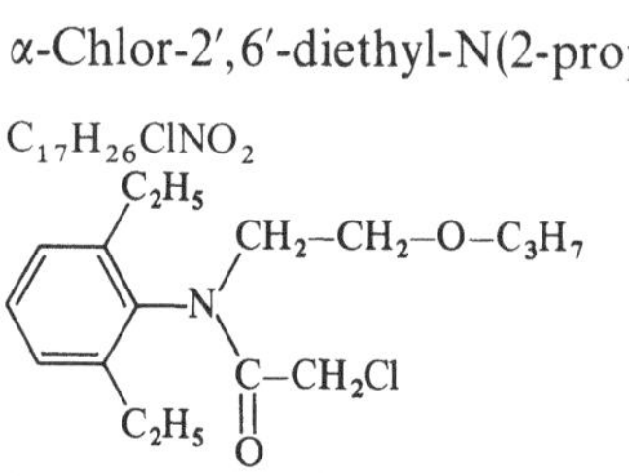

Common name-Vorschlag:
Pretachlor
Rifit® (Ciba-Geigy)

Darstellung:
$ArNH_2 + ClCH_2-CH_2-O-C_3H_7$ usw.

Patente: DOS 2328340, 4.6.73/20.12.73, Ciba-Geigy; CH Prior. 6.6.72 u. 30.3.73; US 3937730 u. 4022611; FR 2187227; GB 1438311 u. 1438312 u.a.

90

(1-Methylethyl)-N-chloracetyl-N-(6-ethyl-2-methyl-phenyl)-glycinat

$C_{16}H_{22}ClNO_3$

H 26910 (Hercules)
Diese Verbindung unterscheidet sich nur in der Estergruppe von Antor® derselben Firma.

Darstellung:

Selektiv in Baumwolle

Patente: DOS 2311897, 9.3.72/4.10.73, Hercules; BE 796263; DOS 2212268 u. 14.3.72/28.9.72, Sumitomo Chem. Co.

Lit.: G.D.Wills: South Weed Sci. Soc. *29*, 110–114 (1976); Chem. Abstr. *85* [11], Ref. *73327* (1976)

2-Chlor-N-(2,3-dimethylphenyl)-N-(1-methylethyl)-acetamid

$C_{13}H_{18}ClNO$

Common name-Vorschlag: Xylachlor
Versuchsprod. Combat®
AC 206784, CL 206784
Amer. Cyanamid Co.

Wasserlöslichkeit: 310 ppm.
Keine mutagene Wirkung

Anwendung gegen Gräser in der Voraussaat und pre-emergence. Selektiv in einigen Bohnensorten, Baumwolle, Erdnüssen, Mais, Reis, Sojabohnen und Weizen. Die Verbindung weicht sowohl in der Stellung der Alkyl-Gruppen am Phenyl-Rest (meist 2,6) wie in der Substitution am N (ohne O in der Alkylen-Kette) von den meisten Chloracetanilid-Herbiziden ab.

Patente: DOS 2632437, 19.7.76/3.2.77, Amer. Cyanamid Co.; US 4021483; CH 603552; FR 2318861; GB 1544482 usw.

Lit.: Weed Abstr. *29* [5], Ref. 1462 (1980)

Offensichtlich ist die Chloracetyl-Gruppe nicht notwendiger Bestandteil eines N-substituierten Anilids wie das nachfolgende Versuchsprodukt zeigt:

N-(3,4-Dichlorphenyl)-N-(ethoxymethyl)-cyclopropancarboxamid

$C_{13}H_{15}Cl_2NO_2$

R 33669
Stauffer Chem. Co.

Patente: DOS 2422721, 10.5.74/5.12.74, Stauffer Chem. Co.; US 4088687, 3140705, 4144250 u.a.; FR 2229685.
Lit.: Proc. Northeast. Weed Sci. Soc. *34*, 431 (1980); Chem. Abstr. *88*, Ref. 131860 (1978)

Die bei den Thiolcarbamaten bekannte Möglichkeit Herbizid-Schäden an Kulturen durch sog. Antidots oder Safener zu vermindern hat nun auch bei den Chloracetaniliden zu einem praktisch verwertbaren Erfolg geführt:

α-(Cyanmethoxyimino)-phenylacetonitril   GGA 43089, Ciba-Geigy 1977

$C_{10}H_7N_3O$

Safener für

Dual®

GGA selbst hat keine herbizide Wirksamkeit, schützt aber wichtige Kulturen wie z.B. Sorghum-Sämlinge, Reis, Weizen vor Schäden durch Dual, wenn eine Vorbehandlung mit dem Safener vorgenommen wird. Aufwandmengen: 1,25–1,5 g/kg Saat. Toxizität LD_{50}: 2277 mg/kg p.o. Ratte (techn. Subst.).
Patente: DOS 2639405, 1.9.76/17.3.77, Ciba-Geigy; CH Prior. 4.9.75; US 4070389; FR 2322861; BE 845827 u.a.

Lit.: Techn. Information der Ciba-Geigy 15.12.77; s.a. unter 5.3.2

An dieser Stelle sei auf eine immer wieder zu beobachtende Tatsache hingewiesen, daß zahlreiche Herbizide nach geringfügiger Abwandlung gut Fungizide ergeben (vgl. Bd. 4).

CGA 29212 ist ein systemisches Fungizid

Patent: DOS 2643477; 2350944 u.a. (Näheres siehe Bd. 6, Fungizide).

92

Die vielfachen Variationsmöglichkeiten, z. B. durch Abänderung der Acyl-Gruppe oder Variation der Substituenten am N haben zu weiteren herbiziden Versuchsprodukten geführt.

2-Chlor-N-(2,6-dimethylphenyl)-N-[(3,5-dimethyl-1H-pyrazol-1-yl)-methyl]-acetamid

$C_{16}H_{20}ClN_3O$

Versuchsprod. Lab 114253 BASF

LD_{50}: 10100 mg/kg Ratte für techn. Produkt. Pre-emergence Anwendung gegen Gräser und Unkräuter in Baumwolle, Raps, Busch-Bohnen, Erdbeeren, Kartoffeln, Sojabohnen usw.

Patente: DOS 2648008 v. 23.10.76/3.5.78, BASF; BE 860043; FR 2368476; JA 78/53651 u. a.
Lit.: B. Würzer, K. Eiken: Proc. EWRS, Intern. Symp. Mainz 10.–12. Okt. 1976, pg. 411–420; Weed Abstr. *29* [5], Ref. 1463 (1980)

Ähnlich ist ein weiteres Herbizid der BASF:

2-Chlor-N-(2,6-dimethylphenyl)-N-(1H-pyrazol-1-ylmethyl)-acetamid

$C_{14}H_{16}ClN_3O$

Common name: Metazolachlor Versuchsprod. BAS 47900 H Butisan® neu

Wasserlöslichkeit: 0,2 g in 100 g
LD_{50}: 2150 mg/kg Ratte für das technische Produkt. Vorauflauf-Anwendung sowie in Vorsaateinarbeitung. Selektiv in Winter- und Sommerraps, Sojabohnen und Zuckerrüben gegen Gräser (Flughafer). Wirksam auch gegen einige Unkräuter.

Darstellung:

Patente: DOS 2648008, 23.10.76/3.5.78, BASF; BE 860043; FR 2368476; JA 78/53651; NL 77/11543; ZA 77/6278 u. a.
Lit.: B. Würzer, K. Eiken: Proc. EWRS Intern. Symp. Mainz 10.–12. Okt. 1979, S. 411–420, 421–428. – B. H. Henk et al.: Proc. EWRS Intern. Symp. Mainz, 10.–12. Okt. 1979, S. 429–430; Weed Abstr. *29* [6], Ref. 1463 (1980)

S-{2-[(4-Chlorphenyl)-(1-methylethyl)-amino]-2-oxo-ethyl}-O,O-
dimethyldithiophosphat

$C_{13}H_{19}ClNO_3PS_2$

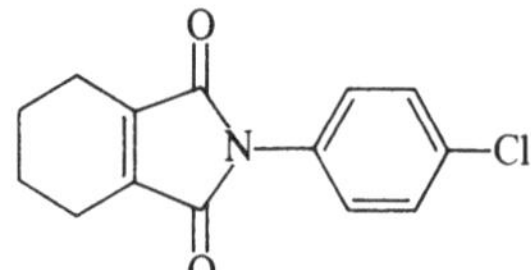

Common name: Anilinofos HOE 30374
Versuchsprodukt der Hoechst AG

Patente: DOS 2821509, 17.5.78/22.11.79, Hoechst AG; GB 78/4448, Gulf Oil Corp.

5.12.4c *Acylanilide von Dicarbonsäuren die zur Imid-Bildung befähigt sind* (Bd. 5, S. 259)

Waren seither nur offene Anilido-carbonsäuren spez. der Phthalsäure als Handels- und Versuchsprodukte bekannt, ist jetzt erstmalig ein cyclisches N-Phenyl-imid als Versuchsprodukt registriert:

N-(4-Chlorphenyl)-3,4-tetrahydrophthalimid

$C_{14}H_{22}ClNO_2$

MK 616
(Mitsubishi Chem. Co. 1976)

Selektiv in Reis

Patente: DOS 2165651, 30.12.71/19.4.73, Mitsubishi; FR 2119703; NL 71/17690.
Lit.: K. Matsui et al.: Chem. Regulat. Plants *9*, 45–50 (1974); Weed Abstr. *25* [3], Ref. 862 (1976). –
H. Ohta et al.: Agric. Biol. Chem. *40* [4], 745–751 (1976)

Auch cyclische N-Phenyl-imide von heterocyclischen Dicarbonsäuren sind als Versuchsprodukte bekannt geworden, als nach der Auffindung von MK 616 (siehe zuvor) systematisch cyclische N-Phenyl-Imide untersucht wurden.

Patente: US 3958976, 3.2.75/25.5.76, Du Pont; DOS 2526358, 12.6.75/8.1.76; NL 75/07233.
Lit.: H. Ohta et al.: Agric. Biol. Chem. *40* [4], 745–751 (1976)

Dicarbonsäure-Amide, Anilide und N-Phenylimide sind seit vielen Jahren Gegenstand zahlreicher Patente und seit 1976 sind auch einige cyclische N-Phenylimide als herbizide Versuchsprodukte bekannt geworden (siehe zuvor). Die Fa. Mitsubishi/Japan, welche derartige Verbindungen besonders intensiv und mit Erfolg bearbeitet hat, veröffentlichte 1978 eine sehr umfassende Zusammenstellung zwischen chemischer Konstitution und herbizide Wirksamkeit.
Lit.: K. Wakabayashi, K. Matsuya, H. Ohta, T. Jikihara: Adv. Perticide Sci., IUPCA Zürich, 1978, Part 2 (Ed.: H. Geissbühler) Pergamon Press, Oxford 1979

Es wurden folgende Verbindungstypen untersucht:

I II III

In vieler Hinsicht fallen bekannte Herbizide im weitesten Sinn unter dieses
Schema, z. B.

Lenacil
Bd. 5, S. 321

Terbacil
Bd. 5, S. 322

BAS 2580 H (s. Bd. 5, S. 325 und zahlreiche weitere Verbindungen auf den Seiten 321–325).

Biologisch besteht eine große Ähnlichkeit mit den Gräserherbiziden (Diphenyloxy-
den) CNP (Bd. 5, S. 78) und Nitrofen (Bd. 5, S. 74 u. 78). Als wichtigste Ergebnisse
sind herauszustellen: Die Alkylen-Gruppe in I bewirkt hohe Herbizidität; sie darf
aber nur 3 bis 5 C-Atome umfassen. Ein Ersatz einer Methylen-Gruppe durch
Heteroatome führt zu inaktiven Verbindungen. Das gleiche gilt für die Einführung
weiterer C–C-Doppelbindungen. Die cyclischen Methylen-Gruppen können ent-
fallen und durch eine Methyl-Substitution am C_1 und C_2 ersetzt werden. Höhere
Alkyl-Gruppen stören den planaren Charakter der Verbindung, der als Voraussetzung für eine herbizide Wirksamkeit angesehen wird. Die C–C-Doppelbindung ist
notwendig, sie darf nur zwischen C_1 und C_2 liegen. Die C-Atome 1 und 2 können
ohne Wirkungseinbuße durch ein oder zwei N-Atome ersetzt werden. Der Imid-
Ring ist essentiell. Ersatz von

durch

führt zu Wirkungseinbuße. Wird beim Ersatz von C_1 und C_2 durch N in II auch
noch C ~ C durch CH_2 ersetzt, fällt die Wirkung ab. Sind X und Y nicht O sondern
S steigt die Wirksamkeit erheblich. Y kann auch $=N—CO—C_2H_5$ sein und gibt

95

hohe Wirksamkeit. R im Phenyl-Ring führt zur höchsten Wirksamkeit wenn R=Cl in 4-Stellung steht. Cl kann durch Br oder I (weniger gut durch F) ersetzt werden. Substitution durch —O—CH$_3$ ergibt etwas geringere Wirksamkeit. Ersatz durch

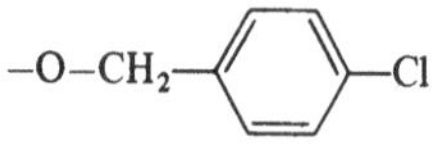

$$-O-CH_2-\langle\;\rangle-Cl$$

erhöht die Wirksamkeit. Cl ist durch SCN ersetzbar. Bei Cl in 4-Stellung kann ohne Einbuße der Wirksamkeit die 2-Stellung durch Halogen oder CH$_3$ substituiert werden. Ein bestimmtes Verhältnis zwischen Hydrophilie und Hydrophobie im Gesamtmolekül wird für wichtig erachtet.

5.12.4 *Acyl-Verbindungen heterocyclischer Amine* (Bd. 5, S. 262)

Acyl-Verbindungen von Heterocyclen, bei denen die N—C—R-Gruppe Glied des Ringes ist, siehe die entsprechenden Heterocyclen.

N-(5-Brom-2-thiazolyl)-butanamid

C$_7$H$_9$BrN$_2$OS

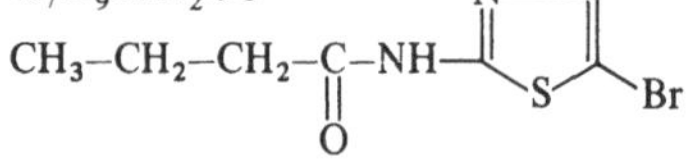

CH$_3$—CH$_2$—CH$_2$—C—NH—[thiazolyl]—Br

Versuchsprod. BABT der ICI

Vorwiegend Wuchsverzögerung in Weizen, (herbizide Wirksamkeit?)

Patente: GB 1276663, 24.12.68/7.6.72, ICI; ZA 69/8643; eine isomere Verbindung hat die Sankyo patentiert: JA 73/27467, 21.4.65/22.8.73.
Lit.: P.Stamp, G.Geisler: Z. Pflanzenernähr. Bodenkd. *1976* [6], 705, 715

Acyl-Derivate von 3-Amino-1,2-benzisothiazolen der Formel:

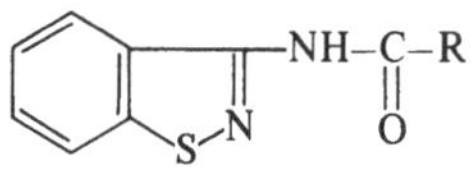

wurden als Herbizide getestet und erwiesen sich als hochwirksam.
Lit.: A.Bellotti et al.: Ateneo Parmense, Acta Naturalia *14* [3], 365–378 (1978); Weed Abstr. *29* [7], Ref. 2227 (1980)

5.12.6 *Sulfonsäure-anilide* (Bd. 5, S. 264)

Im Hauptband 5 wurden einige Trifluormethan-Sulfonsäureanilide als Herbizide aufgeführt. Ebenso wurde schon kurz über *Zusammenhänge zwischen chemischer Konstitution und herbizider Wirksamkeit* berichtet. Eine größere zusammenfassende neuere Arbeit aus dem Arbeitskreis der 3-M-Company weist auf weitere Zusammenhänge hin.
Lit.: T.L.Fridinger: Adv. Pesticide Sci., IUPAC Zürich 1978, Part 2 (Ed.: H.Geissbühler) p. 261, Pergamon Press, Oxford 1979

$\overset{x}{S}O_2$ kann durch SO und selbst S bei nur geringem Wirkungsabfall ersetzt werden. Ersatz von SO_2 durch O führt zu starkem Wirkungsverlust. Die CH_3-Gruppe in 2-Stellung ist essentiell. Geringer Wirkungsabfall bei Ersatz durch C_2H_5, starker Abfall bei höheren Alkyl-Gruppen. Ersatz von CH_3 in 2-Stellung durch H, F, Cl oder NO_2 gibt nur noch gute Wirksamkeit gegen Gräser. 2,5-Dimethyl-Gruppen und CH_3 in 2-Stellung mit Cl in 6-Stellung ergeben nur geringen Wirkungsverlust.

In ⬡$-\overset{x}{S}O_2$ soll die 4-Stellung nicht substituiert sein; F statt H zeigt den geringsten Wirkungsabfall. Im CF_3SO_2NH-Anteil ist CF_3 nur mit Wirkungsabfall durch CF_2HSO_2NH ersetzbar. $C_6H_5SO_2NH$ ergibt starken Wirkungsverlust. Verbindungen wie

zeigen sehr gute Hemmwirkung gegen Gräser-Wachstum.

5.13 Quartäre heterocyclische Ammonium-Verbindungen (Bd. 5, S. 275–283)

1-(2-Iod-1-cyclohexyl)-pyridinium-nitrat

$C_{11}H_{16}JNO$

Selektiv in Bohnen und Flachs.

Die herbizide Wirksamkeit der Verbindung gegen *Avena fatua* scheint eng zusammenzuhängen mit dem Redoxpotential. In der Pflanze wird Iod abgespalten. Anorganische und organische Iod-Verbindungen sollen allgemein gegen *Avena fatua* (Flughafer) wirksam sein. So zeigt z. B. Iodoform eine herbizide Wirksamkeit, aber ohne die Selektivität des Jod-Pyridiniumnitrats zu haben.
Die folgende Verbindung zeigt aus einer Reihe weiterer untersuchter Verbindungen gute Eigenschaften als Wuchsverzögerer, soll aber in höherer Konzentration auch herbizid wirksam sein:

Lit.: G. A. Hüppi et al.: Experientia *32*, 37–38 (1976)

R. Wegler und L. Eue

1-(2-Propenyl)-1-(3,7-dimethyloctyl)-piperidinium-bromid

$C_{18}H_{36}BrN$

Common name:
Piproctanylium-bromid
Handelsname: Alden®, Hoffmann-La Roche
1976

5.14 Heterocyclen

5.14.2 *Heterocyclische 5-Ring-Verbindungen*

a) *5-Ring-Verbindungen mit 1 × O* (Bd. 5, S. 288)

Ein neues Versuchsprodukt stellt eine Abwandlung des schon beschriebenen Nortron® (Ethofumesate) der Fisons Ltd dar.

(2,3-Dihydro-3,5-dimethyl-5-benzofuranyl)-ethansulfonat

$C_{12}H_{16}O_4S$

Versuchsprod. NC 20484
(Fisons Ltd.)

Bodenherbizid gegen Gräser und
Unkräuter in Baumwolle, Tabak, Erdnüssen

Patente: BE 863471, 5.8.77/31.8.78, Fisons Ltd. = DOS 2803991.
Lit.: Techn. Information der Fisons Pest-Control, s. a. Ethofumesate Bd. 5, S. 288

Darstellung:

über Zwischenstufen
z.B. eines Esters

b) *5-Ring-Verbindungen mit 1 × N* (Bd. 5, S. 289)

In erster Linie sind aus dieser Verbindungsklasse wieder die Pyrrolidone wirksam.

3-Chlor-4-chlormethyl-1-(3-trifluormethylphenyl)-2-pyrrolidon

$C_{12}H_{10}Cl_2F_3NO$

Versuchsprod. R 40244
(Stauffer Chem. Co.)

Pre- und post-emergence Anwendung in Mais
gegen Unkräuter spez. gegen *Chenopodium album*
und *Galium aparine.*

Patente: DOS 2612731, 25.3.76/7.10.76, Stauffer Chem. Co.; US Prior. 23.3.75; US 4110105;
GB 1522869; FR 2305434; BE 839977; NL 76/3193; JA 77/89666.
Lit.: Chem. Abstr. *90*, [21], Ref. 163217 (1979)

Darstellung:

I: R = C-CH$_3$ und R^1 = H;   II: R = C-C$_6$H$_5$ und R^1 = H
      ‖   ‖
      O   O

III: R = C-CH$_2$-O-⟨⟩ und R^1 = H   I, II und III sind wissenschaftliche Präparate.
       ‖
       O

Die Verbindungen sind Bodenherbizide in Baumwolle und Mais gegen *Amaranthus sp.* und Hühnerhirse, III ist aber in Baumwolle nicht ausreichend selektiv

Lit.: Chem. Abstr. *91* (19), Ref. 152632 (1979), Orig. russisch

e) *5-Ringverbindungen mit* $1 \times N + 1 \times S$ (Bd. 5, S. 290)

Der Benzthiazol-Rest kommt in wichtigen herbiziden Harnstoffen vor, der Thiazol-Ring zeigt aber auch schon bei einfacher Substitution herbizide Eigenschaften.

Lit.: Chem. Abstr. *85* [23], Ref. 172661 (1976)

Benzthiazolin-2-thion

$C_7H_5NS_2$

N-(3-Methyl-4-thiazolin-2-yliden)-2,4-xylidin

$C_{12}H_{14}N_2S$

Common name-Vorschlag:
Xymiazole
Versuchsprod. MG 07, Nitrokemia Ipartelk Ungarn und Ciba-Geigy

Darstellung:

Akarizid mit herbizider Wirksamkeit (BE 627278, 18.1.63/28.7.63).
Patente: DOS 1218210, 18.1.63/6.6.66, Ciba; s. a. DOS 2619724, 4.5.76/18.11.76 (Verwendung als Acaricid).
Lit.: R. M. Immler et al.: Proc. Brit. Insect. Fungicide Conf. 21.–24.11.1977, Brighton, 2, 383–396 (1977)

g) *5-Ring-Verbindungen mit* $2 \times N$ (Bd. 5, S. 290)

a) *Pyrazole (s. a. heterocyclisch-aliphatische Carbonsäuren Bd. 5, S. 218)*

Pyrazol-Verbindungen waren bis 1976 als Herbizide kaum bekannt. Eine Ausnahme bildet das Difenzoquat (Avenge® der Cyanamid Co. Bd. 5, S. 282). Hier dürfte aber die quartäre Salzstruktur für die Wirksamkeit verantwortlich sein. Unter den heterocyclischen Essigsäuren finden sich zwar ebenfalls Versuchsherbizide, doch ist hier die allgemeine Konstitution (Ar oder Hetero)-CH_2COOH Wirkungsprinzip (Bd. 5, S. 520). Aber es wurden schon in Bd. 5 zahlreiche Patente (siehe Patentergänzung) über Pyrazol-Derivate angeführt (siehe S. 606–608). Dies ließ annehmen, daß bald auch Versuchsprodukte bekannt würden. Am Schluß dieses Abschnitts sind zwei neue Produkte SW 751 und SL 49 angeführt; aber auch in der Patentergänzung dieses Bandes sind wieder zahlreiche Patente über Pyrazole zu verzeichnen, wobei in der Hauptsache japanische Firmen benannt sind. Die Firma Sankyo gab nun einen umfassenden Forschungsbericht über Zusammenhänge zwischen chemischer Konsitution und herbizider Wirksamkeit bekannt.

Lit.: T. Konotsune, K. Kanakubo, T. Yanai: Adv. Pesticide Sci., IUPAC Zürich 1978, Part 2, p. 94 (Ed. M. Geissbühler) Pergamon Press, Oxford 1979

Es wurde gefunden, daß 4-Acylpyrazole spez. 4-Benzoyl-5-hydroxypyrazole bei Pflanzen Chlorose bewirken. Um eine breitere Untersuchung durchführen zu können, wurden Methoden zur Darstellung derartiger Verbindung entwickelt. Das folgende Schema gibt eine Übersicht über bekannte Synthesewege.

Die direkte Darstellung der 4-Acyl-5-hydroxypyrazole (2) verläuft viel schlechter als der Umweg über (5) und dessen Umlagerung zu (2).
Zusammenhänge zwischen herbizider Wirkung und Art der Substituenten zeigt die folgende Tabelle:

R^1	R^2	R^3	Pre-emergence-Wirksamkeit. 10 = höchste Wirksamkeit bei 50 g/a
CH_3	CH_3	CH_3	0
C_6H_5	CH_3	CH_3	0
CH_3	CH_3	$o\text{-}Cl\text{-}C_6H_4$	10!
C_6H_5	CH_3	$o\text{-}Cl\text{-}C_6H_4$	3,6
CH_3	C_6H_5	$o\text{-}Cl\text{-}C_6H_4$	2,4
C_6H_5	C_6H_5	$o\text{-}Cl\text{-}C_6H_4$	2,0
H	CH_3	$o\text{-}Cl\text{-}C_6H_4$	0,4

Den Einfluß der Stellung der Substituenten in R^3 (= Phenyl) zeigen einige Beispiele mit R^1 und $R^2 = CH_3$

		pre-emergence-Wirksamkeit
	2-Cl	8,4
A) $\longrightarrow$	2,4-Cl$_2$	10,0
	2,5-Cl$_2$	7,4
	2,6-Cl$_2$	4,4
	3,5-Cl$_2$	2,4
	2,4-CH$_{3\,2}$	6,0
	3,5-Cl$_2$	6,6
	2-NO$_2$-5-CH$_3$	8,6
	3,5-(NO$_2$)$_2$	1,2
	2-NO$_2$-4-Cl	10,0!

A) $C_{19}H_{16}Cl_2N_2O_2$

Einzelheiten siehe Original

Ansteigende Wirksamkeit von ortho- über meta- nach para- bei elektronenanziehenden Substituenten; aber umgekehrte Reihenfolge bei CH_3 Gruppen.

An der 5-OH-Gruppe (Enol-Form des Pyrazolons) veresterte Verbindungen zeigen durchweg etwa die Wirksamkeit der unveresterten Verbindung (2). Ein Ersatz von OH durch CH_3, NH_2, $NHCH_3$ oder Cl ergibt Unwirksamkeit, S—C—CH$_3$ geringere Wirksamkeit. Die spezielle Verbindung 4-(2,4-Dichlor-benzoyl)-1,3-dimethyl-5-(p-tolyl-oxy)-pyrazol (A) zeigt gute pre-emergence-Wirkung in Reiskulturen. Einarbeitung in den Boden oder post-emergence-Anwendung sind ungeeignet. Andere gleichwirksame Ester sind nicht genügend selektiv in Reiskulturen.

[4-(2,4-Dichlorbenzoyl)-1,3-dimethylpyrazol-5-yl]-4-methylphenylsulfonat
$C_{19}H_{16}Cl_2N_2O_4S$

Versuchsprod. SW 751
Sankyo Co.
Common name: Pyrazolate
Handelsname: Sunbird®

Selektiv in Wasserreis bes. gegen *Sagittaria pygmaea*, bewirkt Chlorose.

Patente: JA 51 106 745, 12.3.75/21.9.76, Sankyo Co.; JA 76/106 738.

R. Wegler und L. Eue

Ein weiteres Versuchsprodukt wurde Ende 1979 bekannt:

{[4-(2,4-Dichlorbenzoyl)-1,3-dimethyl-1H-pyrazol-5-yl]-oxy}-1-phenylethanon

$C_{20}H_{16}Cl_2N_2O_3$

Wirkung wie SW 751

LD_{50}: >1000 mg/kg p.o. Maus

Patente: GB 2002375, 10.11.77/21.2.79; JA 77/96110, 12.8.77/21.2.79; JA 79/41872; s.a. JA 55083752, 18.12.78/24.6.80, Ishihara Sankyo.

Eine Veresterung scheint wie schon erwähnt und nach jüngsten Ergebnissen (1980) nicht notwendig zu sein, wie das folgende Versuchsprodukt beweist:

(2,4-Dichlorphenyl)-(5-hydroxyl-1,3-dimethyl-1H-pyrazol-4-yl)methanon

$C_{12}H_{10}Cl_2N_2O_2$

wissenschaftliche Versuchsprodukte
Sankyo Co.

Bewirkt Chlorose

Patente: DOS 2627223, 18.6.76/30.12.76, Sankyo; JA Prior. 19.6.75; CH 601251; FR 2316235; GB 1481161; US 4070536 u.a.
Lit.: K. Kawakubo et al.: Plant. Physiol. *64*, [5] 774 (1979); Chem. Abstr. *92* [21], Ref. 175648 (1980)

b) *Imidazole* (Bd. 5, S. 292)

Ausgehend von Aminosäuren bzw. einfachen Dipeptiden wurden durch Kondensation mit o-Phenyldiamin Benzimidazol-Derivate hergestellt, welche eine starke Hemmwirkung bei Reissämlingen zeigten.
Lit.: K. Mackawa, J. Ohtani: Agric. Biol. Chem. *41* [5], 811–818 (1977)

$C_{21}H_{26}N_4O$

pre-emergence Wirkung

Ähnliche Hemmwirkung und herbizide pre-emergence Aktivität zeigt das folgende Tryptophan-Derivat:

$C_{17}H_{16}N_4$

Die Kondensation der N-geschützten Aminosäure-Ester geschieht mittels Dicyclohexylcarbodiimid.

Literatur wie zuvor.

Imidazolidinone (Bd. 5, S. 292)

1-(3,4-Dichlorphenyl)-4-methyl-2-imidazolidinon

$C_{10}H_{10}Cl_2N_2O$

Versuchsprod. RU 15086 der Roussel Uclaf

Photosynthese-Hemmer; näheres nicht bekannt. Ringschlußprodukt eines aromatisch-aliphatischen Harnstoffs.

Lit.: D.E.Moreland: Ann. Rev. Plant Physiol. *31*, 597–638 (1980)

h) *5-Ringverbindungen* 2N + 1O, *Oxadiazole* (Bd. 5, S. 296–298)

In Bd. 5 sind aus dieser Verbindungsklasse mehrere Versuchsprodukte sowie ein Handelsprodukt (Ronstar® der Rhone Poulenc) beschrieben. Eine neue Verbindung reiht sich eng an die bekannten Produkte an:

5-tert.-3-[2,4-dichlor-5-(2-propinyloxy)-phenyl]-1,3,4-oxadiazolon-2(3H)on

$C_{15}H_{14}Cl_2N_2O_3$

Versuchsprodukt Rhone Poulenc RP 20630

Anwendung: Pre- und post-emergence in Sojabohnen, Sonnenblumen, Zwiebeln u. a.

I

Patente: DOS 2227012; US 3818026; GB 1345313 u. 1345314; FR 2141442, 2.6.71; BE 789286.
Lit.: W.G.Richardson et al.: Agric. Res. Council-Weed Res. Organization Techn. Rep. [38/40], 41–44, 52 (1976); Weed Abstr. *26* [6], Ref. 1774 (1977)

R. Wegler und L. Eue

Darstellung:

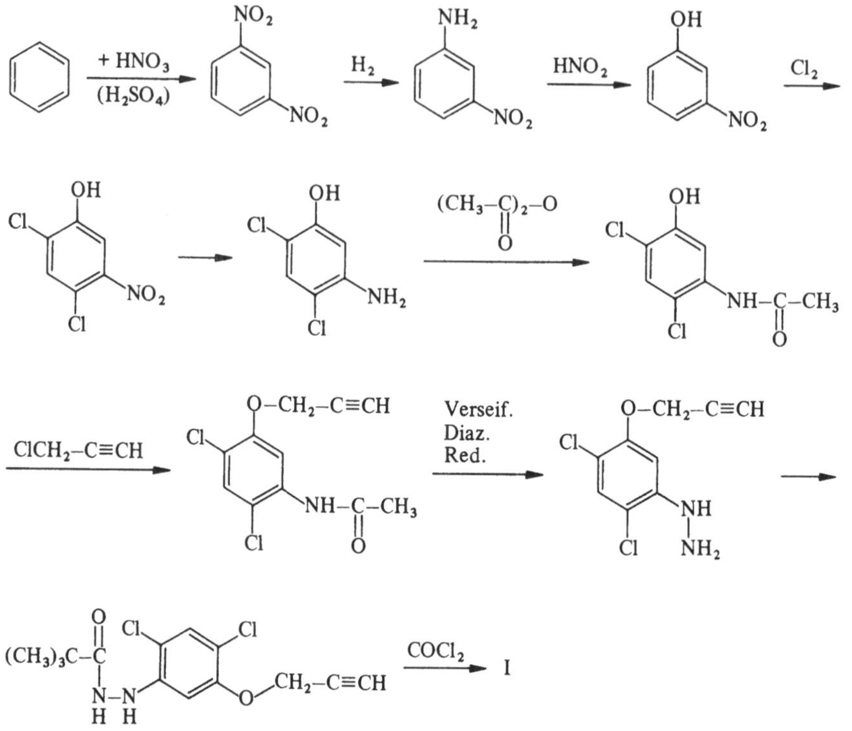

Ähnlich ist RP 20810 der Rhone Poulenc (Bd. 5, S. 297). Die tert. Butyl-Gruppe ist durch *Isopropyl* ersetzt:

$C_{14}H_{12}Cl_2N_2O_3$ Pre-emergence-Anwendung gegen Unkräuter.
Patente: siehe zuvor.
Lit.: Weed Abstr. *26*, [6], Ref. 1774 (1977).

5.14.3 *Heterocyclische 6-Ring- und höhergliedrige- Verbindungen*

a) 6-Ringverbindungen mit $1 \times N$ *(Bd. 5, S. 308)*

2-Chlor-3,5-dijodo-4-pyridinyl-acetat

$C_7H_4ClJ_2N_2O$

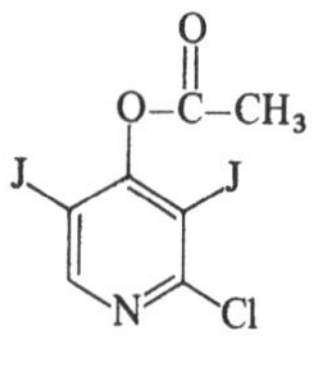

Common name-Vorschlag:
Cliodinate
Versuchsprodukt der Celamerk GmbH

Patente: DOS 2719904, 4.5.77/9.11.78, Celamerk; BE 866699

2-[2,5-Dimethylphenyl-ethylsulfonyl]-pyridin-1-oxid

$C_{15}H_{17}O_3S$

Versuchsprod. UBI S 734
(Uniroyal Inc.)

Gegen Gräser
LD_{50}: 5200 mg/kg Ratte p. o. akut

I

Vorsaat-Anwendung mit Einarbeitung in den Boden bei Sojabohnen, Baumwolle, Erdnüssen.

Darstellung:

R

Patente: DOS 2609204, 5.3.76/7.10.76, Uniroyal; US 3960542; FR 2304292; JA 76/115926 u.a.;
DDR 124876 u. 129731.
Lit.: Farm. Chem. *141*, 11, 65 (1978); Techn. Data Sheed der Fa. Uniroyal

a 1) *6-Ring-Verbindungen mit* $1 \times O$ (Bd. 5, S. 310)

Als Beispiel einer herbizid wirksamen Verbindung aus Penicillium-Stämmen ist Mycaotoxin bekannt geworden.

$C_{13}H_{14}OS$

Mycaotoxin
Versuchsprodukt LD_{50}:
35 mg/kg Maus

Näheres siehe Bd. 6: „Pflanzenschutzpräparate mikrobieller Herkunft".
Lit.: Biologiya *9* [6], 865 (1976); Weed Abstr. *25* [3], Ref. 861 (1976)

R. Wegler und L. Eue

c) *6-Ring-Verbindungen mit* 2N (Bd. 5, S. 311)

6-Chlor-3-phenylpyridazin-4-yl-S-octyl-thiocarbonat

$C_{19}H_{23}ClN_2O_2S$

Common name-Vorschlag: Fenpyrate oder Pyridate. Versuchsprod. CL 11344, Chemie Linz AG.

LD_{50}: >10000 mg/kg Maus p. o. akut; 1800 bzw. 2100 mg/kg männliche bzw. weibliche Ratten p. o. akut. Wasserlöslichkeit: 90 ppm (20 °C?). Post-emergence-Anwendung gegen Unkräuter, selektiv anwendbar in Mais, Weizen und Gerste.

Patente: DOS 2331398, 20.6.73/23.1.75, Chemie Linz; BE 816574; US 3953445.
Lit.: British Crop. Protection Council *2*, 717–722 (1976); Weed Abstr. *26*, Ref. 2757 (1977)

Darstellung:

Aus der in früheren Jahren erfolgreich bearbeiteten Verbindungsklasse der Pyrimidine ist nur ein Versuchsprodukt hervorgegangen.

6-Chlor-5-(ethylthio)-2,4-pyrimidindiamin

$C_6H_9ClN_4S$

Versuchsprodukt der Produits Chimiques Ugine Kuhlmann

Pre- und post-emergence wirksam in Getreide, Mais, Reis und Sorghum gegen Gräser und dikotyle Unkräuter. Wirksam über Wurzeln und Blätter.

Patente: EP 681, 21.7.78/7.2.79; FR Prior. 5.2.78, FR 2398737, 2417507; DDR 140040, JA 79/27586; s. a. JA 78/92789 der Mitsui Toatsu.

6-Ring-Verbindungen 2N *und* 1S (Bd. 5, S. 327)

Im Rahmen von Synthesemöglichkeiten mittels Alkylsulfamidsäurechloriden veröffentlicht die BAS weitere Synthesewege für das Herbizid Basagran®, z. B.

Die Synthesevorschriften für zahlreiche Sulfamidsäurechloride werden angegeben, ebenso ausführliche Zusammenhänge zwischen chemischer Konstitution und herbizider Wirksamkeit von entsprechenden 1,3-Benzothiadiazin-2,2-dioxiden sowie verwandte Verbindungen.

Lit.: G. Hamprecht et al.: Ang. Chem. *93*, 151–163 (1981)

Während Basgran® (BASF), ein bedeutendes Herbizid ist, zeigt eine ähnliche Verbindung, bei der SO_2 und $C=O$ vertauscht sind, keine herbiziden Eigenschaften.

Lit.: Y. Girood et al.: J. Chem. Soc. *1979* [4], 1043

Die bedeutenden Herbizide aus der Klasse der Uracile (Bd. 5, S. 321, 322) haben zu Versuchen geführt, eine $O=C$-Gruppe durch SO_2 zu ersetzen. Derartige Verbindungen lassen sich wie folgt darstellen, ohne daß indessen Verbindungen mit wesentlicher herbizider Wirkung bekannt geworden wären.

Lit.: H. Hansen, K. H. König, W. Rohr: Liebigs Ann. Chem. [7] 950 (1979)

Als Vergleich Sinbar® (Bd. 5, S. 322) $R=C(CH_3)_3$, $R^1=CH_3$

R. Wegler und L. Eue

ROOC–CH$_2$–COOR

\+

HC–OR
‖
O

R^2–C–OR
‖
O

ROOC   COOR
\ /
C
‖
HC–OH

NH$_2$–SO$_2$–NHR^1 →

ROOC   COOR
\ /
C
‖
HC–NH–SO$_2$–NHR^1

1) NaOH
2) HCl

oder

ROOC   COOR
\ /
CH
|
O=C–R^2

\+ NH$_3$ →

ROOC   COOR
\ /
C
‖
R^2–C–NH$_2$

\+ ClSO$_2$NHR^1 →

ROOC   COOR
\ /
C
‖
R^2–C–NH–SO$_2$ NH–R^1

Eine ausreichende herbizide Wirkung ergab sich nicht.

5.14.4 *6-Ring-Verbindungen mit* 3N

a) *Symmetrische Triazine* (Bd. 5, S. 336–354)

Die Klasse der sym. Triazine mit den bedeutendsten Herbiziden wird immer noch
weiter durchleuchtet und abgewandelt. Dabei wurden spezifisch und hoch wirksame Herbizide gefunden:

Ethyl-4-chlor-6-ethylamino-1,3,5-triazin-2-glycidat

C$_9$H$_{14}$ClN$_5$O$_2$

C$_2$H$_5$–NH   Cl

NH–CH$_2$–C–O–C$_2$H$_5$
             ‖
             O

Common name-Vorschlag: Eglinazine-ethyl
MG 06 (Nitrokemia Ipartelepek, Ungarn)

LD$_{50}$: >10000 mg/kg Ratte und Maus
Eglinazin-ethyl wird in Getreide im Vorauflauf-
Verfahren angewendet (3 kg/ha).

Patente: JA 5942/66; weitere Variationen Weed Abstr. 347, Ref. 3128 (1977).

Die freie Säure ist ebenfalls ein Versuchsprodukt. Common name: Eglinazine. Eine ähnliche Verbindung mit $(CH_3)_2CHNH$ statt C_2H_5NH ist ebenfalls beschrieben worden. Common name-Vorschlag: Pyoglinazine (Fa. wie zuvor. Weitere Variationen siehe Weed Abstr. 25347, Ref. 3128 (1976).

Zum Beispiel:

Die früher als allgemeingültig angesehene Regel, daß nur 1,3,5-Triazin-Verbindungen mit 1Cl, $-OCH_3$ oder $-SCH_3$ in 2-Stellung brauchbare herbizide Eigenschaften aufweisen, ist im nachfolgenden Versuchsprodukt durchbrochen. Es fehlen aber bis heute (Januar 1981) Einzelheiten über eine praktische Verwendbarkeit.

(4,6-Bis-[(1-methylethyl)-amino]-1,3,5-triazin-2-yl)-methylcyanamid

$C_{11}H_{19}N_7$

Hersteller: Armn. Agric. Inst., Erewan, UdSSR

Name: Metazine oder Methazin
Anwendung im Kartoffelanbau

Patente: DOS 2657944, 21.12.76/22.6.78; Schweiz 605854; JA 78/92790.
Lit.: Chem. Abstr. *93* [5], Ref. 39337 (1980)

Ein offensichtlich hochwirksames Herbizid ist von der DuPont gefunden worden.

Es wird zusammen mit dem entsprechenden Pyrimidin-Derivat (Pyrimidin statt 1,3,5-Triazin) bei den heterocyclischen Harnstoffen in Abschn. 5-5-5-3 i 2 g besprochen.

R. Wegler und L. Eue

1,3,5-Triazin-2,4-dione (Bd. 5, S. 352–353)

Für das bedeutende Totalherbizid Velpar® (Du Pont) können Synthesewege nachgetragen werden. Darstellung z. B.:

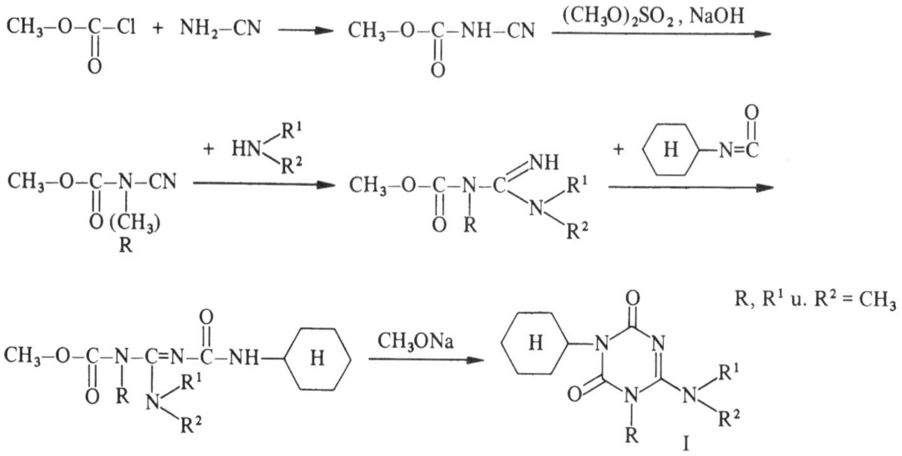

$$CH_3-O-\underset{O}{\overset{\|}{C}}-Cl \;+\; NH_2-CN \longrightarrow CH_3-O-\underset{O}{\overset{\|}{C}}-NH-CN \xrightarrow{(CH_3O)_2SO_2,\;NaOH}$$

R, R¹ u. R² = CH₃

oder

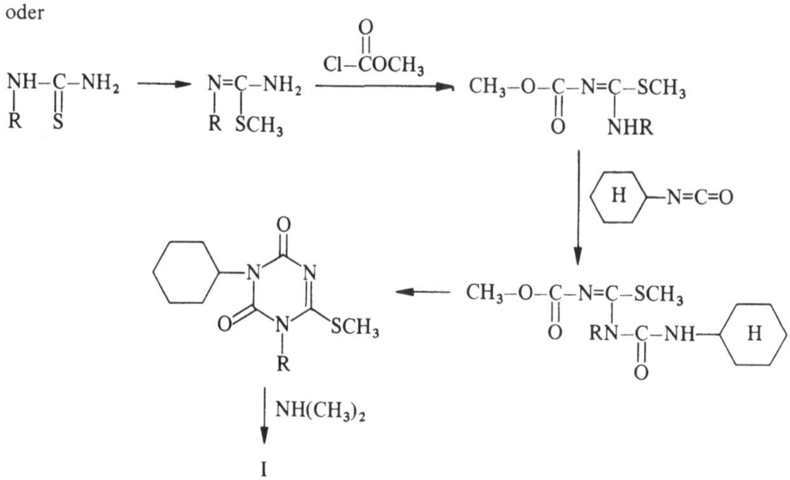

I = Velpar® ; Common name-Vorschlag: Hexazinone, Du Pont.
Patente siehe Bd. 5, S. 352.

Neu ist: 3-(4-Chlorphenyl)-6-methoxy-1,3,5-triazin-2,4(1H,3H)-dion

$C_{10}H_8ClN_3O_3$

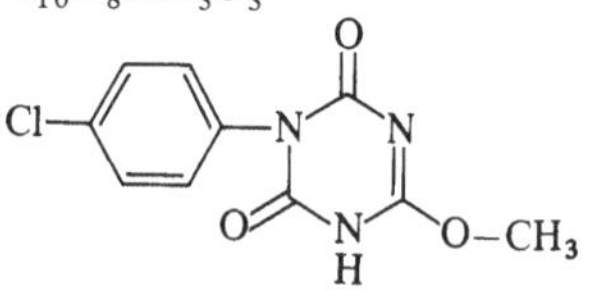

Versuchsprod. DPX 3778 (Du Pont)

xN(CH₂CH₂OH)₃ als Salz
Pflanzenwuchsregulator (und Herbizid)

Patent: US 3951971, 16.10.74/20.4.76, Du Pont.
Lit.: Chem. Abstr. *88* [25], Ref. 100267 (1978)

110

Darstellung:

$$R-C\equiv N \xrightarrow{+\,NH_3} R-\underset{NH_2}{\overset{}{C}}=NH \xrightarrow{+\,R^1N=C=O} R^1-NH-\underset{O}{\overset{}{C}}-NH-\underset{}{\overset{R}{C}}=NH$$

$$\downarrow \; CH_3-O-\underset{O}{\overset{}{C}}Cl$$

$$R^1-NH-\underset{O}{\overset{}{C}}-NH-\underset{}{\overset{R}{C}}=N-\underset{O}{\overset{}{C}}-O-CH_3$$

$R^1 = Cl-\!\!\left\langle\!\!\bigcirc\!\!\right\rangle$

$R = CH_3$

für $R = OCH_3$ entsteht die Verbindung DPX 3778

Darstellungsmöglichkeit:

$$R^1-N=C=O \; + \; \underset{O-CH_3}{\overset{}{NH=C-NH_2}} \longrightarrow$$

$$\downarrow \; + Cl-\underset{O}{\overset{}{C}}-O-CH_3$$

Untersuchungen im Hinblick auf eine gewisse Ähnlichkeit zwischen sym. Triazinen und 2-substituierten 4,6-Pyrimidin-diaminen haben zu einer intensiveren Untersuchung geführt, wobei neue Synthesen für Pyrimidin-Zwischenprodukte gefunden wurden. Herbizide Verbindungen entstanden bis jetzt nicht aus dieser Untersuchung.

$X = Cl, SCH_3$; R^1, R^2, R^3 u. $R^4 = H$, (iso)Alkyl

Lit.: E. William, P. Hymans, P. A. Cruickshank: Abstr. 175th Amer. Chem. Soc. Nat. Meet., Pest., Ref. 18, 1979

5.14.5 *1,2,4-Triazine und Triazinone* (Bd. 5, S. 354–357)

Das Handelsprodukt Goltix® (Bayer AG)[1] hat sich auf Grund seiner sehr guten
Selektivität im Zucker- und Futterrübenanbau als wertvolles Herbizid erwiesen.
Die Synthesen, die zu entspr. Triazinonen bzw. zu Goltix® führen, benutzen leicht
zugängliche Acylhydrazone vom Ketocarbonsäureestern als Ausgangsmateria-
lien:

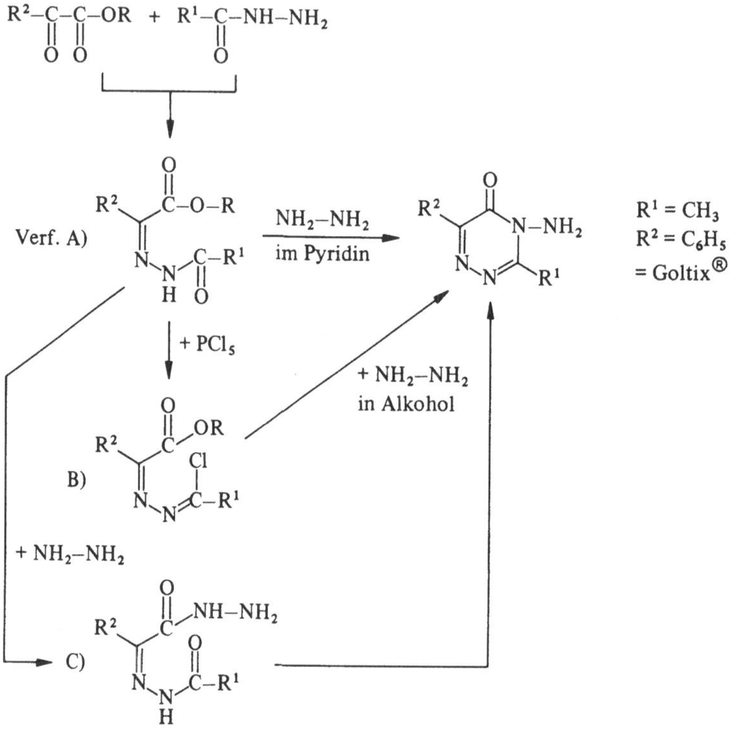

Verfahren A) in heißen Pyridin gibt meist mäßige Ausbeuten.
Verfahren B) gibt in Alkohol beste Ausbeuten. Das Zwischenprodukt der Hydra-
zidchloride ist stabil und gut isolierbar.
Verfahren C) ist nur in einigen Fällen anwendbar.
Möglich ist auch eine Kondensation der Ketocarbonsäure, anstatt der Ester mit
dem Amidrazon. Da die Methylgruppe ($R^1 = CH_3$) mittels N-Chlorsuccinimid
leicht chlorierbar ist, können aus der Chlormethyl-Verbindung mit nucleophilen
Substituenten Derivate hergestellt werden.

Beim Verfahren B) kann an Stelle von Hydrazin ein Amin, Hydroxylamin oder
Monomethylhydrazin eingesetzt werden, was zu neuen Verbindungen und Ersatz
der 4-NH_2-Gruppe führt. Aus den 4-Amin-triazinonen ($R^1 = H$) lassen sich glatt

---

1 Berichtigung zu Bd. 5, S. 356: Handelsprod. Goltix® (nicht Boltis)
Nachtrag: LD_{50}: 1447 mg/kg Maus, 3300 mg/kg Ratte p.o. akut (1977). Das der gleichen Verbindungsklasse entstammende Sencor® (Bayer) ist inzwischen in Kombinationsprodukten enthalten.

Azomethine herstellen, die direkt zu Pyrazolo-triazinonen ringschließen. Letztere können leicht oxydiert werden, z. B.:

Lit.: W.Draber, H.Timmler: Liebigs Ann. Chem. 2206–2221 (1976)

Ein 1,2,4-Triazinonherbizid ohne NH_2 Substituenten wurde von der Fa. Sumitomo gefunden:

3-(Pyridinyl)-1,2,4-triazin-5(2H)-on

$C_8H_6N_4O$

Sumitomo Chem. Co.

Post-emergence-Wirkung gegen Gräser (Blatthirse, Amaranth, Hühnerhirse, grüne Borstenhirse, weißer Gänsefuß).

I

Ausführliche Untersuchungen über chemische Konstitution und Wirksamkeit ergaben: NH_2, OH, COOH in 6-Stellung ergeben totalen Wirkungsverlust. Tert.-Butyl anstelle von 2-Pyridyl führt zu einer Wirkungsverminderung. Notwendig sind die O=C-Gruppierung in 5-Stellung und NH in 4-Stellung.

Darstellung:

Patente: DOS 2602186, 21.1.76/22.7.76, Sumitomo Chem. Co.; JA 86132/76 Prior. 21.1.75 u. 14.2.76; FR 2298276; US 4033752; NL 76/00626
Lit.: O.Kirino et al.: Agric. Biol. Chem. 41, 1093–1094 (1977)

5.14.6 *6-Ring-Verbindungen mit* 4N (Bd. 5, S. 357)

Aus dieser wenig untersuchten Verbindungsklasse der 1,2,4,5-Tetrazine wurden neue, besonders in ihrer Wirkungsweise interessante Vertreter, gefunden:

3-(4-Chlorphenyl)-1,2-dihydro-6-pentafluorethyl-1,2,4,5-tetrazin

$C_{10}H_6ClF_5N_4$

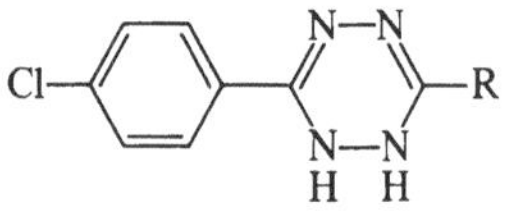

A und B

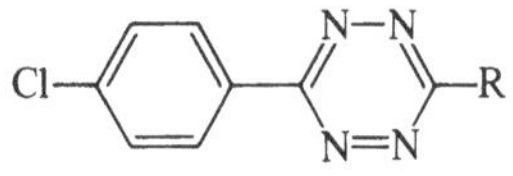

Shell Chem.

$R = C_2F_5$ und C_3F_7

A zeigt post-emergence-Aktivität gegen breitblättrige Unkräuter, weniger gegen Gräser. Aufnahme durch die Blätter. B wird in der Pflanze durch Licht zur Dihydro-Verbindung A reduziert. A und B sind nur unter Lichteinfluß aktiv (Vergleich mit Paraquat). Dementsprechend sind A und B pre-emergence angewandt wenig aktiv. Elektronenabziehende Substituenten stabilisieren den Tetrazin-Ring und erhöhen so die Wirksamkeit.

Patent: US 3860588/89, 23.7.73/14.1.75, Shell Chem. Co.

Lit.: K. H. Pilgram et al.: J. Agric. Food Chem. *25* [4], 888–892 (1977)

Darstellung:

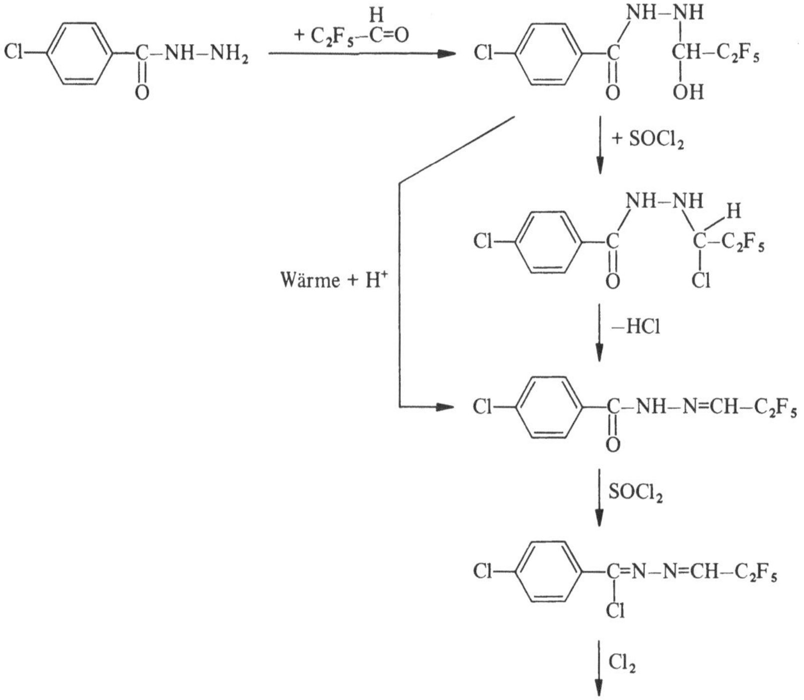

$$Cl-\text{C}_6H_4-\underset{\underset{Cl}{|}}{C}=N-N=\underset{\underset{Cl}{|}}{C}-C_2F_5 \quad \downarrow Cl_2$$

$$\downarrow + NH_2-NH_2$$

(B) $\xleftarrow[\text{Dehydr.}]{\text{FeCl}_3 \text{ oder } H_2O_2}$ (A)

(B)   **(A)**

5.14.7 *Höhere Ring-Verbindungen als 6-Ringe*

Höher-gliedrige Ringverbindungen haben bis 1980 zu keinen Versuchsprodukten geführt, obwohl herbizide Verbindungen bekannt geworden sind.

$$N-CH_2-CH_2-CH(CH_3)_2$$

Lit.: Chem. Abstr. *85* [1], Ref. 1086 (1976)

Lit.: D.P.Clifford et al.: Pest. Sci. *7* [5], 453 (1976)

5.15.4 *Phosphonsäure-ester* (Bd. 5, S. 371)

N-Phosphono-methylglycin

$C_3H_8NO_5P$

$$(HO)_2\overset{O}{\overset{\|}{P}}-CH_2-NH-CH_2-COOH$$

Common name: Glyphosate
Handelsname: Roundup® (Monsanto)

Über dieses Produkt wurde schon in Bd. 5 ausführlich berichtet. Jetzt liegen weitere Daten vor, und die Monsanto hat in großzügiger Weise Einblick in die Geschichte, Entwicklung und Darstellung dieser und ähnlicher Verbindungen gewährt:

Lit.: J.E.Franz: Adv. Pest. Sci., IUPAC, Zürich 1978, Part 2, p. 139–157, Pergamon Press, Oxford 1979; Farm. Chemicals, 15–22 (1977); W.Lanz, Sonderh. 8, Z. Pflanzenkrankh. u. Pflanzensch. 513 (1977)

Das Isopropylamin-Salz ist ein breitwirkendes Herbizid, sehr wenig giftig gegen Säugetiere, Vögel und Fische, Insekten sowie die meisten Bakterien. Post-emergence-Anwendung, systemisch wirkend, pre-emergence in bezug auf die zu schützenden Kulturen; pre-emergence nahezu wirkungslos. Keine Bodenrückstände selbst bei hohen Anwendungsdosen (6 4lbs/acre). Coniferen werden während der Ruhezeit nicht angegriffen, so daß ein Einsatz in Forstkulturen aussichtsreich erscheint.

Wasserlöslichkeit: 1 bis 8% bei 25–100 °C; LD_{50}: 4320 mg/kg Ratte p. o., >7940 mg/kg Kaninchen; EC_{50}: >1000 ppm Fische.

Lit.: Über die Wirkungsweise von Glyphosate in der Pflanze als Synthesehemmer von Flavonoiden und Chlorogensäure siehe: N. Amrhein, I. Schab und M. C. Steinrücken: Naturwissenschaften *67*, 356 (1980)

Allgemeine Darstellungsreaktionen:

A) $(R)_2{>}NH + CH_2O + H{-}\underset{O}{\overset{\|}{P}}(OH)_2 \xrightarrow{HCl} (R)_2N{-}CH_2{-}\underset{O}{\overset{\|}{P}}{<}(OH)_2$

B) $R{-}NH_2 + 2\,CH_2O + 2\,H{-}\underset{O}{\overset{\|}{P}}{<}(OH)_2 \longrightarrow RN\left(CH_2{-}\underset{O}{\overset{\|}{P}}{<}^{OH}_{OH}\right)_2$

C) $HOOC{-}CH_2NH_2 + CH_2O + HP{<}(OH)_2 \xrightarrow{\;} HO\underset{O}{\overset{\|}{C}}{-}CH_2{-}NH{-}CH_2{-}\underset{O}{\overset{\|}{P}}{<}(OH)_2$

(1)

$+ HO\underset{O}{\overset{\|}{C}}{-}CH_2{-}N[CH_2{-}\underset{O}{\overset{\|}{P}}{<}(OH)_2]$

(2)

Die Reaktion (C) verläuft nicht einheitlich, sondern es entstehen mehrere Verbindungen, wobei die Hauptreaktion (B) ist. Die Verbindung (2) ist also leicht darstellbar und als Glyphosine® (Polaris® der Monsanto) im Handel; (2) ist aber vorwiegend ein Wuchsregulator (s. Bd. 5, S. 372). Aber nicht nur die gewünschte Verbindung (1) (Glyphosate) entsteht nur als Beiprodukt von (2), sondern es entstehen noch weitere Nebenprodukte. Diese verdanken ihre Entstehung der reduzierenden Wirkung von Formaldehyd, wobei aus dem Zwischenprodukt, einer Methylol-Verbindung oder einem Methylimin, eine N-Methyl-Verbindung entsteht. Das folgende Schema zeigt die möglichen Reaktionszwischenstufen und die entstehenden Nebenprodukte:

$NH_2{-}CH_2{-}COOH \xrightarrow{+CH_2O} \begin{matrix} CH_2{-}COOH \\ | \\ NH \\ | \\ CH_2{-}OH \end{matrix} \xrightarrow{+CH_2O} \begin{matrix} CH_2{-}COOH \\ | \\ N(CH_2{-}OH)_2 \end{matrix}$

$\downarrow -H_2O$

$\begin{matrix} CH_2{-}COOH \\ | \\ NH \\ | \\ CH_3 \end{matrix} + HCOOH \xleftarrow[+CH_2O]{Red.} \left[\begin{matrix} CH_2{-}COOH \\ | \\ N{=}CH_2 \\[4pt] CH{-}COOH \\ | \\ N{-}CH_3 \end{matrix}\right] \xrightarrow{+\,H\underset{O}{\overset{\|}{P}}(OH)_2} \left[\begin{matrix} CH_2COOH \\ | \\ HN{-}CH_2\underset{O}{\overset{\|}{P}}(OH)_2 \\[4pt] \text{Glyphosate} \\ (1)\;\Big|\;{}^{CH_2O}_{Red.} \\ CH_2COOH \\ | \\ N{-}CH_3 \\ | \\ CH_2\underset{O}{\overset{\|}{P}}(OH)_2 \end{matrix}\right] \xrightarrow[+CH_2O]{+\,\overset{O}{\overset{\|}{H}P(OH)_2}} \begin{matrix} CH_2{-}COOH \\ | \\ N\left[CH_2\underset{O}{\overset{\|}{P}}(OH)_2\right]_2 \\[4pt] \text{Glyphosine} \end{matrix}$

$\begin{matrix} \downarrow +CH_2O \\ +\,H\underset{O}{\overset{\|}{P}}(OH)_2 \end{matrix}$

$\downarrow$ polymere Nebenprodukte

$\begin{matrix} CH_2{-}COOH \\ | \\ N{-}CH_3 \\ | \\ CH_2{-}\underset{O}{\overset{\|}{P}}(OH)_2 \end{matrix}$

Nebenprodukt

$\begin{matrix} CH_2COOH \\ | \\ N{-}CH_3 \\ | \\ CH_2\underset{O}{\overset{\|}{P}}(OH)_2 \end{matrix}$

Nebenprodukt

„Glyphosate" (1) entsteht bei den beschriebenen Reaktionen nicht, wohl aber verschiedene Nebenprodukte. N-methylierte Verbindungen zeigen keine herbizide Wirksamkeit. Mehrere andere Darstellungsmethoden für (1), z.B. mittels $ClCH_2P(OH)_2$, ergeben zwar (1) aber ebenfalls zuviel Nebenprodukte.

Eine andere Reaktionsfolge führt aber zu (1): Wie verschiedene primäre Amine, besonders schwach basische Amine und Amine mit sterischer Behinderung in Nachbarschaft zum Stickstoff, gibt Glycin (siehe Reaktionsschema) leicht ein definiertes 6-Ring-Kondensationsprodukt. Allerdings muß Glycin als Ethyl- oder besser tert.-Butylester angewandt werden. Dieses Trimere läßt sich nun leicht und einheitlich mit einem Phosphonester kondensieren. Der Phosphonester muß als leicht verseifbarer Diphenylphosphonester angewendet werden.

$$3 \text{ x } \underset{\substack{| \\ NH_2}}{\overset{R}{\overset{|}{CH_2}}}-\underset{\overset{||}{O}}{C}-O-C_2H_5 + 3 \text{ x } CH_2O \longrightarrow \text{ 6-Ring-Triazin (R-N, N-R, N-R)}$$

$+ 3 \text{ x } H-\overset{||}{\underset{O}{P}}(OC_6H_5)_2$ in Benzol

100% Ausbeute:

$$\underset{\substack{| \\ \underset{\overset{||}{O}}{CH_2-P(OH)_2}}}{\overset{R}{\overset{|}{NH}}} \quad \xleftarrow[\;HCl\;]{\;NaOH\;} \quad \underset{\substack{| \\ \underset{\overset{||}{O}}{}}}{\overset{R}{\overset{|}{NH}}}-CH_2-\overset{||}{\underset{O}{P}}-(OC_6H_5)_2$$

(1)

48% HBr, 120°C; 3 Stdn.

Ausgehend von $\underset{\overset{|}{H}}{\overset{R}{\overset{|}{N}}}-C(CH_3)_3 + CH_2O + H\overset{||}{\underset{O}{P}}(OC_2H_5)_2$ entsteht $\longrightarrow$ $\underset{\substack{| \\ \underset{\overset{||}{O}}{CH_2-P(OC_2H_5)_2}}}{\overset{R}{\overset{|}{N}}}-C(CH_3)_3$ (A)

sterisch behindertes Amin   einheitlich (A)   (A)

Durch Verseifung und Abspaltung der tert. Butyl-Gruppe aus (A) entsteht ebenfalls in sehr guter Ausbeute (1). Andere Methoden benutzen die Spaltbarkeit eines leicht und einheitlich darstellbaren Kondensationsproduktes (3) mit Wasser zu (1). Ebenso gelingt eine Spaltung mit $H_2SO_4 \cdot SO_3$, durch Oxidation mit H_2O_2 oder durch katalytische Oxidation.

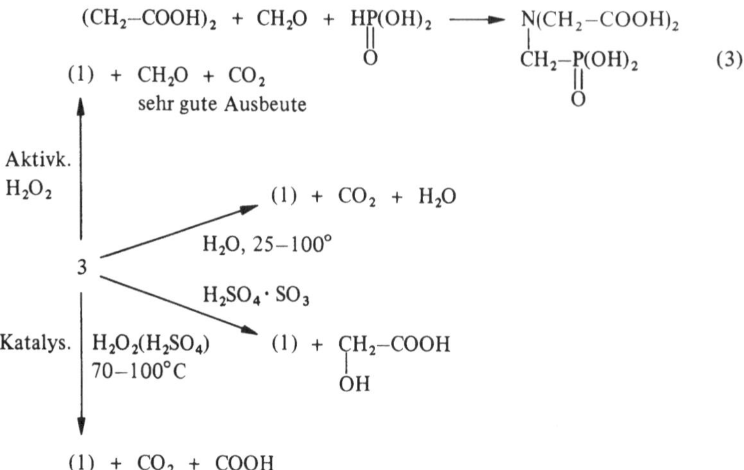

Die dem Glyphosate entsprechende Hydroxylamin-Verbindung ist ebenfalls leicht darstellbar, ihre herbizide Wirksamkeit aber nicht ausreichend. Über zahlreiche weitere Abwandlungsprodukte von (1) zu (2) wurde berichtet. Nur Verbindungen, die zu (1) metabolisieren können, zeigen eine herbizide Wirksamkeit.

Zweite Patentergänzung
September 1976 bis November 1980

Die Forschung auf dem Gebiet der Herbizide ist auf Grund ihrer außergewöhnlichen Bedeutung so rapide gewachsen, daß die neueste Patentliteratur der Aktualität wegen nur in der hier folgenden Form, welche die schnellste Publikation ermöglicht, vorgelegt werden kann.
Die Einteilung entspricht der in Bd. 5 verwendeten. Einige wenige neue Verbindungsklassen mußten eingeführt werden. Im nachfolgenden Text werden nur Verbindungsklassen angeführt, für die neue Patente bzw. Patentanmeldungen vorliegen. Die Zahlen auf der linken Seite bezeichnen die entsprechenden Abschnitts-Nummern in Bd. 5. Die Zahlen auf der rechten Seite weisen auf die dazugehörenden Anfangs-Seitenzahlen im Band 5 hin.

4. Anorganische Herbizide

DOS 2 820 359   10. 5.78/23.11.78   Takala usw. Finnland

Phosphorsäurespritzung vor dem Mähdreschen zur Vernichtung des Blatt-
werks von Nutzpflanzen und von Unkräutern.

JA 52 122621   5. 4.76/15.10.77   Showa Denko

$NaClO_3$ + Polyethylenglycol   Wirksam gegen Gräser.

5. Organische Herbizide

5.1.b   <u>Aliphatische, aromatisch-aliphatische und aromatische Kohlen-
wasserstoffe</u>   Bd. 5/383

US 4 005 153   24.10.63/25. 1.77   Hooker Chem.

Halog ⊢☐⊣ $(Ar)_m$

US 4 018 801   6. 8.70/19. 4.77   Dow Chemical

R u. R^1 = H, Halog., CN, NO_2,
CF_3, C_6H_5-CH_2-O,
(O) -Alkyl

US   4 156 602   30. 6.77/29. 5.79   PPG Ind.

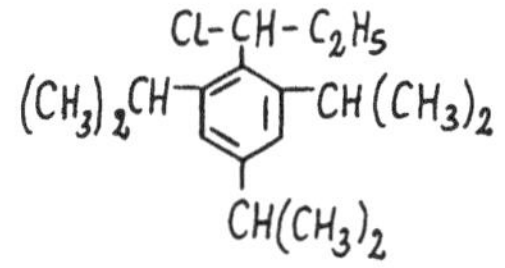

$$(CH_3)_2CH \overset{\displaystyle Cl-CH-C_2H_5}{\underset{\displaystyle CH(CH_3)_2}{\bigcirc}} CH(CH_3)_2$$

5.1.c   <u>Sulfoxide und Sulfone</u>   Bd. 5/385

US   3 984 481   12. 2.76/ 5.10.76   Chevron Res.

$$\left(R-S-CH=CH_2\right)_2 = SO_2$$

$R = C_1-C_6$-Alkyl, subst. Phenyl

Fungizide aber auch Herbizide gegen Wasserpflanzen.

Siehe auch

EP   10 588   27. 8.79/14. 5.80   Ciba-Geigy
EP   11 047   27. 8.79/14. 5.80   Ciba-Geigy

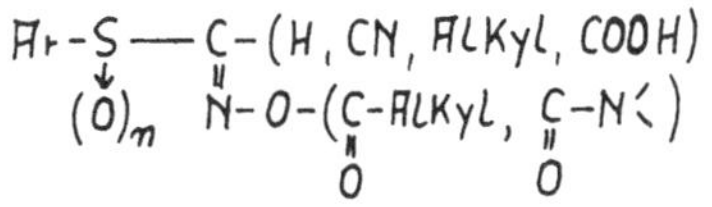

$$\underset{(O)_m}{Ar-S} \overset{O}{\overset{\|}{-}} \underset{\underset{O}{\|}}{C}-(H, CN, Alkyl, COOH) \\ N-O-(C-Alkyl, C-N<)$$

Antidots
Näheres siehe 5.5.2.f. 3

EP   31. 8.78/16. 4.80   Ciba-Geigy

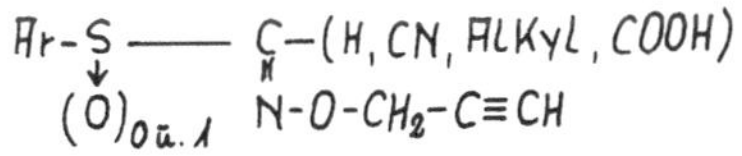

$$\underset{(O)_{0\,\ddot{u}.\,1}}{Ar-S} \overset{O}{\overset{\|}{-}} C-(H, CN, Alkyl, COOH) \\ N-O-CH_2-C\equiv CH$$

Antidots für Herbizide

Ar = Phenyl, Naphthyl, Heterocyclus

US   4 196 152   2. 2.76/ 1. 4.80   Chevron Res.

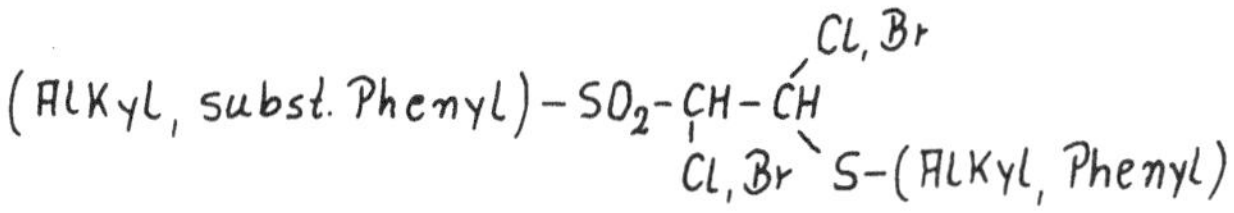

$$(Alkyl, subst. Phenyl)-SO_2-\underset{Cl,Br}{CH}-\overset{Cl,Br}{CH} \\ S-(Alkyl, Phenyl)$$

u.a. auch Wasserherbizide

SU   536 172   19. 2.75/15. 6.77   Leningrad Lensovet

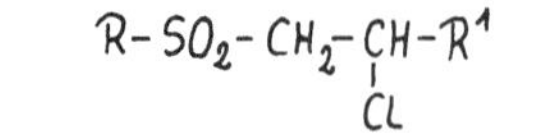

$$R-SO_2-CH_2-\underset{\underset{Cl}{|}}{CH}-R^1$$

R = CH$_3$, C$_6$H$_5$(subst.)
R^1 = C$_6$H$_5$

u.a. auch herbizid wirksam

JA   53 109952   7. 3.77/26. 9.78   Nissan Chem. Ind.

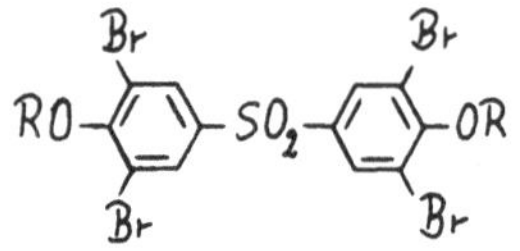

R = H, CH$_2$-$\underset{\underset{O}{\|}}{C}$-OH

5.2 Alkohole, Thiole, Ester, Ether, Acetale, auch cyclische Acetale

5.2.1   Aliphatische Alkohole

    a   Mono-alkohole, Thiole, Ester, Ether und Acetale   Bd. 5/385

US   4 042 371   4. 9.73/16. 8.77   Stauffer Chem.

$$R^1-S-R^2$$

R^1 = R^2 = 3, 3-Dichlorallyl, 3-Chlorallyl
oder 2, 3-Dichlorallyl

Antidots zur Erhöhung der herbiziden Selektivität von Thiocarbamten.

5.2.1 b   Diole und Triole, ihre Ether, offene und cyclische Acetale

Cyclische Acetale siehe auch unter 5 u. 6-Ringen mit   Bd.5/386
2 x O. 5.14.2. 1 und 5.14.3. c

DOS   2 655 843   9.12.76/23. 6.77   Ciba-Geigy
SZ Priorität 12. u. 24.12.75/15.11.76
s.a. DOS 2 655 843 Ciba-Geigy 5.4.2.b

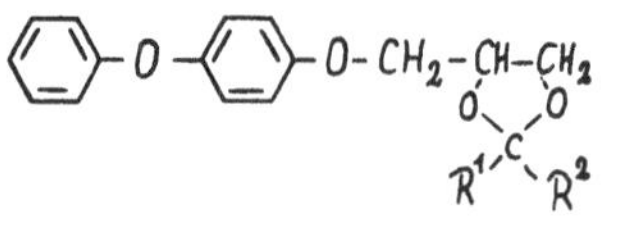

R^1 = H, Alkyl

R^2 = 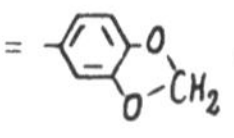 oder subst. Phenyl

DOS   2 739 067   31. 8.76/ 2. 3.78   FMC

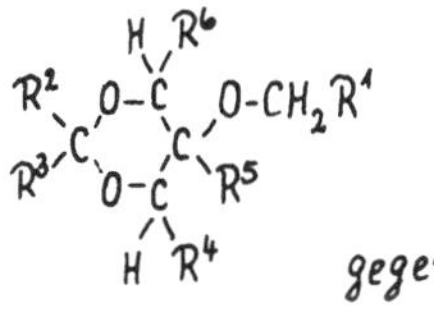

R^1 = Phenyl, Furyl, Pyridyl usw. subst.
R^2 = H, Alkyl, Phenyl
R^3 = H
R^2 + R^3 = $(CH_2)_n$
R^4 = Alkyl, Haloalkyl, Cyanalkyl usw.
R^5 = H, Alkyl
R^6 = H, Alkyl, Halog.-Alkyl, Cyanalkyl

gegen Gräser

DOS   2 842 188 Bayer AG sehr ähnlich dem nachf. EP.

EP   2 214   1. 12.77/13. 6.79   Bayer
=DOS   2 753 556

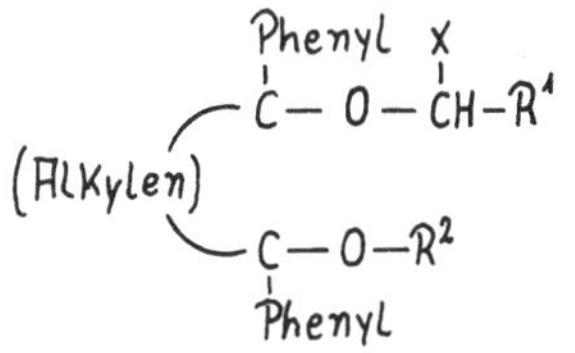

R^1 = subst. Phenyl
R^2 = H, Alk(en, in)yl, Halog. Alkyl
X = H, Alk(en, in)yl, Halog. Alkyl, Phenyl

Beispiel:

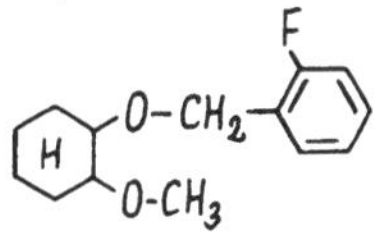

US   4 035 178   31. 8.76/12. 7.77   FMC

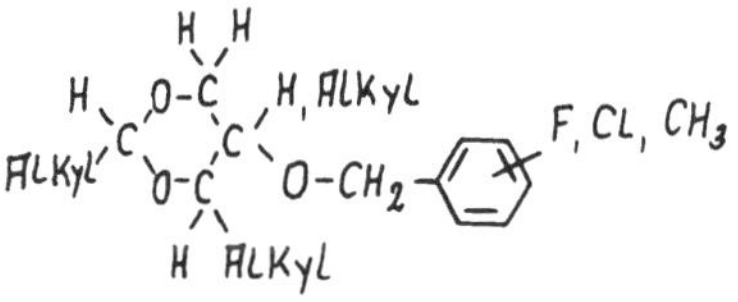

US   4 077 982   19. 1.68/ 7. 3.78   FMC
s.a. US 3 753 678

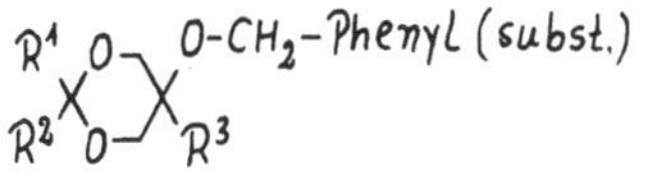

R^1 = H, Alk(en, in)yl, Cycloalkyl, Phenyl usw.
R^2 = H, Alkyl
R^3 = H, Cyanalkyl, Halog. Alkyl

US   4 042 369   21. 6.74/16. 8.77   Shell Oil

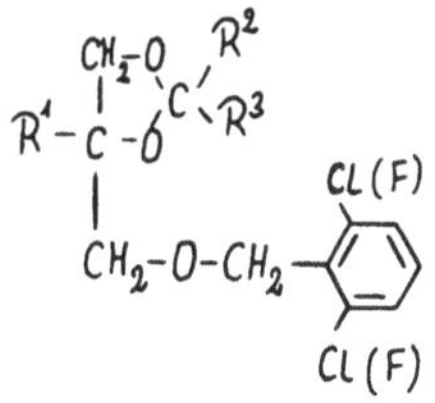

R^1 u. R^2 = CH_3, C_2H_5
R^3 = H, CH_3 u. C_2H_5

spez. gegen Ungräser in Weizen

5.2.2   Aromatisch-aliphatische Alkohole und ihre Ether

a) Benzylalkohole, Phenethylalkohole   Bd. 5/388

DOS   2 753 556   1.12.77/ 7. 6.79   Bayer
= EP   2 214

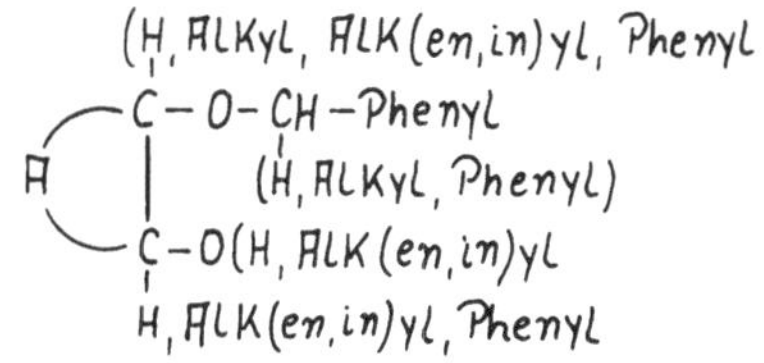

Selektiv gegen Gräser

Beispiele:

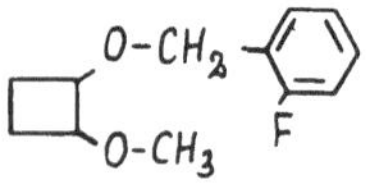

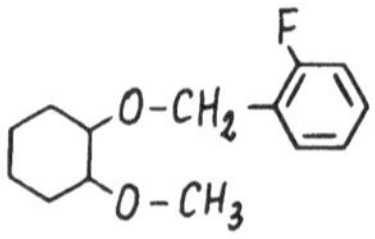

JA   52 079026   25.12.75/ 2. 7.77   Ube Ind.

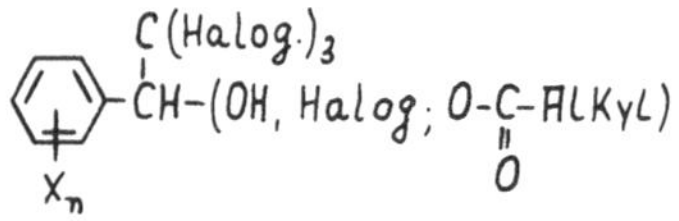

X = Halog, O-Alkyl, $-O\underset{CH_2}{\diagup\diagdown}O-$

124

5.3 Aldehyde und Ketone

5.3.1 Aliphatische und cycloaliphatische Vertreter Bd.5/391

DOS 2 616 756 15. 4.76/28.10.76 Ciba-Geigy

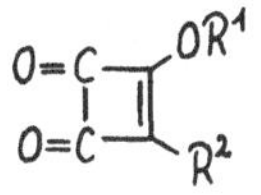

R^1 = Alkyl, Phenyl (subst.), Alk(en,in)yl
N<, CN, NO_2
R^2 = H, OH

Kontakt-herbizid

(Quadratsäure)

Ähnliche Verbindungen aus einem Pilz (fusevium moniliforme) siehe Science 179, 1324 (1973).

Beispiele:

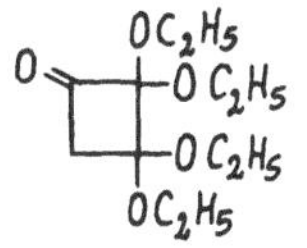

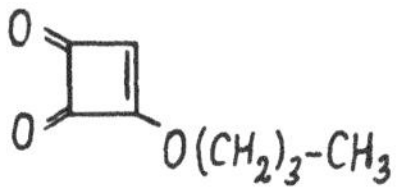

DOS 2 813 341 28. 3.77/ 5.10.78 Union Carbide
US Prior. 28.3.77

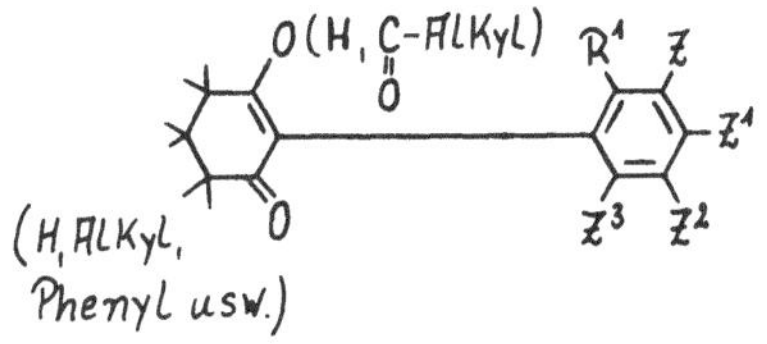

Z-Z^3 = H, Halog.-Alkyl, (O)-Alkyl, CF_3, CN, NO_2, Alkyl-SO_2, NH_2 usw.

unter anderem auch herbizid wirksam

DOS 2 822 304 22. 5.78/30.11.78 Nippon Soda
ähnlich JA 55 028957

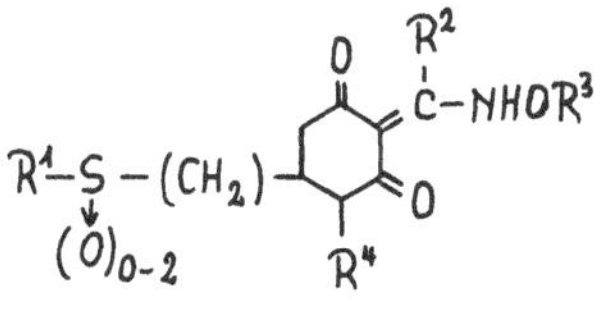

R^1 = Alkyl, Phenyl, Benzyl
R^2 = Alkyl
R^3 = Alk(en)yl
R^4 = H, C(=O)-O-Alkyl

Wirksam gegen Gräser

DDR   123 452   12.12.75/20.12.76   Hesse

X, Y u. Z = H, Cl, Br

Darstellungsverfahren

GB   1 534 275   30. 1.75/29.11.78   Shell Int.

R^1 = H, Alkyl, Aryl, O-Alkyl
                                                usw.
R^2 = R^1
X = Halog., O-Alkyl, N$\diagup$,

$$NH-\underset{\underset{O}{\|}}{C}-O\text{-Alk.}$$

US   3 976 469   16.11.72/24. 8.76   Stauffer

R^1 = Halog subst. Cycloalkyl, $ClCH_2$ usw.
       Phenyl, Benzyl
R^2 = ähnlich R^1

Antidot gegen Herbizidschäden durch Thiolcarbamate.

US   4 091 006   12.12.72/23. 5.78   Union Carbide

R^1 = H, Halogen
R^2 = H, CH_3, C_2H_5, O-CH_3, O-C_2H_5, F,
       Cl, Br, CCl_3, CF_3
R^3 = H, (O)-Alkyl, NO_2, Halog. usw.
R^4 = Halog. CH_3, O-CH_3, O-C_2H_4-
R^5 = H, Alkyl, Halog.-Alkyl, O-Alkyl,
       NH-$\underset{\underset{O}{\|}}{C}$-Alkyl
R^6 = H, Alkyl, O-Alkyl, CF_3, NH-$\underset{\underset{O}{\|}}{C}$-Alkyl

Unter anderem auch Herbizide

126

US   4 089 673   16.11.72/16. 5.78   Stauffer

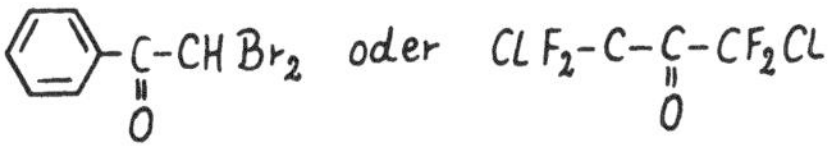

$$\text{Ph}-\overset{\parallel}{\underset{O}{C}}-CHBr_2 \quad oder \quad ClF_2-C-\overset{\parallel}{\underset{O}{C}}-CF_2Cl$$

Antidots gegen Triazinherbizidschäden, auch für Thiocarbamat-Herbizide.

US   4 202 840   29. 9.78/13. 5.80   Stauffer

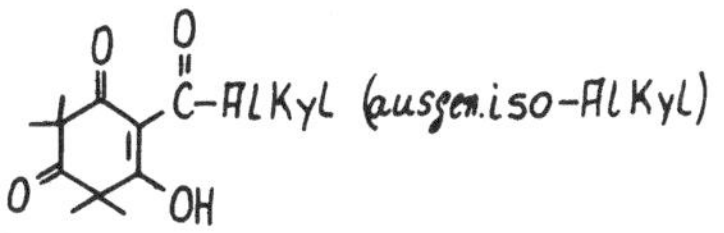

$$\overset{O\ O}{\underset{O\quad OH}{}}\ \overset{\parallel}{C}-ALKyl\ (ausgen.iso-ALKyl)$$

JA   51 125040   23. 8.74/ 1.11.76   Nippon Soda

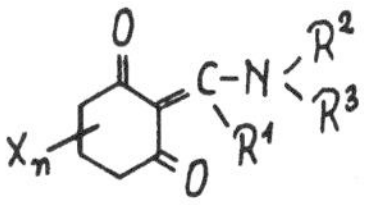

$$X_n \quad \overset{O}{\underset{O}{}}\ C-N{<}\overset{R^2}{\underset{R^3}{}}\ R^1$$

R^1 = Alkyl, Phenyl
R^2 u. R^3 = Alk(en)yl, Phenyl
X = Alkyl, Phenyl, Furyl, Thienyl,
    5,6-Tetramethylen

JA   53 090248   23. 8.74/ 8. 8.78   Nippon Soda

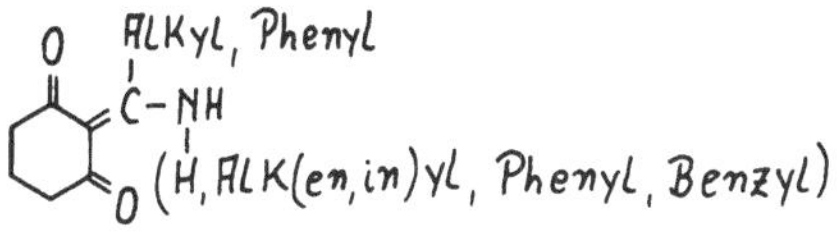

$$\overset{O}{\underset{O}{}}\ \overset{ALKyl,\ Phenyl}{C-NH}\ (H, ALK(en,in)yl,\ Phenyl,\ Benzyl)$$

Gegen Gräser und Unkräuter.

JA   53 130430   18. 4.77/14.11.78   Nippon Soda

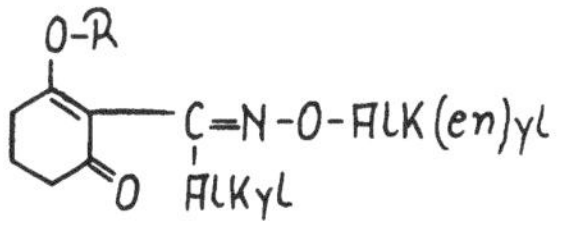

$$\overset{O-R}{\underset{O\ \ ALKyl}{}}\ C=N-O-ALK(en)yl$$

R = Alkyl, Benzyl, $\underset{S}{\overset{\parallel}{P}}(OAlkyl)_2$ usw.

JA   54 019 945   15. 7.77/15. 2.79   Nippon Soda

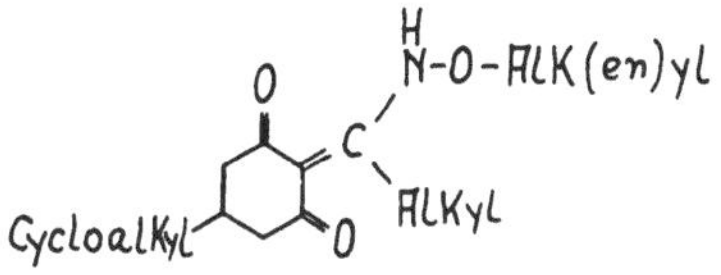

$$Cycloalkyl\ \overset{O}{\underset{O}{}}\ \overset{H}{\underset{ALKyl}{C}}\ N-O-ALK(en)yl$$

EP 3 884 23. 2.78/ 5. 9.79 ICI

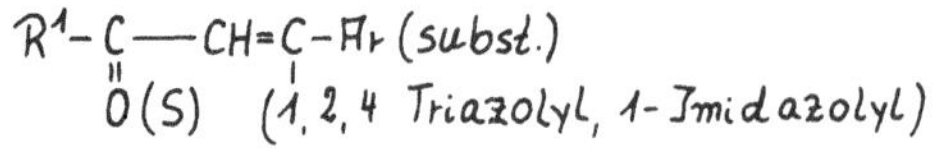

$$R^1-\underset{\underset{O(S)}{\|}}{C}-CH=C-Hr\ (subst.)\quad (1,2,4\ Triazolyl,\ 1-Imidazolyl)$$

R^1 = Alkyl, Phenyl subst.

Breitspektrum-Herbizid

EP 12 158 30. 8.79/25. 6.80 Ciba-Geigy

CH Priorität   1. 9.78

(Sehr umfassendes Patent mit 12 Erfindern)

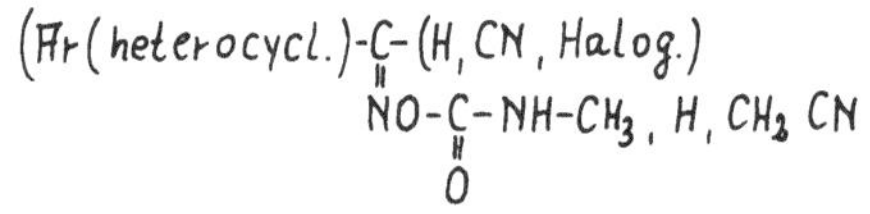

$$\left(Hr\,(heterocycl.)-\underset{\underset{NO-\underset{\underset{O}{\|}}{C}-NH-CH_3,\ H,\ CH_2\,CN}{\|}}{C}-(H,CN,Halog.)\right)$$

Antidots gegen sym. Triazinherbizide, Diphenyletheroxypropionsäuren usw.

US 4 009 022 30.10.75/22. 2.77 Rohm u. Haas

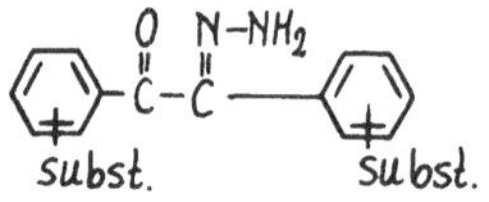

Antagonisten gegen Herbizidschäden durch Triazine

US 4 089 673 16.11.72/16. 5.78 Stauffer

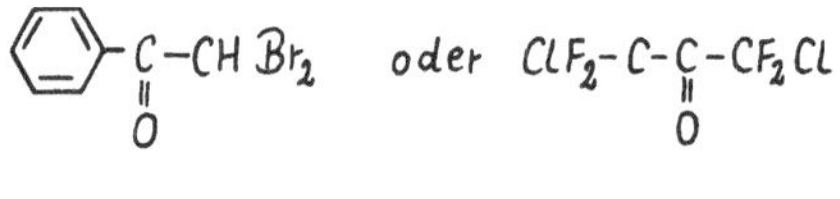

Antidots gegen Herbizidschäden durch sym. Triazine oder Thiocarbamate

US 4 113 465 16.12.75/12. 9.78 ICI

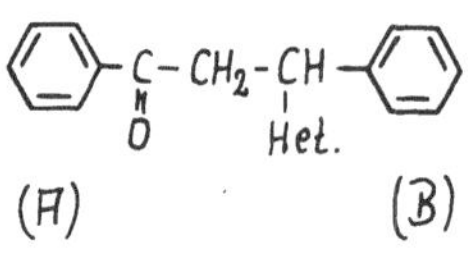

A u. B subst. durch Halogen, Alkyl, CN, NO_2, N$\langle$ usw.

128

| US | 4 123 255 | 3. 1.77/31.10.78 | Chevron Res. |

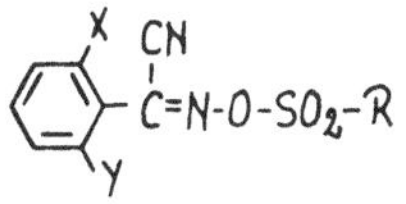

R = Alkyl (subst. d. Halog.)
X u. Y = Cl, Br

| SU | 498 938 | 4. 4.74/30. 4.76 | ASUZB-Plant Biol. |

| ZA | 75 006575 | 17.10.75/ 2. 8.76 | Shionogi |

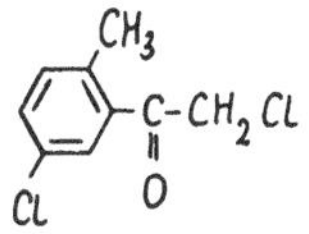

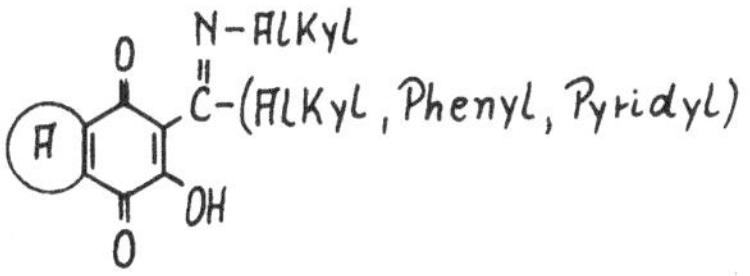

A = Benzo, Pyrido

5.4 Phenolderivate

Bd. 5/398

5.4.1   Phenole, Thiophenole und ihre Ester

| DOS | 2 803 991 | 31. 1.78/10. 8.78 | |
| BE | 863 471 | 5. 2.77/31. 7.78 | Fisons Ltd. |

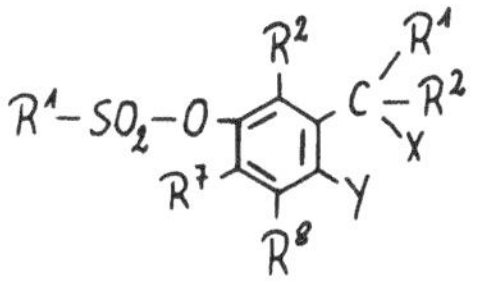

$X = -CH-O-R^4$ mit R^3

$Y = OR^5$; $X + Y = -CH-O$ usw. mit R^3

R^6, R^7, R^8 = H, Halog.-Alkyl, CN, O-Alkyl, Cycloacyl

R^1, R^2, R^3 = H, Alkyl

$R^1 + R^2$ oder $R^2 + R^3$ = zusammen Alkylen

Spez. Beispiel:

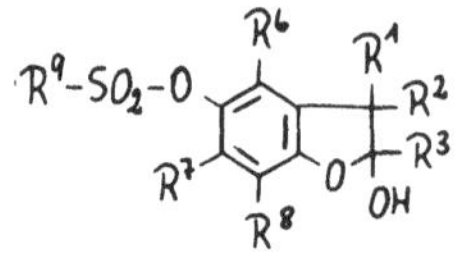

DOS   2 636 452   13. 8. 76 / 16. 2. 78   Amer. Cyanamid

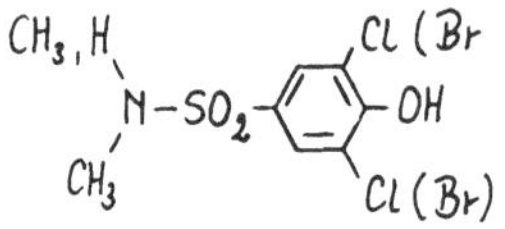

Gegen Wildhafer in Getreide,
Zuckerrüben, Flachs oder
Raps

DOS   2 815 237   8. 4. 78 / 18. 10. 79   Bayer

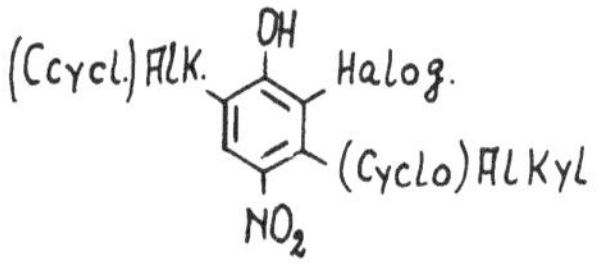

Entblätterungsmittel für
Baumwollpflanzen

EP   542   26. 7. 77 / 7. 2. 79   Bayer

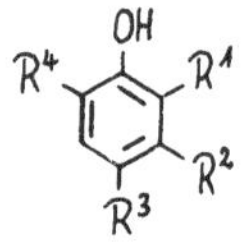

R^1 = H, Cl, CF_3
R^2 = H, O-CH_3
R^3 = Cl, NO_2, CF_3
R^4 = Cl, NO_2, COOH

BE   841 523   7. 5. 75 / 1. 9. 76   Gulf Oil

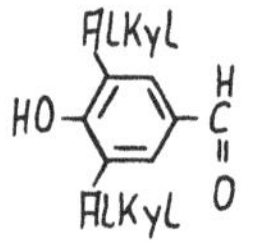

Zwischenprodukt für
Herbizide

DDR   131 125   18. 5.75/ 7. 6.78   VEB Chemiefaserkombinat
                                                            Bitterfeld

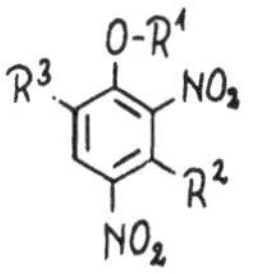

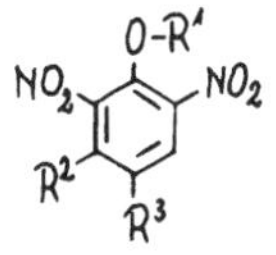

R^1 = H, NH_4 usw.

R^2 = H, CH_3

R^3 = subst. Cyclohexyl, Cyclopentyl, Aralkyl

ü.

DDR·   131 448   25. 5.77/26. 6.78   VEB Chemiefaserkombinat
                                                            Bitterfeld

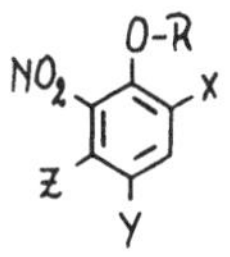

R = H, NH_4

X = Cl, Br

Y = H, Cl, Br

Z = H, Cl

Dessicant oder Defoliant

DDR   131 615   27. 6.77/12. 7.78   VEB Chemiekombinat
                                                            Bitterfeld

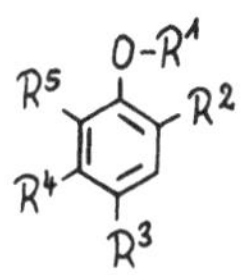

R^1 = H, NH_4, CO-Alkyl, $\overset{\text{O}}{\overset{\|}{C}}$-$CH_2Cl$

R^2 = C_6H_5-CH_2-CH_2-, Cumyl

R^3 u. R^5 = CH_3, C_2H_5, Cl, NO_2

R^4 = H, CH_3, Halogen

FR   2 374 850   24.12.76/25. 8.78   Goffin

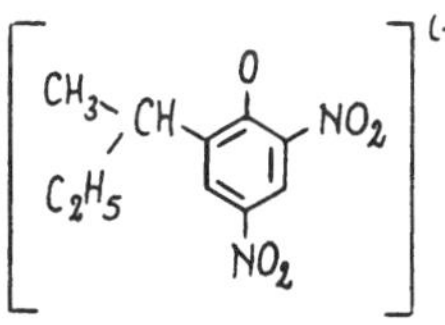

$NH_4^{(+)}$ in überschüssiger NH_3 Lösung

ph >9

US   3 988 350   29.10.71/26.10.76   Gaf Corp.

Phenol-Lactam (Pyrrolidon) Komplexe

US   3 988 351   29.10.71/26.10.76   Gaf Corp.

Phenol-Lactam-Komplexe z.B. N-Methylpyrrolidon,
2,4,5-Trichlorphenol oder Polylactame z.B. Hexamethylen-
bis-(2-1-pyrrolidone)+Phenole. Unter anderem auch herbizid wirksam.

US   4 002 661   13. 4.70/11. 1.77   Minnesota Mining

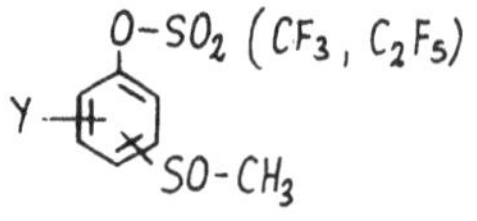

Y = Cl, CH_3, C_2H_5

JA   52 064423   26.11.75/27. 5.77   Mitsui Toatsu

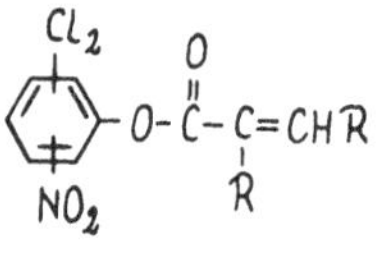

R = H, CH_3

JA   53 142526   19. 5.77/12.12.78   Hokkai Sankyo

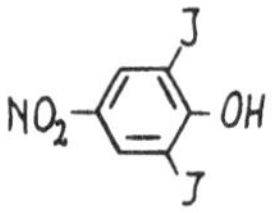

JA   54 002323   7. 6.77/ 9.11.79   Mitsubishi Petrochem.

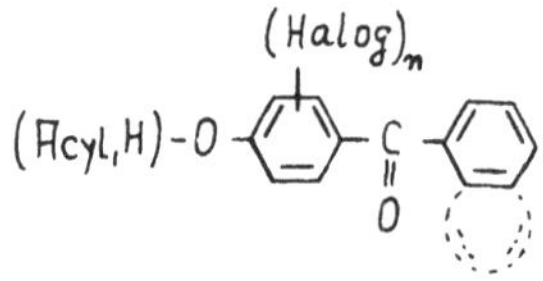

Insektizid-Herbizid

5.4.2   Phenol-ether (O u. S) und Acetale, siehe auch 5.7.3 b 1 - c
          ===================================

    a) <u>Phenolether mit aliphatischen Ethergruppen</u>   Bd. 5/400

DOS   2 738 873   29. 8.77/ 2. 3.78   Eli Lilly
US Prior.   31. 8.76

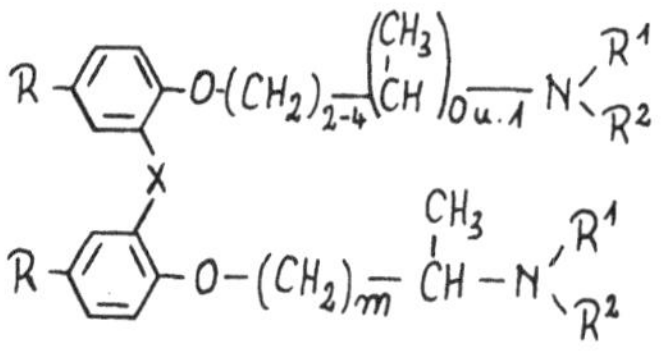

X = CH_2, CH_2-CH_2, CH_2-$\overset{CH_3}{CH}$, S
R = Halogen, NO_2, (O)-Alkyl
R^1 u. R^2 = H, Alkyl,
        CH_2-CH_2-OH

R^2 auch C_6H_{11}, CH_2-⬡,

        CH_2-CH_2-OH

        Herbizid und Algizid

132

DOS   2 641 167   19.10.76/ 5. 5.77   Shell Inst. Res.
GB Prior.   21.10.75

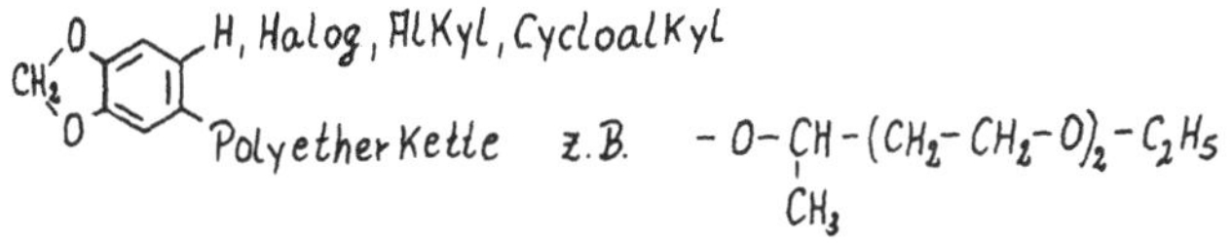

Synergist für bekannte Herbizide gegen wilden Hafer.

DOS   2 738 902   29. 8.77/ 2. 3.78   Eli Lilly u. Co.

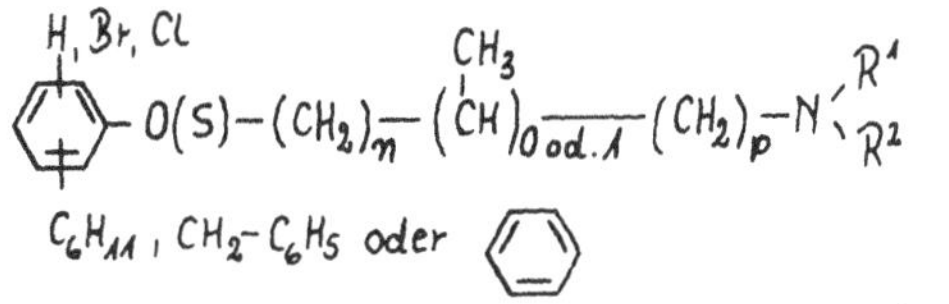

n + p = 2-10
R^1 = H, Alkyl, CH_2-CH_2-OH
R^2 = H, Alkyl, Alkenyl,
        Cycloalkyl, CH_2-CH_2-OH,
        CH_2-C_6H_5, Phenyl,
        1-Adamantyl

Herbizid und Algizid

JA   52 091834   26. 1.76/ 2. 8.77   Kumiai Chem.

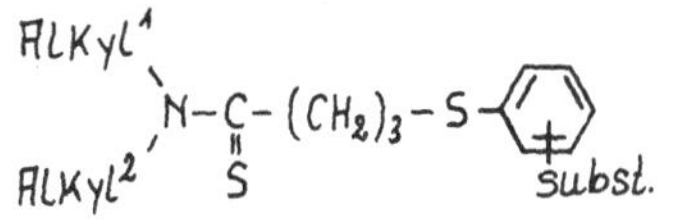

b) <u>Di- u. Triphenylether, Phenyl-benzylether, Di-Benzylether,
   Phenyl-oximether</u>

Nachfolgend werden hier <u>nicht</u> registriert: Ar-O-Ar-O-CH-COOR
                                                         |
                                                      (H, CH_3)
siehe diese unter Phenoxyalkancarbonsäuren 5.7.3 b - c 1

Erfaßt sind aber Verbindungen wie Ar-O-Ar-O-CH_2 R^1 mit

R^1 = Cl, $(CH_2)_n$-CH (z.T. auch COOH umfassend, dann doppelt registriert)

Verbindungen des Typs:

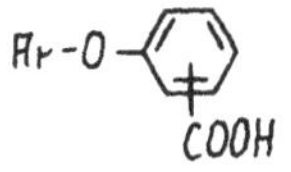

werden hier beschrieben, da offensichtlich
die Diphenylether-Struktur für die herbizide Wirksamkeit maßgebend ist.

DOS   2 366 039   3. 7.73/ 7. 7.77   Bayer

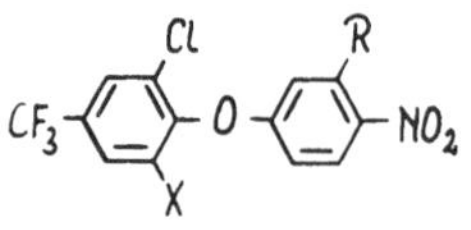

R = Halog., CH_3-S

X = H, Cl

R nicht = Cl wenn X = H

Darstellungsverfahren

DOS   2 366 040   3. 7.73/14. 7.77   Bayer

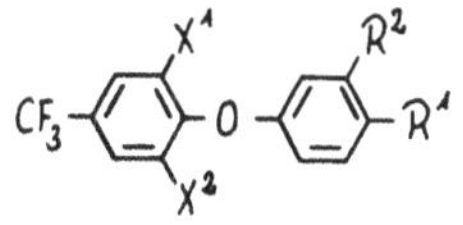

R^1 = S(O)$_{0-2}$-Alkyl, NH_2-C,
                                  $\overset{\|}{S}$

R^2 = H, CH_3
X^1 = H, CH_3
X^2 = H, Halog.

DOS   2 520 815   9. 5.75/18.11.76   Bayer

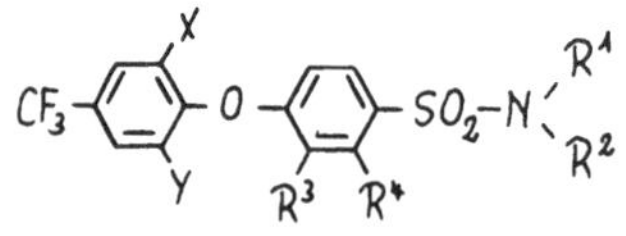

X u. Y = H u. Halog.
R^1 u. R^2 = H, Alkyl
R^3 = H, Halog.
R^4 = H, Halog., Alkyl

z.B.

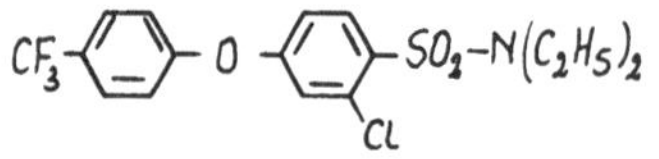

DOS   2 611 695   19. 3.76/29. 9.77   Hoechst
=BE   852 701

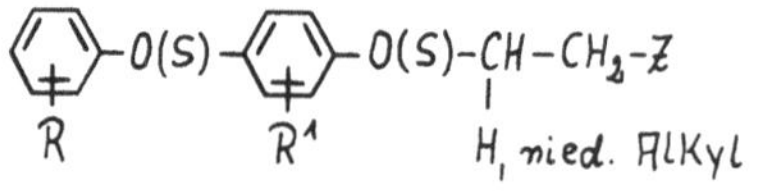

R = Halog., Halogen-Alkyl, Alkyl, Thioalkyl

R^1 = Halog., CH_3, CF_3, O-C-CH_2-O-Ar subst. usw.
                          $\overset{\|}{O}$

Z = H, OH, SH, OR^3,    n = 0, 1 oder 2
                                        R^3 = Alkyl, Cycloalkyl, Alkenyl, Phenyl,
                                                Benzyl, alles Reste ev. subst.
                                        R^4 = Alk(en)yl, Cycloalkyl
                                        R^5 = H, Alkyl
                                        R^4 u. R^5 Teil eines Heteroringes usw.

134

DOS   2 619 489   3. 5.76/18.11.76   Rohm u. Haas

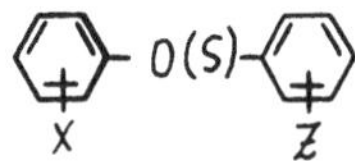

X u. Z = Halog, Alkyl, CF_3,
          NO_2, OAlkyl

Verwendung zur Wuchsdämpfung z. B. bei Tabak; ähnliche Verbindungen als
Herbizide in Verwendung.

DOS   2 632 619   20. 7.76/24. 2.77   Stauffer
US Prior.   23. 7.75

X = Cl, CH_3, n = 0-2
Y = Cl, CF_3
R^1 = H, (O-)Alkyl, S-Alkyl,
       Alkenyl
R^2 = H, O-Alkyl usw.

DOS   2 635 099   4. 8.76/ 9. 2.78   Celamerk
=BE   857 444

R^1 = Halog., R^2 = H, Halogen

DOS   2 639 796   3. 9.75/31. 3.77   Rohm u. Haas

R^1 = H, Halog., CN; R =

S in p-Stellung dann

R^1 = H, Halog, CN,
       NO_2, CF_3,
       (O)-Alkyl usw.

DOS   2 643 438   17. 3.76/ 7. 4.77   Ishihara Ind.

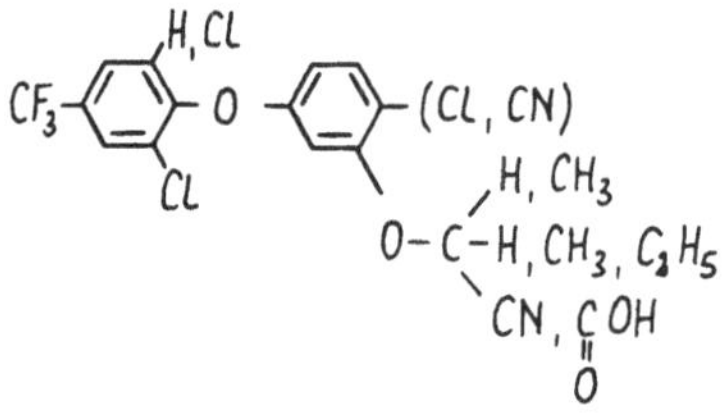

Gegen Unkräuter

DOS   2 644 486   1.10.76/14. 4.77   Mobil Oil
US Prior.   1.10.75

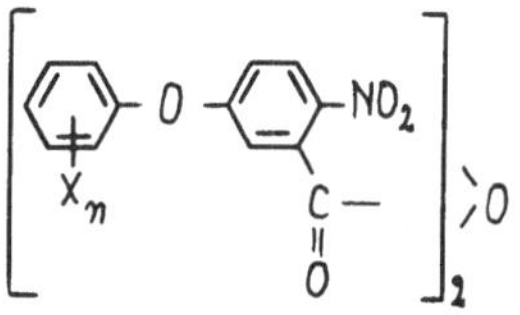

X = Halog., NO_2, Alkyl, CF_3, CN

Entsprechende Ester siehe
US 3 652 645

DOS   2 649 706   29.10.75/ 5. 5.77   Ishihara Sangyo

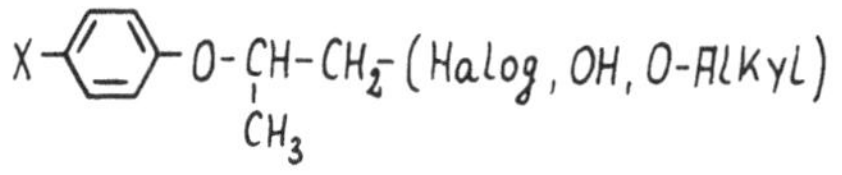

X = Phenyl-O, subst. Phenyl-O

und

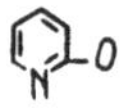

DOS   2 655 843   9.12.76/23. 6.77   Ciba-Geigy
CH Prior.   12.u.24.12.75/15.11.76

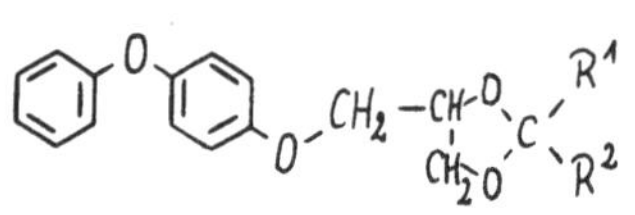

R^1 = H, Alkyl

R^2 =

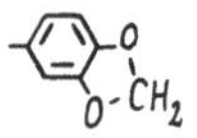

, subst. Phenyl

Insektizid und Herbizid.

DOS   2 655 910   9.12.76/16. 6.77   Ciba-Geigy
CH Prior. 12.12.75; 24.12.75; 1.9.75; 19.11.76

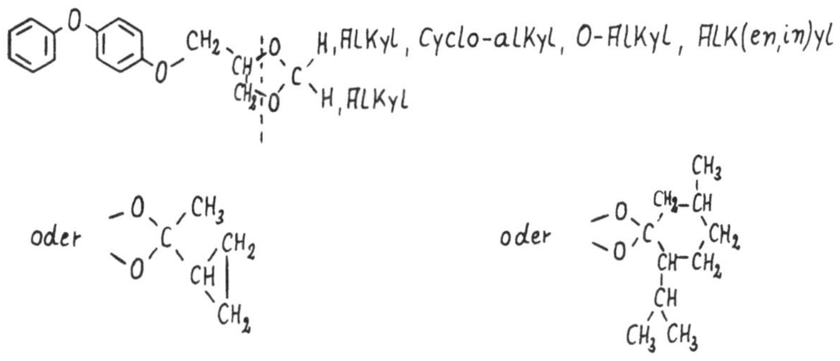

H, Alkyl, Cyclo-alkyl, O-Alkyl, Alk(en,in)yl
H, Alkyl

oder

oder

Herbizide und Insektizide.

DOS   2 720 427   6. 5.77/24.11.77   Sumitomo Chem.
JA Prior. 7.5.76; 13.5.76; 6.10.76; 28.12.76

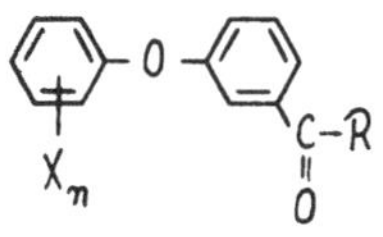

X = Alkyl, O-Alkyl, Halogen; n = 0-5

$R = N$ ⟨ H, Alkyl / Alk(en,in)yl, O-Alkyl usw.

Substitution von X vorwiegend in
3 und 5 Stellung

DOS   2 729 529   30. 6.77/11. 1.79   Celamerk

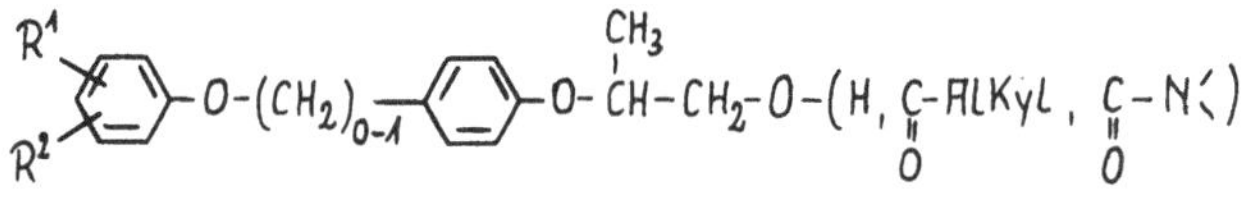

Zusammen mit einem Antidot z.B. $Cl_2CH-C-N$ ⟨
                                        ‖
                                        O

R^1 = Halog; R^2 = H oder Halogen

DOS   2 745 005   6.10.77/19. 4.79   Bayer

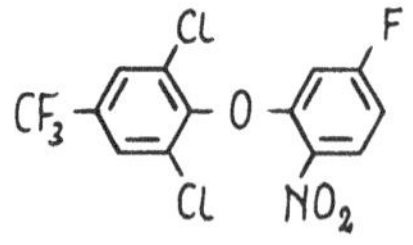

Defoliant spez. für Baumwolle

DOS   2 745 006   6.10.77/19. 4.79   Bayer
=EP   1 427   6.10.77/18. 4.79

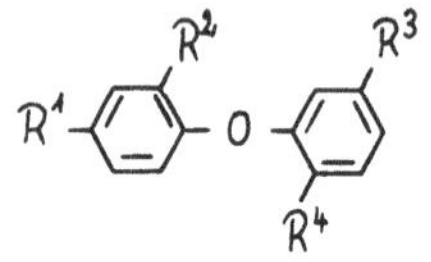

R^1 = H, Halog., Alkyl, CN, NO_2, CF_3,
O-Alkyl, Alkyl-SO_2
R^2 = H, Halog.; R^3 = H, Halog.,
R^4 = CN, NO_2

Defoliant, Desiccant und Herbizid.

DOS   2 748 658   29.10.77/10. 5.79   Hoechst

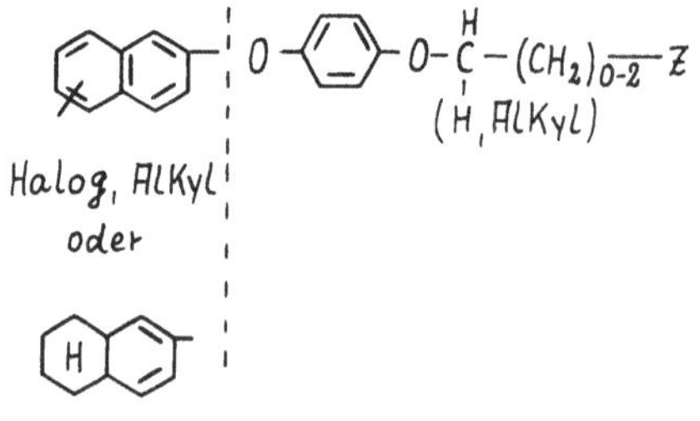

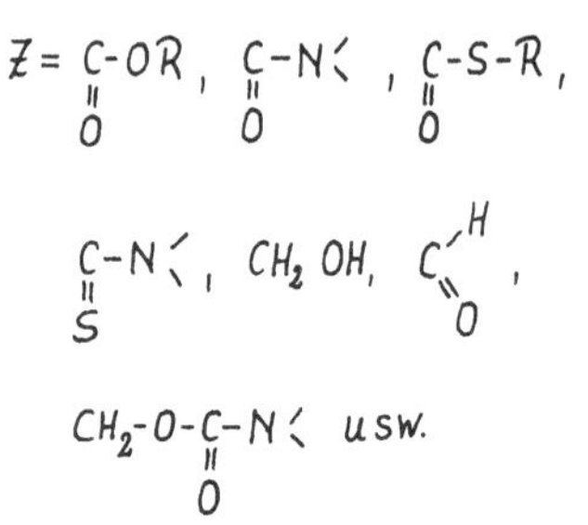

$Z = \underset{O}{\overset{H}{C}}-OR$, $\underset{O}{\overset{H}{C}}-N\langle$, $\underset{O}{\overset{H}{C}}-S-R$,

$\underset{S}{\overset{H}{C}}-N\langle$, $CH_2 OH$, $\underset{O}{\overset{H}{C}}$,

$CH_2-O-\underset{O}{\overset{H}{C}}-N\langle$ usw.

DOS   2 753 900   3.12.77/ 8. 6.78   Rohm und Haas
US Prior.   3.12.76

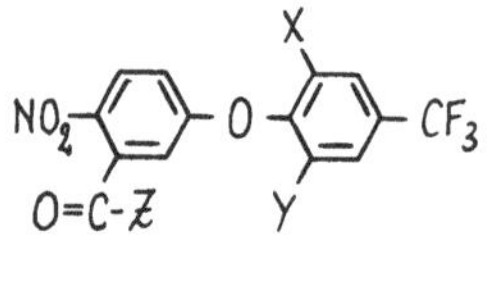

X = H, Halog., CCl_3, CF_3, CN, Alkyl
Y = H, Halog., CCl_3, CF_3, CN, Alkyl
Z = O-Alkenyl, O- Halog-Alkyl, O-Alkinyl,
    Ar-alkenyl

$N\langle$ H, Alkyl / Alk(en, yl)-yl, Aralkenyl

Gegen breitblättrige Unkräuter in Getreide, Mais, Bohnen und Reis.

138

DOS   2 833 021   27. 7.78/ 8. 2.79   Nagai
JA Prior.   28. 7.77
=GB   2 001 635   28. 7.77/ 7. 2.79   Ube Industr.

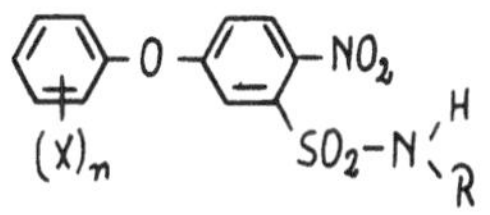

X = Halogen, CF_3
n = 1-3
R = H, $\overset{\text{O}}{\underset{\|}{C}}$-N(CH$_3$)$_2$, Acyl

DOS   2 926 829   5. 7.78/17. 1.80   Mitsui Toatsu Chem.

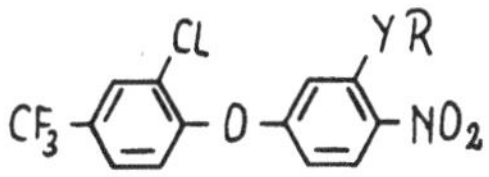

Y = O, S, NH
R = Alk(en, in)yl

Darstellungsverfahren

DOS   2 930 728   28. 7.79/14. 2.80   Sandoz
US Prior.   7. 8.78

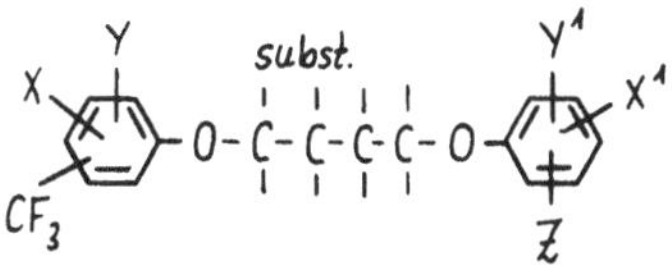

X u. X^1 = H, Halog., CF_3,
        O(S)-Alkyl, NO_2,
        Alkyl, Alkyl-SO_2 usw.

Z = H, CF_3, Halog., Alkyl
Y u. Y^1 = H, Halog, CF_3,
        Alkyl, O(S)-Alkyl usw.

Substituenten an den Zwischengliedern: H, Alkyl, COOR, Halog., Epoxygruppen.

DOS   3 009 153   9. 3.79/11. 9.80   Nippon Kayaku

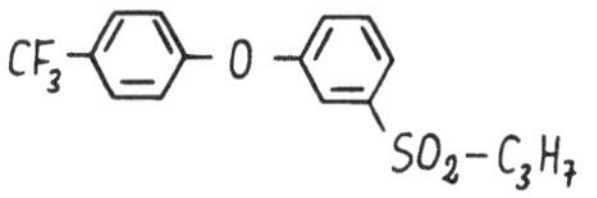

DOS   2 944 783   9.11.79/22. 5.80   Mitsui Toatsu Chem.

DOS   2 946 024   14.11.79/29. 5.80   Ihara Chem. Ind.

EP   359   30. 6.78/24. 1.79   Ciba-Geigy
CH Prior.   7. 7.77

EP   1 427   siehe DOS 2 745 006   Bayer
EP   2 812   2. 1.78/11. 7.79   Ciba-Geigy

140

EP   3 416   19. 1.78/ 8. 8.79   ICI
GB   19.1.78; 30.5.78; 21.12.78

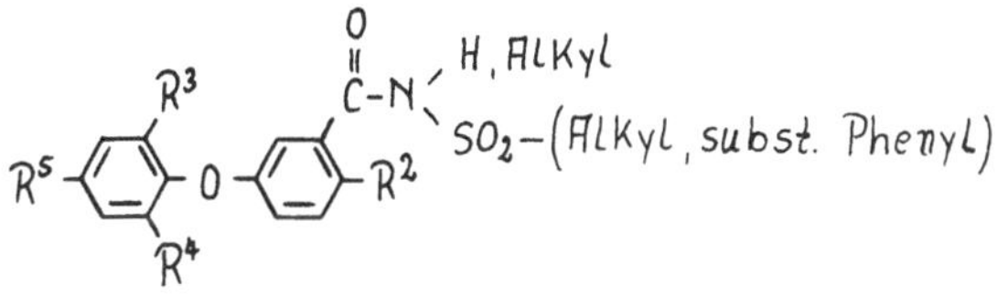

R^2 = H, Halog., NO_2
R^3 = H, Halog, Alkyl, CF_3, CN
R^4 = H, Halog, Alkyl, CF_3, CN
R^5 = H, Halog, Alkyl, CF_3, CN
R^6 = H, Alkyl

EP   7 471   2. 7.79/ 6. 2.80   Ciba-Geigy
CH Prior.   4. 7.78

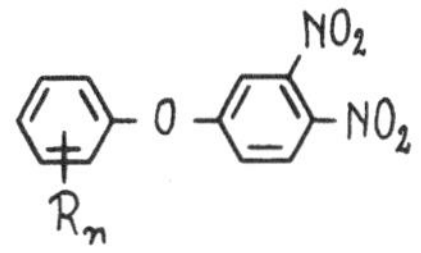

R = Halog., CF_3, NO_2, CN, $\underset{O(S)}{\overset{\|}{C}}-NH_2$

n = 0-3

EP   3 877   5. 2.79/ 5. 9.79   ICI
GB Prior.   15. 2.78

Z = Halog., Cl(F)-Alkyl
Y = H, Halog., Cl(F)-Alkyl

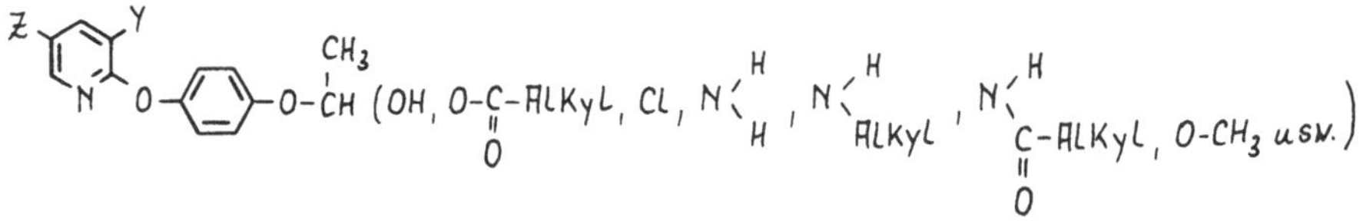

EP   7 482   4. 7.79/ 6. 2.80   Celamerk
=DOS   2 831 262   15. 7.78/31. 1.80   Celamerk

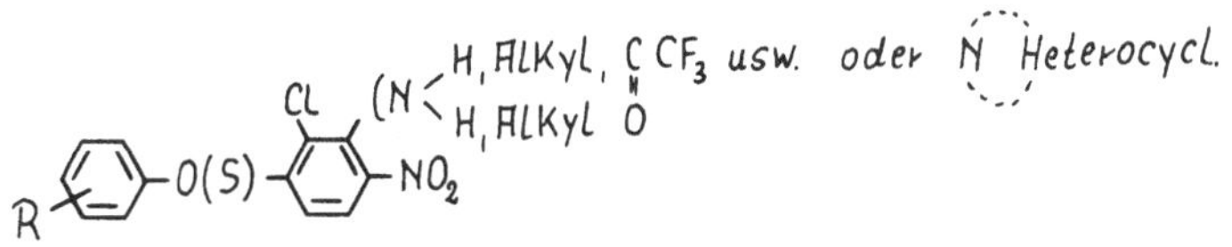

R = H, Halog., CH_3, $O-CH_3$,
    NO_2

EP   13 660   9. 1.79/23. 7.80   Ciba-Geigy

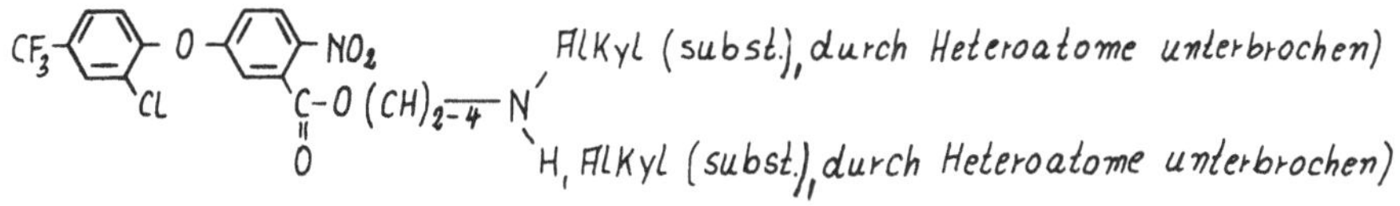

EP   14 684   31. 1.80/20. 8.80   Ciba-Geigy
CH Prior.   6. 2.79

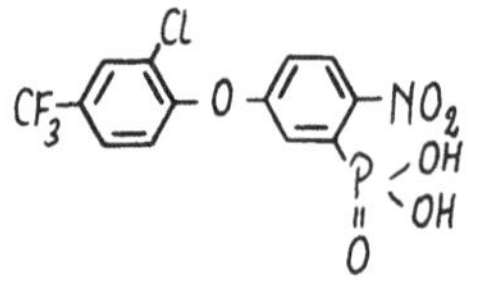

und ähnliche Verbindungen

US   3 976 470   23. 7.75/24. 8.76   Stauffer

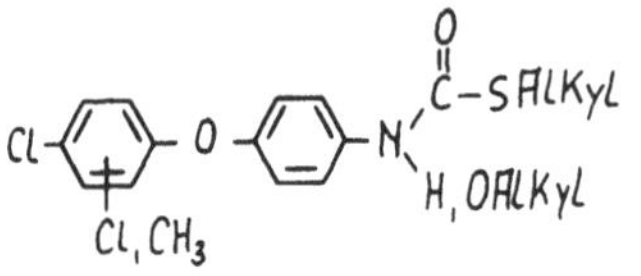

selektiv in Reis

US   3 979 437   25. 4.69/ 7. 9.76   Mobil Oil

Siehe damit zusammenhängene Patente
US   3 941 830, 3 784 635 u. 3 652 645

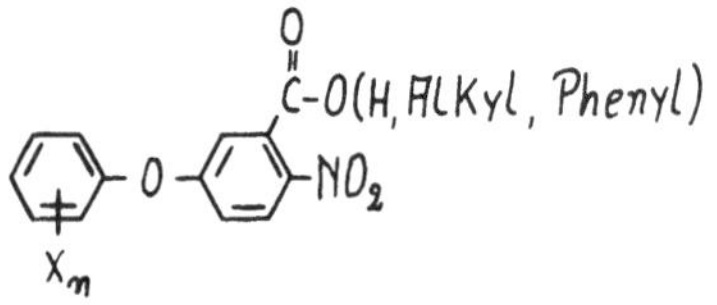

X = Halog, CF_3
n = 2-5

US   3 983 168   25. 4.69/28. 9.76   Mobil Oil

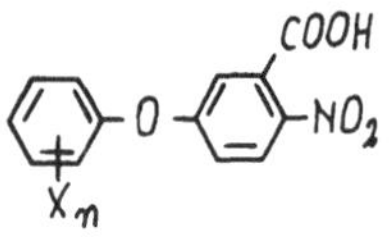

X = Halogen
n = 1-5

US   4 001 005   7. 6.74/ 4. 1.77   Mobil Oil

Siehe auch US 3 907 866

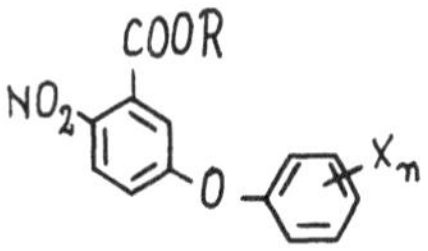

X = Halog.,
n = 1-3
R = (Cyclo)Alkyl usw.

US   4 002 662   25. 4.69/11. 1.77   Mobil Oil

R = OH, O-Alkyl

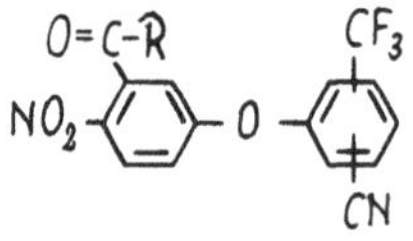

142

US   4 031 131   29. 9.75/21. 6.77   Rohm u. Haas

US   4 042 369   21. 6.74/16. 8.77   Shell Oil Co.

R^1 u. R^2 = CH_3, C_2H_5

R^3 = H, CH_3, C_2H_5

Spezif. gegen Gräser in Weizen.

US   4 063 929   14. 3.72/20.12.77   Rohm u. Haas

Siehe auch US 3 928 416 u. 3 798 276

US   4 029 493   4. 9.70/14. 6.77   Mobil Oil

Gräserherbizid

US   4 093 446   14. 3.72/ 6. 6.78   Rohm u. Haas

X = H, Halogen, CF_3, Alkyl
Y = H, Halogen, CF_3
Z = O-Alkyl

US   4 112 002   21. 1.77/ 5. 9.78   Gaf Corp.

Fungizid und Herbizid

US   4 164 408   25. 4.69/14. 8.79   Mobil Oil
US   4 164 409   25. 4.69/14. 8.79   Mobil Oil
US   4 164 410   25. 4.69/14. 8.79   Mobil Oil

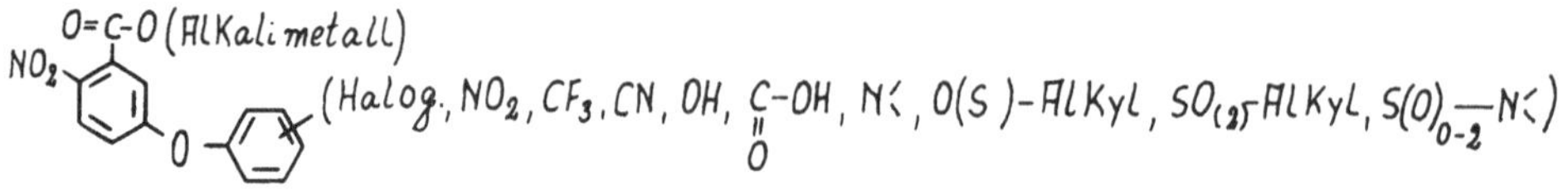

US   4 191 554   3.10.77/ 4. 3.80   Du Pont

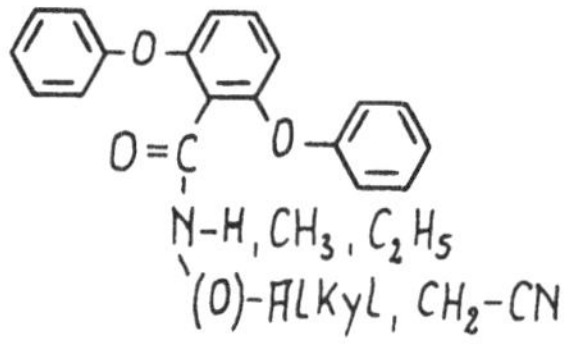

US   4 208 205   3.10.77/17. 6.80   Du Pont

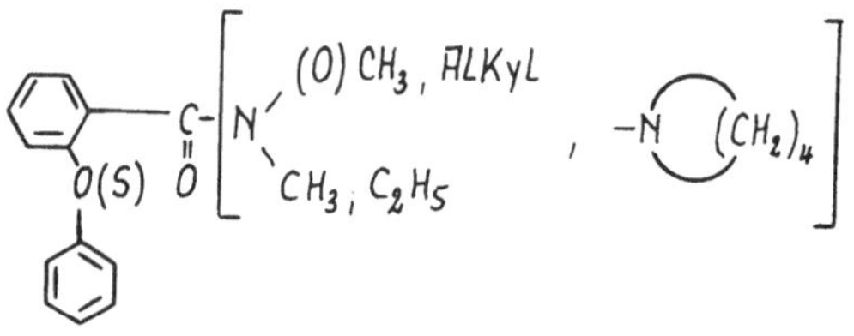

US   4 220 468   14. 3.72/ 2. 9.80   Rohm u. Haas

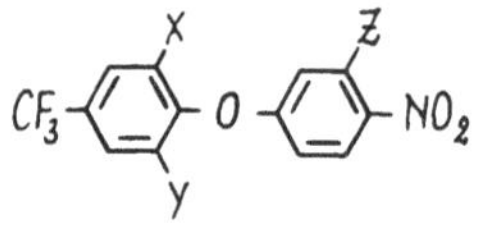

X u. Y = H, Halog., CF_3, Alkyl

Z = Alkyl, verzweigt d. Alk(en, in)yl

RD   193 039   20. 4.80/10. 5.80   Anonym

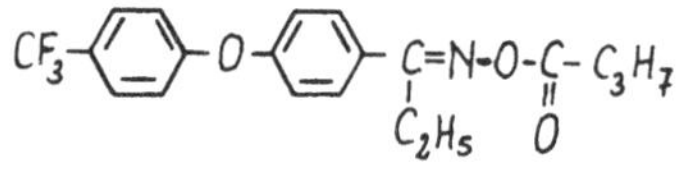

Gegen Gräser

144

BE   474 578   2. 3.78/ 3. 9.79   Philagro

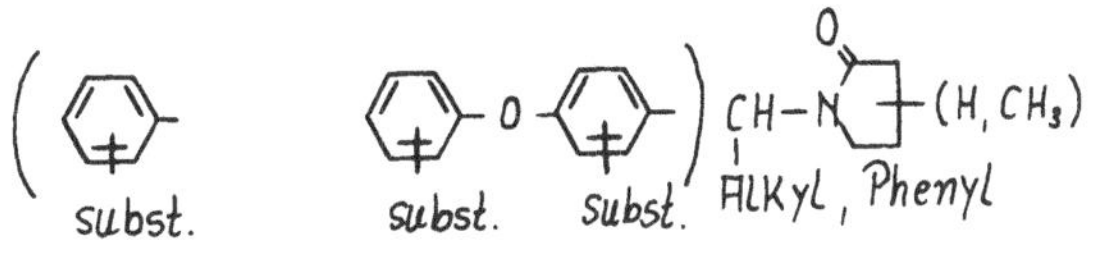

BE   870 068   30. 8.78/28. 2.79   Ciba-Geigy

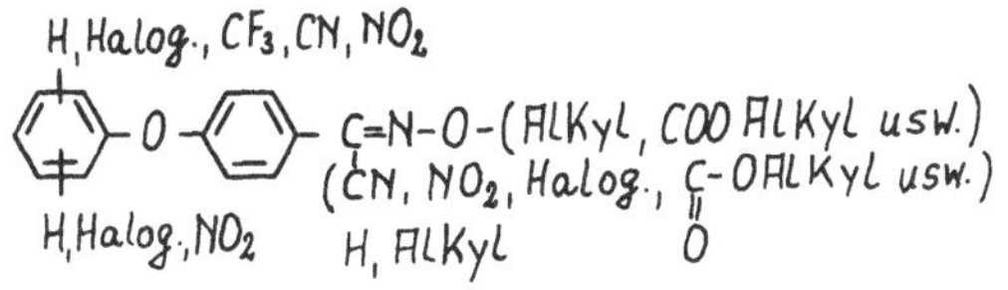

GB   1 543 964   8. 4.76/11. 4.79   ICI

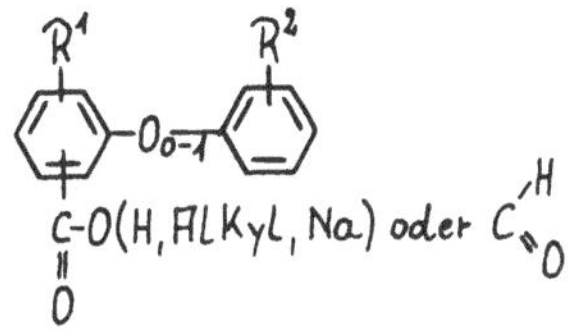

R^1 u. R^2 = H, Halog.,
                (O)-Alkyl, CF$_3$, CN,
                C-Alkyl, NH$_2$, OH
                $\overset{\|}{O}$

GB   1 543 965   8. 4.76/19. 4.79   ICI

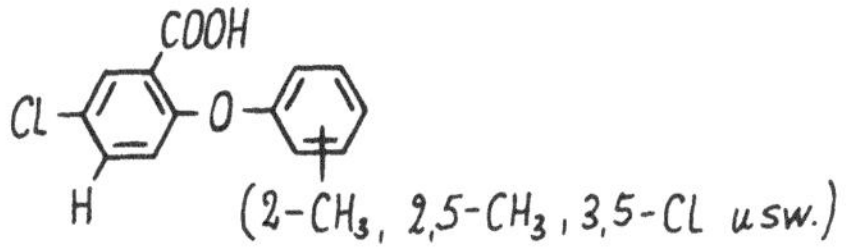

GB   2 035 309   10.10.78/18. 6.80   Shell Int. Res.

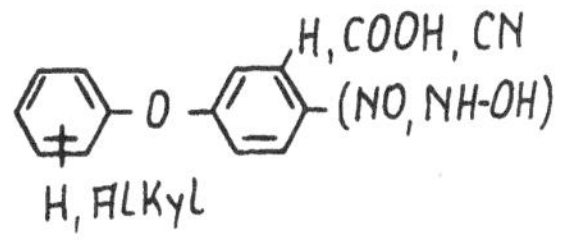

NL   75/14299   9.12.75/13. 6.77   Ishihara Sangyo

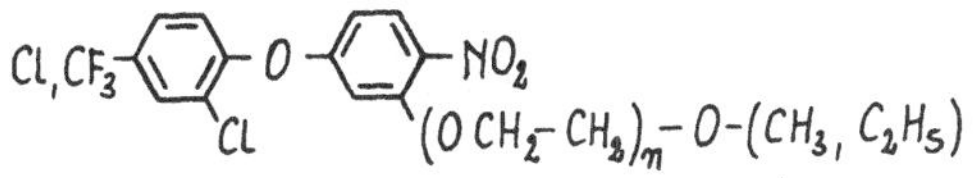

NL   77/05037   7. 5.76/ 9.11.77   Sumitomo Chem.

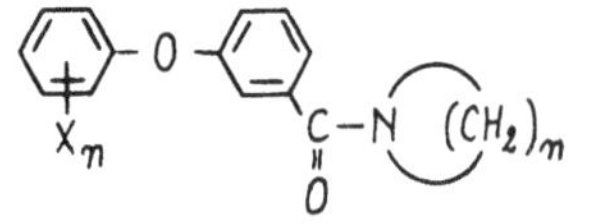

X = Halogen, (O-Alkyl)

Gegen Ungräser und Unkräuter
im Reis.

X_n —benzene— O —benzene— C(=O)—N(CH$_2$)$_n$

JA   50 107134   4. 2.74/23. 8.75   Mitsui Toatsu Chem.

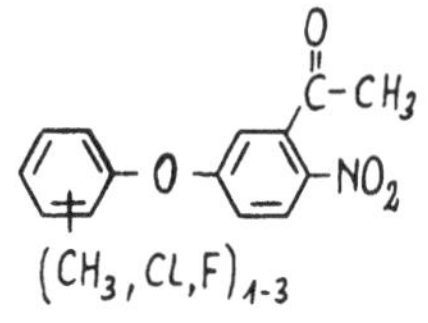

$(CH_3, Cl, F)_{1-3}$ —benzene— O —benzene (C(=O)-CH$_3$, NO$_2$)

JA   50 117927   5. 3.74/16. 9.75   Mitsui Toatsu Chem.

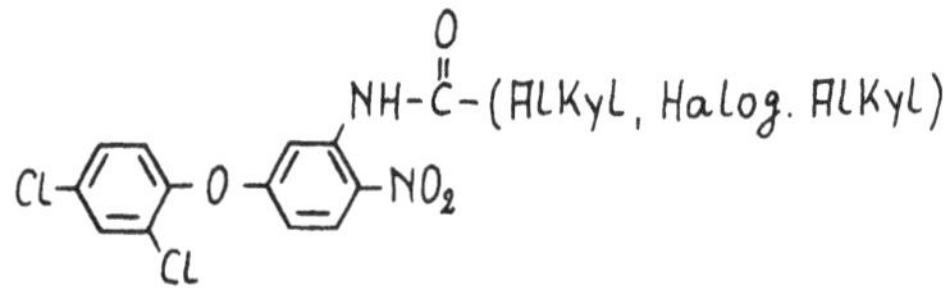

Cl, Cl —benzene— O —benzene (NH-C(=O)-(Alkyl, Halog. Alkyl), NO$_2$)

JA   50 129737   5. 4.74/14.10.75   Ishihara Min.

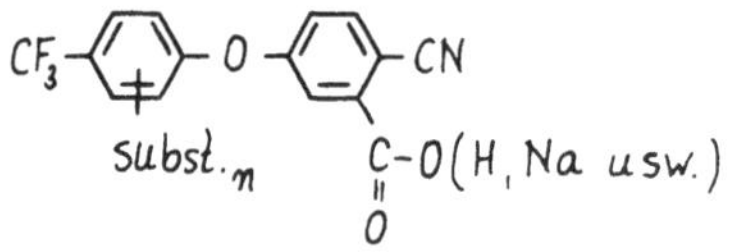

CF_3 —benzene (subst.$_n$)— O —benzene (CN, C(=O)-O(H, Na usw.))

JA   50 132125   12. 4.74/20.10.75   Ishihara Min.

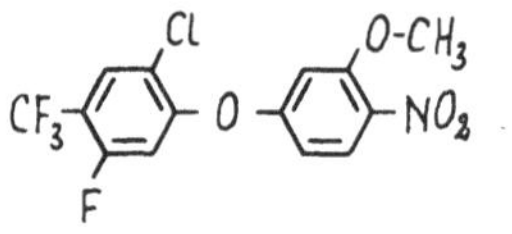

Bodenherbizid

CF_3, F —benzene (Cl)— O —benzene (O-CH$_3$, NO$_2$)

JA   50 154216   31. 5.74/12.12.75   Fujisawa

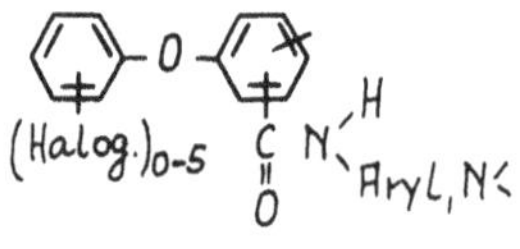

(Halog.)$_{0-5}$ —benzene— O —benzene— C(=O)—N(H, Aryl, N<)

JA   50 160 427   19. 6.74/25.12.75   Fujisawa Pharm.

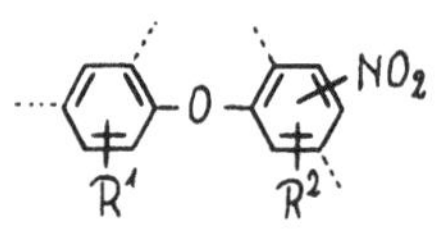

R^1 = Halog., Alkenyl, Phenyl, CN
R^2 = C(=O)-OH, C(=O)-O-Alkyl usw.

R^1 —benzene— O —benzene (NO$_2$)— R^2

- - - - bevorzugte Stellungen f. Substituenten

146

JA   51 098327   25. 2.75/30. 8.76   Mitsui Toatsu Chem.

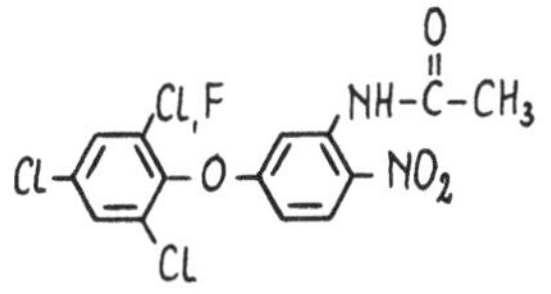

JA   51 104030   5. 3.75/14. 9.76   Mitsui Toatsu Chem.

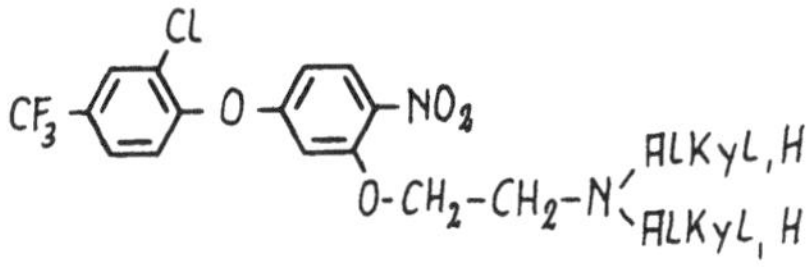

Gegen Gräser im Reis, auch
gegen einige Unkräuter.

JA   51 128426   28. 4.75/ 9.11.76   Mitsui Toatsu Chem.
JA   51 128429   28. 4.75/ 9.11.76   Mitsui Toatsu Chem.
JA   51 128431   28. 4.75/ 9.11.76   Mitsui Toatsu Chem.

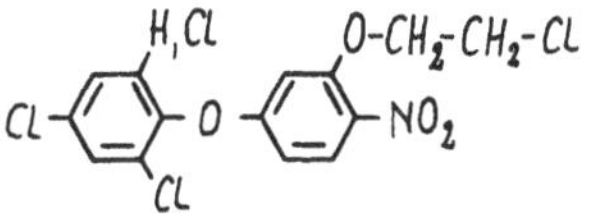

In Kombination mit anderen
Herbiziden.

JA   51 133422   16. 5.75/19.11.76   Mitsui Toatsu Chem.

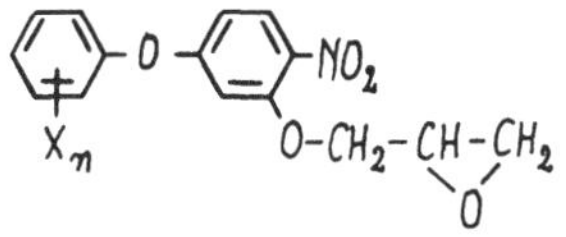

X = Halog., nied. Alkyl
n = 2 oder 3

JA   51 141826   3. 6.75/ 7.12.76   Mitsui Toatsu Chem.

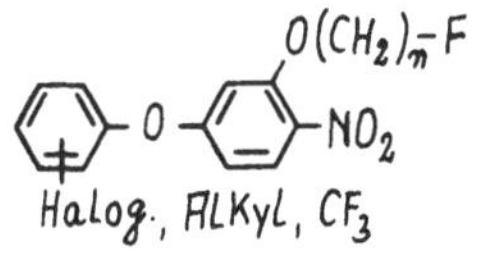

JA   52 001020   24. 6.75/ 6. 1.77   Mitsui Toatsu Chem.

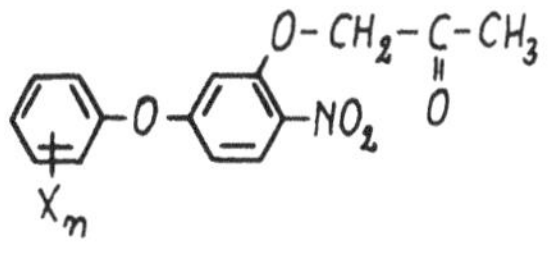

X = Halog., Alkyl, CF_3
n = 1-3

JA   52 012928   21. 7.75/31. 1.77   Mitsubishi

$$Cl-\underset{\underset{O}{\|}}{C}-N\underset{H,\,ALKYL}{\diagdown}\text{—}\underset{subst.}{\diagup}\!\!\!\bigcirc\!\!\!-O-\!\!\!\bigcirc\!\!\!\underset{subst.}{-}NO_2$$

JA   52 015822   29. 7.75/ 5. 2.77   Ishihara Sangyo

$$CF_3\!\!-\!\!\bigcirc\!\!\!\underset{Cl}{-}O-\!\!\!\bigcirc\!\!\!\underset{O(CH_2-CH_2)_{\overline{n}}-O-ALKYL}{-}NO_2$$

JA   52 028934   27. 8.75/ 4. 3.77   Mitsui Toatsu Chem.

                                                       R = Cl, F

$$ALKYL-\underset{(O)_{0-2}}{\overset{\uparrow}{S}}\!\!-\!\!\overset{(R)_{0-2}}{\bigcirc}\!\!-O-\!\!\!\bigcirc\!\!\!\underset{O-CH_2-CH_2\,(Cl,F)}{-}NO_2$$

JA   52 041228   27. 9.75/30. 3.77   Ishihara Sangyo

                                                       Selektiv in Reis.

$$Cl-\!\!\bigcirc\!\!\!\underset{Cl}{-}O-\!\!\!\bigcirc\!\!\!\underset{(OC_2H_4)_{\overline{n}}-O-ALKYL}{-}NO_2$$

JA   52 047917   8.10.75/16. 4.77   Ishihara Sangyo

                                                       n = 2-4

$$CF_3\!\!-\!\!\bigcirc\!\!\!\underset{Cl}{-}O-\!\!\!\bigcirc\!\!\!\underset{(O-CH_2-CH_2)_{\overline{n}}-O-ALKYL}{-}Cl,\,CN$$

JA   52 057320   4.11.75/11. 5.77   Mitsui Toatsu Chem.

$$(H,Cl,CH_3,SO_2)\!\!-\!\!\overset{}{\underset{(Cl,F,CH_3)_{0-2}}{\bigcirc}}\!\!-O-\!\!\!\bigcirc\!\!\!\underset{O-CH_2-CF_3}{-}NO_2$$

148

JA 52 072816   16.12.75/17. 6.77   Mitsui Toatsu Chem.

CF_3—[ring: Cl]—O—[ring: $O-CH_2-CH_2-O-CH_2-CH_2-O-Alkyl$, NO_2]

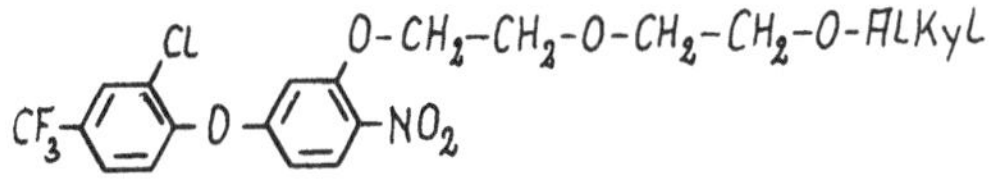

JA 52 072817   16.12.75/17. 6.77   Mitsui Toatsu Chem.

$Halog.$, CF_3—[ring: Cl]—O—[ring: $O-CH_2-CH_2-Halog.$, $H, Halog., CN$]

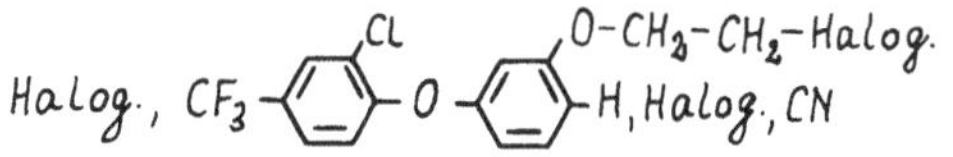

JA 52 125144   13. 4.76/20.10.77   Chugai Pharm.

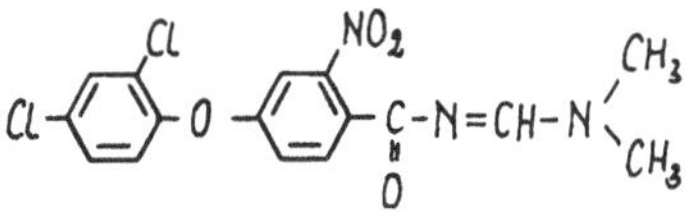

Cl—[ring: Cl]—O—[ring: NO_2, $\overset{\overset{}{C}}{\underset{O}{}}-N=CH-N\overset{CH_3}{\underset{CH_3}{}}$]

JA **53** 059637   11.11.76/29. 5.78   Mitsui Toatsu Chem.

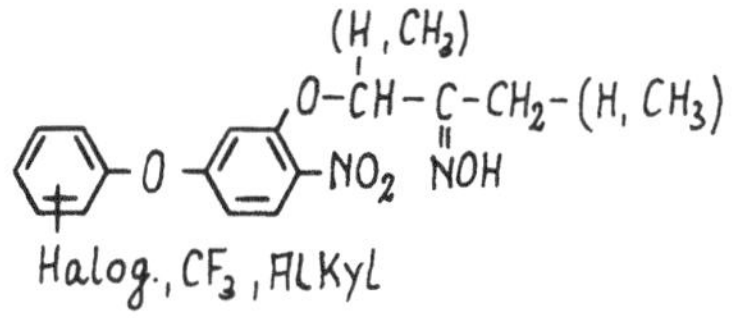

[ring]—O—[ring: $O-\overset{(H,CH_3)}{CH}-\underset{NOH}{C}-CH_2-(H, CH_3)$, NO_2]

$Halog., CF_3, Alkyl$

JA 53 092762   25. 1.77/15. 8.77   Sumitomo Chem.

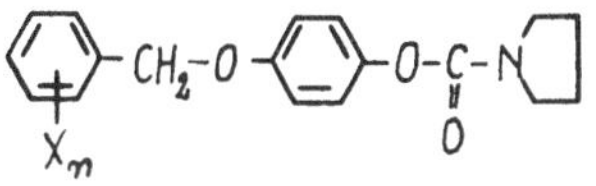

X = (O)-Alkyl, Halog., NO_2,
CN, S-Alkyl

[ring: X_m]—CH_2-O—[ring]—$O-\underset{O}{C}-N$[ring]

JA 53 095932   31. 1.77/22. 8.78   Nitron Noyaku

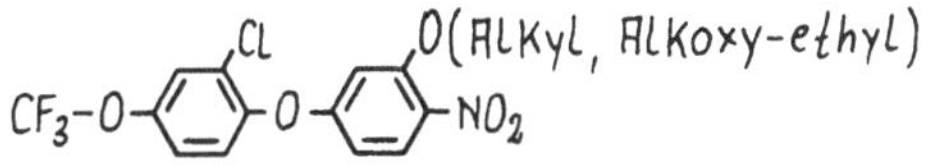

CF_3-O—[ring: Cl]—O—[ring: $O(Alkyl, Alkoxy-ethyl)$, NO_2]

JA 50 123824   15. 3.74/20. 9.75   Mitsui Toatsu Chem.

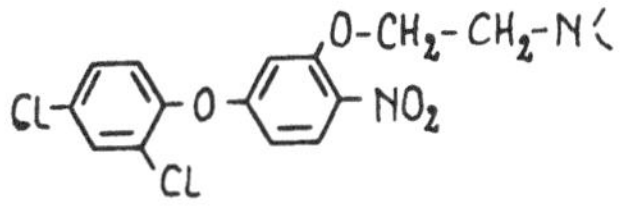

Cl—[ring]—O—[ring: $O-CH_2-CH_2-N<$, NO_2], Cl

JA   53 111027   8. 3.77/28. 9.78   Sumitomo Chem.

Z = CN, Halog., (O)-Alkyl, NO_2,
X u. Y = (S)-Alkyl, $S(O)_{1-2}$-Alkyl, OH,
O-Alkyl, CF_3, SO_2-N$<$ CH$_3$ / CH$_3$,
C-OH usw. ($\overset{\parallel}{O}$)

Gräserherbizid

JA   53 116378   16. 3.77/11.10.78   Sumitomo Chem.

JA   54 005033   15. 6.77/16. 1.79   Mitsui Toatsu Chem.

Gegen Gräser wirksam.

JA   54 032427   18. 8.77/ 9. 3.79   Mitsui Toatsu Chem.

JA   54 039034   29. 8.77/24. 3.79   Mitsui Toatsu Chem.

Gegen Gräser

JA   54 095527   10. 1.78/28. 7.79   Mitsui Toatsu Chem.

150

JA   54 095530   10. 1.78/28. 7.79   Matsushita

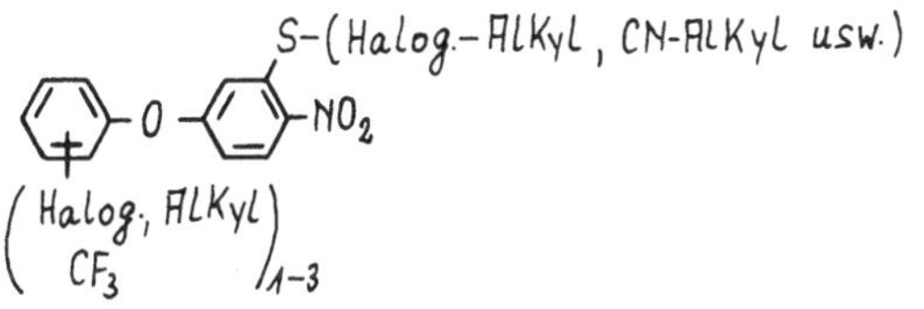

JA   54 141767   25. 4.78/ 5.11.79   Mitsui Toatsu Chem.

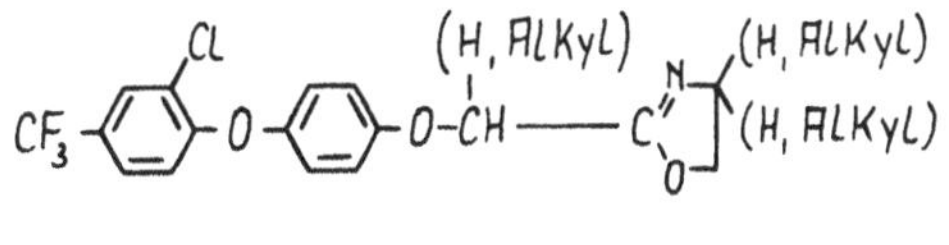

                                                      Gräserherbizid

JA   55 035016   2. 9.78/11. 2.80   Nitron Noyaku

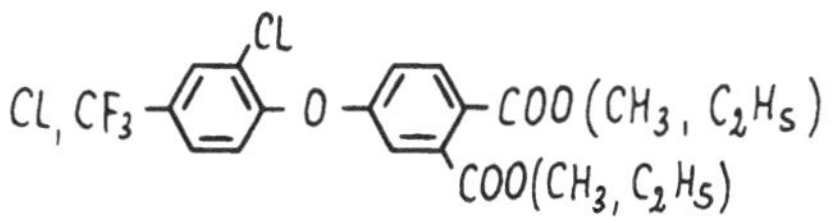

JA   55 064567   10.11.78/15. 5.80   Ube Ind.

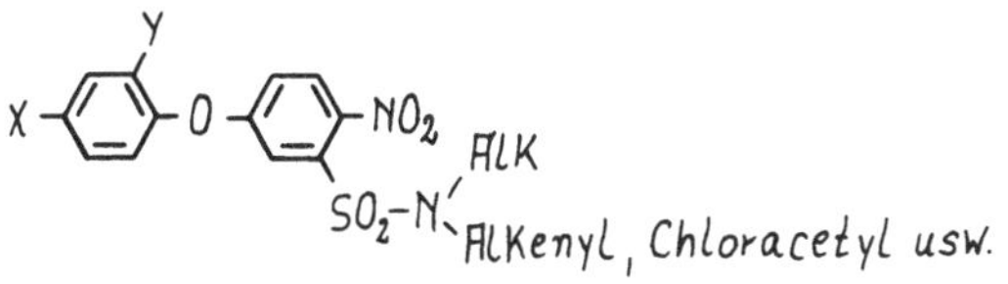

                                                      X u. Y = Halog., d. andere
                                                              CF_3

JA   77 046286   28. 5.74/24.11.77   Sumitomo Chem.

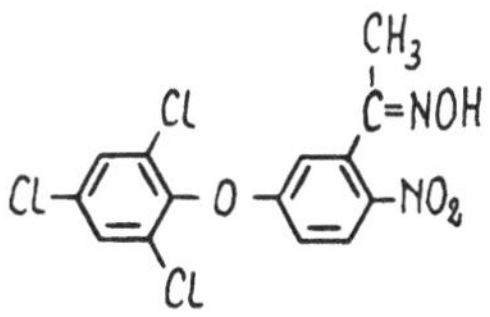

                                                      Selektiv in Reis

JA   80 008482   23. 8.68/ 4. 3.80   Hodogaya
JA   80 008483   27. 9.68/ 4. 3.80   Hodogaya

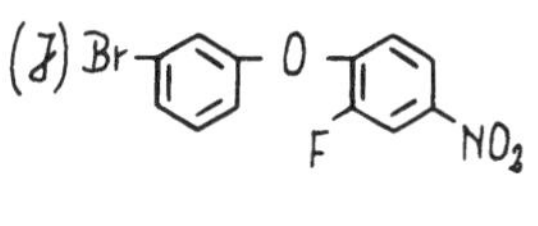

                                                      X u. Y = Halog., d. andere
                                                              CF_3

Z.T. unter 5.4.2.b miterfaßt, siehe auch
5.7.3.f 2

DOS   2 640 730   10. 9.76/16. 3.78   Hoechst
= BE   858 618   13. 3.78   Hoechst

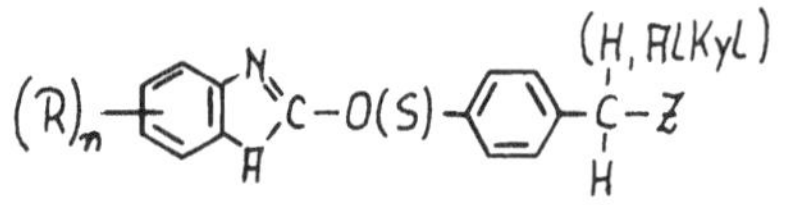

$(R)_n$—[benzazole]C-O(S)—[phenyl]—$\overset{(H, Alkyl)}{\underset{H}{C}}$-Z

Z = $\underset{O}{\overset{\|}{C}}$-O(Alkyl, H), $\underset{O}{\overset{\|}{C}}$-S-Alkyl, $\underset{O}{\overset{\|}{C}}$-N$<$, $\underset{O}{\overset{\|}{C}}$-N-N$<\atop H$, $\underset{S}{\overset{\|}{C}}$-NH$_2$, CN, CH$_2$-OH,

CH$_2$O-$\underset{O}{\overset{\|}{C}}$-Alkyl, CH$_2$-O-$\underset{O}{\overset{\|}{C}}$-N$<$, CH$_2$-O-SO$_2$-(Alkyl, Phenyl)

R = Halogen, NO$_2$, CN, NH$_2$, Alkyl, CF$_3$, O-Alkyl usw.; n = 0-4
A = O, S, NH, N-Alkyl

DOS   2 649 706   29.10.75/ 5. 5.77   Ishihara Sangyo

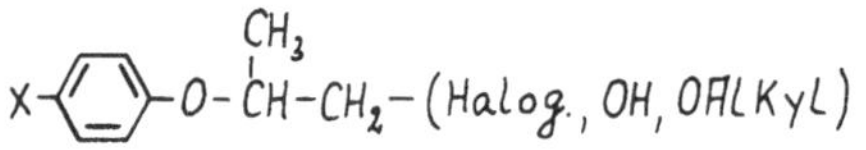

X—[phenyl]—O-$\overset{CH_3}{\underset{}{CH}}$-CH$_2$—(Halog., OH, OAlkyl)

X = Cl—[phenyl]—O- ,   Cl—[phenyl(Cl)]—O- , CF$_3$—[phenyl(Cl)]—O- ; [Cl-pyridyl]O- usw.

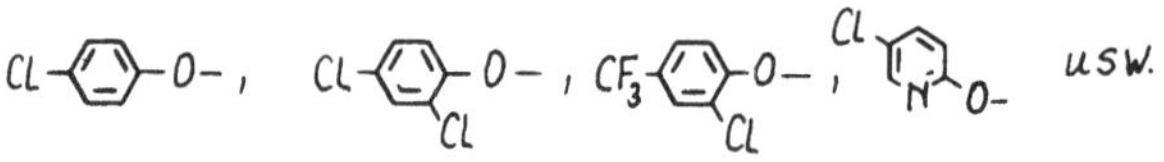

DOS   2 709 144   3. 3.77/ 7. 9.78   Bayer

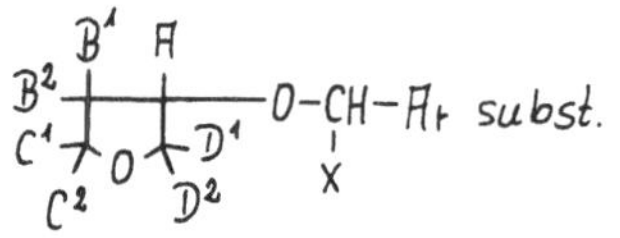

[bicyclic B^1,B^2,C^1,C^2,D^1,D^2,A ring]—O-CH-A$_r$ subst.
                                                           |
                                                           X

X = H, Alk(en, in)yl, subst.
    Phenyl
A = H, Alk(en, in)yl, Cyclo-
    Alkyl, Phenyl
B^1, B^2, C^1, C^2, D^1, D^2 = H,
    (Halog.)-Alkyl, Alkoxy-
    alkyl, subst. Phenyl

DOS   2 724 675   Bayer
= EP   02 (GE Prior. 1.6.77/20.12.78   Bayer

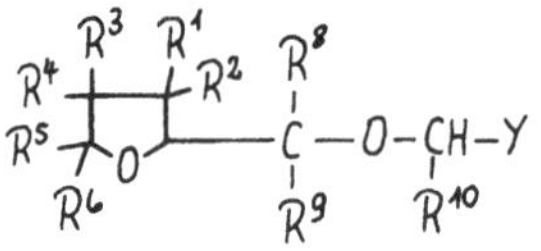

[bicyclic R^1–R^6 ring]—$\underset{R^9}{\overset{R^8}{C}}$—O-$\overset{}{\underset{R^{10}}{CH}}$-Y

R^1-R^6 = H, Alkyl, Phenyl usw.
R^5-R^6 auch Glieder eines
         Ringsystems
R^8 u. R^9 = H, Alk(en, in)yl,
         O-Alkyl, Cycloalkyl,
         Phenyl, Benzyl

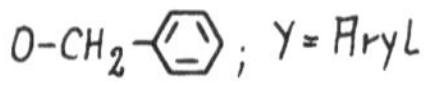

O-CH$_2$—[phenyl] ; Y = Aryl

152

Beispiel: [Struktur: Tetrahydrofuranyl-CH$_2$-O-CH$_2$-Phenyl mit F] Gräserherbizid

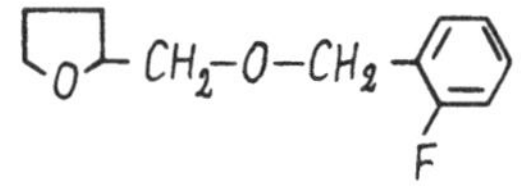

DOS   2 724 677   1. 6.77/14.12.78   Bayer

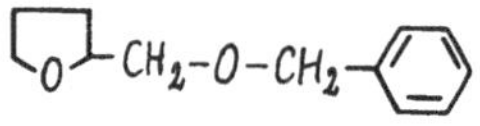

[Struktur: Tetrahydrofuranyl-CH$_2$-O-CH$_2$-Phenyl]   Gräserherbizid

DOS   2 909 816   16. 3.78/27. 9.79   Ciba-Geigy

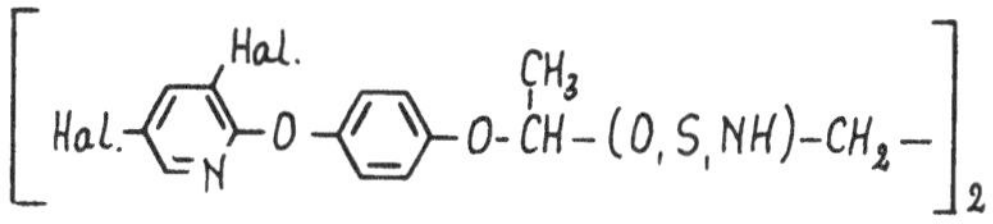

$$\left[\text{Hal.} \underset{\text{Hal.}}{\text{Pyridinyl}} - O - \text{Phenyl} - O - \underset{CH_3}{CH} - (O, S, NH) - CH_2 - \right]_2$$

DOS   2 921 567   28. 5.79/24. 1.80   Kumiai Chem.
JA Prior. 14.u.15.7.78, 14.10.78

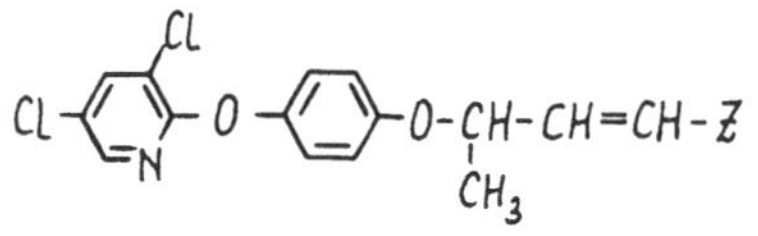

[Struktur: Cl$_2$-Pyridinyl-O-Phenyl-O-CH(CH$_3$)-CH=CH-Z]

Z = CH$_2$-OH, $\underset{O}{\overset{\parallel}{C}}$-OR, $\underset{S}{\overset{\parallel}{C}}$-SR usw.

DOS   2 923 371   9. 6.79/13.12.79   Ishihara Sangyo

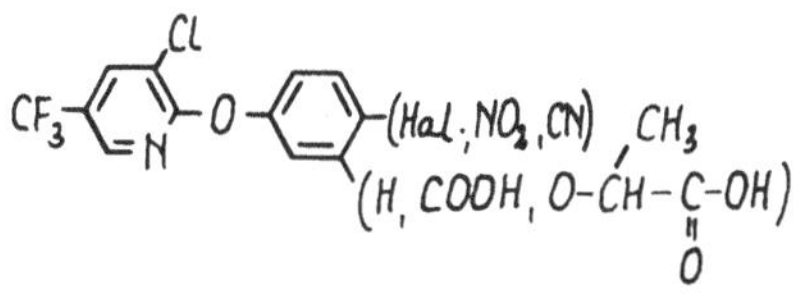

[Struktur: CF$_3$-, Cl-Pyridinyl-O-Phenyl-(Hal.,NO$_2$,CN), (H, COOH, O-CH(CH$_3$)-C(=O)-OH)]

EP   02   1. 6.78/20.12.78   Bayer
siehe DOS 2 724 675 zuvor

EP   924   17. 8.78/ 7. 3.79   Hoechst
=DOS   2 738 963   30. 8.77

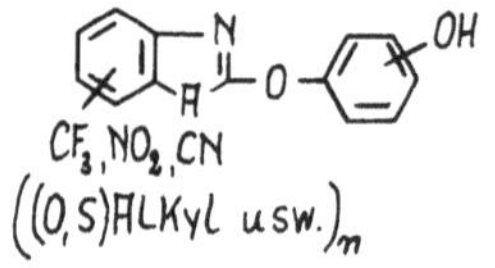

$R = O, S, -N-$
$ Alkyl$

(structure: benzoxazole with CF$_3$, NO$_2$, CN, ((O,S)Alkyl usw.)$_n$ substituents, linked via O to phenyl-OH)

EP   1 187   12. 9.78/21. 3.79   ICI
=DOS   2 820 032   8. 5.78

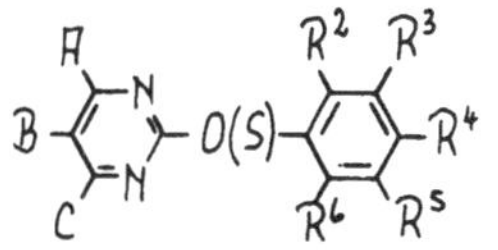

(structure: pyrimidine with A, B, C substituents, O(S) linked to benzene ring with R^2, R^3, R^4, R^5, R^6)

A, B, C = H, Halog., NO$_2$, CN,
 CNS, C-OR,
 $\overset{\|}{}$
 O
 SO$_2$-Alkyl, Alk(en)yl,
 O(S)-Alkyl, Phenyl
 subst.
R^2-R^6 = H, Halog., NO$_2$, HC-,
 $\overset{\|}{}$
 O

 SO$_3$H, CN usw.

EP   2 812   2. 1.78/11. 7.79   Ciba-Geigy
CH Prior.   2. 1.78

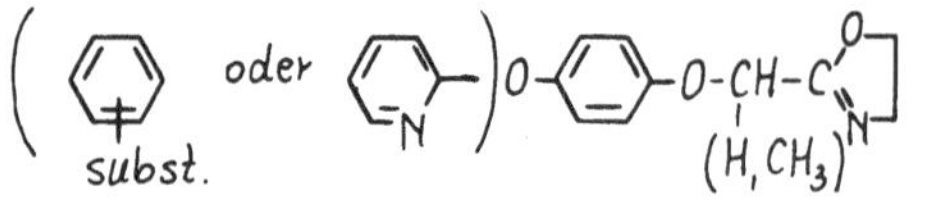

(structure: phenyl subst. oder pyridin-2-yl — O — phenyl — O-CH-C(=O) with (H,CH$_3$) and N ring)

EP   3 877   5. 2.79/ 5. 9.79   ICI
GB Prior.   15. 2.78

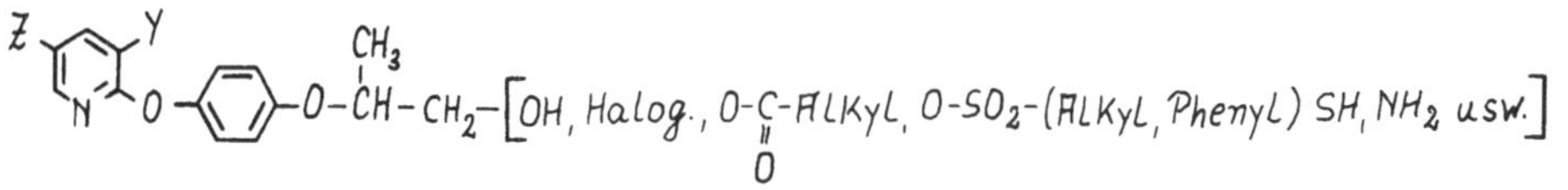

(structure: Z, Y substituted pyridine — O — phenyl — O-CH(CH$_3$)-CH$_2$—[OH, Halog., O-C(=O)-Alkyl, O-SO$_2$-(Alkyl, Phenyl) SH, NH$_2$ usw.])

Z = Halog., Cl(F)-Alkyl
Y = H, Halog., Cl(F)-Alkyl

EP   7 790   21. 6.79/20. 2.80   BASF
=DOS   2 819 289   4. 7.78

(structure: pyrazole with R^1, R^2, R^3, R^4)

Nähere Einzelheiten siehe
5.14.2 g 1

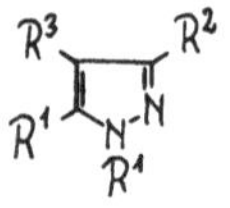

R^2 = O-Alkyl, O-Aryl,
 O-Heterocycl., S-Alkyl

154

EP   8 192   31. 7.79/20. 2.80   ICI, Australien
AU Prior.   8. 8.78

W u. X = O oder S

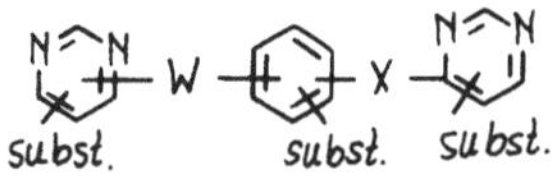

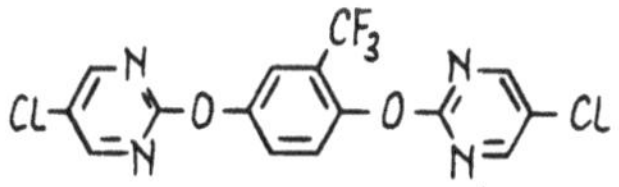

EP   6 608   26. 6.79/ 9. 1.80   Ciba-Geigy
CH Prior.   29. 6.78

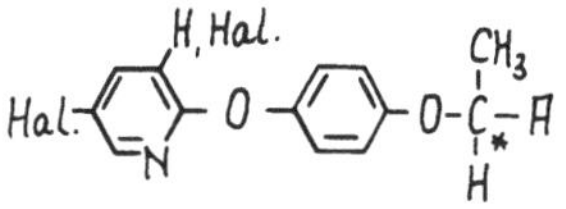

$A = CN$, $C\!-\!O(H, Alkyl)$, $C\!-\!N<$, $C\!-\!N$

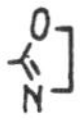

R = Isomeres

US   4 019 892   31.12.75/26. 4.77   Shell Oil

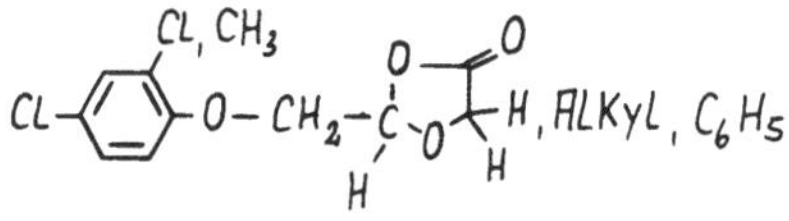

US   4 045 208   23. 7.75/30. 8.77   Stauffer

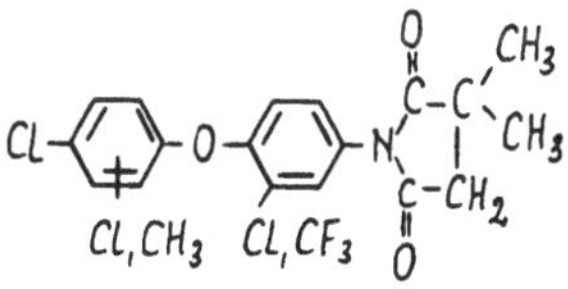

post emerg. gegen Unkräuter

US   4 062 670   31.12.75/13.12.77   Shell Oil

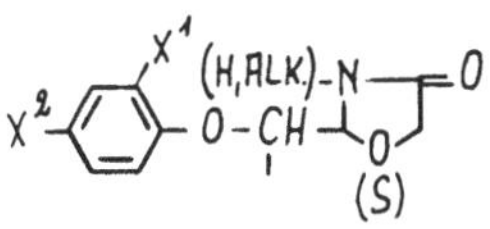

X^1 = H, Halog., CN, (O, S)-Alkyl, Alkyl
X^2 = H, Halog., (O)-Alkyl

| US | 4 180 395 | 11. 4.74/25.12.79 | Dow Chem. |

R = CH_2-O-C-Alkyl, CH_2-O-⟨⟩ ,
$\quad\quad$ O

CH_2-O-CH_2-CH_2-O-Alkyl, C-N< ,
$\quad\quad\quad\quad\quad\quad\quad\quad\quad\quad\quad$ O

C-N-CH_2-COOH usw.
O H

| JA | 55 115873 | 27. 2.79/ 6. 9.80 | Sankyo |

X = Halogen

| JA | 52 072821 | 15.12.75/17. 6.77 | Mitsubishi Petroch. |

R = Halog., nied. Alkyl
R^1 = H, Halog., (O)-Alkyl, Halog.-Alkyl
R^2 u. R^3 = H, F

| JA | 52 139034 | 13. 5.76/19.11.77 | Ishihara Sangyo |

Selektiv gegen Ungräser

| JA | 53 111063 | 8. 3.77/28. 9.78 | Nippon Soda |

Gräserherbizid

| JA | 55 007250 | 3. 7.78/19. 1.80 | Sankyo |

R^1 u. R^2 = Alk(en)yl

156

JA   55 092368   29.12.78/12. 7.80   Ishihara Sangyo

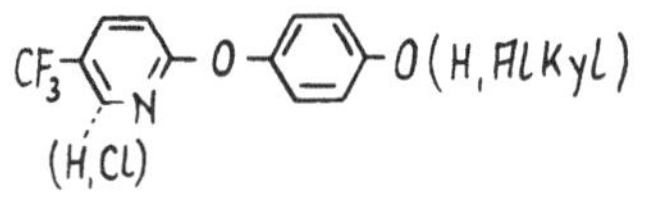

Darstellungsverfahren

CF₃ pyridine ring -O- benzene ring -O(H,Alkyl) (H,Cl) on pyridine N

JA   80 007403   6. 5.70/25. 2.80   Ishihara Sangyo

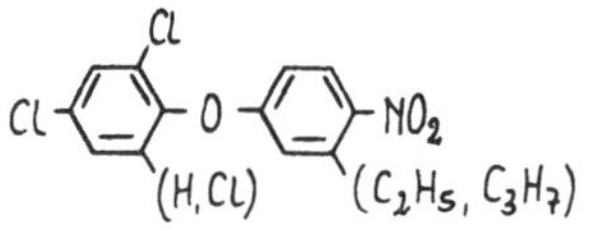

Cl substituted benzene -O- benzene -NO₂ structure (H,Cl) (C₂H₅, C₃H₇)

5.5 Kohlensäure und Thiokohlensäure-Derivate   Bd. 5/408

5.5.1   Kohlensäure und Thiokohlensäure-diester

DOS   2 837 204   25. 8.78/ 6. 3.80   Ciba-Geigy

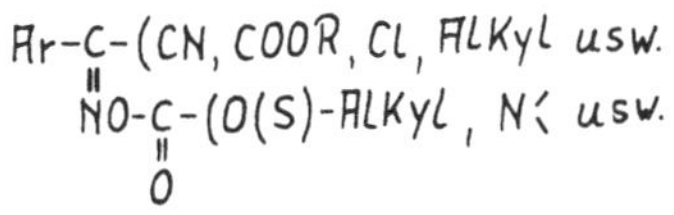

Structure: Ar-C-(CN, COOR, Cl, Alkyl usw.) / NO-C-(O(S)-Alkyl, N< usw.) with =O

EP   10 588   27. 8.79/14. 5.80   Ciba-Geigy
EP   11 047   27. 8.79/14. 5.80   Ciba-Geigy

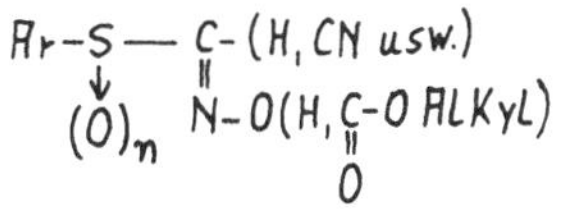

Structure: Ar-S—C-(H, CN usw.) / (O)ₙ N-O(H, C-O Alkyl) with =O

Näheres siehe unter 5.5.2.f 3

US   4 022 609   28. 8.70/10. 5.77   PPG Industr.

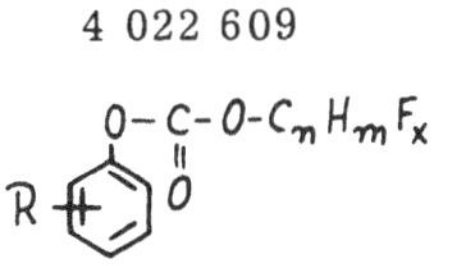

Structure: R-benzene with O-C-O-CₙHₘFₓ (C=O)

R = Alk(en)yl, Halog. - Alk(en)yl,
S-Alkyl, Cycloalkyl

US   4 204 857   24. 1.79/27. 5.80   Olin Corp.

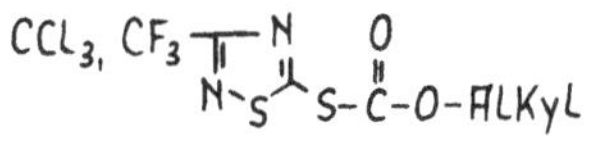

Structure: CCl₃, CF₃ triazine-thio ring, N-S-C(=O)-S-C-O-Alkyl

a 1) <u>Carbamidsäureester aliphatischer und cycloaliphatischer Amine</u>
      <u>mit Alkoholen, Phenolen usw.</u>

DOS   2 644 504   1.10.76/ 2. 6.77   Stauffer
US Prior.   2.10.75 u. 13. 9.76

$$\text{Phenyl}(X_n)-SO_2-N(H)-\underset{O}{\overset{\|}{C}}-O\{\text{Alkyl d. Cl subst., Alk(en,in)yl, } N(C_2H_5)_2, CH_2-NH-\underset{O}{\overset{\|}{C}}-CF_3, CH_2-S-\text{Phenyl}-Cl,$$

$$N=C\begin{smallmatrix}CH_3\\CH_3\end{smallmatrix}\}$$

X = H, Br, Cl, O-CH$_3$, CH$_3$, CF$_3$; n = 0-3

Antidot gegenThiocarbamat-Schäden.

DOS   2 837 204   25. 8.78/ 6. 3.80   Ciba-Geigy

$$Ar-\underset{N-O-\underset{O}{\overset{\|}{C}}}{\overset{\|}{C}}-(CN, COOR)\left(N\begin{smallmatrix}Alkyl\\(H,Alkyl)\end{smallmatrix}\right)\begin{pmatrix}NH-Ar\ usw.\\O-\underset{O}{\overset{\|}{C}}-Alkyl\end{pmatrix}$$

Antidots für sym.Triazin-
Herbizide u. Diphenylethoxy-
propionsäurederivate

EP   11 670   30.11.78/11. 8.80   Bayer

$$\text{Cycloheptatrienyl}\begin{cases}R(=H, \underset{O}{\overset{\|}{C}}-Alkyl, \underset{O}{\overset{\|}{C}}-O-Alkyl)\\N(H)-R^1(=\underset{O}{\overset{\|}{C}}-S-Alkyl, \underset{O}{\overset{\|}{C}}-N\langle\ usw.)\end{cases}$$

EP   9 555   20. 7.79/16. 4.80   BASF
=DOS 2 832 974   27. 7.78

$$\text{Norbornyl}-N(\underset{C_2H_5}{})-\underset{O}{\overset{\|}{C}}-S-C_2H_5 \ + \ \text{Halog. Alkyl } \underset{O}{\overset{\|}{C}}-N\begin{smallmatrix}Alk(en)yl\\Alk(en)yl\end{smallmatrix} \quad \text{als Safener}$$

US   4 171 449   21. 5.75/16.10.79   Exxon Res.

$$\text{Alkylen}-\left[O-\underset{O}{\overset{\|}{C}}-N-\overset{|}{\underset{(H,Cl)}{}}\ \overset{|}{C}-\overset{|}{C}=\overset{|}{C}-\overset{|}{C}-Cl\right]_2 \quad u.a.\ auch\ Herbizide$$

158

b) <u>S-Verbindungen</u>

<u>Mono u. Dithiocarbamidsäureester sec. aliphatischer und cyclo-
aliphatischer Amine (z.T. unter a.1 miterfaßt).</u>

Verbindungen der Formel

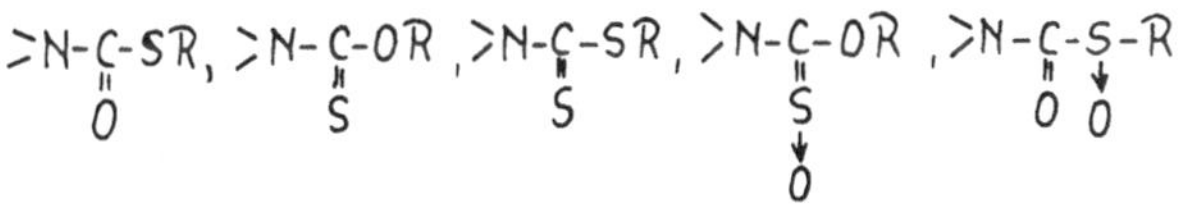

$$>N-\underset{\underset{O}{\parallel}}{C}-SR,\ >N-\underset{\underset{S}{\parallel}}{C}-OR,\ >N-\underset{\underset{S}{\parallel}}{C}-SR,\ >N-\underset{\underset{\overset{\downarrow}{O}}{S}}{C}-OR,\ >N-\underset{\underset{O}{\parallel}}{C}-\underset{\underset{O}{\downarrow}}{S}-R$$

US	4 059 609	4. 6.76/22.11.77	PPG Ind.

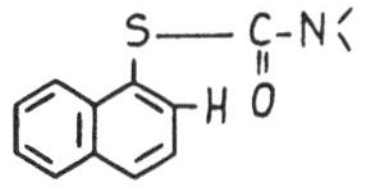

Substituenten in der 1. und 2. Stellung vertauschbar, unter anderem auch herbizid wirksam.

US	4 086 265	1.12.76/25. 4.78	Minnesota University

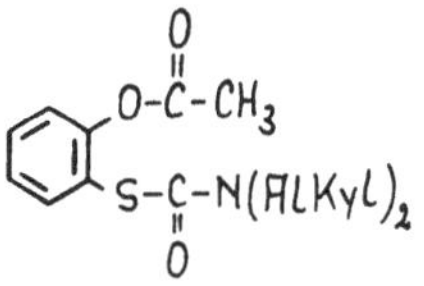

Darstellungsverfahren

BE	850 872	28. 1.76/28. 7.77	Montedison

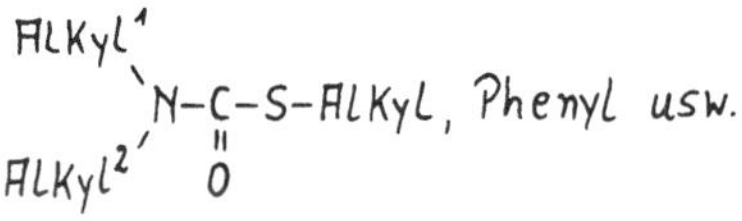

Neues Darstellungsverfahren

BE	858 348	3. 9.76/ 2. 3.78	Stauffer

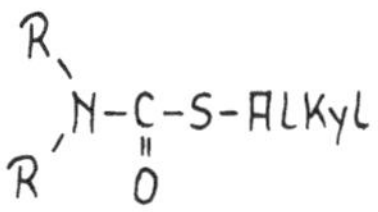

Verbessertes Darstellungs-fahren

R = Alkyl

BE	859 097	4.10.76/28. 3.78	Stauffer

Neues Darstellungsverfahren aus COS, Amin u. Alkylchlorid.

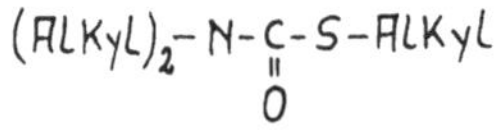

DDR   128 673   29. 9.75/30.11.77   Stauffer

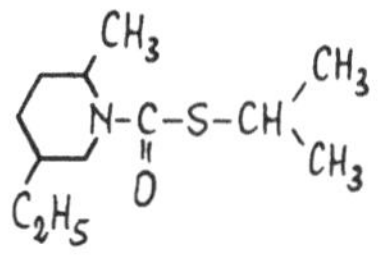

Verbessertes Darstellungsverfahren.

JA   52 125136   10. 4.76/20.10.77

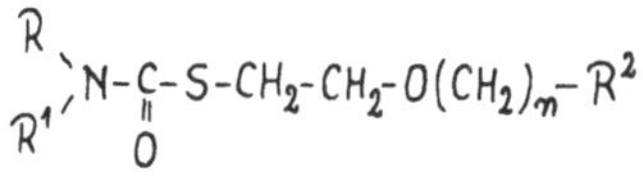

R u. R^1 = Alkyl oder zusammen
Heteroring
R^2 = subst. Phenyl

JA   53 024022   13. 8.76/ 6. 3.78   Asahi Chem.

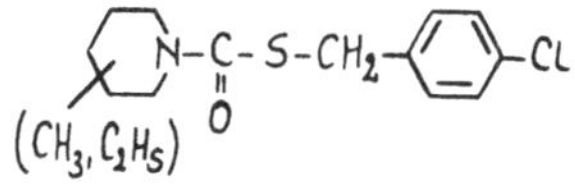

Gegen Ungräser und breitbl.
Unkräuter

JA   53 024023   16. 8.76/ 6. 3.78   Asahi Chem.

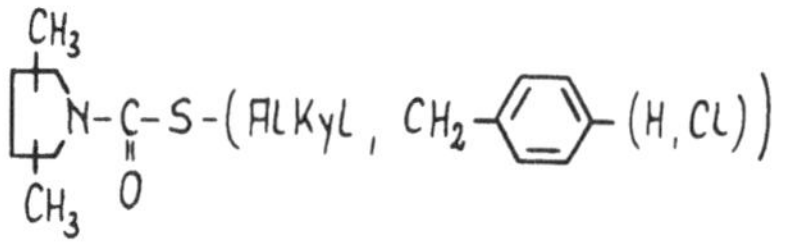

JA   53 024024   17. 8.76/ 6. 3.78   Asahi Chem.

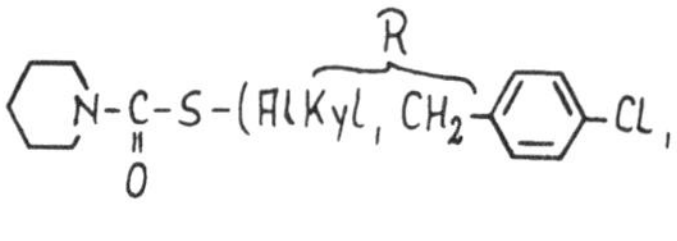

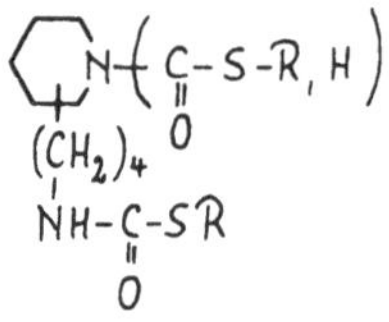

JA   53 069829   2.12.76/21. 6.78   Tokyo Org. Chem.

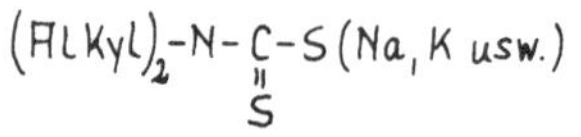

Gegen Unkräuter in
Wasser-Reis

JA   54 145653   4. 5.78/14.11.79   Kuraray

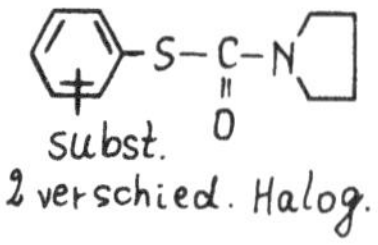

JA   54 145654   4. 5.78/14.11.79   Kuraray

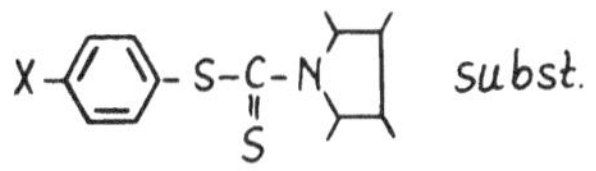

X = Halog., CH_3, $O\text{-}CH_3$,
CF_3, NO_2

JA   54 145679   4. 5.78/14.11.79   Kuraray

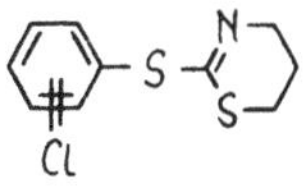

JA   54 145680 wie zuvor Br, F, J, CH_3 statt Cl

JA   55 059161   28.10.78/ 2. 5.80   Ihara Chem.

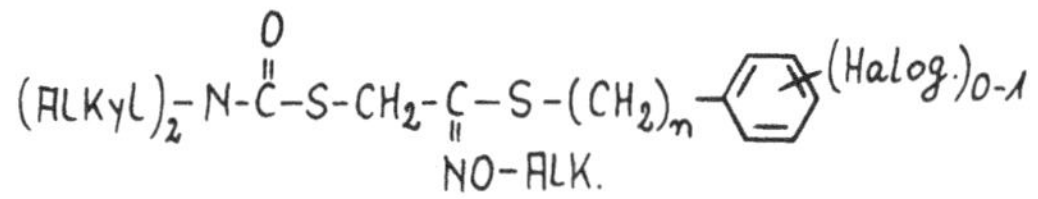

n = 0 oder 1

c)   Benzylthiole und Phenyl-ethyl-thiole als Esterkomponente
     (Patente umfassen z.T. auch aliphatische Thiole)   Bd. 5/415

DOS   2 636 620   13. 8.76/17. 2.77   Kumiai Chem.
JA Prior.   15. 8.75

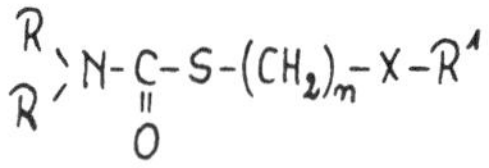

R = niedr. Alkyl
R^1 = subst. Phenyl mit
       Cl, (O)-Alkyl, NO_2, CF_3
n = 0-3; X = O oder S

Herbizid, Acarizid

DOS   2 644 446   1.10.76/14. 4.77   Stauffer
US Prior.   2.10.75 u. 17.9.76

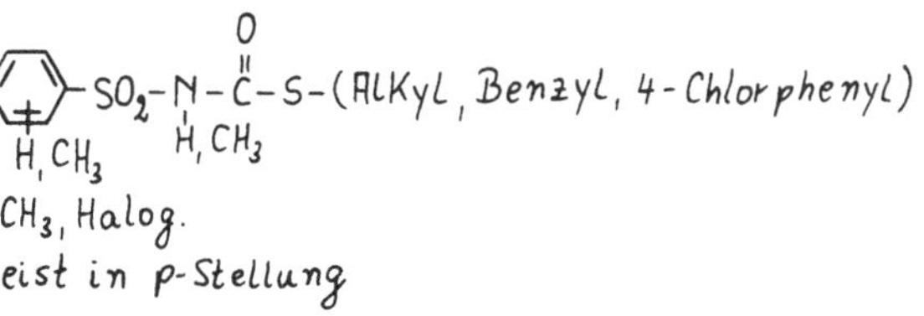

Antidots gegen Thiocarbamatschäden.

DOS   2 703 123   26. 1.77/11. 8.77   Kumiai Chem.
JA Prior.   30. 1.76

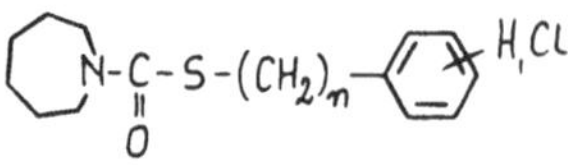

n = 2 oder 3

DOS   2 745 009   6.10.77/12. 4.79   Esza Kmagyarovszagi

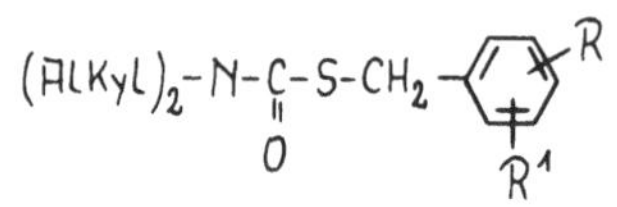

R u. R^1 = H, Halog.,
        (O)-Alkyl, S-Alkyl

Neues Darstellungsverfahren

BE   859 599   11.10.77/ 1. 2.78   Eszakmagyar.

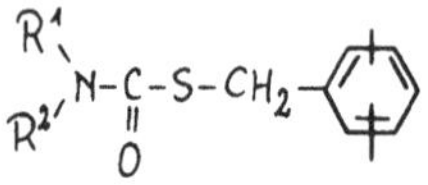

Herstellungsverfahren

BE   865 594   1. 4.77/ 2.10.78   Montedison

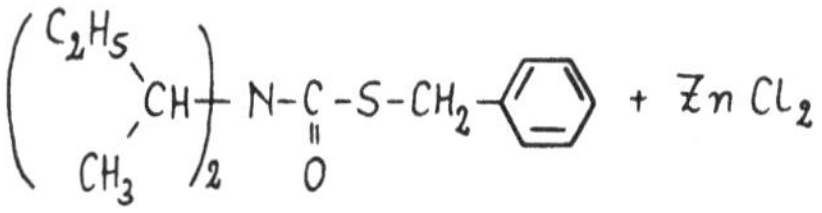

JA   50 063135   1. 5.73/29. 5.75   Showa Denko

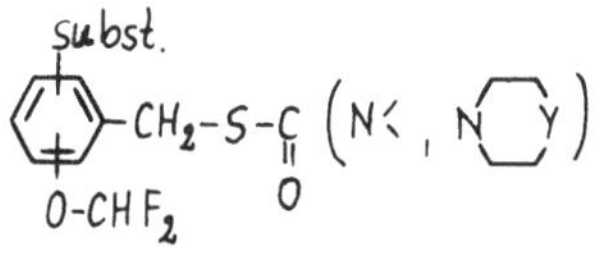

Y = CH_2, O

JA   50 129740   28. 3.74/14.10.75   Kumiai Chem.

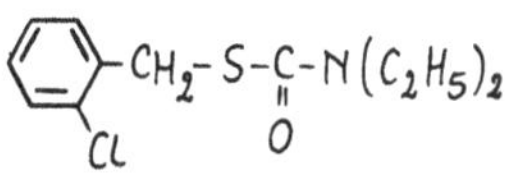

JA   51 098330   12. 2.75/30. 8.76   Sankyo

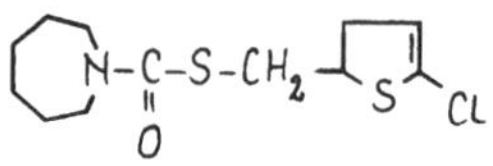

Selektiv in Reis.

162

JA   51 098331   25. 2.75/30. 8.76   Mitsubishi Petroch.

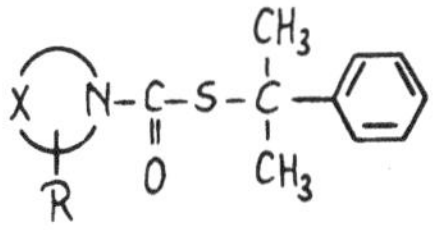

$X = (CH_2)_{4-6}$, $CH_2\text{-}CH_2\text{-}O\text{-}CH_2\text{-}CH_2$

R = H, nied. Alkyl

Selektiv in Reis

JA   51 148023   13. 6.75/18.12.76   Mitsui Toatsu Chem.

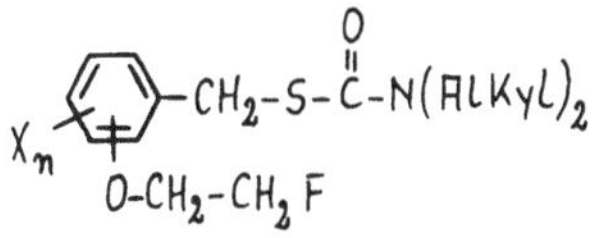

$X = Cl$, $(O)\text{-}CH_3$

n = 0-2

JA   52 070023   10.12.75/10. 6.77   Mokko Chem.

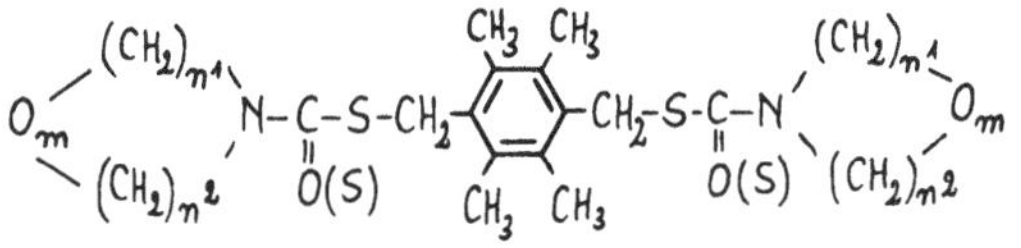

n^1 u. n^2 = 2 od. 3

m = 0 oder 1

JA   52 130914   23. 4.76/ 2.11.77   Kumiai Chem.

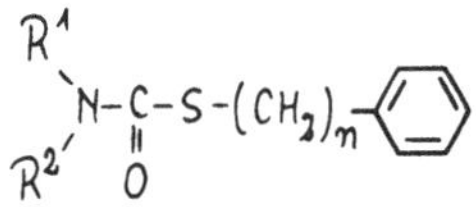

R^1 u. R^2 = nied. Alkyl

n = 2 oder 3

Bodenherbizid

JA   52 142086   24. 5.76/26.11.77   Kumiai Chem.

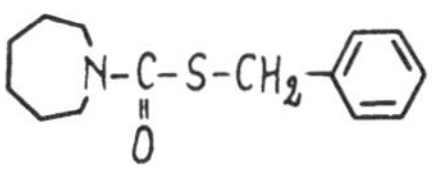

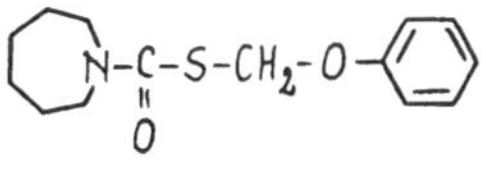

JA   52 153935   11. 6.76/21.12.77   Kyowa Fermentation

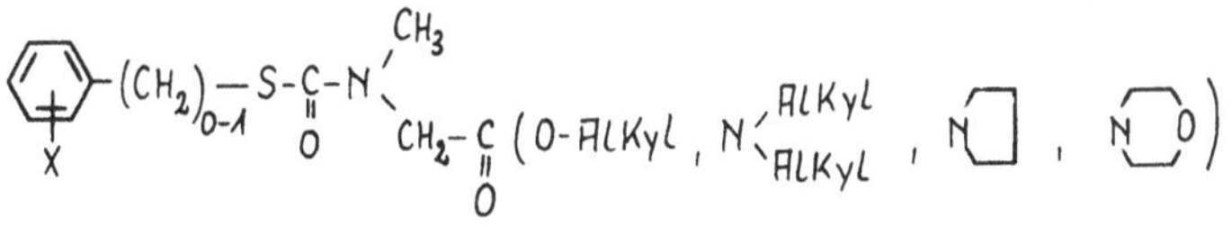

X = (O)-Alkyl, Halog., OH, NO_2

| JA | 53 111022 | 8. 3.77/28. 9.78 | Kumiai Chem. |

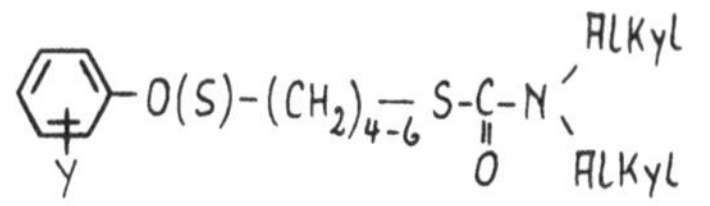

$$\text{(C}_6\text{H}_4\text{)}-O(S)-(CH_2)_{4-6}-S-\underset{\underset{O}{\|}}{C}-N\underset{Alkyl}{\overset{Alkyl}{<}}$$

Y = Halog., (O)-CH_3, CF_3,
S-CH_3, NO_2

| JA | 56 028693 | 21. 7.69/20. 8.76 | Ishihara Sangyo |

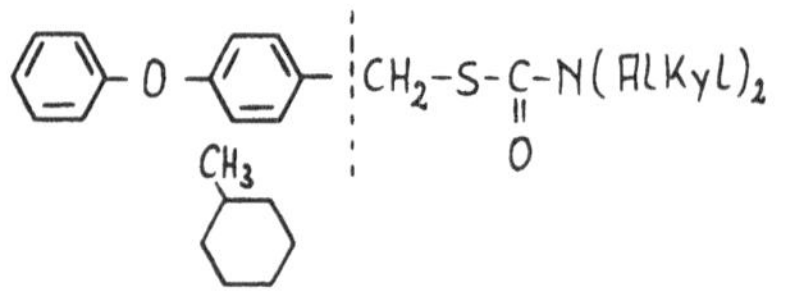

$$(C_6H_5)-O-(C_6H_3)(CH_3)-CH_2-S-\underset{\underset{O}{\|}}{C}-N(Alkyl)_2$$

| JA | 56 034890 | 6. 3.70/29. 9.76 | Kumiai Chem. |

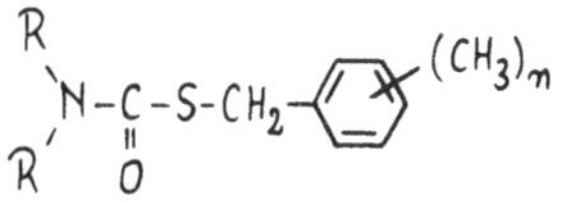

$$R,R'{-}N-\underset{\underset{O}{\|}}{C}-S-CH_2-(C_6H_3)(CH_3)_m$$

n = 3
R u. R^1 = Alkyl, Alkenyl, H, Cyclohexyl

Selektiv in Reis.

| JA | 77 009678 | 00.00.74/17. 3.77 | Nihon Tokushu Noyak |

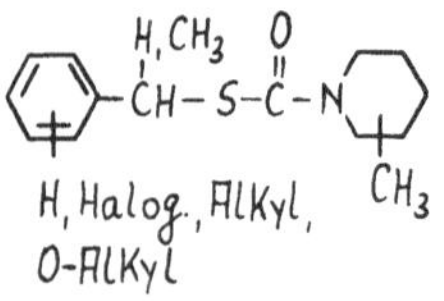

$$(C_6H_3)-\underset{H,CH_3}{\underset{|}{C}H}-S-\underset{\underset{O}{\|}}{C}-N(CH_3)\text{-piperidyl}$$

H, Halog., Alkyl,
O-Alkyl

d) <u>Heterocyclische Alkohole und Thiole als Esterkomponente</u>   Bd.5/417

| DOS
=BE | 2 633 790
857 260 | 28. 7.76/ 2. 2.78 | BASF |

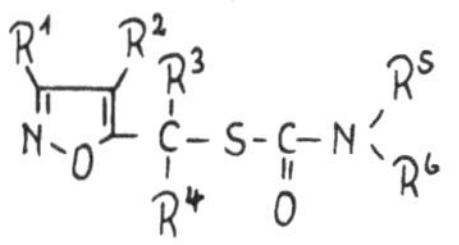

R^1 = H, Alkyl, Cycloalkyl, Aralkyl
R^2 = H, Alkyl, Aryl
R^3 u. R^4 = H, Alkyl
R^5 u. R^6 = Alk(en)yl, Cyclo-alkyl, Glieder(CH_2) eines 4-7 Ringes

DOS   2 823 384   29. 5.78/ 7.12.78   Philagro
=BE   867 636
FR Prior.   31. 5.77

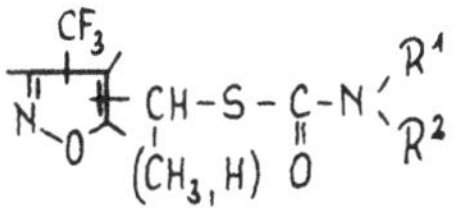

$$\text{N-O} \overset{CF_3}{\underset{(CH_3, H)}{\cdots}} CH - S - \overset{O}{\underset{||}{C}} - N \overset{R^1}{\underset{R^2}{<}}$$

R^1 u. R^2 = H, Alk(en, in)yl, Cycloalkyl, O-Alkyl usw.

Glieder eines N-haltigen 5 oder 6-Ring-systems.

JA   53 095962   28. 1.77/22. 8.78   Nihon Noyaku

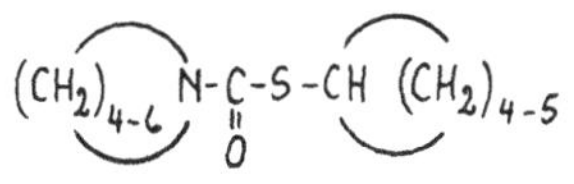

Selektiv in Reis

$$(CH_2)_{4-6} \text{N} - \overset{O}{\underset{||}{C}} - S - CH (CH_2)_{4-5}$$

JA   54 005972   13. 6.77/17. 1.79   Nissan

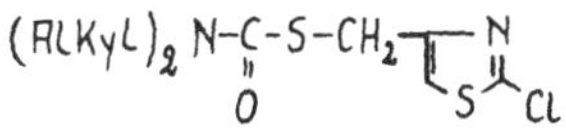

$$(Alkyl)_2 \, N - \overset{O}{\underset{||}{C}} - S - CH_2 - \underset{S}{\overset{N}{\diagdown}} Cl$$

e)   <u>Oxidationsprodukte von Thiolcarbamaten</u>   Bd. 5/418

DOS   2 712 760   23. 3.77/20.10.77   Stauffer
US Prior.   29. 3. u. 28. 4.76
=BE   852 933

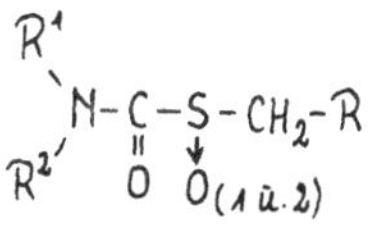

$$\overset{R^1}{\underset{R^2}{>}} N - \overset{O}{\underset{||}{C}} - \overset{\downarrow}{\underset{O_{(1 u. 2)}}{S}} - CH_2 - R$$

R = tert. Butyl, Cyclo(propyl, butyl, pentyl)
R^1 u. R^2 = Alkyl, Alkenyl, Cyclopropyl, Phenyl, Benzyl usw. oder zusammen subst. Alkylengruppe

US   4 081 468   8. 9.76/28. 3.78   Stauffer

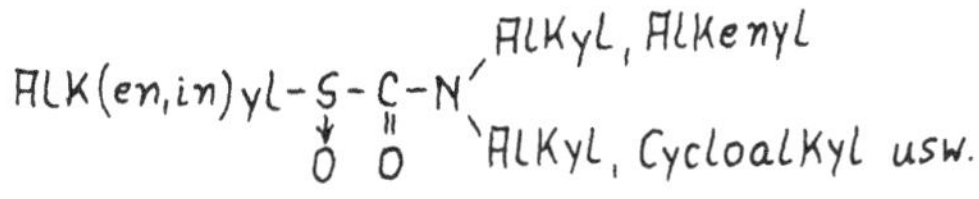

$$Alk(en, in)yl - \overset{\downarrow}{\underset{O}{S}} - \overset{O}{\underset{||}{C}} - N \overset{Alkyl, Alkenyl}{\underset{Alkyl, Cycloalkyl \ usw.}{<}}$$

US   4 117 010   8. 9.76/26. 9.78   Stauffer

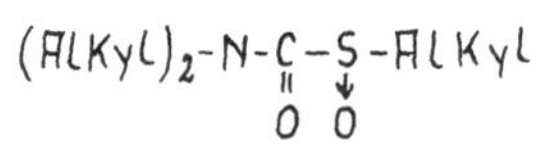

Stabilisiert durch Phenole mit verzweigten Alkyl subst. in 2 u. 6-Stellung.

$$(Alkyl)_2 - N - \overset{O}{\underset{||}{C}} - \overset{\downarrow}{\underset{O}{S}} - Alkyl$$

JA   53 137964   9. 5.77/ 1.12.78   Nissan Chem.

$(CH_2)\!\!\!\diagdown\!\!\!\underset{(CH_3)}{\diagup}\!\!N\!-\!\underset{\underset{O}{\|}}{C}\!-\!S\!-\!CH_2\!-\!\overset{N-N}{\underset{S}{\diagdown\!\!\diagup}}\!-\!Cl$

Pre emergence Wirkung,
selektiv in Reis.

f)   <u>Carbamidsäureester primärer u. sec. aromatischer Amine</u>   Bd.5/420

f 1)   <u>Aliphatische und cycloaliphatische Alkohole als Esterkomponente</u>

Z.Teil überschneiden Patente mehrere Abschnitte z.B. f (primäre Amine) mit
f 4 sec. Amine sowie f 1 aliph. Alkohole als Ester-Komponente mit Phenolen
als Ester-Komponente f 2. Zur sicheren Übersicht sind also f 1- f 5 durchzusehen.

DOS   2 617 917   23. 4.75/28.10.76   Anic Spa

$Ar\,NH-\underset{\underset{O}{\|}}{C}-S-CH_3$ *Darstellung aus* $Ar\,NH_2 + CO + CH_3-S-S-CH_3$

DOS   2 619 733   30. 4.76/17.11.77   Schering AG, Berlin

X = Halog., (O)-Alkyl
R = Alkyl

Gegen breitbl. Unkräuter

X_n-substituiertes Benzol mit CN und $NH-\underset{\underset{O}{\|}}{C}-OR$

DOS   2 725 893   18. 6.76/ 9. 6.77   Gulf Oil Corp.

Barban

Cl-substituiertes Benzol $-NH-\underset{\underset{O}{\|}}{C}-O-CH_2-C\equiv C-CH_2\,Cl$

       + $C_{12}\!-\!C_{18}\!-$ *Fettalkohole mit 10-40 Mol Ethylenoxid*

DOS   2 839 973   14. 9.78/ 3. 4.80   BASF

HO-substituiertes Benzol (H, Alkyl) $-N-\underset{\underset{O}{\|}}{C}-O(S)-$ Alkyl, Aryl usw.

Halog.,
NO_2, CN, Ar usw.

DOS   2 855 699   22.12.78/28. 6.79   Sumitomo
JA Prior.   22.12.77

R^1 = Alk(en)yl, Hal.-Alkyl,
CN-Alkyl, Cycloalkyl,
O-Alk(en)yl, S-Alk(en)yl

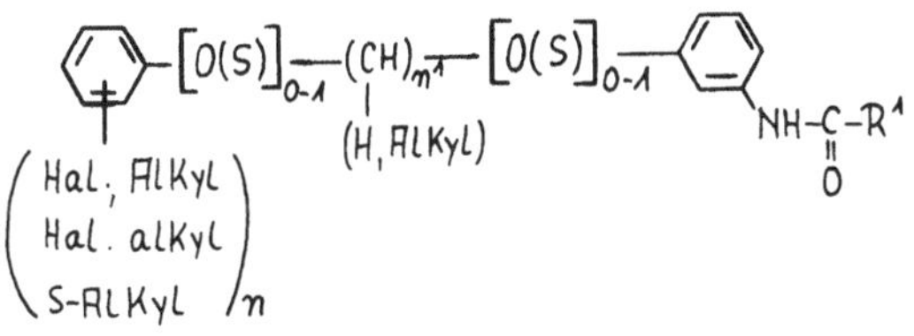

(Hal., Alkyl / Hal. alkyl / S-Alkyl)$_n$ —$\bigcirc$— [O(S)]$_{0-1}$—(CH)$_{m-1}$—[O(S)]$_{0-1}$—$\bigcirc$—NH–C(=O)–R^1

(H, Alkyl)

z.B.

Cl—$\bigcirc$(Cl)—O(CH$_2$)$_8$–O—$\bigcirc$—NH–C(=O)–O–C$_2$H$_5$

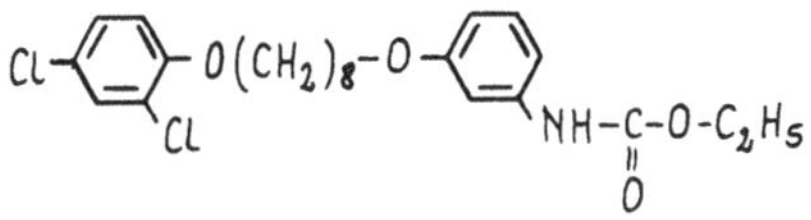

EP   030   1. 6.78/20.12.78   BASF

X = H, 6-F, 6-CF$_3$, 2-CH$_3$ usw.

Beisp.   X—$\bigcirc$(NO$_2$, NH$_2$)—N(H, CH$_2$O Alkyl)–C(=O(S))–O(CH$_3$, subst. Phenyl usw.)

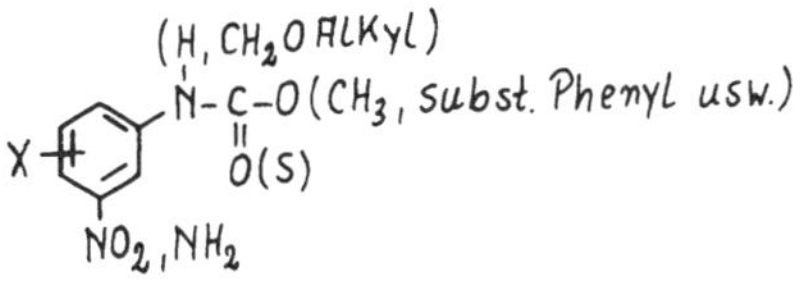

US   3 976 470   23. 7.75/24. 8.76   Stauffer

Selektiv in Reis

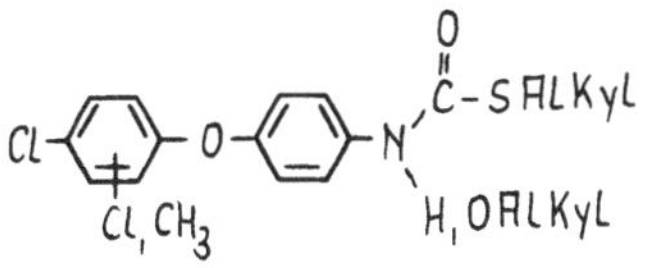

Cl—$\bigcirc$(Cl, CH$_3$)—O—$\bigcirc$—N(H, O Alkyl)–C(=O)–S Alkyl

US   4 193 787   31. 8.78/18. 3.80   Staufffer

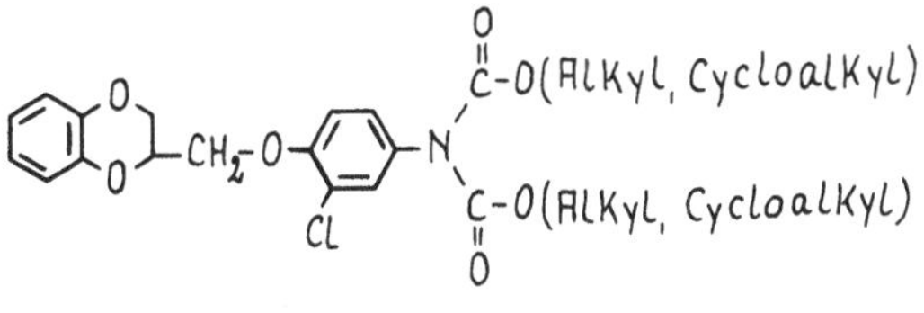

$\bigcirc$(O,O-Dioxan)–CH$_2$–O—$\bigcirc$(Cl)—N[C(=O)–O(Alkyl, Cycloalkyl)][C(=O)–O(Alkyl, Cycloalkyl)]

US   4 204 858   23. 7.75/27. 5.80   Stauffer

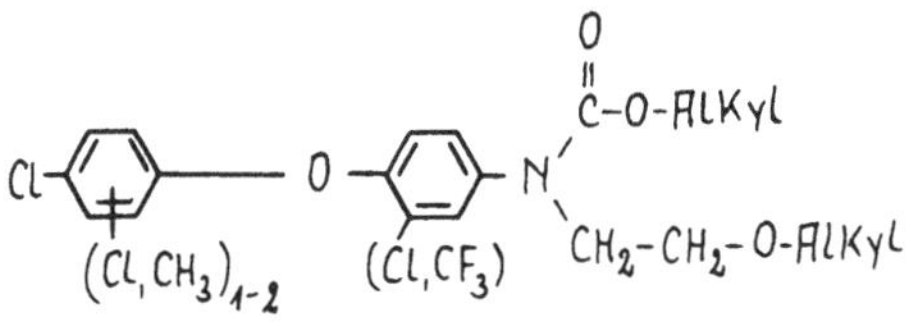

Cl—$\bigcirc$(Cl, CH$_3$)$_{1-2}$—O—$\bigcirc$(Cl, CF$_3$)—N[C(=O)–O–Alkyl][CH$_2$–CH$_2$–O–Alkyl]

FR   2 297 840   16. **1**.75/17. 9.76   Aries

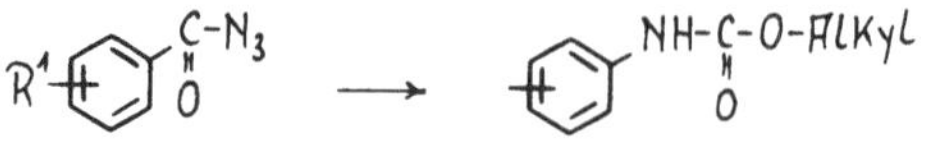

                                              Bekannte Herbizide

$$R'\!-\!\boxed{}\!-\!\overset{\overset{\displaystyle C-N_3}{\|}}{\underset{O}{}} \longrightarrow \boxed{}\!-\!NH\!-\!\overset{\overset{\displaystyle }{\|}}{\underset{O}{C}}\!-\!O\!-\!Alkyl$$

BE   854 128   30. 4.76/31.10.77   Schering

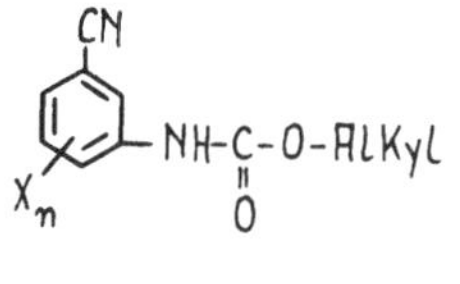

                                              X = Alkyl, O-Alkyl, Halog.

The structure bears a CN group and X_n substituent on a benzene ring, with $-NH-\underset{O}{\overset{\|}{C}}-O-Alkyl$.

                                                n = 0 und 1

pre und postemergence Herbizide.

BE   855 812   18. 6.76/19.12.77   Gulf Oil Corp.

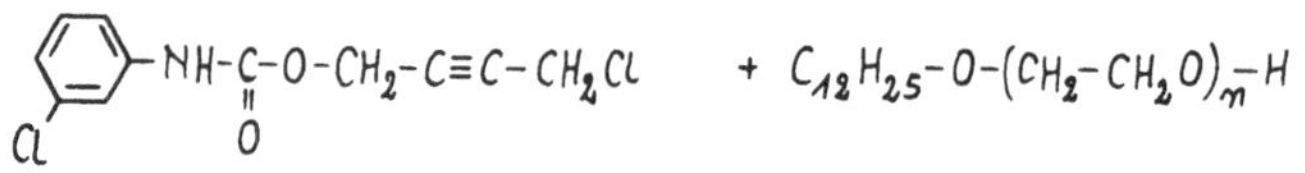

A Cl-substituted phenyl bearing $-NH-\underset{O}{\overset{\|}{C}}-O-CH_2-C\equiv C-CH_2Cl \;+\; C_{12}H_{25}-O-(CH_2-CH_2O)_n^-H$

JA   52 139721   18. 5.76/21.11.77   Mitsui Toatsu Chem.

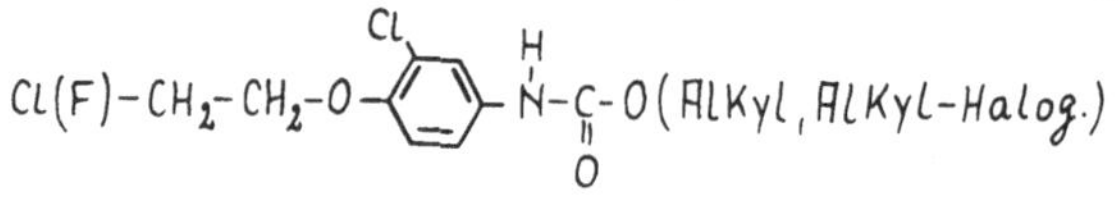

$Cl(F)-CH_2-CH_2-O-$ (Cl-substituted phenyl) $-\overset{H}{N}-\underset{O}{\overset{\|}{C}}-O(Alkyl, Alkyl-Halog.)$

JA   54 005942   14. 6.77/17. 1.79   Sangyo

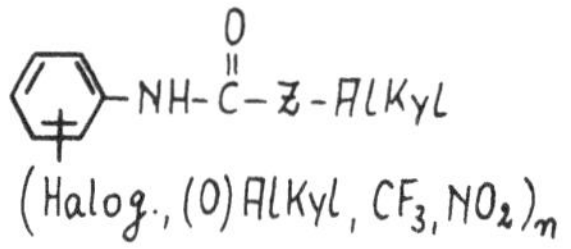

A phenyl bearing $-NH-\overset{\overset{\displaystyle O}{\|}}{C}-Z-Alkyl$, ring substituted $(Halog., (O)Alkyl, CF_3, NO_2)_n$ Z = O

JA   55 094353   9. 1.79/17. 7.80   Mitsui Toatsu Chem.

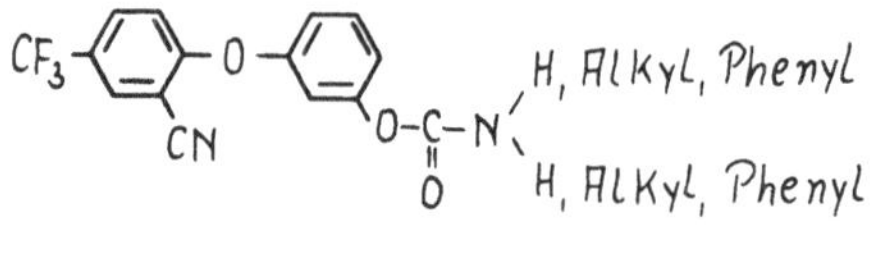

CF_3- (phenyl with CN) $-O-$ (phenyl) $-O-\underset{O}{\overset{\|}{C}}-N\big\langle\; \begin{matrix} H, Alkyl, Phenyl \\ H, Alkyl, Phenyl \end{matrix}$

JA   57 044370   14.12.70/ 8.11.77   Kumiai Chem.

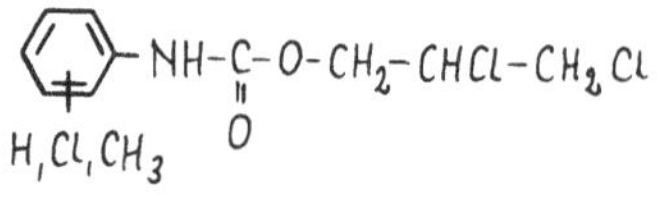

A phenyl bearing $-NH-\underset{O}{\overset{\|}{C}}-O-CH_2-CHCl-CH_2Cl$, ring substituent H, Cl, CH_3

DOS   2 851 766   30.11.78/19. 6.80   Bayer

$$R^1-O(S)\underset{\underset{O}{|}}{\overset{\overset{O}{\|}}{C}}-\underset{R}{\overset{|}{N}}-\overset{\overset{O}{\|}}{C}-(CH_2)_n-O-\langle\text{phenyl}\rangle$$

R = Alkyl, Cycloalkyl u.
    subst. Aryl
R^1 = (Cyclo)Alkyl oder <u>Aryl</u>

EP   13 759   27.12.79/ 6. 8.80   BASF
=DOS   2 901 626   17. 1.79

$$\text{subst.} \; \langle\text{aryl}\rangle - N \begin{cases} SCFCl_2 \\ \underset{O}{\overset{O}{C}}-O-\langle\text{aryl}\rangle - N \begin{cases} SCFCl_2 \\ \underset{O}{\overset{O}{C}}-O\text{-Alkyl (subst.)} \end{cases} \end{cases}$$

f 3) <u>Oxime als Esterkomponente</u>

DOS   2 934 353   24. 8.79/ 6. 3.80   Sumitomo

$$\underset{(R^1)_n}{\langle\text{aryl}\rangle}-O(S)-\underset{(H,CH_3)}{CH}-(CH_2)_n-O(S)-\langle\text{aryl}\rangle-NH-\overset{\overset{O}{\|}}{C}-O-N=C\begin{cases}(O)-CH_3\\CH_3\end{cases}$$

R^1 = Halog., S-CH$_3$, CF$_3$, $\underset{O}{\overset{O}{\diagdown}}CH_2$ usw., O-Alkyl

EP   10 588   27. 8.79/14. 5.80   Ciba-Geigy
CH Prior.   28. 8.78
EP   11 047   27. 8.79/14. 5.80   Ciba-Geigy
CH Prior.   28. 8.78

Die beiden Patente umfassen mehrere Verbindungsklassen.

$$Ar-SO_n-\underset{\underset{\underset{O}{\|}}{NO(H, Alkyl, CH_2COOR, \overset{\overset{O}{\|}}{C}-Alkyl, \overset{\overset{O}{\|}}{C}-N\langle\; usw.}}{\overset{\overset{O}{\|}}{C}(H, CN, Halog., CHOH, Alkyl\; usw.)}}$$

Ar = Phenyl, Heterocycl.

$$Ar(SO)_n-\underset{NO\{C-Alkyl, C-N-NH_2, C-N-C-N\langle, C-CH_2-C-Ar, C-N\langle\; usw.\}}{\overset{\overset{O}{\|}}{C}(CN, H, Alkyl, COOH\; usw.)}$$

Ar = Phenyl, heterocycl. Rest usw.

Die Sulfone beider Patente wirken als Antidots für Herbizide aus der Klasse
der sym. Triazine und der Diphenyletheroxypropionsäuren.

US   4 003 912   11. 6.73/18. 1.77   Monsanto

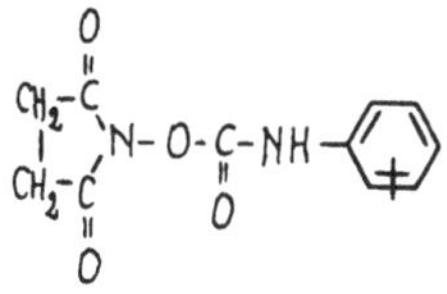

g)   Carbamidsäureester aromatischer Amine mit einer weiteren
     herbizid wirksamen Gruppierung   Bd.5/424

DOS   2 530 908   10. 7.75/13. 1.77   Wcesojusny

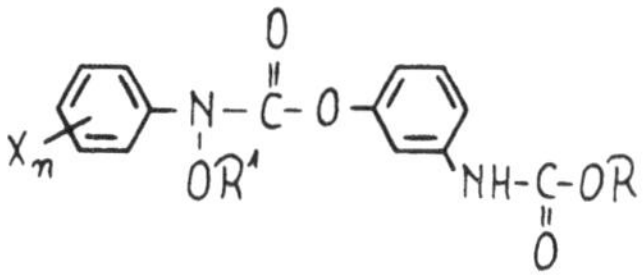

X = H, Halog., (O)-CH_3, NO_2
n = 1-2
R = Alkyl, (subst.)-Phenyl
R^1 = H, Acyl, C-N$\lessdot$

DOS   2 557 552   18.12.75/30. 6.77   Schering
=BE   849 571

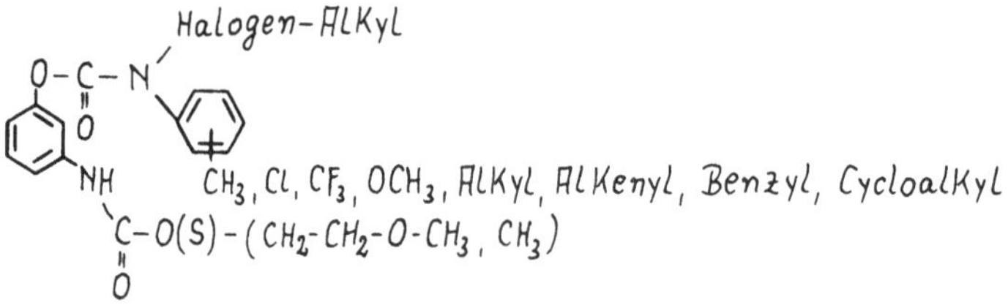

CH_3, Cl, CF_3, OCH_3, Alkyl, Alkenyl, Benzyl, Cycloalkyl

DOS   2 608 473   27. 2.76/ 1. 9.77   Schering

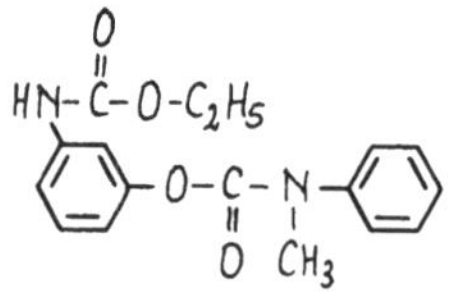

                                                     als selectives Herbizid in Baumwolle

DOS   2 630 418   2. 7.76/ 5. 1.78   Schering
=BE   856 382

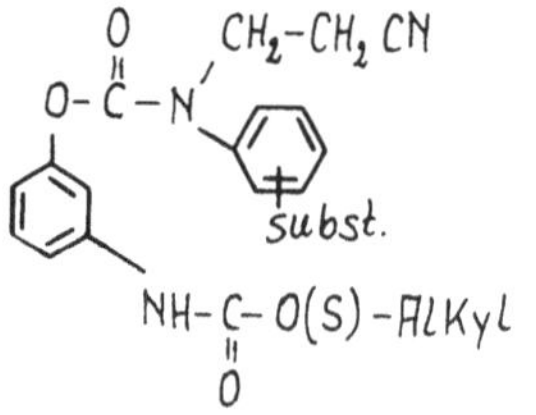

                                                     Selektiv in Baumwolle

170

DOS   2 650 796   3.11.76/11. 5.78   Schering
=BE   860 437

O-C(=O)-N((Halog)-Alkyl / subst. Phenyl, Cyclohexyl)
(phenyl ring) NH-C(=O)-O-CH_2-C≡CH

DOS   2 651 526   9.11.76/18. 5.78   Schering
=BE   860 648

O-C(=O)-N(Alk(en)yl, Halog.-Alkyl / subst. Phenyl)
(phenyl ring) NH-C(=O)-O-CH_2-CH(CH_3)(CH_3)

DOS   2 703 838   31. 1.77/10. 8.78   BASF

(phenyl ring) NH-C(=O)-O[(phenyl)-(F,CH_3)(H,CH_3) , (cyclohexyl) H]
NH-C(=O)-O-(CH_3, C_2H_5)

DOS   2 725 074   3. 6.77/21.12.78   BASF

X u. Y = F/F; F/Cl, Cl/F

X,Y-(phenyl)-NH-C(=O)-O-(phenyl)-NH-C(=O)-OCH_3

DOS   2 732 848   18. 7.77/ 8. 2.79   Schering
=BE   869 074   18. 7.77/18. 1.79   Schering

Selektiv in Baumwolle,
Kartoffeln, Reis, Karotten usw

O-C(=O)-N—(phenyl)-(O)-Alkyl , O Alkyl
(phenyl ring) N(H)-C(=O)-O(S)-Alk(en,in)yl

DOS   2 819 748   2. 5.78/15.11.79   Schering

Darstellungspat.

O-C(=O)-N(C_2H_5)(phenyl)
(phenyl ring) NH-C(=O)-OCH_3

| DOS | 2 844 806 | 11.10.78/24. 4.80 | Schering |
| =NL | 79/06019 | 12.10.78/15. 4.80 | Schering |

R^1 = Alk(en, in)yl

u. andere Substit.

NH–C–ALK(en,in)yl, Cyclopropyl, CH₃

| DOS | 2 843 691 | 4.10.78/24. 4.80 | Schering |

R^1 = H, Alk(en)yl
R^2 = subst. Phenyl, Benzyl, Cyclohexyl
R^3 = Alk(en, in)yl, Chloralkyl

| DOS | 2 846 625 | 26.10.78/ 8. 5.80 | BASF |

=EP   10 692

Postemergence gegen Unkräuter in Baumwolle, Reis, Getreide, Mais.

| DOS | 2 901 626 | 17. 1.79/31. 7.80 | BASF |

| DOS | 2 901 658 | 15. 1.79/24. 7.80 | Schering |

| EP | 00 030 | | |
| =DOS | 2 725 146 | 3. 6.77/14.12.78 | BASF |

172

DOS 2 934 353   25. 8.78/ 6. 3.80   Sumitomo Chem.

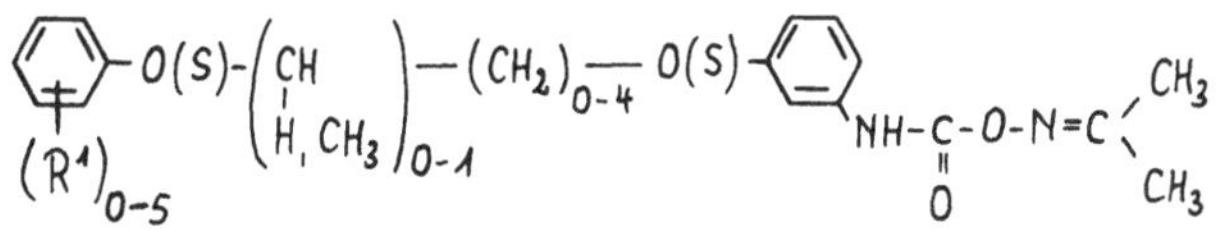

$$\text{(R}^1\text{)}_{0-5}\text{—Phenyl—O(S)—}\left(\underset{H, CH_3}{\overset{CH}{|}}\right)_{0-1}\text{—(CH}_2\text{)}_{0-4}\text{—O(S)—Phenyl—NH—}\underset{O}{\overset{||}{C}}\text{—O—N=C}\overset{CH_3}{\underset{CH_3}{\diagdown}}$$

US   3 997 325   24. 5.73/14.12.76   Amer. Cyanamid
siehe US 3 964 285

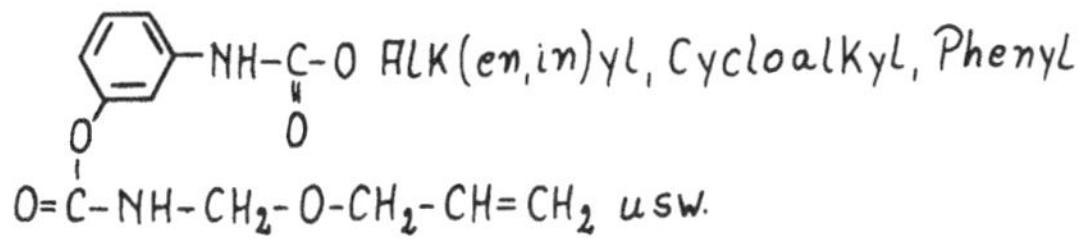

$$\text{Phenyl—NH—}\underset{O}{\overset{||}{C}}\text{—O}\ ALK(en,in)yl,\ Cycloalkyl,\ Phenyl$$
$$O=\underset{|}{\overset{}{C}}\text{—NH—CH}_2\text{—O—CH}_2\text{—CH=CH}_2\ usw.$$

BE   858 059   28. 8.76/24. 2.78   BASF

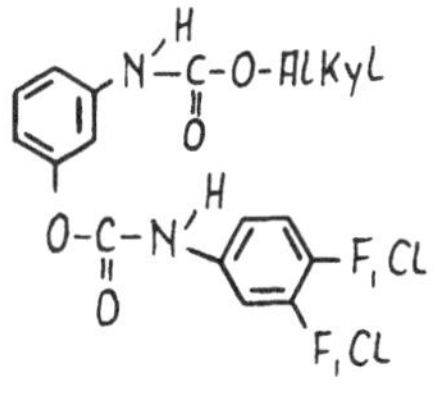

Selektiv in Soja, Zucker-
rüben

$$\text{Phenyl—N}\overset{H}{\diagup}\underset{O}{\overset{||}{C}}\text{—O—Alkyl}$$
$$\text{O—}\underset{O}{\overset{||}{C}}\text{—N}\overset{H}{\diagup}\text{Phenyl}\underset{F,Cl}{\overset{F,Cl}{}}$$

BE   875 991   2. 5.78/ 5.11.79   Schering

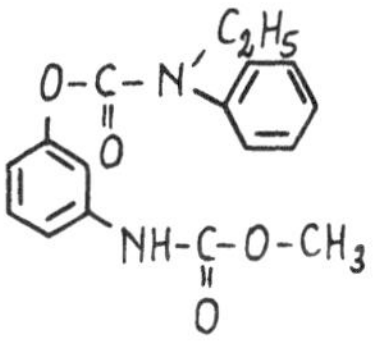

Rübenherbizid

$$\text{O—}\underset{O}{\overset{||}{C}}\text{—N}\overset{C_2H_5}{\diagup}\text{Phenyl}$$
$$\text{NH—}\underset{O}{\overset{||}{C}}\text{—O—CH}_3$$

NL   79/05935   4.10.78/ 9. 4.80   Schering

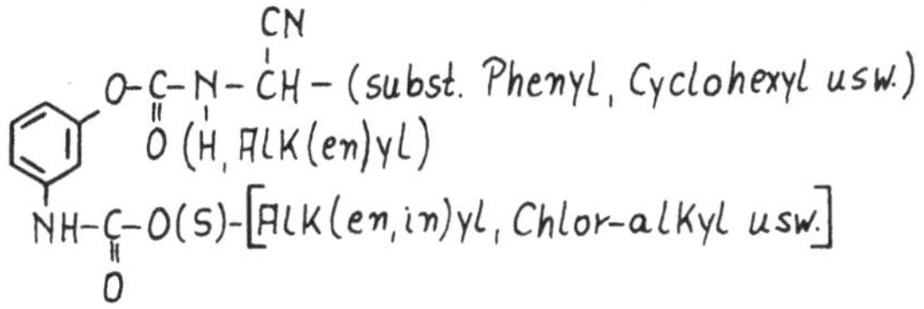

$$\text{O—}\underset{O\ (H,\ ALK(en)yl)}{\overset{||}{C}}\text{—N—}\underset{}{\overset{CN}{\underset{|}{CH}}}\text{—(subst. Phenyl, Cyclohexyl usw.)}$$
$$\text{NH—}\underset{O}{\overset{||}{C}}\text{—O(S)—}[ALK(en,in)yl,\ Chlor\text{-}alkyl\ usw.]$$

SU   471 785   16. 5.73/16. 2.77   Baskakov

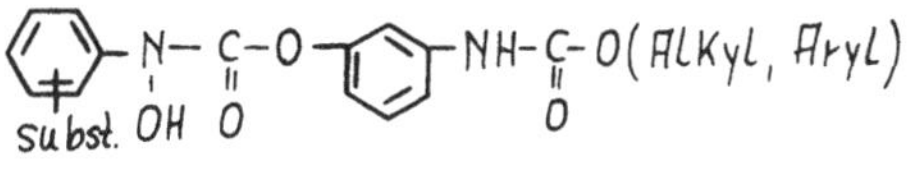

$$\text{Phenyl}\underset{subst.\ OH}{}\text{—N—}\underset{O}{\overset{||}{C}}\text{—O—Phenyl—NH—}\underset{O}{\overset{||}{C}}\text{—O(Alkyl, Aryl)}$$

JA 51 08851   14. 2.79/21. 8.80   Mitsui Toatsu Chem.

$$Cl\text{-}pyridyl(Cl)\text{-}O\text{-}C_6H_3\text{-}O\text{-}\underset{O}{\overset{\|}{C}}\text{-}N(\text{subst. Phenyl})_2$$

<u>g)</u> **g)** Carbamidsäureester heterocyclischer Amine

BE   882 627   5. 4.79/ 3.10.80   Schering

$$O\text{-}\underset{O}{\overset{\|}{C}}\text{-}N\underset{NH\text{-}\underset{O}{\overset{\|}{C}}\text{-}O(S)\text{-}ALK(en,in)yl}{\overset{CH_2\text{-}C\equiv CH}{\diagdown}}$$

<u>h 1)</u> **h 1)** <u>Bicyclische und heterocyclische Carbamidsäureester</u>   Bd. 5/428

   Siehe auch bei 5.5.2 b

Carbamidsäureester bei denen die Gruppierung

$$-\underset{|}{N}\text{-}\underset{\|}{C}\text{-}OR \quad \text{Teil eines Ringsystems ist, werden hier miterfaßt.}$$

DOS 2 632 676   16. 7.76/19. 1.78   Schering

                Gegen Ungräser

$$\text{Ring}\text{-}N\text{-}\underset{O}{\overset{\|}{C}}\text{-}S\text{-}C_2H_5$$

DOS 2 632 915   19. 7.76/26. 1.78   Schering

                Gräserherbizid z. B. gegen
                Flughafer

$$\text{(bicyclisches Ringsystem)}\text{-}N\text{-}\underset{O(S)}{\overset{\|}{C}}\text{-}S(O)\text{-}(Alkyl, Cycloalkyl, Phenyl)$$

174

US   4 207 089   24. 1.79/10. 6.80   Olin Corp.

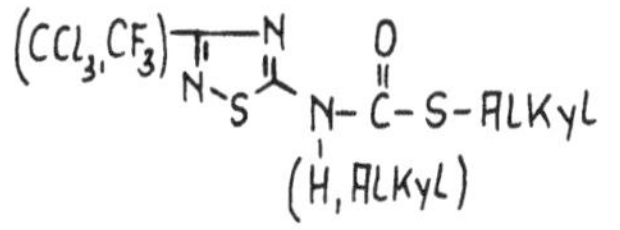

JA   52 072820   15.12.75/17. 6.77   Mitsubishi Petrochem.

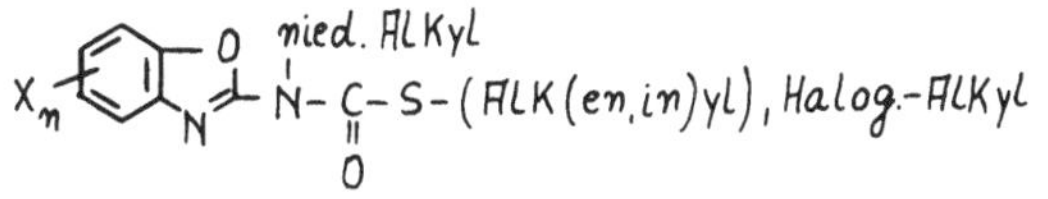

X = Halog., Alkyl
n = 0-4

JA   52 128221   20. 4.76/27.10.77   Mitsubishi Petrochem.

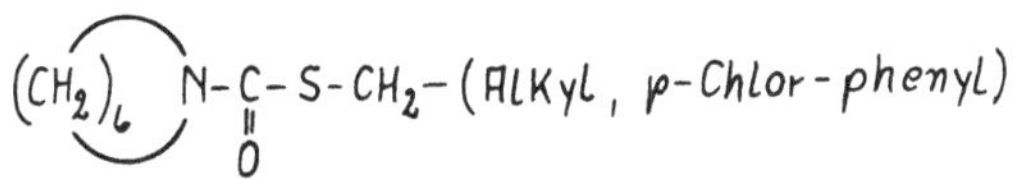

wird in seiner Wirkung verstärkt durch

Alkyl-NH-C-S-(Alkyl, subst. Benzyl)
         ‖
         O

JA   53 024022   13. 8.76/ 6. 3.78   Asahi Chem.

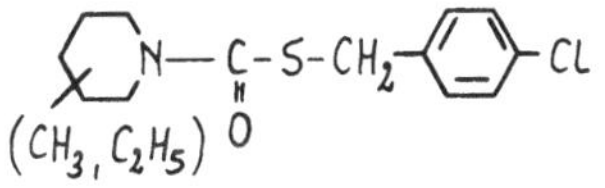

Gegen Ungräser und breitbl.
Unkräuter

JA   53 024023   16. 8.76/ 6. 3.78   Asahi Chem.

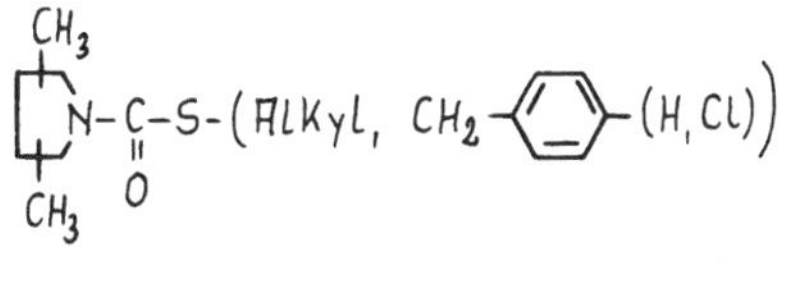

JA   53 024024   17. 8.76/ 6. 3.78   Asahi Chem.

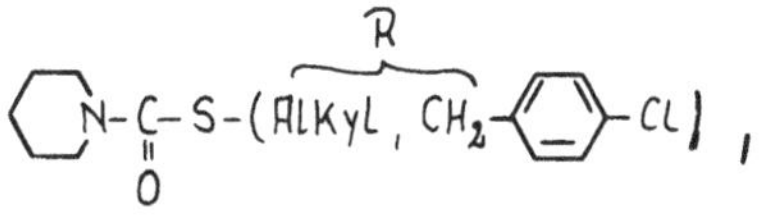

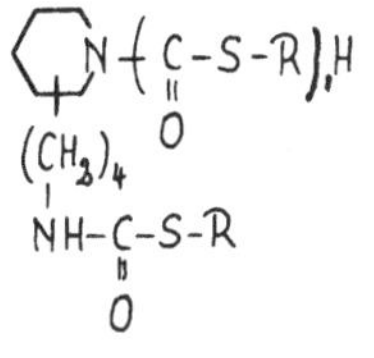

DOS   2 644 446   1.10.76/14. 4.79   Stauffer
US Prior.   2.10.75 u. 17. 9.76

$$\underset{\substack{H,CH_3,\\ O\text{-}CH_3,\,Cl}}{\bigoplus} - SO_2 - \underset{(H,CH_3)}{N} - \overset{O}{\underset{\|}{C}} - S - (Alkyl,\ Benzyl,\ 4\text{-}Chlorphenyl)$$

meist in p-Stellung

Antidots gegen Thiocarbamat-Schäden.

DOS   2 644 504   1.10.76/ 2. 6.77   Stauffer
US Prior.   2.10.75 u. 13. 9.76

n = 0-3
X = H, Cl, Br, CH_3, CF_3,
    O-CH_3, CH_2-N-C-CF_3,

$$\underset{(X)_n}{\bigoplus} - SO_2 - NH - \overset{O}{\underset{\|}{C}} - (O\text{-}Chlor\text{-}alkyl,\ Alk(en,in)yl,$$

$N\langle\substack{C_2H_5\\C_2H_5}$

$CH_2\text{-}S\text{-}\bigcirc\text{-}Cl,$ $N{=}C\langle\substack{CH_3\\CH_3}$

Antidot gegen Thiocarbamatschäden

DOS   2 735 001   3. 8.77/ 9. 2.78   May and Baker GB
GB Prior.   5. 8.76

$$NH_2\text{-}\bigcirc\text{-}SO_2\text{-}NH\text{-}\overset{O}{\underset{\|}{C}}\text{-}O\text{-}CH_3 \quad als\ Na\text{-}Salz \quad (Asulam)$$

$$+\ NH_2\text{-}\overset{O}{\underset{\|}{C}}\text{-}N\langle\substack{H\\(H,CH_3)} \quad zur\ Wirkungsverstärkung$$

5.5.3 Harnstoffe   Bd. 5/431
       ==========

a) <u>Aliphatische Harnstoffe und Thioharnstoffe (z.T. auch aromatisch-
   aliphatische Harnstoffe)</u>

US   3 982 019   17. 7.68/21. 9.76   Velsicol Chem.
Div.   3 895 061

$$\substack{subst.\ Aryl,\ Alk(en)yl\\ \\ H,\ Alk(en)yl}\ N\text{-}\overset{O}{\underset{\|}{C}}\text{-}N\langle\substack{H\\CF_3} \qquad u.a.\ auch\ Herbizide$$

US   3 989 727   6. 8.77/ 2.11.76   Monsanto

$$O=P-CH-N-C-N\begin{smallmatrix}H\\CH_3\end{smallmatrix}$$

(structure: $O=\overset{OH}{\underset{OH,\,C_2H_5,\,Naphthyl}{P}}-\overset{R}{\underset{}{CH}}-\overset{CH_3}{\underset{}{N}}-\overset{O(S)}{\underset{}{C}}-N\begin{smallmatrix}H\\CH_3\end{smallmatrix}$)

R = C_1-C_5-Alkyl, Alkenyl,

CH_2-S-C_2H_5, [structure] , C_6H_5

US   4 066 440   30. 7.73/ 3. 1.78   PPG Ind.

$$R-NH-\underset{O}{\overset{}{C}}-NH-R^1$$

R u. R^1 = Alkinyl

US   4 107 436   21.10.75/18. 8.78   Dow Chem.

(structure with morpholine: $O\underset{}{\bigcirc}N-\underset{O}{\overset{}{C}}-N\begin{smallmatrix}H\\CH-CH-OH\\R^1\;\;R^2\end{smallmatrix}$)

R^1 u. R^2 = H, Alkyl, Phenyl

DDR   131 646   8. 6.77/12. 7.78   VEB Chemiekomb. Bitterfeld

$$CCl_3-\underset{OH}{\overset{}{CH}}-NH-\underset{O}{\overset{}{C}}-NH-\underset{OH}{\overset{}{CH}}-CCl_3$$ *u.a. auch Herbizid wirksam*

b) <u>Cycloaliphatische Harnstoffe</u>   Bd.5/432

   b 1) <u>Monocyclische Harnstoffe (s.a. unter a)</u>

EP   11 670   30.11.78/11. 8.80   Bayer
=DOS   2 851 755

(structure: cycloheptatriene ring with R (=H, $\underset{O}{\overset{}{C}}$-Alkyl, $\underset{O}{\overset{}{C}}$-OAlkyl) and NH-$R^1$)

R^1 = $\underset{O}{\overset{}{C}}-N\begin{smallmatrix}H\\H\end{smallmatrix}$, $\underset{O}{\overset{}{C}}-N\begin{smallmatrix}Alkyl\\Alkyl\end{smallmatrix}$,

$\underset{O}{\overset{}{C}}-SR$

EP   15 605   19. 2.80/17. 9.80   Shell Int. Research

(structure: cyclohexene ring with C_3H_7, O-CH_3, $\underset{H\;\;O}{N}-\overset{}{C}-NH-CH_3$)

b 2) Bicyclische Ringe enthaltende Harnstoffe Bd.5/432

US 4 191 830 25. 1.74/ 4. 3.80 American Cyanamid

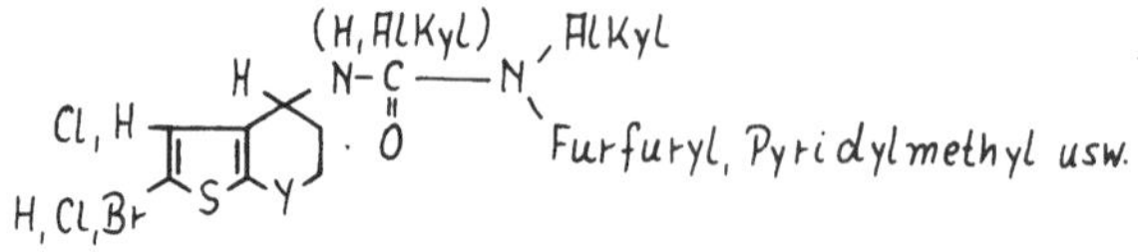

Y = CH_2, C(=O), CHOH usw.

c) Harnstoffe des Benzylamins z.T. in früheren Patente miterfaßt.

Bd.5/434

DOS 2 659 404 29.12.76/14. 7.77 Showa Denko
AU Prior. 30.12.75

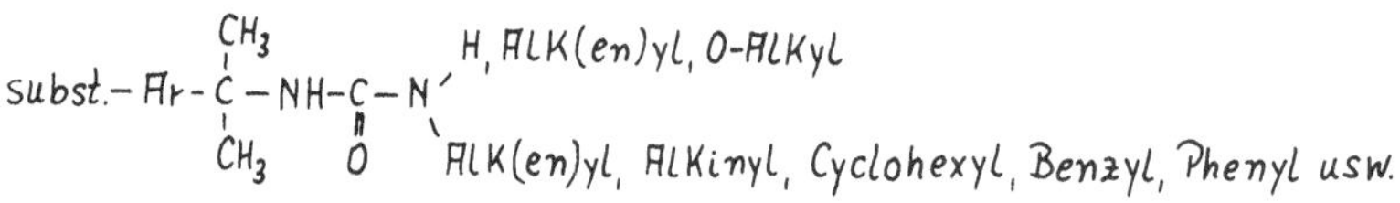

US 4 107 436 21.10.75/18. 8.78 Dow Chem.

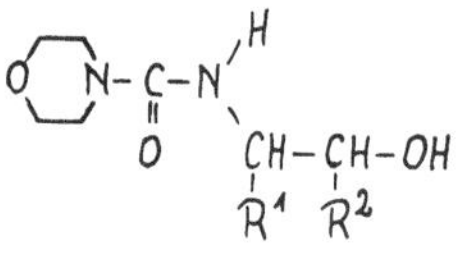

R^1 u. R^2 = H, Alkyl, Phenyl

JA 53 081621 27.12.76/19. 7.78 Showa Denko KG

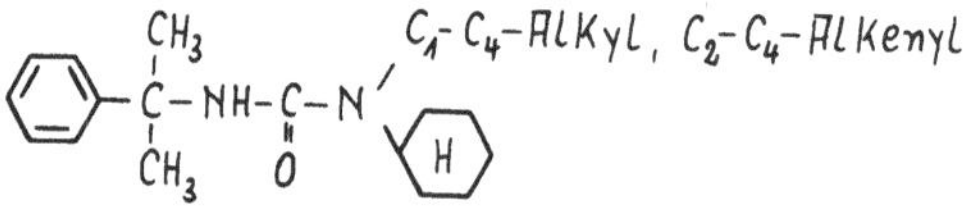

d 1) Aromatisch-aliphatische Harnstoffe und Thioharnstoffe primärer und sekundärer aromatischer Amine. Z.T. in 5.5.3 a-c miterfaßt.

DOS 2 501 729 17. 1.75/22. 7.76 Celamerck

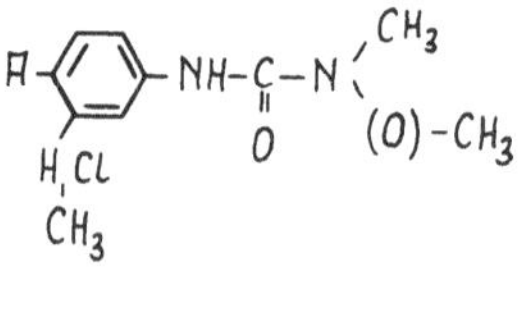

A = CCl_3-CHCl-, CCl_2=CCl-

DOS 2 538 178   27. 8.75/10. 3.77   Bayer

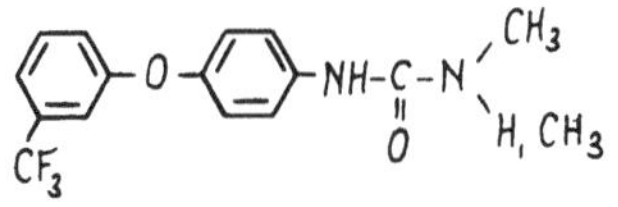

DOS 2 542 468   24. 9.75/ 7. 4.77   Bayer

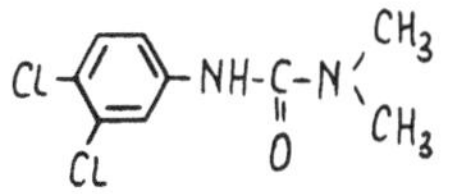

Anwendung als Herbizid gegen
Gräser in Mais.

DOS 2 558 078   19.12.75/23. 6.77   Schering

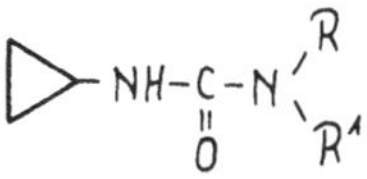

R = H, Alkyl, Aromaten
R^1 = subst. Phenyl

DOS 2 624 094   29. 5.75/ 2.12.76   Sumitomo

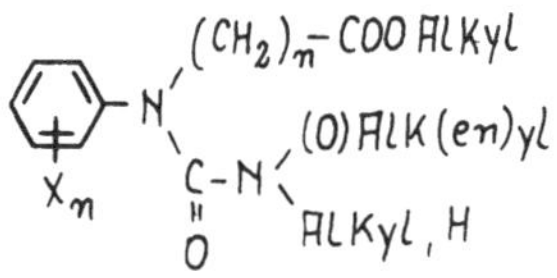

X = Halog., Alkyl, O-Alkyl,
    CF_3
n = 0-2

DOS 2 624 822   31. 5.76/15.12.77   Schering

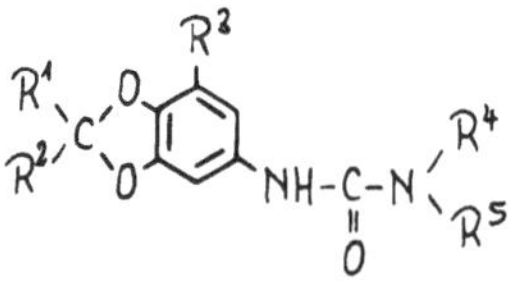

R^1 u. R^2 = H, Alkyl, Aryl,
                Heterocycl.
R^1 u. R^2 Glieder eines
Ringsystems
R^3 = H, Alkyl
R^4 u. R^5 = H, Alk(en)yl

Gegen Ungräser und Unkräuter.

DOS 2 711 230   16. 3.76/29. 9.77   Sumitomo

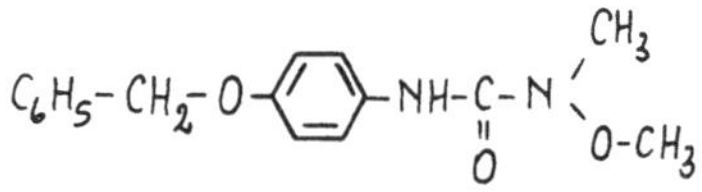

Selektiv gegen Ungräser und
Unkräuter in Sojabohnen

DOS   2 744 169   30. 9.76/ 6. 4.78   Sumitomo

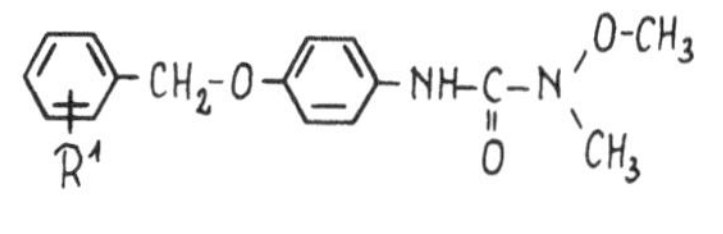

R^1 = NO_2, CN, CF_3

DOS   2 739 444   1. 9.77/ 8. 3.79   Amer. Cyanamid

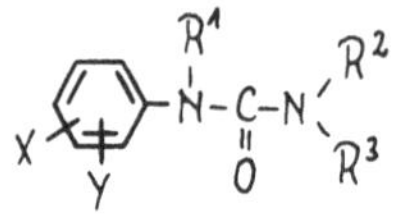

R^1, R^2 u. R^3 = H, (O)-Alkyl
X = H, Halog, (O)-Alkyl, NO_2,
    Halog.-Alkyl

Y = -C-O-C-O-   z.B. ...

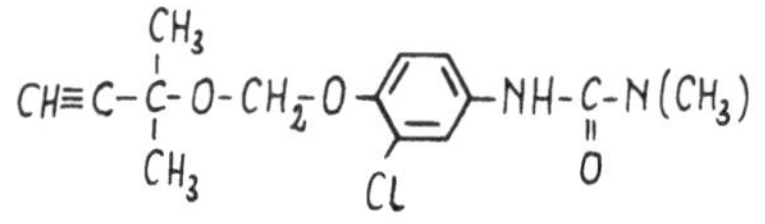

DOS   2 800 111   3. 1.78/13. 7.78   Aziende Colori Naz.
IT Prior. 7.1.77 u. 14.2.77

Darstellungsverfahren

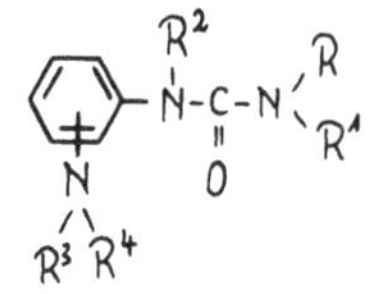

R u. R^1 = H, Alkyl, Cyclo-
          alkyl, Aryl, Benzyl
R^2 = H, Alkyl, Cycloalkyl, Aryl,
      Benzyl
R^3 u. R^4 = H, Alkyl, Cyclo-
          alkyl, Aryl, Benzyl
Glieder eines Ringsystems.

DOS   2 804 739   2. 2.78/ 9. 8.79   Schering
=BE   873 906   2. 2.78/ 2. 8.79   Schering

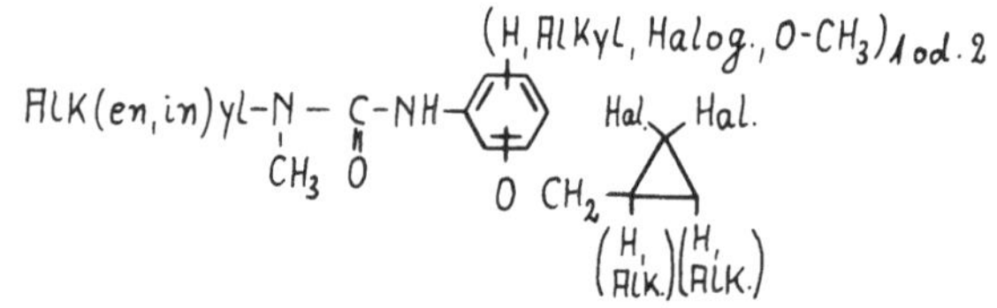

DOS   2 809 035   2. 3.78/ 7. 9.78   Sumitomo
JA Prior. 3.3.77 ähnlich JA 54 098326

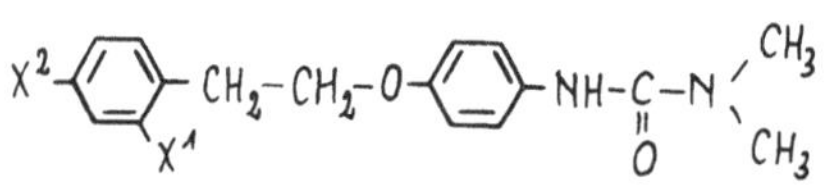

X^1 = H, CH_3
X^2 = (O)-Alkyl (C_1-C_5)

Selektiv in Reis, Sojabohnen, Erdnüssen, Mais.

DOS   2 826 531   17. 6.77/ 4. 1.79   Hokko Chem.

$$CH_3-CH=CH-CH_3-O-\text{[Ph]}-NH-\underset{\underset{O}{\|}}{C}-N(CH_3)_2$$

Spezifisch gegen Echinochloa-crus-galli var. oryzicola in Reiskulturen.

DOS   2 828 417   28. 6.77/ 4. 1.79   Sumitomo

R^1, R^2, R^3 = H, Halog., Alkyl, O-Alkyl, CF_3
Z = Alkylen (O oder S enthaltend)

Selektiv in Reis

DOS   2 846 723   20.10.78/ 3. 5.79   Sumitomo
JA Prior. 26.10.77; 28.2.78 u. 1.3.78
siehe auch DOS 2 744 169, 2 809 035 u. 2 828 417 zuvor referiert, ferner
DOS 2 909 828 13.3.79/20.3.79   Sumitomo

DOS   2 853 791   13.12.78/21. 6.79   Sumitomo
JA Prior.   13.12.77

DOS   2 901 659   15. 1.79/24. 7.80   Schering

R^1 = 1-3 x Alk(en, in)yl, Halog., O-Alkyl,

$-O-\text{[Ph]}$, NO_2 usw.
R^2 = H, Alkyl

EP   1 556   16. 5.78/ 2. 5.79   Stauffer Chem.

EP   3 835   21. 2.79/ 5. 9.79   Sumitomo
JA Prior.   54 115345   24. 2.78/ 7. 9.79   Sumitomo
            54 115344   22. 2.78   (H statt CH_3 am N^1)

$$CH_3\text{—}\underset{}{\bigcirc}\text{—}CH_2\text{—}CH_2\text{—}O\text{—}\underset{}{\bigcirc}\text{—}\underbrace{\left[NH\text{—}\underset{\underset{O}{\|}}{C}\text{—}N\diagdown\begin{matrix}CH_3\\O\text{-}CH_3\end{matrix}\right]}_{Z}$$

Beansprucht auch

Z = N=C=O usw.

als Zwischenprodukt

EP   11 179   9.11.78/28. 5.80   Bayer
=DOS   2 848 531   9.11.78/29. 5.80   Bayer

Z = CF_2, CF_2-CH_2, CF_2-CHF

oder CF_2-$CFCl$

$$Z\diagdown\underset{O}{\overset{O}{}}\bigcirc\text{—}NH\text{—}\underset{\underset{O}{\|}}{C}\text{—}N\diagdown\begin{matrix}H, ALKyl\\ALKyl, ALK(en,in)yl, O\text{-}ALKyl\ usw.\end{matrix}$$

US   3 982 019   17. 7.68/21. 9.76   Velsicol Chem.
Div.   3 895 061

$$\begin{matrix}subst.\ Aryl, ALK(en)yl\diagdown\\H, ALK(en)yl\diagup\end{matrix}N\text{—}\underset{\underset{O}{\|}}{C}\text{—}N\diagdown\begin{matrix}H\\CF_3\end{matrix}$$

US   3 988 300   31. 5.73/26.10.76   American Cyanamid

A=B= H u. p-C_6H_5-$\underset{\underset{O}{\|}}{C}$,

m-C_6H_5-$\underset{\underset{O}{\|}}{C}$

$$B\bigcirc\text{—}\underset{\underset{O}{\|}}{C}\text{—}\overset{A}{\bigcirc}\text{—}NH\text{—}\underset{\underset{O}{\|}}{C}\text{—}N\diagdown\begin{matrix}CH_3\\CH_3\end{matrix}$$

US   4 026 697   31. 5.73/31. 5.77   American Cyanamid

X u. X^1 = H, Halog., CH_3,
CF_3, O-CH_3

$$X\bigcirc\text{—}\underset{\underset{O}{\|}}{C}\text{—}\overset{X^1}{\bigcirc}\text{—}NH\ \underset{\underset{O}{\|}}{C}\ N\diagdown\begin{matrix}H, (O) ALKyl, ALK(en,in)yl, Cycloalkyl\\ALKyl, Heterocycl. usw.\end{matrix}$$

X^1 in m oder p Stellung (dann $\overset{}{\bigcirc}\underset{X}{}\text{—}\underset{\underset{O}{\|}}{C}$ in m-Stellung)

US   4 090 864   31. 5.77/23. 5.78   Stauffer Chem.

X u. Y = H, Cl, CH_3, CF_3

$$X\underset{Y}{\bigcirc}\text{—}NH\text{—}\underset{\underset{O}{\|}}{C}\text{—}CH_2\text{—}S\text{—}CH_2\text{—}\underset{\underset{CH_3}{|}}{N}\text{—}\underset{\underset{O}{\|}}{C}\text{—}NH\underset{Y}{\bigcirc}\text{—}X$$

182

US   4 102 673   30. 8.68/25. 7.78   Hercules

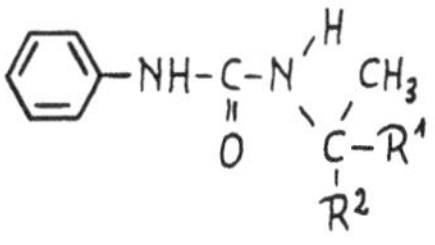

R^1 = H; R^2 = iso.,sec.,tert.-Butyl

R^1 = CH$_3$; R^2 = CH$<$CH$_3$, CH$_3$

US   4 111 682   13. 1.75/ 5. 9.78   Stauffer Chem.

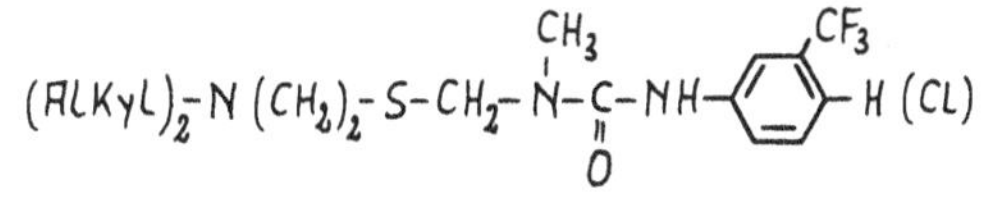

(Alkyl)$_2$-N (CH$_2$)$_2$-S-CH$_2$-N-C-NH-⟨C$_6$H$_3$(CF$_3$)⟩-H (Cl)   (N-CH$_3$, C=O)

US   4 111 683   16. 3.71/ 5. 9.78   Chevron Res.

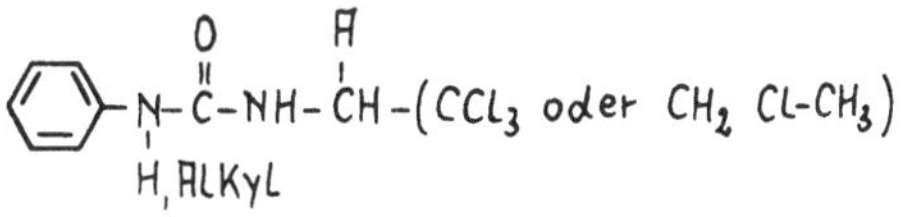

⟨C$_6$H$_5$⟩-N-C-NH-CH-(CCl$_3$ oder CH$_2$Cl-CH$_3$)   (N with H, Alkyl; C=O; CH with A)

A = (O, S-H, Alkyl)

CA   107 840   16. 3.73/27. 5.80   ITT Ind.

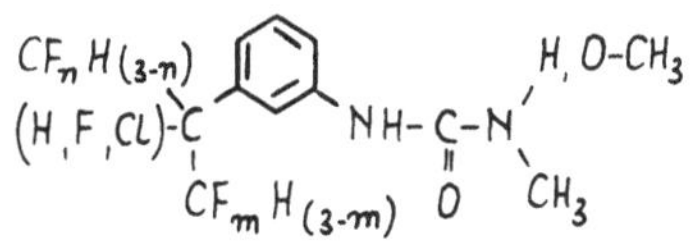

(Formel stimmt nicht mit
Beschr. überein:
m-Stellung, Formel p-Stellung)

[Structure: CF$_n$H$_{(3-n)}$, (H,F,Cl)-C, CF$_m$H$_{(3-m)}$ substituent on benzene ring; NH-C-N with H,O-CH$_3$ and CH$_3$; C=O]

CA   1 024 527   21.12.73/17. 1.78   Ciba-Geigy

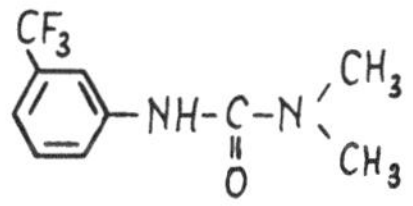

Darstellungsverfahren

[Structure: ⟨C$_6$H$_4$(CF$_3$)⟩-NH-C-N(CH$_3$)(CH$_3$); C=O]

BE   838 616   18. 2.75/16. 8.76   Roussel Uclaf

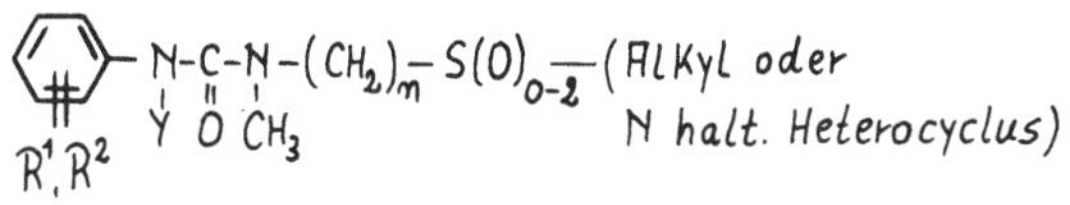

Y = H, Alkyl, S-Heterocycl.
R^1 u. R^2 = H, Cl, NO$_2$, CF$_3$,
(O)Alkyl

[Structure: ⟨C$_6$H$_3$ with R^1,R^2⟩-N-C-N-(CH$_2$)$_m$-S(O)$_{0-2}$-(Alkyl oder N halt. Heterocyclus); Y, O, CH$_3$]

BE   845 528   27. 8.75/28. 2.77   Bayer

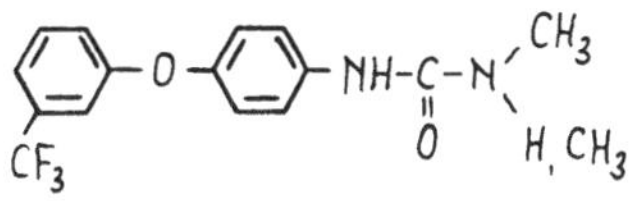

[Structure: ⟨C$_6$H$_4$(CF$_3$)⟩-O-⟨C$_6$H$_4$⟩-NH-C-N(CH$_3$)(H,CH$_3$); C=O]

| BE | 849 570 | 19.12.75/17. 6.77 | Schering |

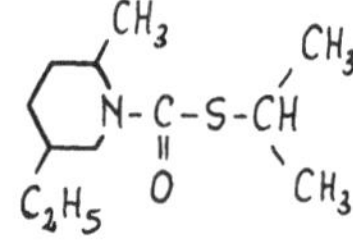

| BE | 855 218 | 31. 5.76/30.11.77 | Schering |

R^1 u. R^2 = H, Alkyl, subst.
Aryl, subst. Hetero-
cycl.

| BE | 864 029 | 3. 3.77/16. 8.78 | Sumitomo |

R^1 = H, CH_3
R^2 = C_{1-5}-(O)-Alkyl

| DDR | 134 714 | 29. 9.75/21. 3.79 | Stauffer Chem. |

cis u. trans Isomerengemisch

Selektiv in Baumwolle

| GB | 2 016 010 | 13. 3.78/19. 9.79 | Sumitomo |

R = Halog., CF_3, CN,
O(S)-Alkyl usw.

| GB | 2 027 699 | 25. 7.78/27. 2.80 | Philagro |
| =BE | 878 130 | 26. 2.79/ 7. 2.80 | Philagro |

NE   78 01458   13. 2.68/30. 6.78   Ciba-Geigy

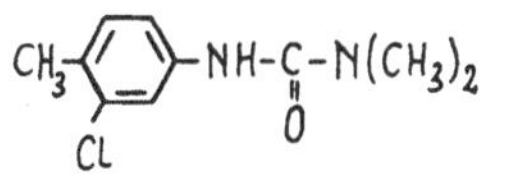

Selektiv in Weizen und Gerste gegen Gräser

NE   78 10700   26.10.77/ 1. 5.79   Sumitomo

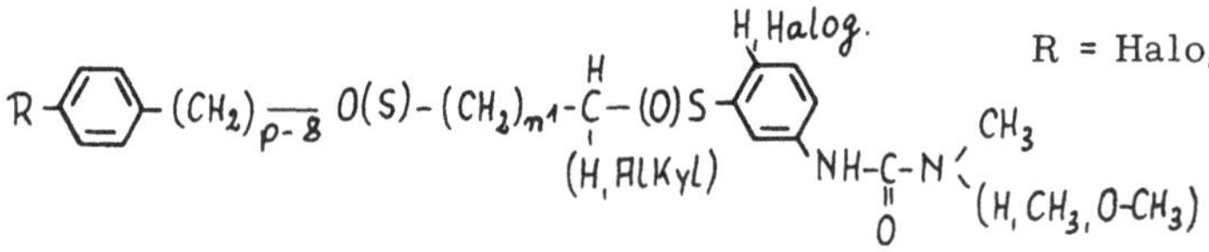

R = Halog., CF$_3$, (O(S))-Alkyl

SU   556 716   2. 2.76/26.10.77   Stauffer

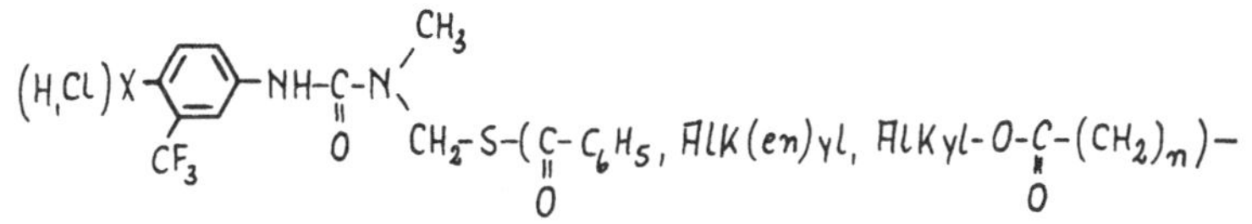

JA   51 106734   13. 3.73/21. 9.76   Nihon Tokushu Noyak

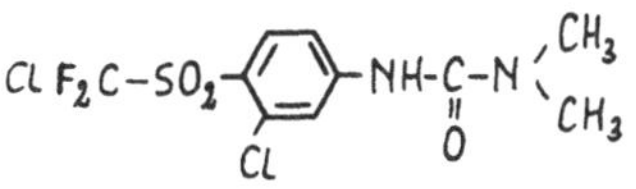

Selektiv in Reis gegen Unkräuter und Ungräser

JA   52 113943   17. 3.76/24. 9.77   Mitsubishi

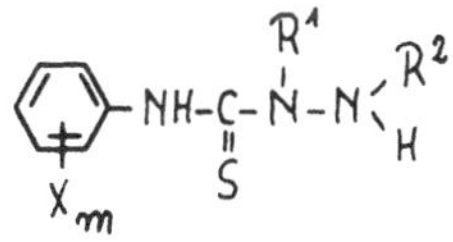

R^1 u. R^2 = Alkyl
X = Halogen, Alkyl, O-Alkyl, CF$_3$ etc.

In Reis gegen Barngard Gras

JA   52 154523   17. 6.76/22.12.77   Kyowa Fermentation

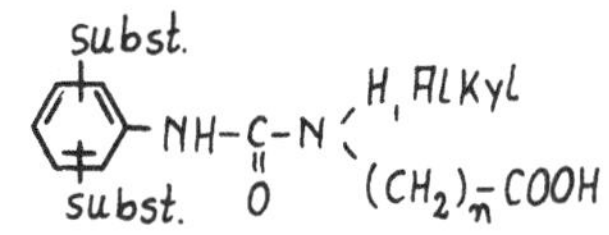

Bekanntes Herbizid, Zusatz von aliphatischen Di- und Polyaminen zur Verminderung von Herbizidschäden

JA   53 018733   5. 8.76/21. 2.78   Kyowa Fermentation

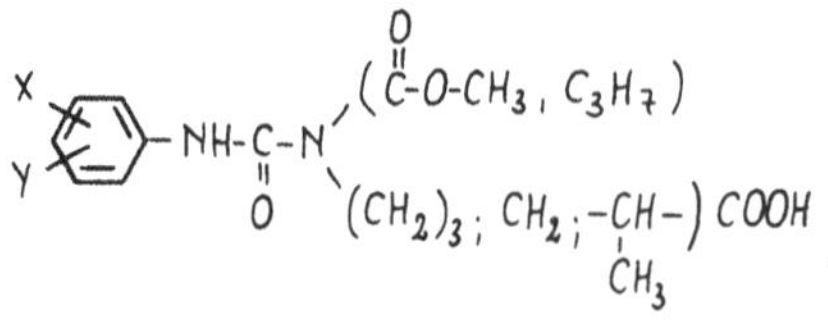

X u. Y = H, Halogen, NO_2,
       CH_3-SO_2, CF_3

Antidot für Harnstoff-Herbizide.

JA   53 029930   27. 8.76/20. 3.78   Ube Industries

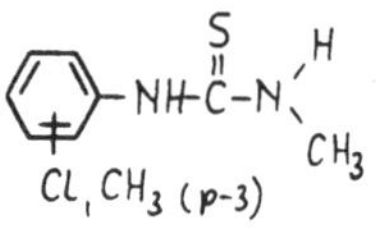

Gegen Barnyard Gras in Reis

JA   53 094035   24. 1.77/17. 8.77   Kyowa Fermentation

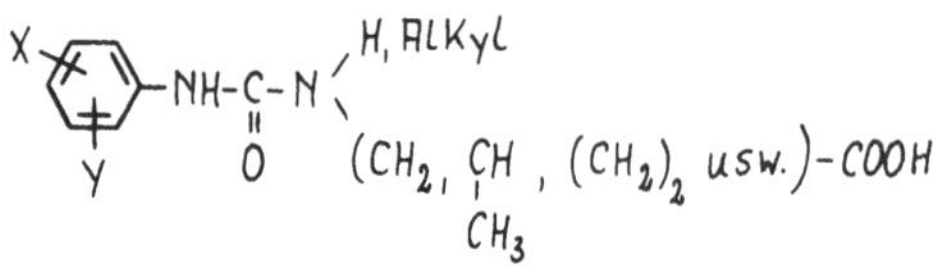

JA   53 104730   18. 2.77/12. 9.78   Kumiai Chem.

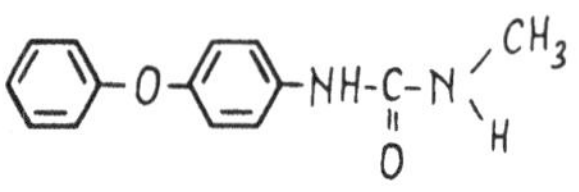

Selektive Anwendung in
Erbsen

JA   53 104731   18. 2.77/12. 9.78   Kumiai Chem.

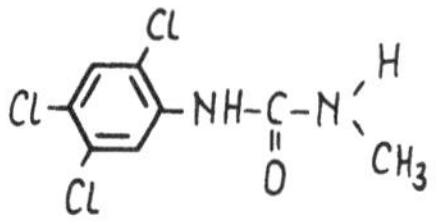

Selektiv post emerg. in
Erbsen

JA   54 014938   5. 7.77/ 3. 2.79   Kumiai Chem.

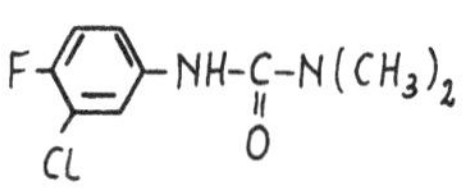

Gegen Gräser in Reis und
Getreide

JA   54 022341   19. 7.77/20. 2.79   Mokko Chem.

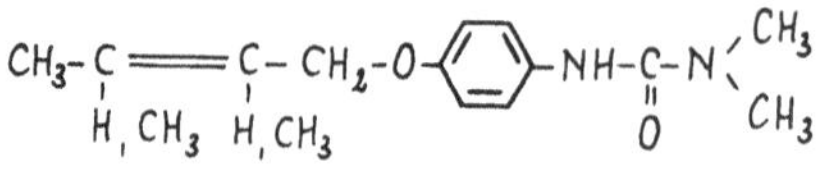

186

JA   54 061149   20.10.77/17. 5.79   Kumiai Chem.

Gräserherbizid in Reis

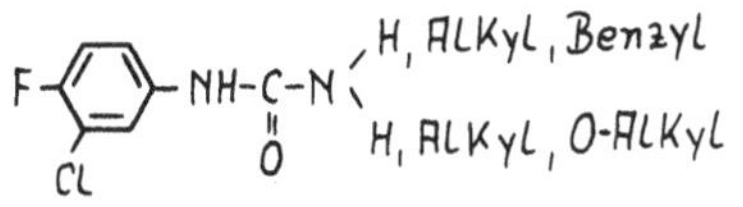

JA   54 098326   20. 1.78/ 3. 8.79   Sumitomo
siehe auch 54 098327 u. 54 098328

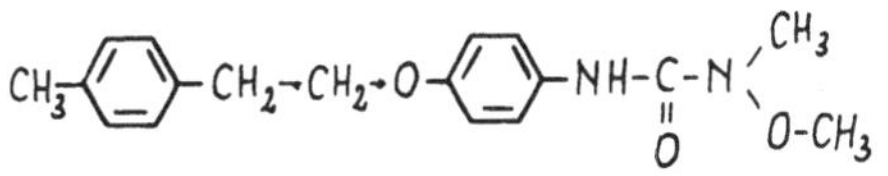

In Kombination mit z.B. Cyclohexanon-herbiziden und anderen Herbiziden.

JA   54 103837   1. 2.78/15. 8.79   Sumitomo

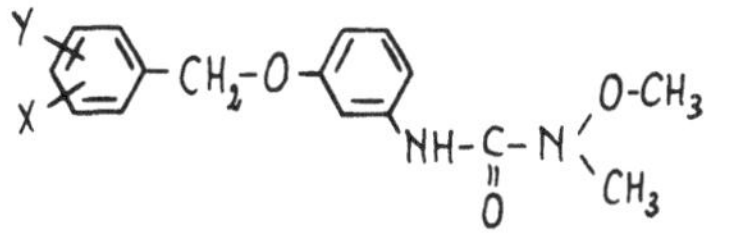

X u. Y = Halog., CN, NO_2,
(O)-Alkyl usw.

Darstellung

JA   54 103838   2. 2.78/15. 8.79   Sumitomo Chem.

Formel wie zuvor. Anwendung als Herbizide.

JA   54 138540   18. 4.78/27.10.79   Sumitomo

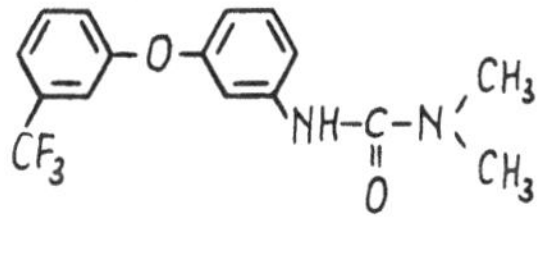

Selektives post emergence
Herbizid in Soyabohnen, Reis,
Baumwolle

JA   55 004354   28. 6.76/12. 1.80   Kumiai Chem.

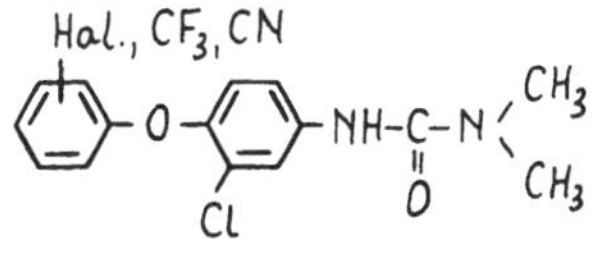

JA   55 053259   17.10.78/18. 4.80   Sumitomo

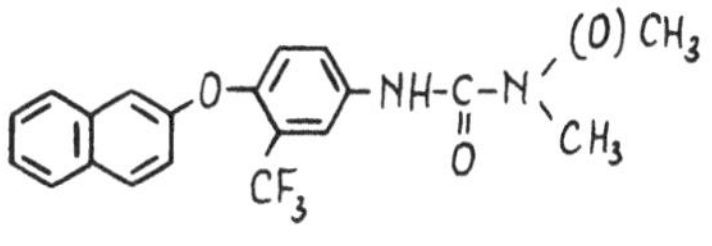

| JA | 55 076855 | 6.12.78/10. 6.80 | Sumitomo |

Selektiv in Reis

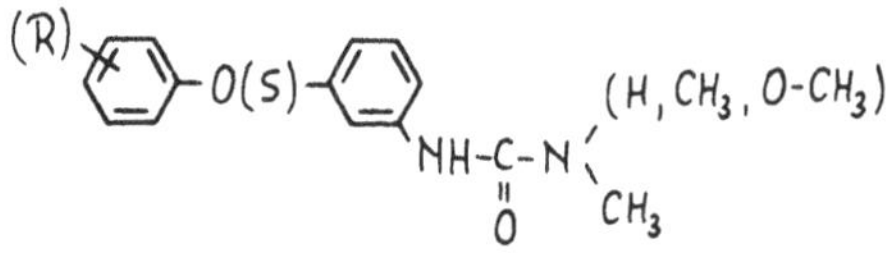

$$\text{(R)} \bigcirc\!\!\!-O(S)-\bigcirc\!\!\!-NH\!-\!\underset{\underset{O}{\|}}{C}\!-\!N\!\!\begin{smallmatrix}(H,CH_3,O\text{-}CH_3)\\CH_3\end{smallmatrix}$$

| JA | 55 102554 | 31. 1.79/ 5. 8.80 | Sumitomo |

$$R\!-\!Z\!-\!Y\!-\!\underset{\underset{X}{|}}{\bigcirc}\!-\!NH\!-\!\underset{\underset{O}{\|}}{C}\!-\!N\!\!\begin{smallmatrix}H,(O)\text{-}CH_3\\CH_3\end{smallmatrix}$$
$$(H, Halog.)$$

R = Cyclohex(en)yl
Z = Alkylen, ev. O enthaltend
Y = aus Pat. nicht ersichtlich!

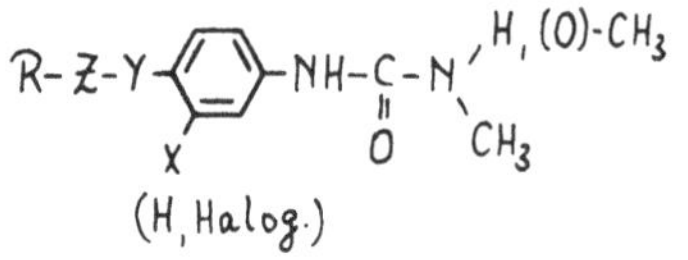

d 2) Aromatisch-aliphatische Harnstoffe am N^2 substituiert entfällt da
durch d1 miterfaßt.

d 3) <u>Aromatisch aliphatische Harnstoffe zusätzlich eine herbizid wirksame Gruppierung enthaltend</u>   Bd. 5/449

Siehe auch z. T. mitbeansprucht bei aromatischen Carbamidsäureestern.
5.5.2 g

| DOS | 2 634 455 | 29. 7.76/ 2. 2.78 | Schering |
| =BE | 857 317 | | |

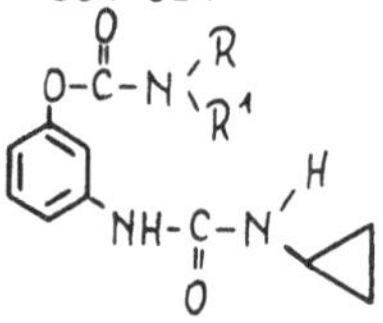

R u. R^1 = H, Alkyl, Cycloalkyl,
Aryl (subst.), Hetero-
cyclus
R^1 u. R^2 = CH_2-Glieder eines
Ringsystems

| DOS | 2 730 325 | 1. 7.77/11. 1.79 | Schering |

R = Alk(en)yl, Cycloalkyl
R^1 = H, CH_3
R^2 = CN-Alkyl, O-Alkyl,
S-Alkyl
R^3 = H, Alkyl, Alkoxy-alkyl
usw.

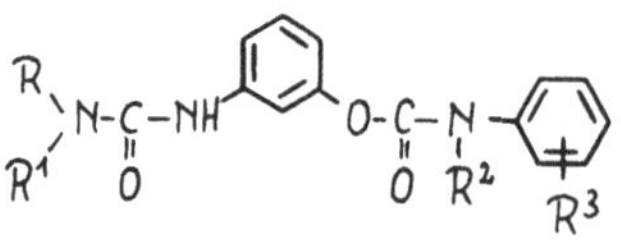

US   4 013 450   2.10.70/22. 3.77   Monsanto
siehe US 3 867 426 (Hauptpatent)

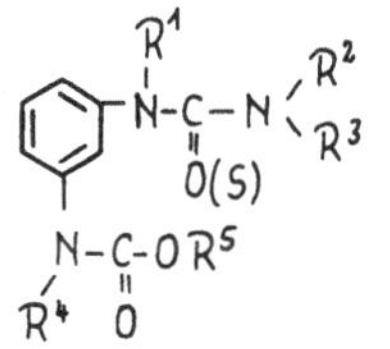

R^1 = H, Alk(en)yl, C_2-C_6
R^2 = Alk(en, in)yl, Cycloalk(en)yl,
      O-Alkyl
Phenyl subst. d. Alkyl, NO_2,
Halogen

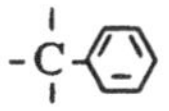

-C-⟨⟩ (substituiert;

-C-C-⟨⟩ (substituiert

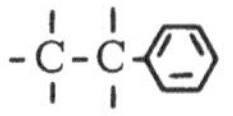

R^4 = H, Alk(en)yl; R^5 = Alk(en)yl, Cycloalk(en)yl, subst. Phenyl

US   4 046 808   24. 8.72/ 6. 9.77   Amer. Cyanamid

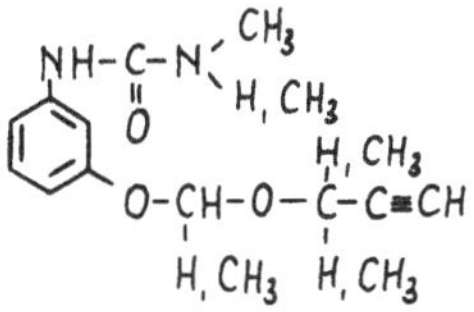

FR   2 319 628   31. 7.75/ 1. 4.77   Roussel Uclaf

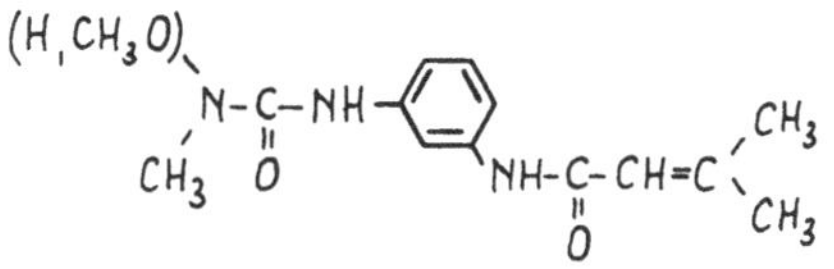

Selektiv in Getreide

d 4) Aromatisch-aliphatische N^1 oder N^2-Oxyharnstoffe und N-Thioharnstoffe

Bd. 5/455

DOS   2 711 230   15. 3.77/29. 9.77   Sumitomo
JA Prior.   16. 3.76

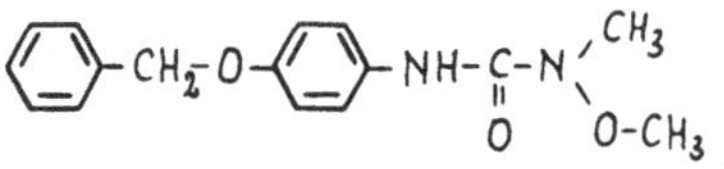

Gegen Gräser und Unkräuter
selektiv in Sojabohnen

DOS   2 739 349   1. 9.77/15. 3.79   BASF
=EP   1 225   28. 8.78/ 1. 9.77   BASF

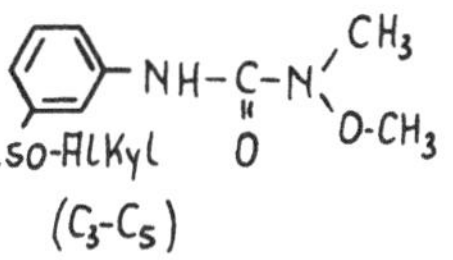

Selektiv in Getreide gegen
Unkräuter und Ungräser

DOS   2 744 169   30. 9.77/ 6. 4.78   Sumitomo
JA Prior.   30. 9.76

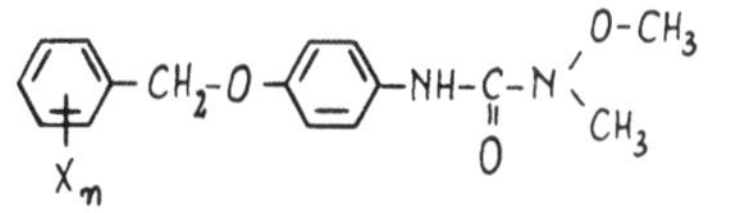

X = Halogen, bes. F, CN,
    CF_3, (O)-Alkyl
n = 1-5

US   4 200 450   16. 3.71/26. 6.78   Chevron Res.

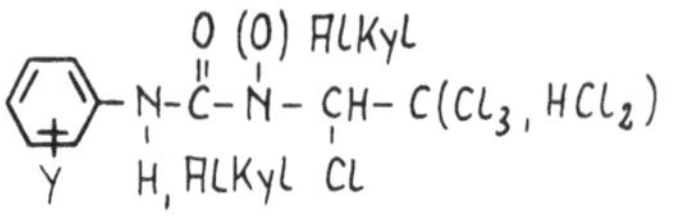

Y = Halog., (O)-Alkyl, CF_3

Gegen Unkräuter und Gräser

BE   846 029   3. 3.77/16. 8.78   Sumitomo

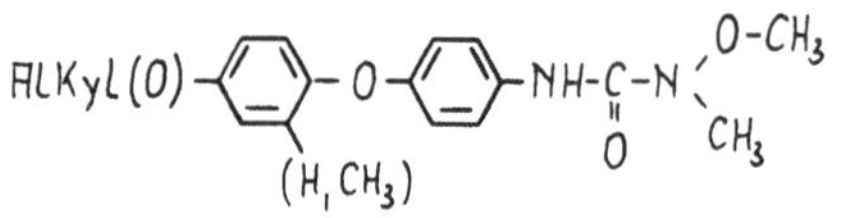

JA   53 112845   14. 3.77/ 2.10.78   Chugai Pharmac.

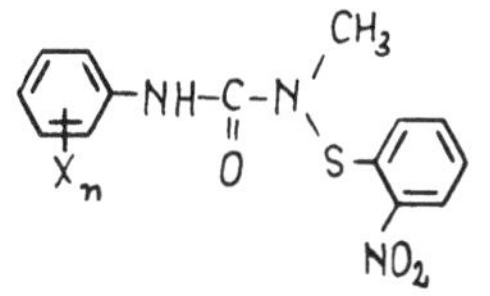

X = Halog, Alkyl, O-Alkyl,
    CF_3, NO_2
n = 1-3

JA   54 151947   19. 5.78/29.11.79   Sumitomo

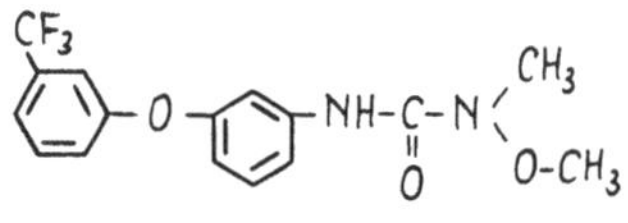

Gegen Gräser und Unkräuter

190

e) <u>N-Acyl,-Sulfonyl-Harnstoffe, Acylimide, Biurete, Iso-Harnstoff-Ether</u>   Bd. 5/460

Aliphatische und aromatische Vertreter

DOS   2 558 839   27.12.75/22. 7.76   Sandoz
CH Prior.   15. 1.75 u. 27.11.75

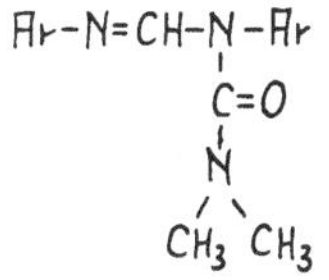

$$Ar-N=CH-N-Ar$$
$$|$$
$$C=O$$
$$|$$
$$N$$
$$/ \backslash$$
$$CH_3 \quad CH_3$$

Ar = subst. Phenyl

Selektiv in Kartoffelkulturen
Mais gegen Unkräuter und
einige Ungräser.

DOS   2 610 804   17. 3.75/30. 9.76   Stauffer Chem.

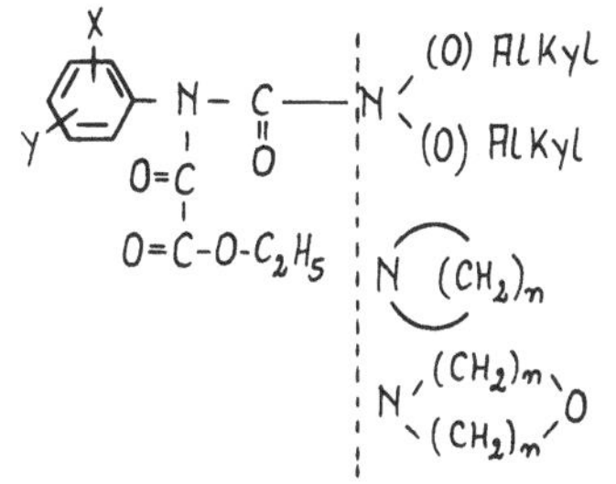

X u. Y = Cl, Br, CF_3

DOS   2 612 080   22. 3.76/14.10.76   Roussel Uclaf
FR Prior.   26. 3.75
FR   2 341 567
BE   840 000

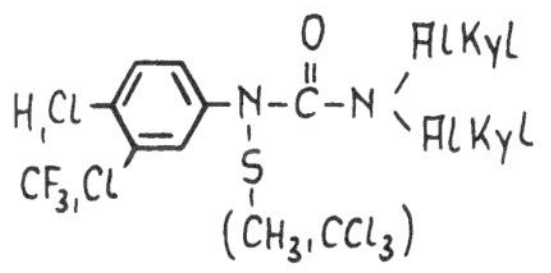

DOS   2 644 504   1.10.76/ 2. 6.77   Stauffer Chem.

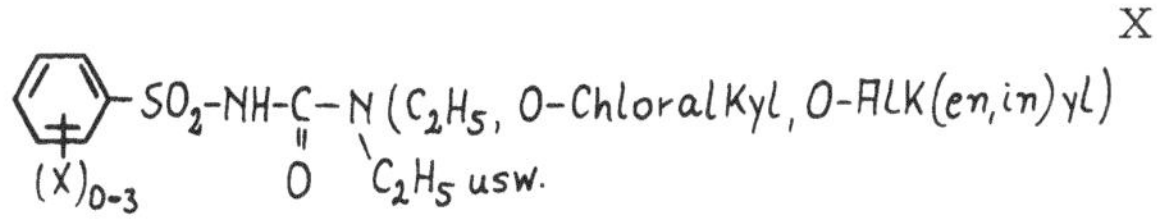

X = H, Cl, Br, CF_3, $O-CH_3$,
$CH_2-N-C-CF_3$
$\quad\ \ H \ \overset{\|}{O}$

DOS   2 720 161   5. 5.77/18.11.78   Chem. Fabrik Kalk

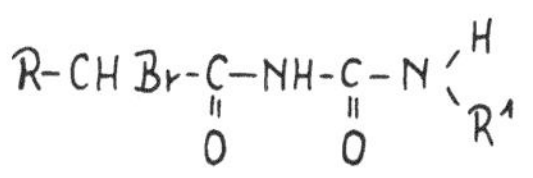

$$R-CH\,Br-C-NH-C-N \begin{smallmatrix} H \\ \diagdown R^1 \end{smallmatrix}$$
$$\quad\quad\quad \overset{\|}{O}\quad\ \overset{\|}{O}$$

R u. R^1 = H, Alkyl

DOS   2 727 545   18. 6.77/ 4. 1.79   Celamerk

X u. Y = H, Alkyl, O-Alkyl,
        Cl, Br, CF$_3$

DOS   2 926 480   30. 6.79/24. 1.80   Duphar Internat. Res.

US   3 996 275   7. 7.71/ 7.12.76   Monsanto
Siehe auch
US   3 860 634, 3 903 153 u. 3 956 384

X = Alkanoyl, NH$_2$
R^1 u. R^2 = Alk(en, in)yl oder
                C$_6$H$_5$-CH$_2$
R^3 u. R^4 = Alkyl, C$_6$H$_5$-CH$_2$
                subst. d. Halog.

US   3 997 342   9.11.70/14.12.76   Chevron Res. Co.
Siehe auch
US   3 928 407 u. 3 853 966

R, R^1 u. R^2 = H, Alkyl,
                Cycloalkyl, Aryl
R^3 = Alkyl, Aryl

US   4 002 629   11. 4.73/11. 1.77   Hercules

R^1 = H, Alkyl

R^2 =

R^3 = Alkyl, Allyl, Phenyl (subst.)
R^4 = Alkyl, Allyl, Cyclohexyl,
       Phenyl (subst.)

US   4 004 915   7.12.72/25. 1.77   Du Pont
Siehe auch
US   3 823 179 u. 3 882 160 (N-Carbamoylamidinocarbamate)

R^1 = subst. Alkyl, z. B. durch
       O(S)Alkyl, Alk(en, in)yl
R^2 = Alkyl; R^3 = H, CH$_3$, C$_2$H$_5$
R^4 = Alk(en, in)yl; R^5 = O-Alkyl

192

(Struktur: Phenylring subst. durch Cl, Br, CF₃ mit $O=C-\overset{\overset{O}{\|}}{C}-O-C_2H_5$ und $N-\overset{\overset{O}{\|}}{C}-N<\overset{ALKyl}{(O)\,ALKyl}$)

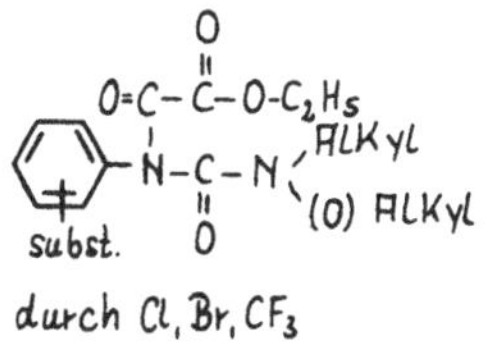

subst. durch Cl, Br, CF₃

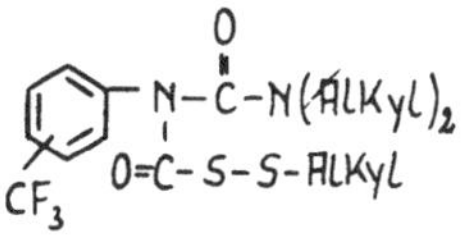

(Struktur: Phenylring mit CF₃, $N-\overset{\overset{O}{\|}}{C}-N(ALKyl)_2$ und $O=C-S-S-ALKyl$)

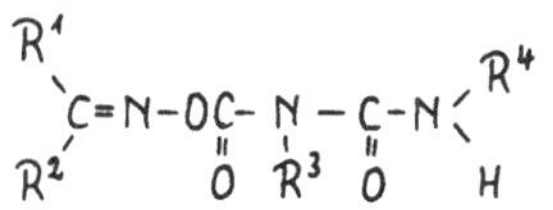

R^4 = Alkyl, Allyl, Cyclohexyl, subst. Phenyl

(Struktur: $\overset{R^1}{\underset{R^2}{}}C=N-O\overset{}{C}-N-\overset{}{C}-N<\overset{R^4}{H}$, mit O, R^3, O)

R^1 = H, Alkyl
R^2 = Alkyl, Phenyl, 2-Furanyl
R^1+R^2 = zus. Cycloalkyl
R^3 = Alkyl, Allyl, subst. Phenyl

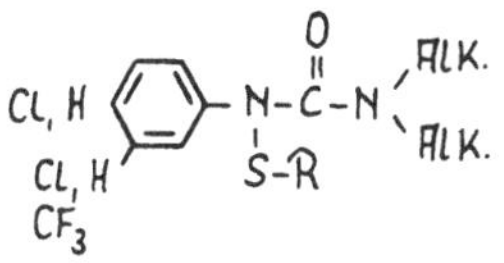

R = C_1-C_6 Alkyl subst. d. Cl
z.B. CCl_3

(Struktur: Phenylring mit Cl, H; Cl, H; CF₃ und $N-\overset{\overset{O}{\|}}{C}-N<\overset{ALK.}{ALK.}$, $S-R$)

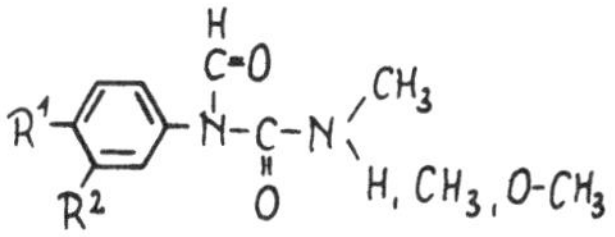

R^1 = $CH<\overset{CH_3}{CH_3}$, F, Cl, Br,
CF₃ usw.
R^2 = Alkyl, Halog., CF₃, O-Alkyl

(Struktur: Phenylring mit R^1, R^2, $\overset{H}{C=O}$ und $N-\overset{\overset{O}{\|}}{C}-N<\overset{CH_3}{H, CH_3, O-CH_3}$)

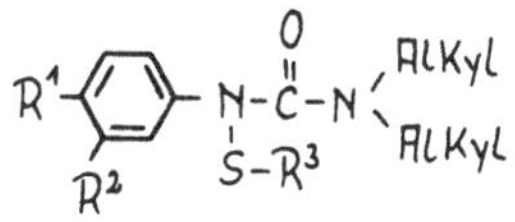

R^1 u. R^2 = H, Cl, CF₃ (1 x)
R^3 = Alkyl, Halog.-alkyl

pre u. post emergence Herb.
gegen Gräser u. Unkräuter
einschl. Flughafer in Getreide.

(Struktur: Phenylring mit R^1, R^2 und $N-\overset{\overset{O}{\|}}{C}-N<\overset{ALKyl}{ALKyl}$, $S-R^3$)

JA   51 136825   22. 5.75/26.11.76   Nihon Tokushu

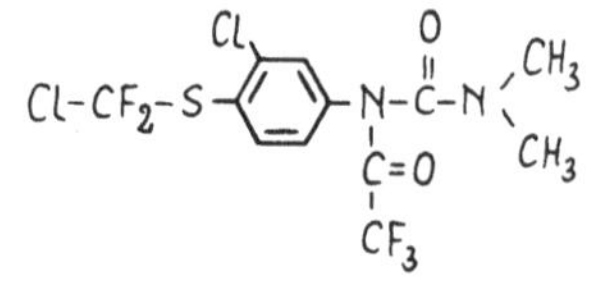

$$Cl\text{-}CF_2\text{-}S\text{—}\underset{\underset{CF_3}{|}}{\underset{C=O}{|}}\overset{}{N}\text{-}\overset{\overset{O}{\|}}{C}\text{-}N\overset{CH_3}{\underset{CH_3}{<}}$$

f) <u>Semicarbazide, Carbodihydrazide und die entsprechenden Thio-</u>
<u>verbindungen</u>   Bd.5/466

Aliphatische und aromatische Vertreter

EP   9 324   14. 9.78/ 2. 4.80   Gulf Oil
=BE   878 668   14. 9.78/ 7. 3.80   Gulf Oil

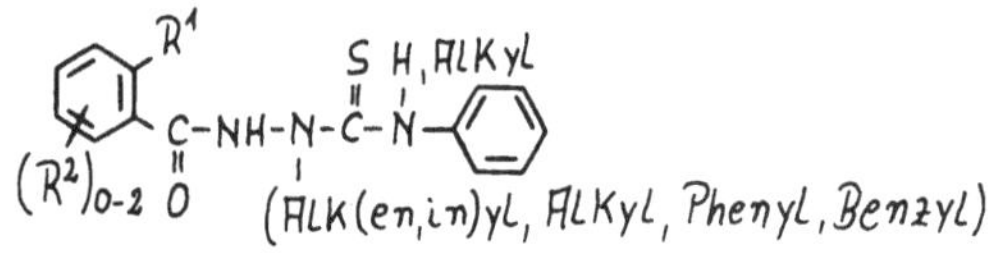

R^1 = C-OH, C-N$<$, C-Alkyl (each with $\|$ O)

R^2 = Alkyl, NO_2, CF_3, Halog.,

US   4 099 956   25. 8.72/11. 7.78   Shell Oil Co.

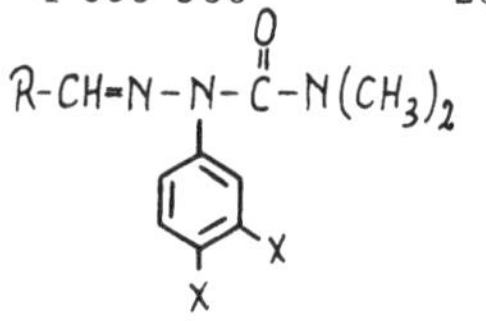

$$R\text{-}CH\text{=}N\text{-}N\text{-}\overset{\overset{O}{\|}}{C}\text{-}N(CH_3)_2$$

R = H, Alk(en)yl
X = Halog., O-Alkyl, CF_3 usw.

BE   847 331   15.10.75/15. 4.77   Bayer

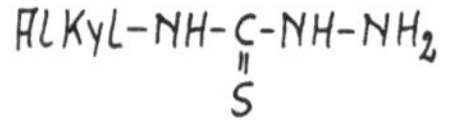

$$Alkyl\text{-}NH\text{-}\underset{\underset{S}{\|}}{C}\text{-}NH\text{-}NH_2$$

Darstellungsverfahren

Zwischenprodukt f. Herbizide.

GB   2 011 904   9. 1.78/18. 7.79   Shell Int. Res.

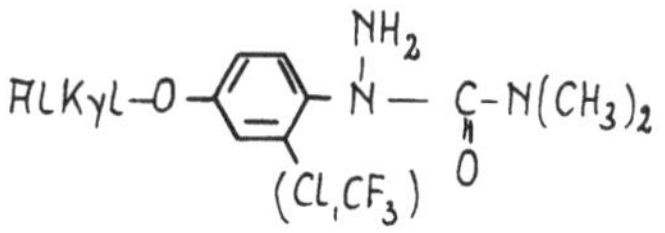

$$Alkyl\text{-}O\text{—}\underset{(Cl,CF_3)}{}\text{—}\overset{\overset{NH_2}{|}}{N}\text{—}\underset{\underset{O}{\|}}{C}\text{-}N(CH_3)_2$$

f 1) Aromatisch-aromatische Harnstoffe (neu), z.T. schon miterfaßt

SU 502 628 4. 1.71/30.10.76 As Uab Veg. Mater

Selektiv in Baumwolle und Getreide.

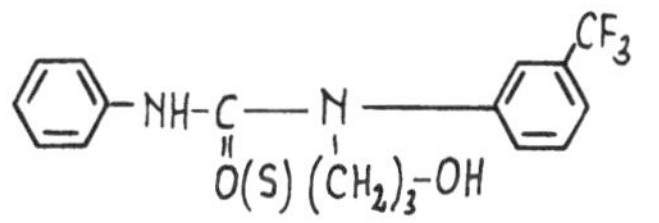

JA 53 046937 7.10.76/27. 4.78 Ube Industr.

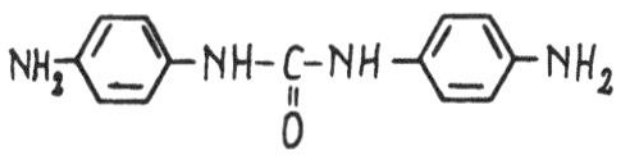

g) Aromatisch-heterocyclische Harnstoffe mit N^1 als Glied eines Ringsystems, z.T. unter d 1 und 5.12.1 a miterfaßt Bd.5/467

(Nicht unter g) fallend: ArNHC(=O)-NH-Heterocycl.

DOS 2 532 645 22. 7.75/10. 2.77 Consort. f. elektrochem. Industrie

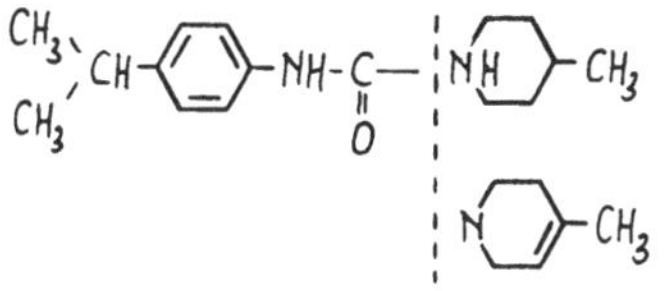

EP 10 770 2.11.79/14. 5.80 Shudo et al.

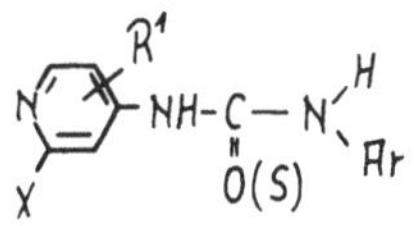

X = Halog., OH, O(S)-Alkyl, NH_2, CF_3, R-C(=O)-NH

R^1 = H, Halog., Alkyl

US 4 065 291 18. 6.76/27.12.77 Chevron Res.

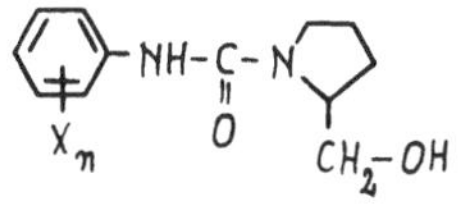

X = H, Halog., CCl_3, CF_3, CN, (O)-Alkyl

n = 1-3

US   4 220 465   1. 2.77/ 2. 9.80   Du Pont

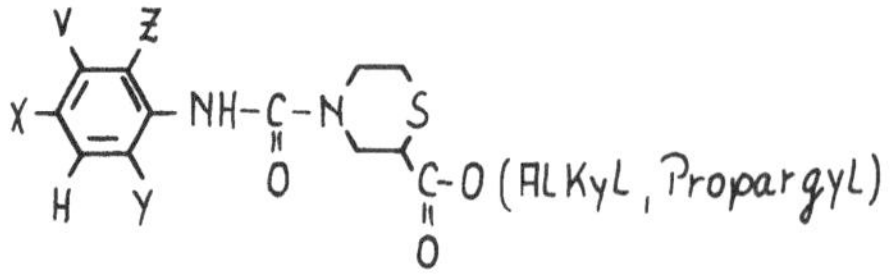

$X = H, Halog, CN, (O)-CH_3, NO_2$
$Y = H, Halog., CH_3$
$Z = H, F, CCH_3,$
      $F, Cl, O-CH_3$
$V = H, F, Cl, O-Alkyl$

JA   52 061225   11.11.75/20. 5.77   Sumitomo

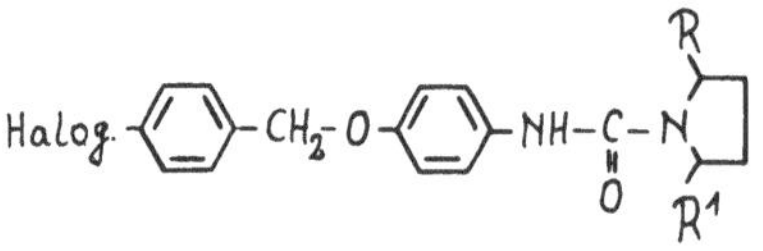

R u. R^1 = H oder Methyl

JA   53 065 879   24.11.76/12. 6.78   Nippon Soda

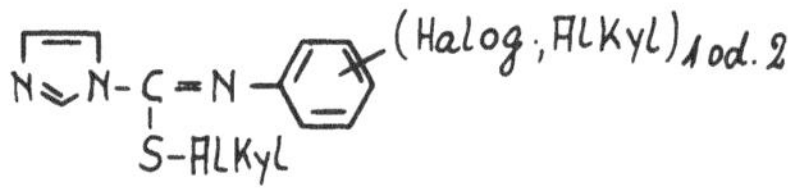

Selektiv in Reis

h)   <u>Aromatisch-aliphatische Harnstoffe mit einem heterocyclischen Substituenten am Aromaten</u>   Bd.5/470

US   3 991 068   22. 1.74/ 9.11.76   Dow Chem.

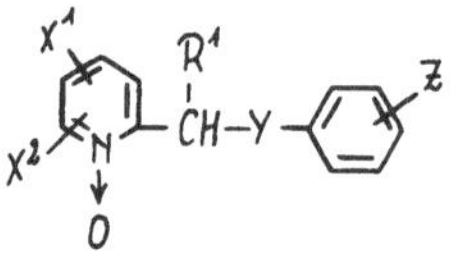

X^1 u. X^2 = Cl, Br, F, CF_3,
                    NH_2, CN, CCl_3 usw.
R^1 = H, CH_3
$Y = O, S, SO_2$
$Z = NH-C-N(Alkyl)_2$
        ‖
        O (S)

US   4 003 906   22. 1.74/18. 1.77   Dow Chem.

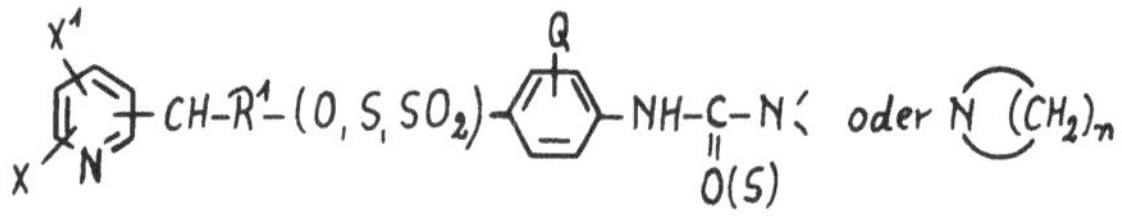

$X = F, Cl, Br$
$X^1 = CCl_3, NH_2, CF_3, CN,$
        $O(S)CH_3$ usw.
R^1 = H, CH_3
$Q = Alkyl, Halogen$

196

US   4 028 092   22. 1.74/ 7. 6.77   Dow Chem.

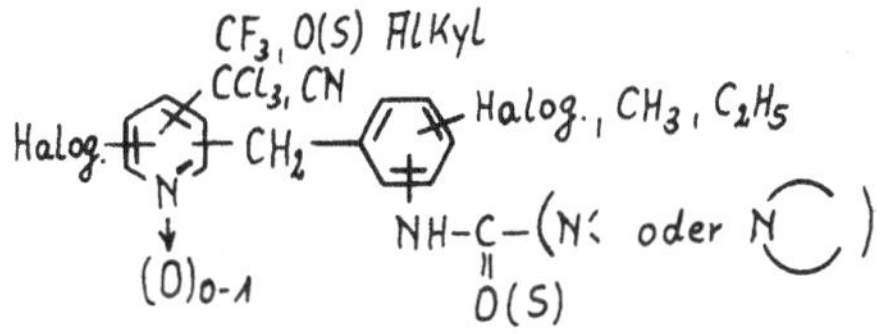

$CF_3, O(S)$ Alkyl
CCl_3, CN
Halog. — CH$_2$ — Halog., CH_3, C_2H_5
N
NH-C-(N< oder N◯)
(O)$_{0-1}$
O(S)

US   4 226 612   2. 7.79/ 7.10.80   Shell Oil Co.

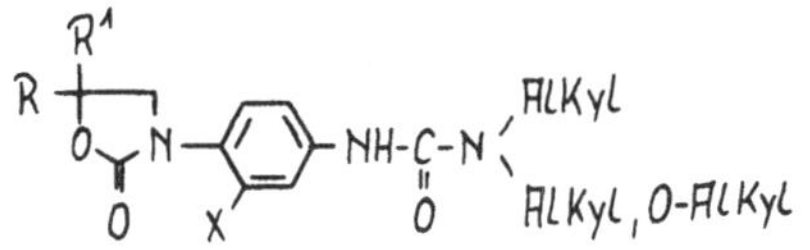

R u. R^1 = H, Alkyl usw.

R^1
R
O   N — NH-C-N< Alkyl
O   X   O   Alkyl, O-Alkyl

<u>i)   Heterocyclisch-aliphatische Harnstoffe</u>   Bd.5/471

"Überbrückte" Harnstoffe werden hier und bei den entsprechenden
heterocyclischen Verbindungen erfaßt.

<u>i 1) 5-Ringe mit 1 x O oder S</u>

US   4 156 670   25. 1.74/29. 5.79   American Cyanamid

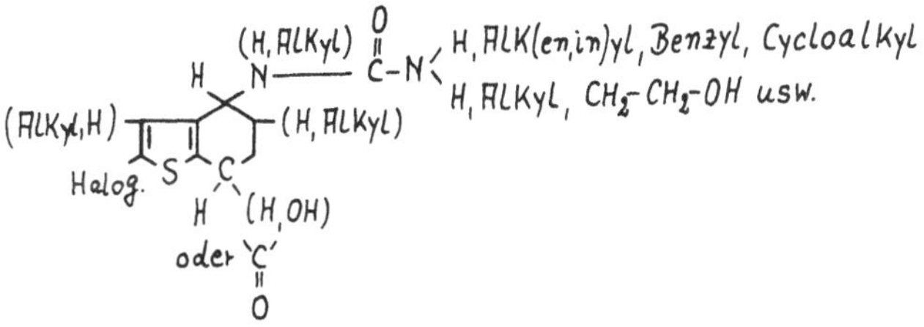

(H,Alkyl) O
H   N——— C-N< H, Alk(en,in)yl, Benzyl, Cycloalkyl
(Alkyl,H) — (H, Alkyl)   H, Alkyl, CH_2-CH_2-OH usw.
Halog. S   C
H   (H,OH)
oder C
O

US   4 154 740   25. 1.74/15. 5.79   American Cyanamid

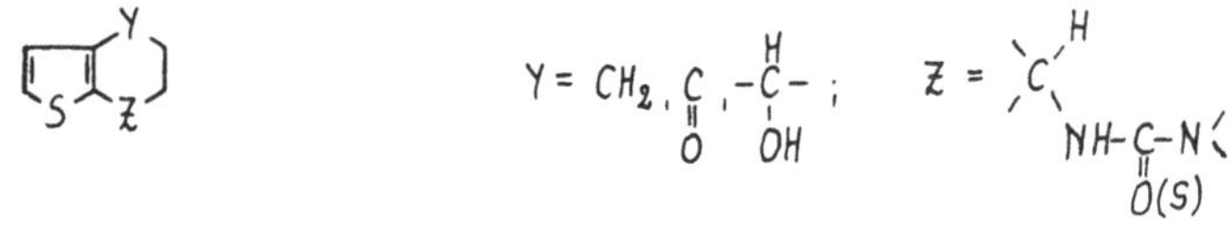

Y   H   H
C   Y = CH$_2$, C, -C- ;   Z = C
S   Z   O   OH   NH-C-N<
O(S)

BE   866 581   28. 4.77/14. 8.78   Shionogi

Spezielle Darstellung bekannter
heterocycl. Harnstoffe

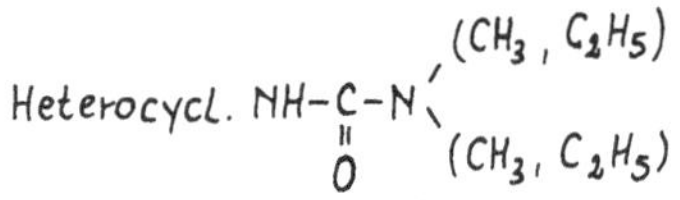

(CH$_3$, C_2H_5)
Heterocycl. NH-C-N<
O   (CH$_3$, C_2H_5)

EP   2 881   12.10.78/11. 7.79   Fisons
GB Prior.   2.11.77

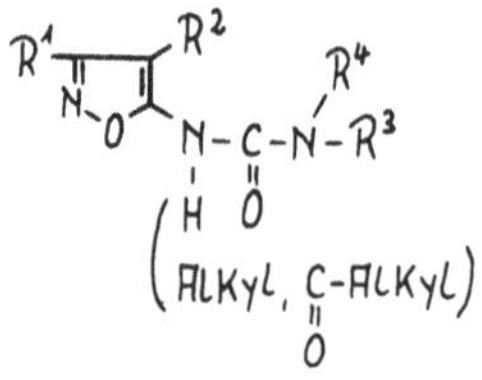

R^1 = H, Halog., NO_2, NH_2, Alkyl, Aryl,
Aralkyl, $\overset{\text{O}}{\overset{\|}{C}}$-O-Alkyl, O-Alkyl

R^2 = CN, NH_2, $\overset{\text{O}}{\overset{\|}{C}}$-OH, $\overset{\text{O}}{\overset{\|}{C}}$-N$\langle$, $\overset{\text{O}}{\overset{\|}{C}}$-N$(CH_2)_n$

R^3 = H, (Cyclo)-Alkyl
R^4 = H, (Cyclo)-Alkyl, O-Alkyl, Aryl
R^3+R^4 = $(CH_2)_n$-Glieder eines Ringes

US   4 212 981   23. 7.74/15. 7.80   Shionogi

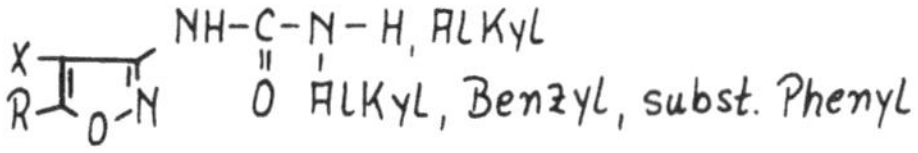

N Glied eines Heterocyclus z.B.
Morpholino usw.

X = H, Halog.
R = H, Halog., Phenyl
R + X = 3-5-Alkylen

JA   53 086033   00.00.77/29. 7.78   Shionogi

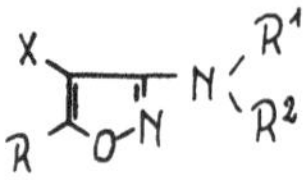

R^1 = H, (Cyclo)-Alkyl, Aryl, Aralkyl,
Alk(en, in)yl
R^2 = H, Alkyl, $\overset{\text{O}}{\overset{\|}{C}}$-N$\langle$, $\overset{\text{O}}{\overset{\|}{C}}$-Alkyl, $\overset{\text{O}}{\overset{\|}{C}}$-N$)$

X = H, Alkyl
R = H, Alkyl, Cycloalkyl, Aryl
X u. R = Glieder eines Ringes (CH=CH)

JA   54 046779   19. 9.77/12. 4.79   Shionogi

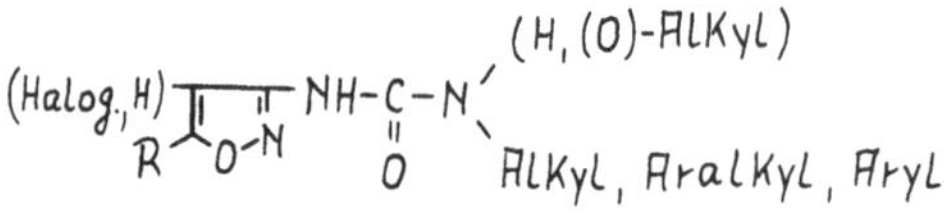

R = H, Alkyl, Aryl

JA   54 059272   00.00.77/12. 5.79   Shionogi

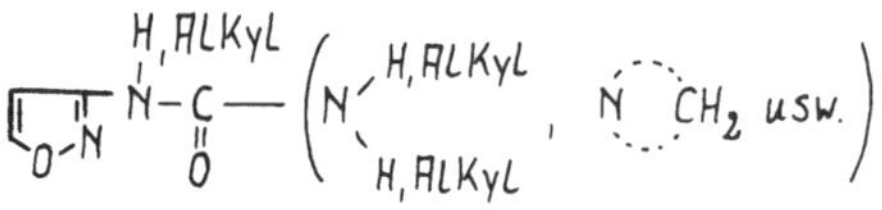

198

DOS   2 724 614   1. 6.77/15.12.77   Velsicol Chem.
=US   4 097 485
US Prior.   1. 6.76

X = Br, Alkyl-SO$_2$

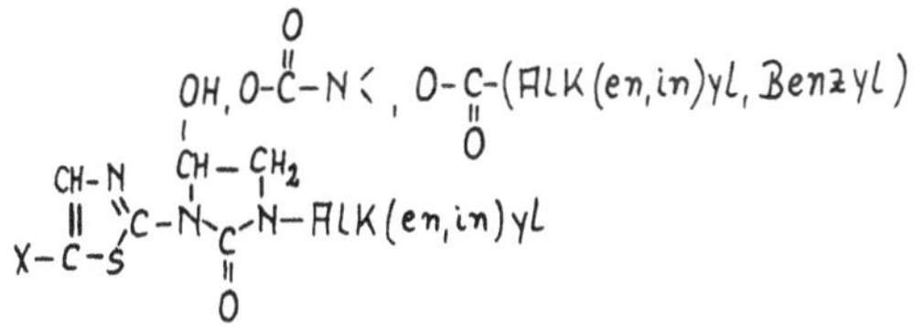

[Strukturformel]

DOS   2 818 947   28. 4.78/ 2.11.78   Shionogi

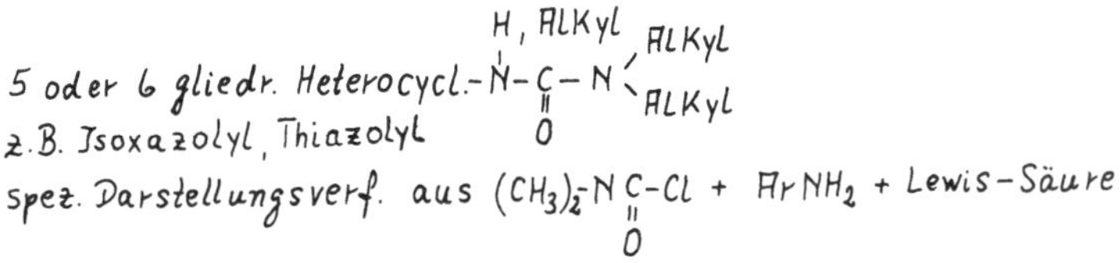

5 oder 6 gliedr. Heterocycl.-N-C-N (H, Alkyl / Alkyl, Alkyl) , mit =O

spez. Darstellungsverf. aus (CH$_3$)$_2$N-C-Cl + ArNH$_2$ + Lewis–Säure, mit =O

US   4 086 240   1. 6.76/25. 4.78   Velsicol Chem.

R = F, Cl, Alkyl-SO$_2$

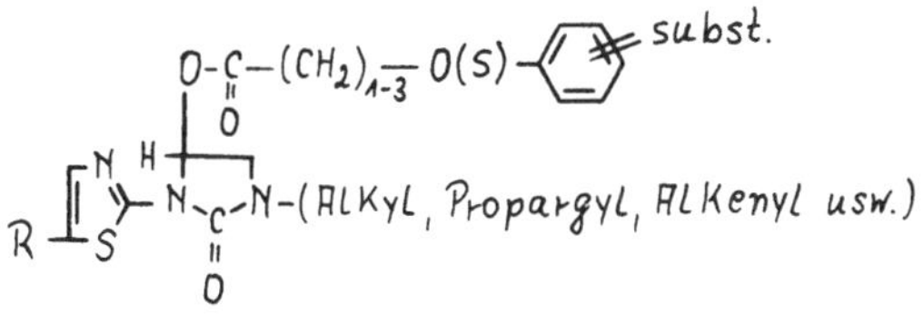

[Strukturformel] O-C-(CH$_2$)$_{1-3}$ O(S)-〈 〉 subst. ; -N-C-N-(Alkyl, Propargyl, Alkenyl usw.)

US   4 097 485   17. 6.76/28. 6.78   Velsicol Chem.

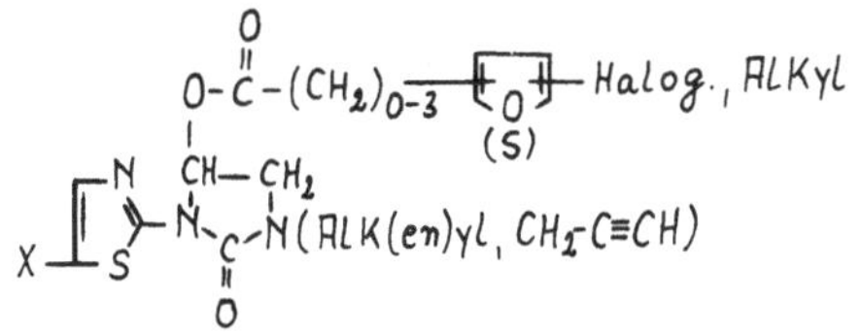

[Strukturformel] O-C-(CH$_2$)$_{0-3}$ O(S) Halog., Alkyl ; -N-C-N(Alk(en)yl, CH$_2$-C≡CH)

X = Cl, Br, F, Alkyl-SO$_2$

SU   509 583   30. 9.74/ 4. 6.76   Melnikov NN

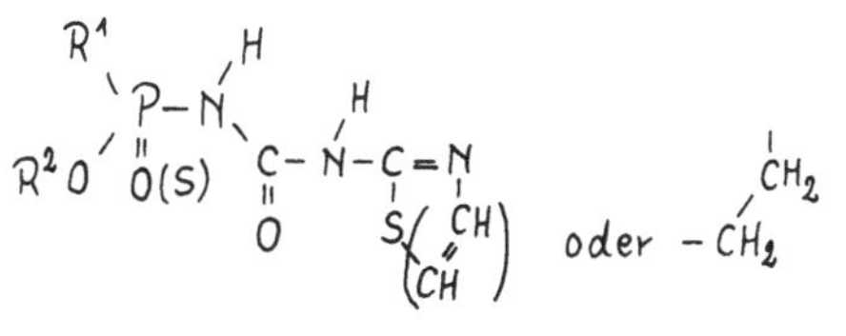

[Strukturformel]

i 3.3) <u>Benzthiazolyl-Harnstoffe</u>

DOS   2 527 394   19. 6.75/30.12.76   Bayer

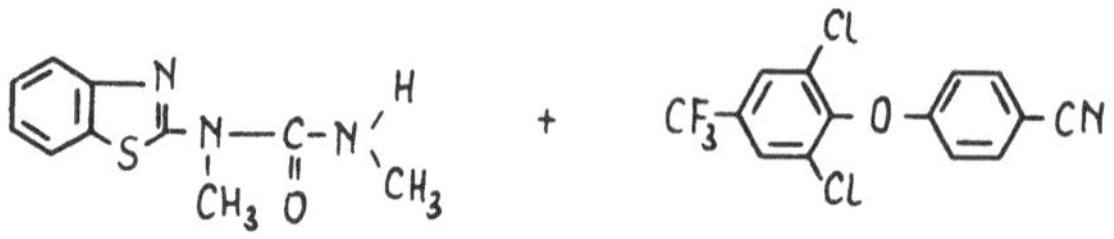

Selektiv in Getreide.

DOS   2 624 823   31. 5.76/15.12.77   Schering

B^+ = Na, K, Li

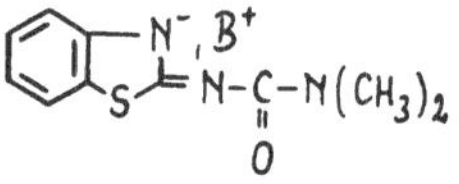

DOS   2 644 425   30. 9.76/ 6. 4.78   Schering

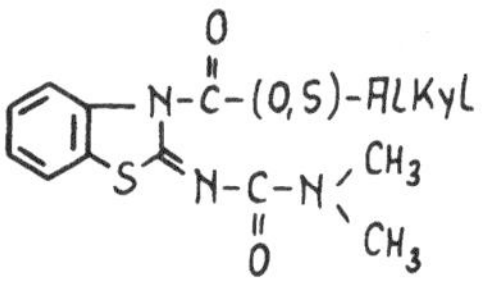

DOS   2 722 949   20. 5.77/ 1.12.77   Velsicol Chem.
=US   4 054 574 u. 4 055 569
US Prior. 20. u. 24.5.76
ähnlich BE 882 010 v.2.3.79/16.6.80

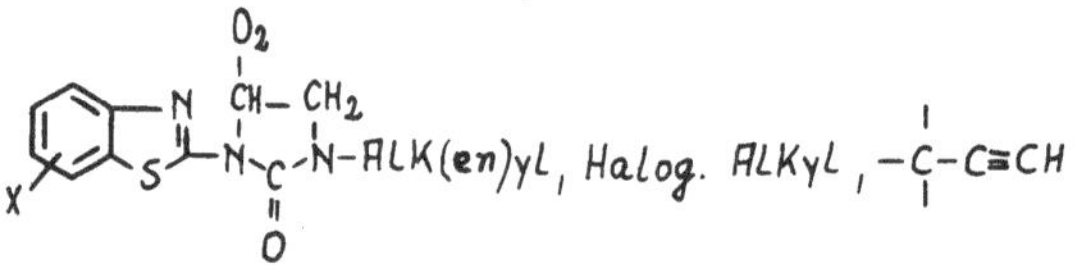

Q = OH, N$\langle$, O-C-Alk(en, in)yl, CH$_2$-

X = Alkyl, Halogen, O-Alkyl

200

DOS   3 007 556   28. 2.80/11. 9.80   Velsicol Chem.
US Prior.   2. 5.79

$(X)_n$—[ring]—N=C(S)—N—C(OH)H—C(=O)—N(Alk(en)yl, Halog.-Alkyl)

X = (O)-Alkyl, S-Alkyl, CN, N(Alkyl)(Alkyl), CN usw.; n = 0-4

US   4 029 491   29. 4.76/14. 6.77   Velsicol Chem.

X—[ring]—N=C(S)—N(—C(=O)—N-R)—CH(OH)—CH₂

X = Alkyl, Halog.-Alkyl, O-Alkyl, Halog., CF₃

R = Alk(en)yl, Halog.-Alkyl, CH≡C-C(H,Alkyl)(H,Alkyl)

US   4 045 446   17. 6.76/30. 8.77   Velsicol Chem.

subst.—[ring]—N=C(S)—N(—C(=O)—N Alk(en,in)yl)—CH(—O-C(=O)—[pyridyl]—subst.)—CH₂

US   4 086 241   14. 6.76/25. 4.78   Velsicol Chem.

subst.—[ring]—N=C(S)—N(H)—N(—C(=O)—N Alkyl)—O-C(=O)-(CH₂)₀₋₃—[ring (S)]—subst.

BE   855 217   31. 5.76/30.11.77   Schering

[[benzothiazolium N⁺]—N-C(=O)-N(CH₃)₂] Alkali⁺

Zerfällt in Wasser in bekannte und patentierte Herbizide.

DOS   2 847 449   2.11.78/16. 5.79   PPG Ind.
=BE   871 762   4.11.77/ 3. 5.79   PPG Ind.
US Prior.   4.11.77

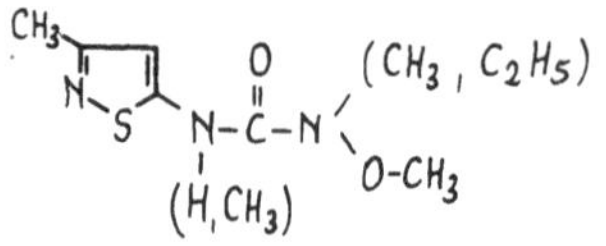

US   4 032 321   18. 6.76/28. 6.77   FMC

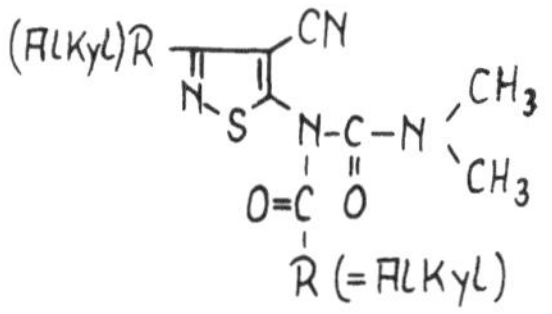

R = z.B. $CH(CH_3)_2$

US   4 057 415   18. 6.76/ 8.11.77   FMC

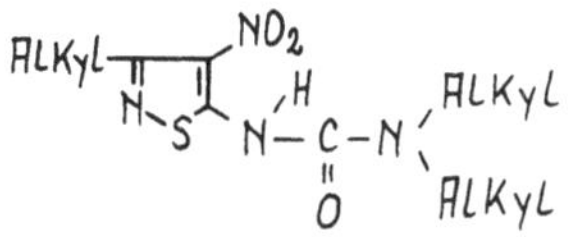

pre and post-emergence
Herbizide, selektiv in Mais

US   4 057 416   18. 6.76/ 8.11.77   FMC

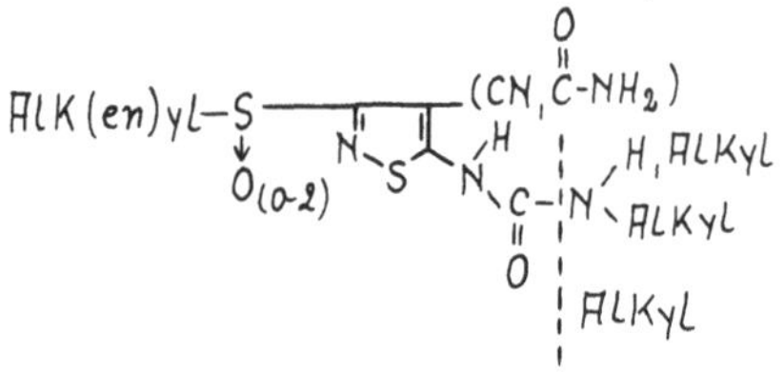

selektive in Mais

US   4 059 433   18. 6.76/22.11.77   FMC

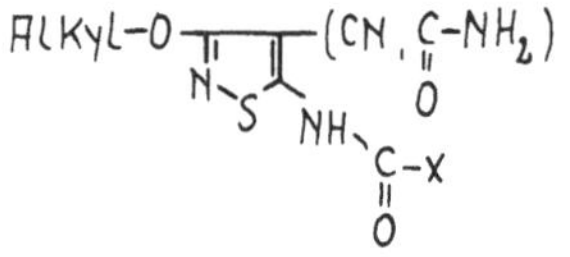

X = Alkyl, $N<$

US   4 075 001   18. 6.76/21. 2.78   FMC

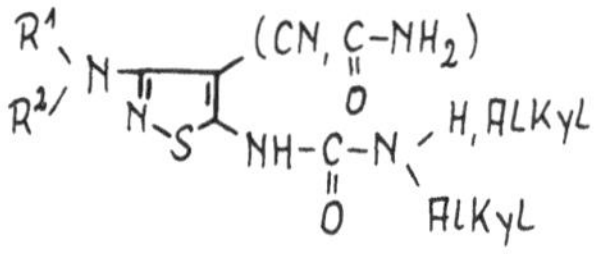

R^1 u. R^2 = Alkyl, $(CH_2)_4$

US   4 152 516   9. 2.78/ 1. 5.79   PPG Ind.

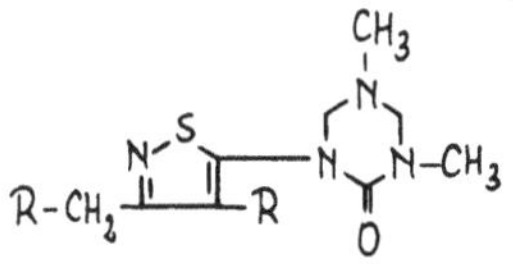

R = H, Alkyl, Phenyl (subst.)

i 3.5) <u>5-Ringe mit 2 x N + 1 x O oder S</u>

i 3.5.1) <u>1,2,3-Thiadiazolyl-Harnstoffe</u>   Bd.5/477

DOS   2 506 690   14. 2.75/ 2. 9.76   Schering

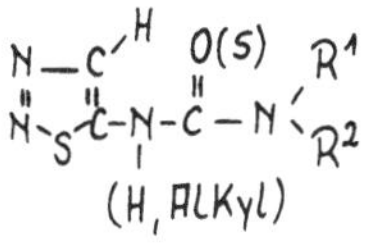

R^1 = H, (O,S-Alkyl)
R^2 = Alkyl, (O,S)-Alkyl,
        Cycloaliphat, subst.
        Phenyl
R^1 u. R^2 = Rest des Piperidins,
        Morpholins usw.

Entblätterungsmittel.

DOS   2 619 861   3. 5.76/24.11.77   Schering

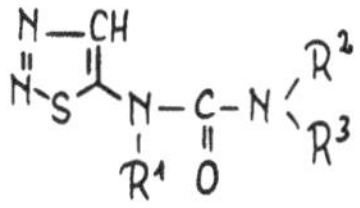

R^1 = H, Alkyl
R^2 = H, Alkyl(ev.durch O oder
        S unterbrochen)
R^3 = heterocycl. Rest Pyridyl

Wuchsverzögerer u. Entblätterungsmittel.

DOS   2 646 712   14.10.76/20. 4.78   Schering

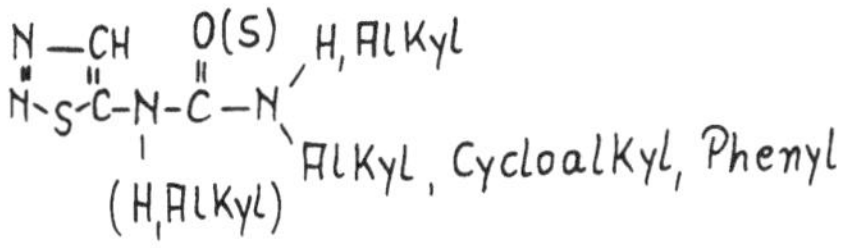

Wuchsstoffe und Entblätterungsmittel.

DOS 2 716 324   7. 4.77/12.10.78   Schering
=BE   865 805

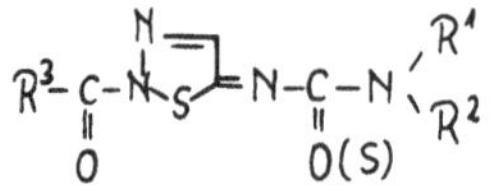

$$R^3-\underset{O}{\underset{\|}{C}}-N-\overset{N}{\underset{S}{\underset{\|}{\ }}}=N-\underset{O(S)}{\underset{\|}{C}}-N\underset{R^2}{\overset{R^1}{<}}$$

R^1 = H, Alkyl usw.
R^2 = Alkyl, Cycloalkyl,
    Aryl (subst.)
R^1 u. R^2 = Glieder des Morpho-
    lino-, Piperidino usw.
    Restes
R^3 = H, Alkyl, Cycloalkyl,
    Aryl, N

DOS 2 719 810   28. 4.77/ 2.11.78   Schering
=BE   866 556   28. 4.77/30.10.78   Schering

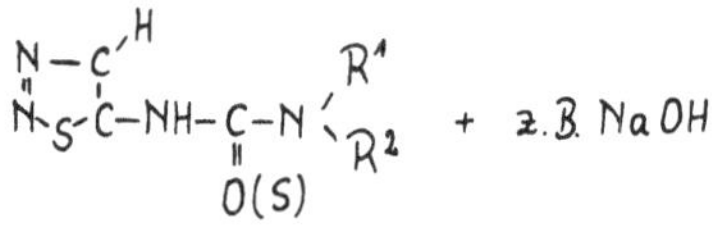

$$\underset{N\diagdown_S}{\overset{N-C\diagup^H}{}}-C-NH-\underset{O(S)}{\underset{\|}{C}}-N\underset{R^2}{\overset{R^1}{<}} \quad + \text{ z.B. NaOH}$$

R^1 = H, Alkyl (ev. durch O oder S unterbrochen)
R^2 = Alkyl (ev. durch O oder S unterbrochen), Aryl, (subst.), Cycloalkyl

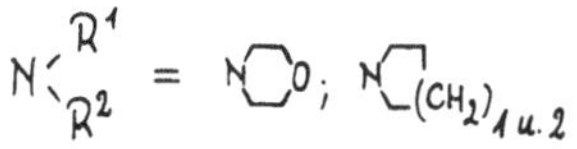

$$N\underset{R^2}{\overset{R^1}{<}} \ = \ N\bigcirc O \ ; \ N\bigcirc (CH_2)_{1\,u.\,2}$$

Wuchsregulatoren und Defoliant.

DOS 2 848 330   3.11.78/14. 5.80   Schering

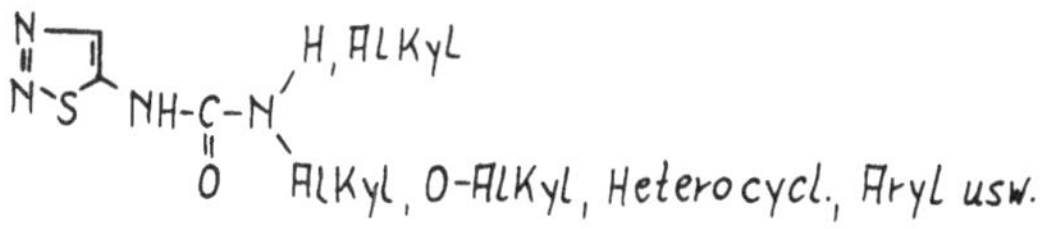

$$\underset{N\diagdown_S}{\overset{N\diagup}{}}NH-\underset{O}{\underset{\|}{C}}-N\diagup^{H,\,Alkyl}_{\diagdown Alkyl,\ O-Alkyl,\ Heterocycl.,\ Aryl\ usw.}$$

204

DOS   2 548 847   31.10.75/28.10.76   Velsicol Chem.
US Prior. 14.4. u. 30.4.75, 25.4. u. 13.1 .75

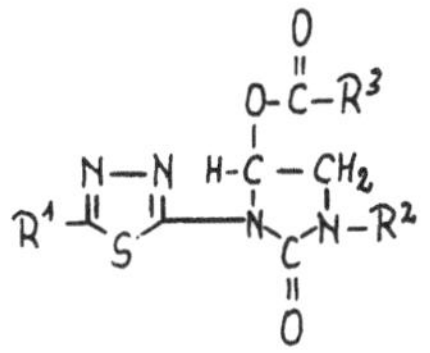

R^1 = Alk(en)yl, Cycloalkyl, Halog.-Alkyl,
    O(S)-Alkyl, Alkylsulfonyl, Alkyl-
    sulfinyl
R^2 = Alk(en)yl, Halog.-Alkyl,

$-\overset{|}{\underset{|}{C}}-C\equiv CH$, $-(CH_2)_n$

R^3 = Alk(en,in)yl, Cycloalkyl

DOS   2 601 987   16. 1.76/21. 7.77   Schering
=BE   850 394

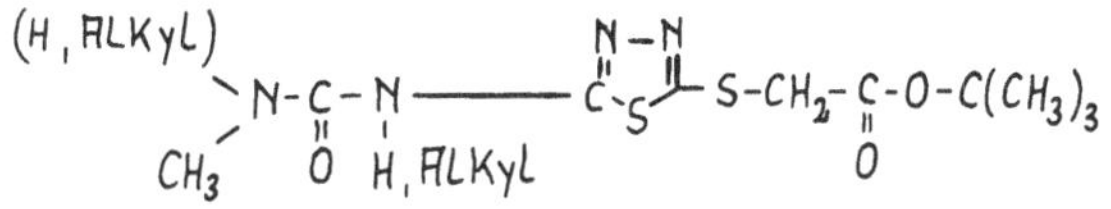

DOS   2 601 988   16. 1.76/21. 7.77   Schering
=BE   850 395

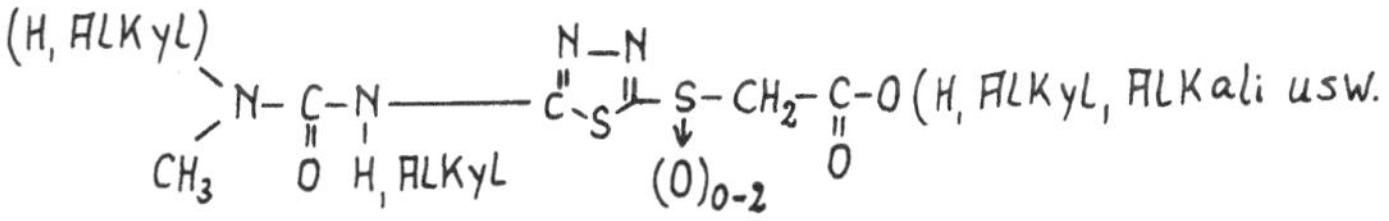

DOS   2 614 831   6. 4.76/20.10.77   Bayer
=BE   853 304

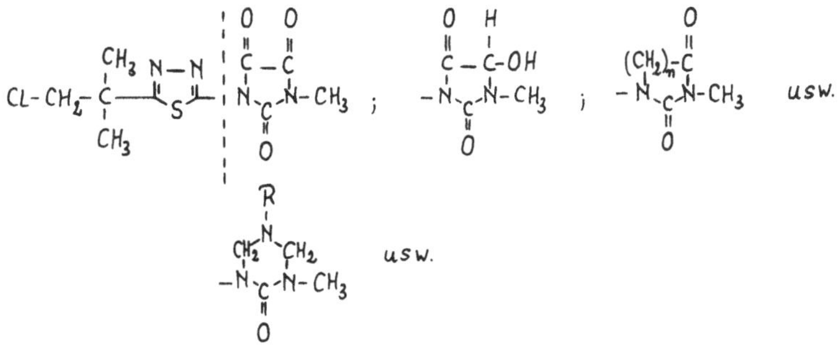

DOS   2 614 842   6. 4.76/20.10.77   Bayer
=BE   853 327

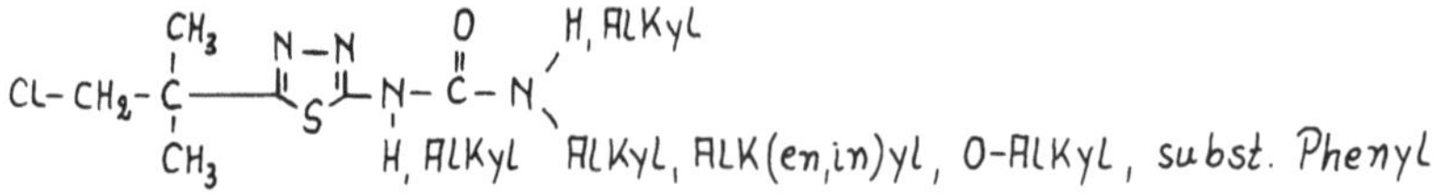

$$Cl-CH_2-\underset{CH_3}{\overset{CH_3}{C}}-\underset{S}{\overset{N-N}{C}}-\underset{H,\,Alkyl}{N}-\overset{O}{\overset{\|}{C}}-N\begin{smallmatrix}H,\,Alkyl\\ Alkyl,\,Alk(en,in)yl,\,O-Alkyl,\,subst.\,Phenyl\end{smallmatrix}$$

DOS   2 621 647   13. 5.76/ 1.12.77   Schering AG
=BE   854 614

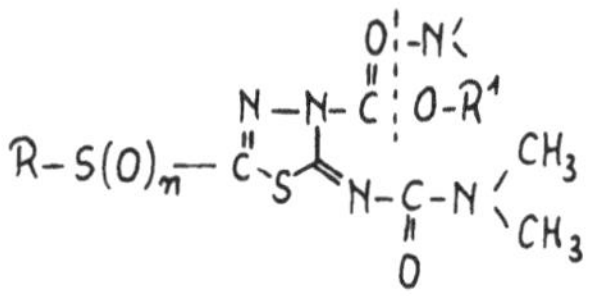

R u. R^1 = Alkyl, Halog. -
        Alkyl

$$R-S(O)_n-\underset{S}{\overset{N-N}{C}}\!=\!\underset{N-\underset{O}{\overset{\|}{C}}-N\begin{smallmatrix}CH_3\\CH_3\end{smallmatrix}}{\overset{O\text{-}N<}{\overset{\|}{C}}\,O-R^1}$$

DOS   2 623 657   24. 5.76/ 8.12.77   Schering
=BE   851 600

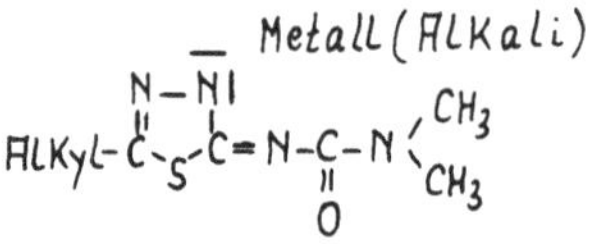

$$\overline{\quad Metall\,(Alkali)\quad}$$
$$Alkyl-\underset{S}{\overset{N-N}{C}}\,C\!=\!N-\underset{O}{\overset{\|}{C}}-N\begin{smallmatrix}CH_3\\CH_3\end{smallmatrix}$$

DOS   2 638 319   25. 8.76/10. 3.77   Eli Lilly
US   3 990 881   25. 8.75/ 9.11.76

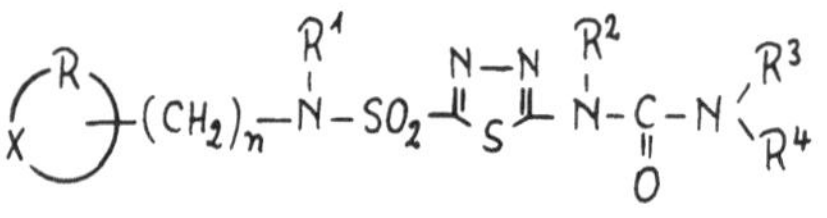

R = Alkylen; X = O, S, NH, $\geq$N

n = 0 oder 1; R^1 = H, CH_3, C_2H_5
R^2 u. R^3 = H oder
                C_1-C_3-Alkyl
R^4 = C_1-C_3-Alkyl oder
        O-Alkyl

$$\underset{X}{\overset{R}{\bigcirc}}-(CH_2)_n-\overset{R^1}{N}-SO_2-\underset{S}{\overset{N-N}{C}}-\overset{R^2}{N}-\underset{O}{\overset{\|}{C}}-N\begin{smallmatrix}R^3\\R^4\end{smallmatrix}$$

Siehe auch
GB   1 290 223, 1 230 432, 1 254 468, 1 340 267; US 3 726 892

alle Patente mit Substituenten in 5-Stellung.

DOS   2 644 426   30. 9.76/ 6. 4.78   Schering
=BE   859 267   30. 9.76/30. 3.78   Schering

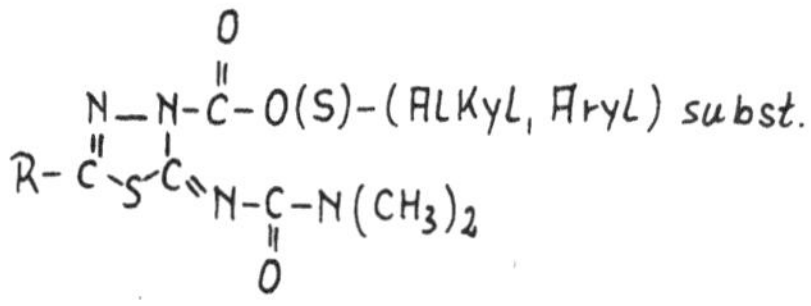

$$R-\underset{S}{\overset{N-N-\overset{O}{\overset{\|}{C}}-O(S)-(Alkyl,\,Aryl)\,subst.}{C}}\,C\!=\!N-\underset{O}{\overset{\|}{C}}-N(CH_3)_2$$

DOS   2 650 362   3.11.76/26. 5.77   Eli Lilly
GB Prior.   7.11.75

R = CF$_3$, C(CH$_3$)$_3$, C$_2$H$_5$-SO$_2$

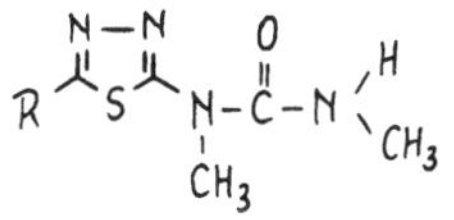

zusammen mit anderen Herbiziden wie z.B.

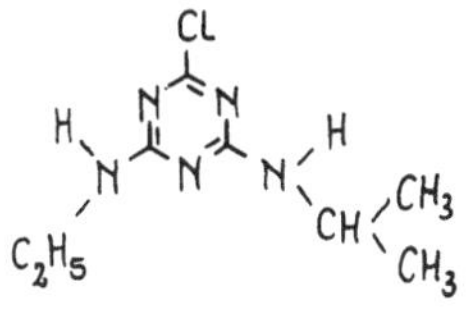

Selektiv in Zuckerrohr.

DOS   2 708 243   25. 2.77/22. 9.77   Velsicol Chem.
US Prior.   10. 3.76 u. 12. 3.76

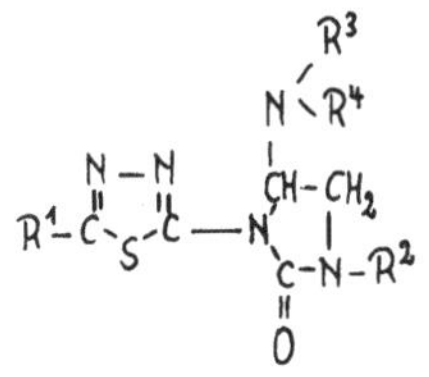

R^1 = (Cyclo)-Alk(en, in)yl
       (O, S, SO, SO$_2$)-Alkyl
R^2 = Alk(en, in)yl, Halog.-
       Alkyl

$$-\overset{|}{\underset{|}{C}}-C \equiv CH$$

R^3 u. R^4 = H, Alkyl, Alkenyl,
       OH, O-Alkyl oder Teil
       eines Ringsystems

DOS   2 724 674   1. 6.77/ 7.12.78   Bayer

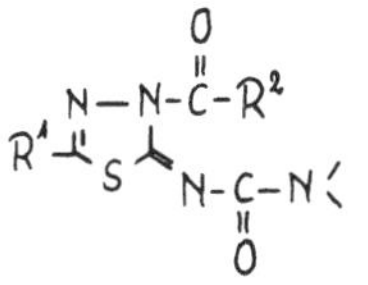

R^1 = (O)-Alkyl, Alkyl-S-
                     (O)$_{0-2}$
R^2 = Alkyl, O-Alkyl, Aryl,
       Aralkyl usw.

DOS   2 728 607   24. 6.77/ 5. 1.78   Gulf Oil
US Prior.   30. 6.76
=BE   856 254

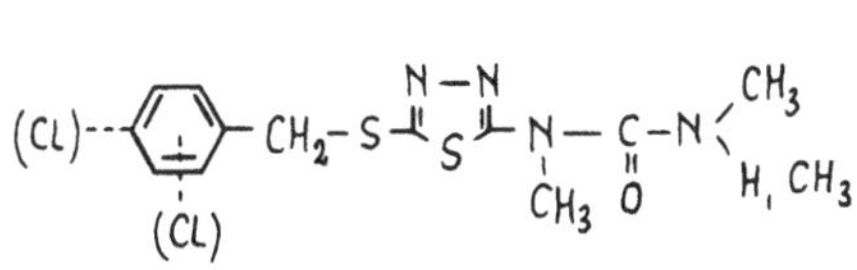

Gegen breitblättrige Unkräuter
in Sojakulturen.

DOS   2 728 610   24. 6.77/ 5. 1.78   Gulf Oil
=BE   856 255   30. 6.76/29.12.77

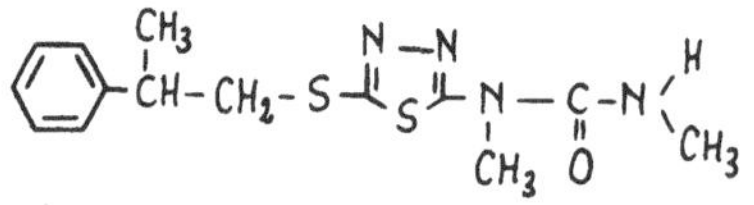

$$\text{Phenyl-CH(CH}_3\text{)-CH}_2\text{-S-C(=S)(N{-}N)-N(CH}_3\text{)-C(=O)-NH-CH}_3$$

Selektiv in Sojabohnen und
Erdnußpflanzungen.

DOS   2 728 612   24. 6.77/ 5. 1.78   Gulf Oil Corp.
=BE   856 256
         862 734   25. 8.77/ 6. 7.78   Gulf Oil Corp.
US Prior.   30. 6.76

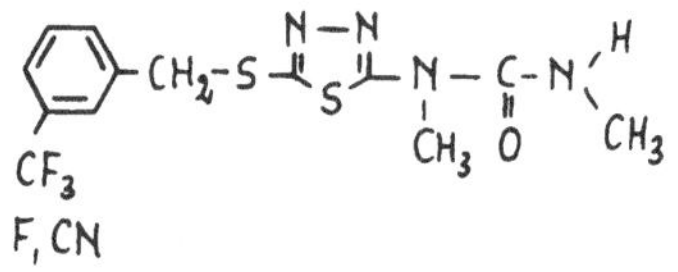

In Sojakulturen post emergence
Anwendung.

$$\text{Phenyl(CF}_3\text{, F, CN)-CH}_2\text{-S-C(=S)(N{-}N)-N(CH}_3\text{)-C(=O)-NH-CH}_3$$

DOS   2 808 362   27. 2.78/ 1. 3.79   Gulf Oil

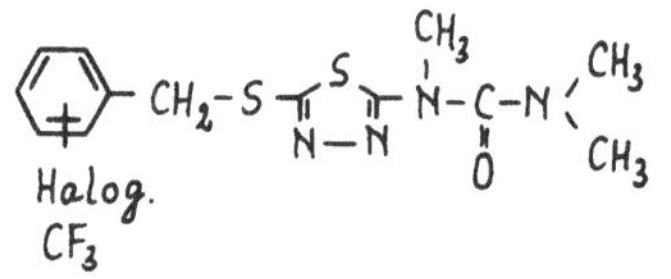

Selektiv gegen Unkräuter.

$$\text{Phenyl(Halog., CF}_3\text{)-CH}_2\text{-S-C(S,N{-}N)-N(CH}_3\text{)-C(=O)-N(CH}_3\text{)}_2$$

US   3 984 228   2. 2.72/ 5.10.76   Velsicol Chem.

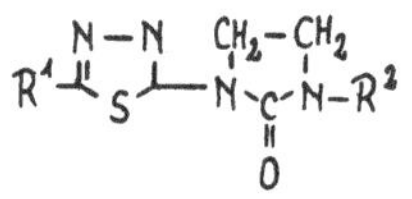

$$R^1 = CH \big\langle {}^{CH_3}_{CH_3}, \; C(CH_3)_3, \; CF_3$$

$$R^2 = \text{nied. Alkyl}$$

$$R^1\text{-C(N{-}N)(S)-N(CH}_2\text{-CH}_2\text{)(N-R}^2\text{)-C=O}$$

US   3 988 143   1.11.74/23.10.75   Velsicol Chem.

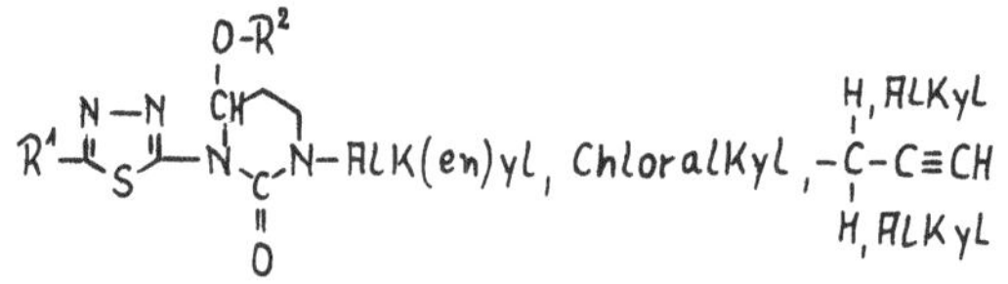

$$R^1 = (O, S, SO_2)\text{-Alkyl, Alkenyl, } CF_3$$

$$R^2 = C_1\text{-}C_5\text{-Alkyl}$$

$$R^1\text{-C(N{-}N)(S)-N(C(O-R}^2\text{)H}_2\text{)-C(=O)-N-ALK(en)yl, Chloralkyl, -C-C≡CH}$$

US   3 990 879   26.12.74/ 9.11.76   Eli Lilly

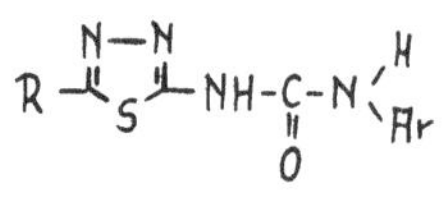

$$R = C_1\text{-}C_2\text{-Halog.-Alkyl, Phenyl, NO}_2\text{, Halog. } C_2\text{-}C_{13}\text{-SO}_2 \text{ usw.}$$

Wasserherbizid.

$$R\text{-C(N{-}N)(S)-NH-C(=O)-N}\big\langle{}^{H}_{Ar}$$

208

| US | 3 990 882 | 31.10.75/ 9.11.76 | Velsicol Chem. |

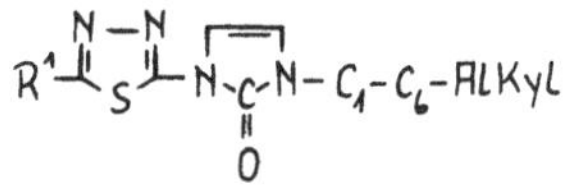

R^1 = C_1-C_2-Alkyl, Chlor-alkyl, O(S)-Alkyl, $S(O)_{1 \text{ u. } 2}$-Alkyl, CF_3

| US | 4 007 031 | 38. 3.74/ 8. 2.77 | Velsicol Chem. |

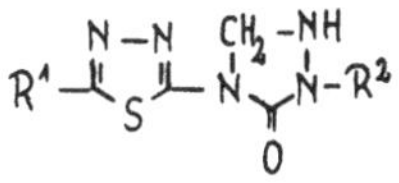

R^1 = Alk(en)yl, O,S-Alkyl, $S(O)_{1 \text{ u. } 2}$-Alkyl, CF_3, Cycloalkyl
R^2 = niedr. Alkyl

| US | 4 012 223 | 15. 2.74/15. 3.77 | Velsicol Chem. |

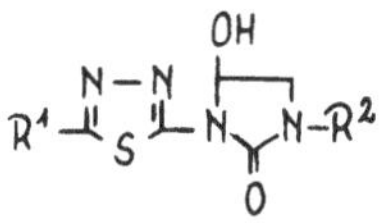

R^1 = C_3-C_7-Cyclo-Alkyl (ev. subst. durch (O)-Alkyl, Cl, Br, OH
R^2 = nied. Alkyl

| US | 4 008 068 | 8. 3.76/12. 2.77 | Velsicol Chem. |

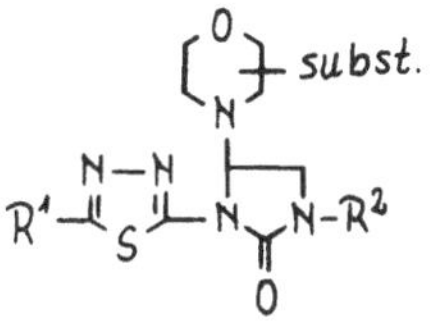

R^1 = Alk(en, in)yl, $S(O)_n$-Alkyl, CF_3, Halog.-Alkyl
n = 0-2
R^2 = Alkyl, Halog.-Alkyl, Propargyl

| US | 4 023 957 | 26.12.73/17. 5.77 | Velsicol Chem. |

| US | 4 033 753 | 5. 4.76/ 5. 7.77 | Velsicol Chem. |

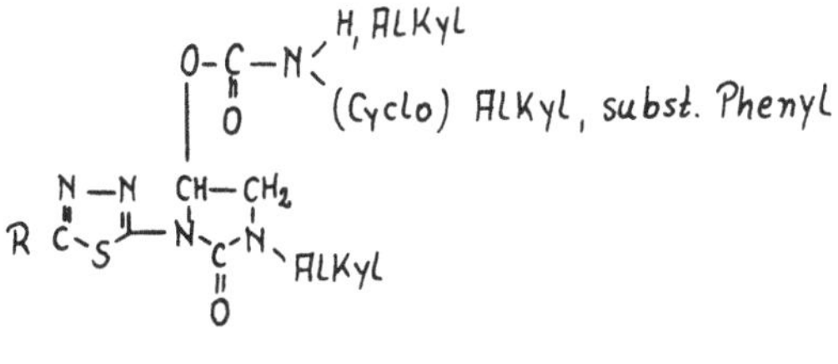

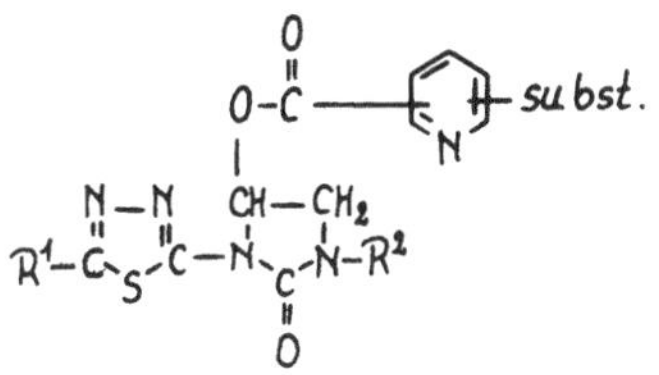

R^1 = (Cyclo)-Alkyl, O(S)-Alkyl, S-Alkyl
$\downarrow$
$(O)_{1 \text{ u. } 2}$
R^2 = (Halog.)-Alkyl, Alkenyl,
$C-C\equiv CH$

209

US   4 036 848   15. 3.76/19. 7.77   Velsicol Chem.

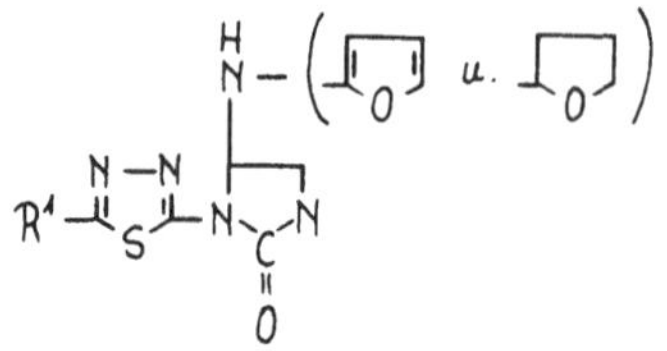

R^1 = Alk(en)yl, O-Alkyl,
       Halog.-Alkyl, Alkyl-SO$_2$
       usw.
R^2 = Alk(en)yl, Propargyl,
       Halog.-Alkyl

US   4 040 812   18. 6.75/ 9. 8.77   Velsicol Chem.

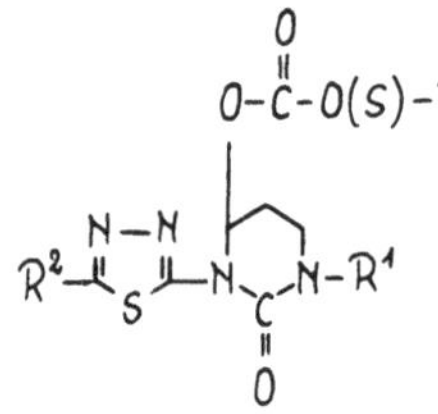

R = Alkyl, Halog.-Alkyl,
    Alkenyl, Alkinyl, Cyclo-
    alkyl
R^1 = Alkyl, Chloralkyl, Alkenyl
    u. -C-C≡CH
R^2 = Alkyl, Halog.-Alkyl,
    Alkenyl, CF$_3$, O-Alkyl,
    Cycloalkyl, S(O)$_{0-2}$-Alkyl
    usw.

US   4 052 192   18. 3.76/ 4.10.77   Velsicol Chem.

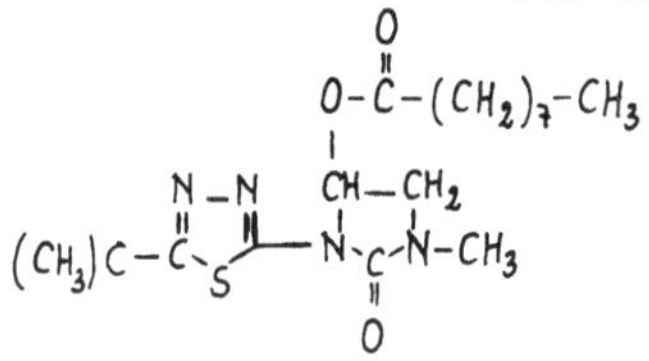

US   4 052 193   18. 3.76/ 4.10.77   Velsicol Chem.

wie zuvor   O-C-(CH$_2$)$_6$-CH$_3$

US   4 053 208   18. 9.74/11.10.77   Velsicol Chem.

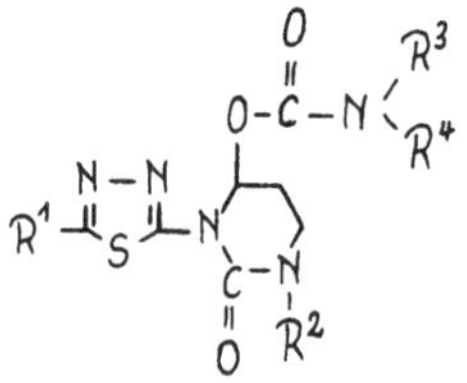

R^1 = Alk(en)yl, CF$_3$,
    O(S)-Alkyl, Alkyl-SO$_2$- usw.
R^2 = Alk(en)yl, Chloralkyl,
    -C-C≡CH
R^3 u. R^4 = H, Alkyl, Cyclo-
    alkyl, subst. Phenyl

210

US   4 053 480   26. 4.76/11.10.7   Velsicol Chem.

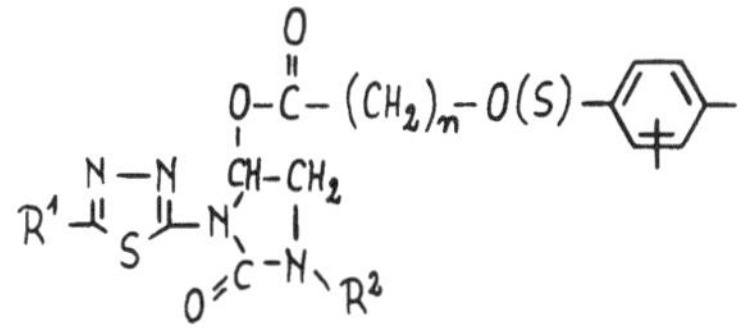

R^1 = Alkyl, Cycloalkyl, Alkenyl, O(S)-Alkyl

R^2 = Alk(en)yl, Halog.-Alkyl

subst. d. Cl, Alkyl, O-Alkyl, NO_2, CN

$(O)_{0-2}$

US   4 056 382   11.12.74/ 1.11.77   Eli Lilly

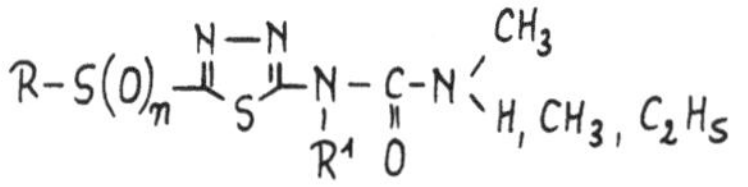

R = Alkyl, Ph-CH_2-,
Ph-CH_2-CH_2-

R^1 = CH_3, C_2H_5

Wasserherbizid.

US   4 063 924   29.10.76/20.12.77   Velsicol Chem.

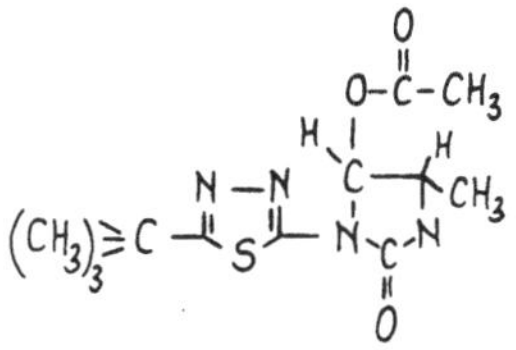

US   4 086 077   17. 4.70/25. 4.78   Gulf Res.

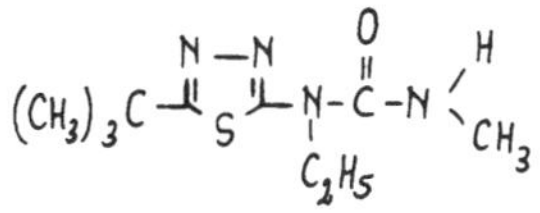

US   4 086 238   21. 6.76/25. 4.78   Velsicol Chem.

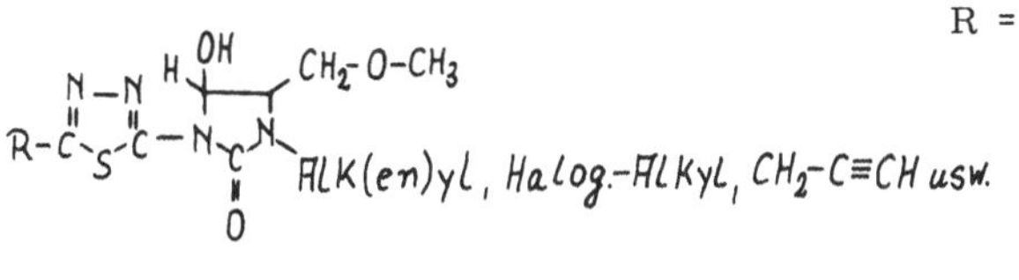

R = Alkyl, Cycloalkyl,
S-Alkyl, Alkyl-$SO_{(1\ u.\ 2)}$

CH_2-C$\equiv$CH usw.

US   4 097 486   29.10.76/28. 6.78   Velsicol Chem.

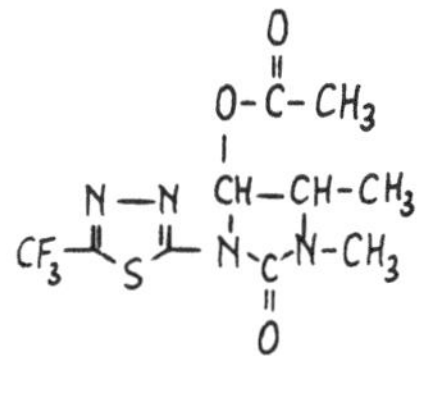

US   4 111 949   25. 3.76/ 5. 9.78   Velsicol Chem.

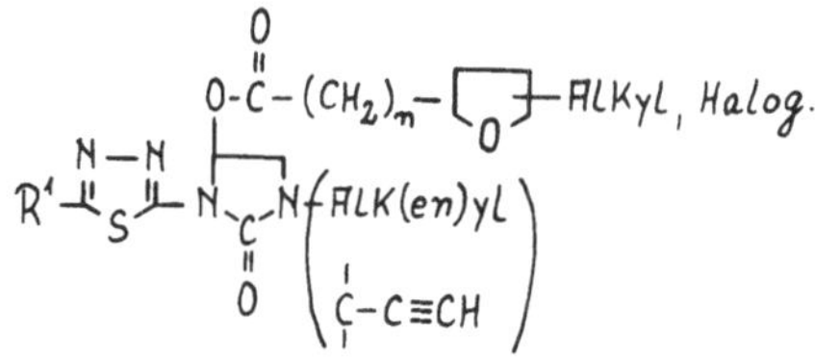

R^1 = Alk(en)yl, Halog.-Alkyl,
Cycloalkyl, O(S)-Alkyl,
$SO_{(1\ u.\ 2)}$-Alkyl

US   4 165 229   28. 6.73/21. 8.79   Eli Lilly

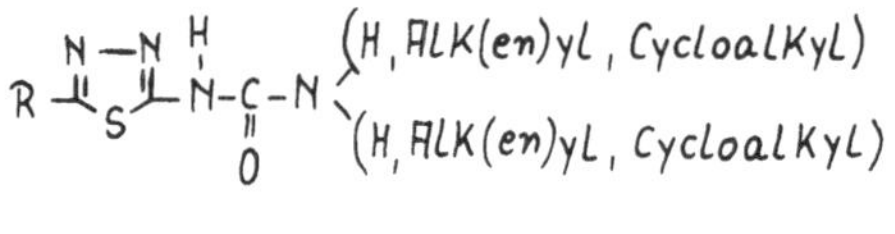

R = Alkyl-S, Alkyl-S(O)$_{1-2}$,
(Alkyl)$_2$-NS(O)$_{1\ u.\ 2}$ usw.

US   4 175 081   20.11.79/ 1. 2.80   Mobil Oil

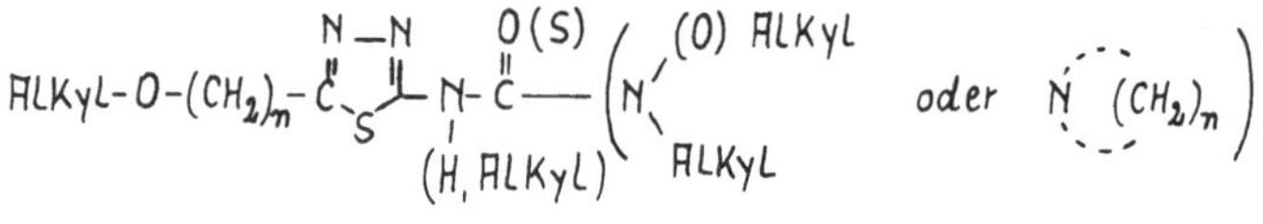

US   4 182 712   1. 2.68/ 8. 1.80   Mobil Oil

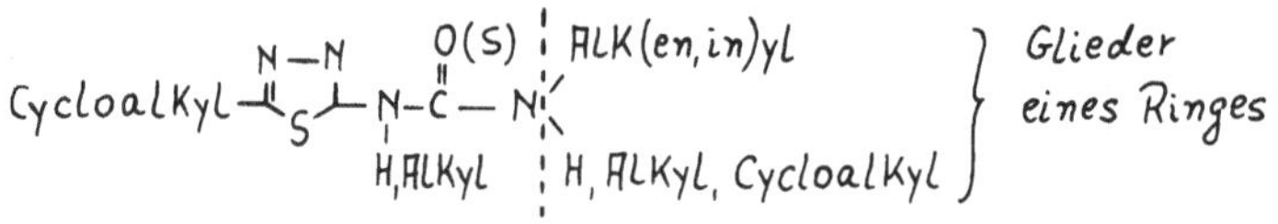

Glieder
eines Ringes

US   4 218 236   18. 6.79/19. 8.80   PPG Ind.

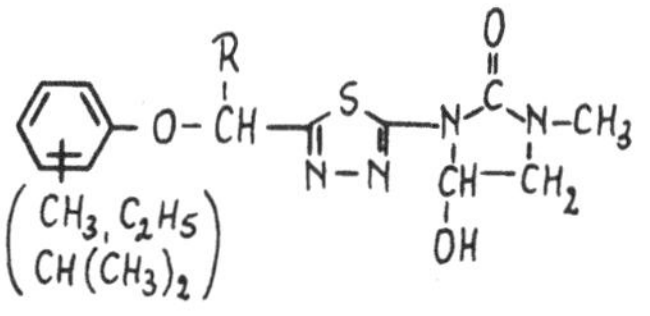

R = Alk(en, in)yl, C_2H_4 Cl usw.

BE   852 149   10. 3.76/ 1. 7.77   Velsicol Chem.

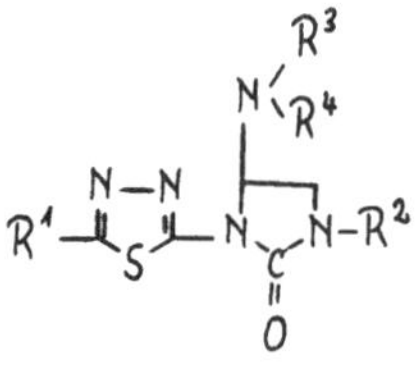

R^1 = Alk(en)yl, Cycloalkyl, O(S)-Alkyl,
S(O)$_{1\ u.\ 2}$-Alkyl
R^2 = Alk(en)yl, Alkyl-Halogenid, C-C≡CH
$R^3 + R^4$ = Alk(en)yl, Cycloalkyl, Phenyl,
Benzyl usw.

212

| BE | 854 971 | 24. 5.76/24.11.77 | Schering |

$$\left[R \overset{\displaystyle N-N}{\underset{S}{\diagup\!\!\diagdown}} N-\underset{\underset{O}{\|}}{C}-N(CH_3)_2 \right] B^+$$

R = Alkyl

B = Li, Na, K

(zerfällt wohl imBoden in bekanntes Harnstoffherbizid)

| BE | 862 734 | 25. 8.77/ 6. 7.78 | Gulf Oil |

X = Br, F, CN, CF_3

$$\underset{X}{\diagup\!\!\diagdown}-CH_2-S-\overset{\displaystyle N-N}{\underset{S}{\diagup\!\!\diagdown}}-\underset{CH_3}{N}-\underset{O}{\overset{\|}{C}}-\underset{CH_3}{N}-(H,CH_3)$$

| JA | 51 088967 | 23. 2.71/ 4. 8.76 | Rikasaku Res. Lab. |

R = CH_3-C_3H_7

Darstellungsverfahren

$$R-\overset{\displaystyle N-N}{\underset{S}{\diagup\!\!\diagdown}}-NH-\underset{\underset{O}{\|}}{C}-N(CH_3)_2$$

| JA | 52 028938 | 25. 8.75/ 4. 3.77 | Eli Lilly |

$$Cl-\underset{(H,Cl)}{CH}-\overset{CH_2-(H,Cl)}{\underset{CH_2-(H,Cl)}{C}}-\overset{\displaystyle N-N}{\underset{S}{\diagup\!\!\diagdown}}-\underset{(H,CH_3,C_2H_5)}{N}-\overset{\overset{O}{\|}}{C}-N\diagup^{H,(O)-CH_3}_{\diagdown H,CH_3,\ 2,2\text{-}Dimethoxy\text{-}ethyl}$$

i 5.4) 5-Ringe mit 2 x N + 1 x S

1, 2, 4-Thiadiazolyl-Harnstoffe

| US | 4 224 449 | 8. 8.79/23. 9.80 | Olin Corp. |

R^1 u. R^2 = H, Alkyl, Phenyl,
C-Phenyl
$\overset{\|}{O}$
wenigstens 1 x H

$$(CCl_3, CF_3)\overset{\displaystyle N-N}{\underset{S}{\diagup\!\!\diagdown}}-\underset{(Alkyl,\ Phenyl)}{N}-\overset{\overset{O}{\|}}{C}-\underset{(Alkyl,\ Phenyl)}{N}-\overset{\overset{O}{\|}}{C}-N\diagup^{R^1}_{\diagdown R^2}$$

5.5.3 k 1-3 Heterocyclisch-aliphatische Harnstoffe. Heterocyclen mit 6-Ringen 1 x N oder 1 x O bzw. S usw. Bd. 5/491

(Harnstoffe der Formel $ArNH\underset{\overset{\|}{O}}{C}-N\begin{smallmatrix}H \\ Heterocycl.\end{smallmatrix}$ siehe 5.5.3 g)

DOS 2 843 722 6.10.78/19. 4.79 Shudo, Koiohi usw.

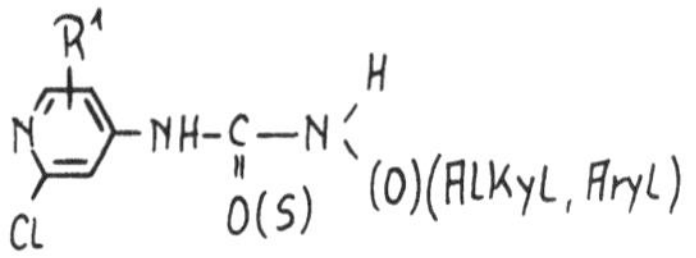

Vorwiegend Pflanzenwuchs-regulatoren.

DOS 2 853 587 12. 2.78/28. 6.79 Philagro S.A.
=BE 872 689 12.12.77/12. 6.79 Philagro S.A.

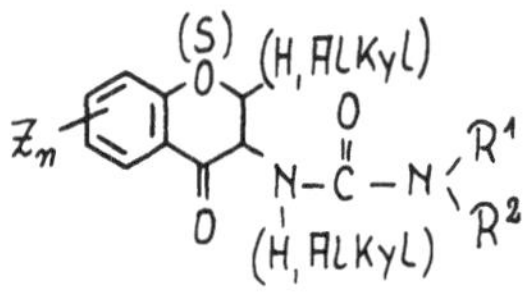

R^1 u. R^2 = H, (O)-Alkyl, Alk(en, in)yl,

$N\begin{smallmatrix}R^1 \\ R^2\end{smallmatrix}$ = Glied eines 5 oder 6-Ringes

Z = Alkyl, Cycloalkyl, O-Alkyl, Phenyl

DOS 3 009 683 13. 3.80/25. 9.80 Sumitomo
JA Prior. 13.3. u. 31.8.79

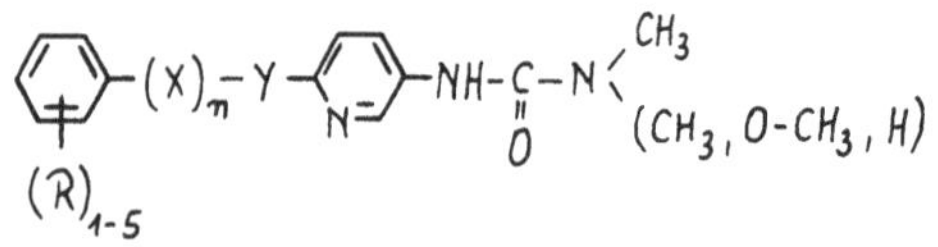

R = CH_3, O-CH_3, CF_3, Halog.
X = verzw. Alkylen
Y = O oder S
n = 0-5

GB 2 001 316 21. 7.77/31. 1.79 Shell Int. Res.

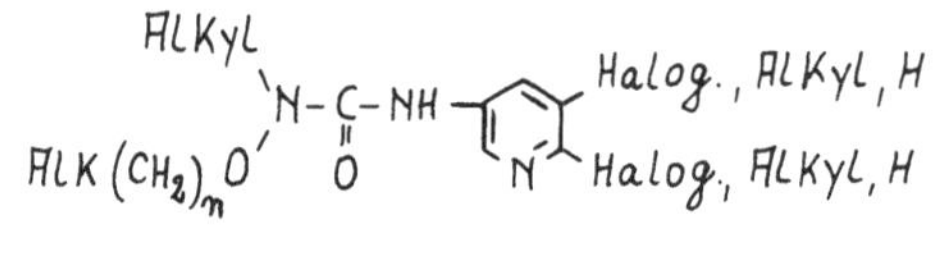

5.5.3. k.4 <u>N[1]-Acyl- u. Sulfonyl-Verbindungen heterocyclisch-</u>
<u>aliphatischer Harnstoffe (neue Verbindungsklasse)</u>

DOS   2 715 786   7. 4.77/13.10.77   Du Pont
US Prior. 7.4.76, 23.2.77

$R^1-SO_2-NH-\overset{O(S)}{\underset{\|}{C}}-N\overset{H}{\underset{R}{\diagdown}}$

$R = $ (Pyrimidinyl) oder (Triazinyl)

$R^1 = $ (Phenyl) , (Thienyl/Furyl) , (Naphthyl) (O)

z. B. $Cl-$(Phenyl, $O-CH_3$)$-SO_2-NH-\overset{O}{\underset{\|}{C}}-N-$(Triazin, CH_3, CH_3)

sehr umfassendes Patent mit 247 Seiten.
=BE   853 374   7. 4.76/ 7.10.77   Du Pont

EP   1 485   18. 9.78/18. 4.79   Du Pont
US   19.4.77 u. 4.8.78

$R^1-SO_2-NH-\overset{}{\underset{O(S)}{\underset{\|}{C}}}-NH-$(Pyrimidin, X, Y)

X u. Y = H, Cl, Br, (O)-CH$_3$,
         O-C$_2$H$_5$

$R^1 = $ (Phenyl) H, NO$_2$, Halog., (O) CH$_3$
         subst. wie zuvor

oder R^1 = 2-Furyl, 2-Thienyl, 1-Naphthyl

EP   1 514   6.10.78/11. 7.79   Du Pont
=GB   2 015 503   2. 3.78/12. 9.79   Du Pont

(subst. (Phenyl) oder 2-Thienyl)$-SO_2-N-\overset{O}{\underset{\|}{C}}-N-$(Triazin, Z, X, Y)
H,(O)-Alkyl / H,Alkyl

X = Cl, CH$_3$, C$_2$H$_5$, -O-Alkyl,
    CF$_3$, -CH$_2$-O-Alkyl
Y = (O)-CH$_3$
Z = CH oder N

EP   1 515   6.10.78/18. 4.79   Du Pont
US Prior.   6.10.77

(Phenyl, R^1, R^2, H, H)$-SO_2-NH-\overset{O}{\underset{\|}{C}}-NH-$(Pyrimidin/Triazin, X, Z, Y)

R^1 = H, Halog., (O)-CH$_3$, NO$_2$,
R^2 = NH-$\overset{O}{\underset{\|}{C}}$-O-Alkyl,

       NH-$\overset{O}{\underset{\|}{C}}$-Alkyl, NH-$\overset{O}{\underset{\|}{C}}$-N<
X = CH$_3$, O-(CH$_3$, C$_2$H$_5$)
Y = CH$_3$, O-CH$_3$
Z =   CH, N

EP   5 986   29. 5.79/12.12.79   Du Pont
US Prior.   30. 5.78

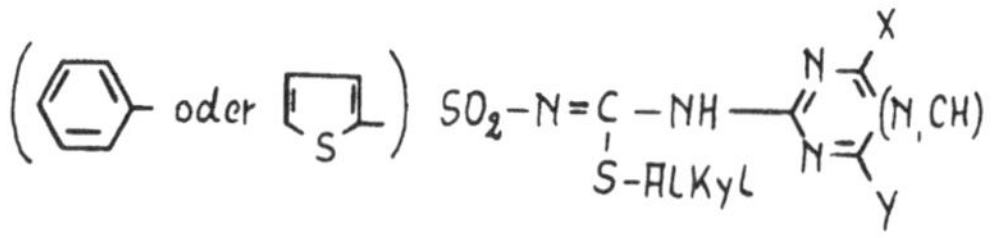

$$\left(\text{C}_6\text{H}_5\text{- oder }\underset{S}{\square}\right)SO_2\text{-N}=\overset{|}{\underset{S\text{-Alkyl}}{C}}\text{-NH} \longrightarrow \text{Triazin (N,CH)}\begin{smallmatrix}X\\Y\end{smallmatrix}$$

X = CH_3, $O\text{-}CH_3$, $O\text{-}C_2H_5$, $CH_2\text{-}O\text{-}CH_3$

Y = CH_3, $O\text{-}CH_3$

EP   7 687   29. 5.79/ 6. 2.80   Du Pont
US Prior. 30.5.78, 30.11.78, 1.3.79, 13.4.79

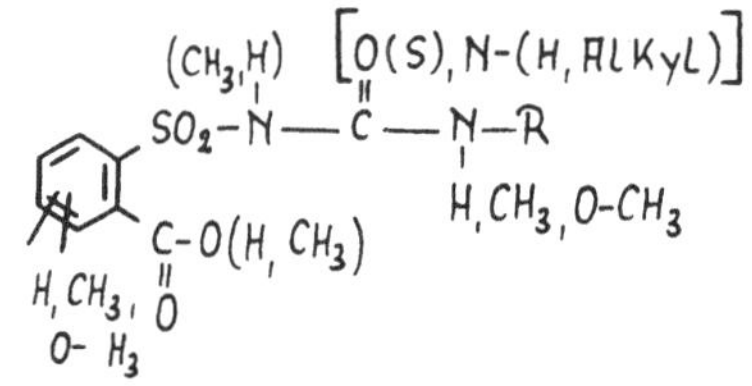

$$\underset{\substack{H,CH_3,\\O\text{-}H_3}}{\text{Phenyl}}\;\underset{C\text{-}O(H,CH_3)}{\overset{(CH_3,H)}{SO_2\text{-N}\text{-}\overset{O(S),N\text{-}(H,Alkyl)}{C}\text{-N-R}}}\quad \underset{H,CH_3,O\text{-}CH_3}{}$$

$$R = \text{Triazinyl} \;\; \ddot{u}. \;\; \text{Pyrimidinyl}$$

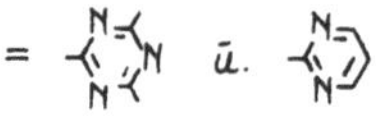

Darstellung ähnl. Verb. ohne COOR siehe NE 121 788 (1966), siehe auch
FR 1 468 747; Chem. Abstr. 53,18052 g (1959). Sehr umfassendes Patent 215 Seiten.

EP   9 419   26. 9.79/ 2. 4.80   Du Pont
US Prior. 7.9.78 u. 7.8.79

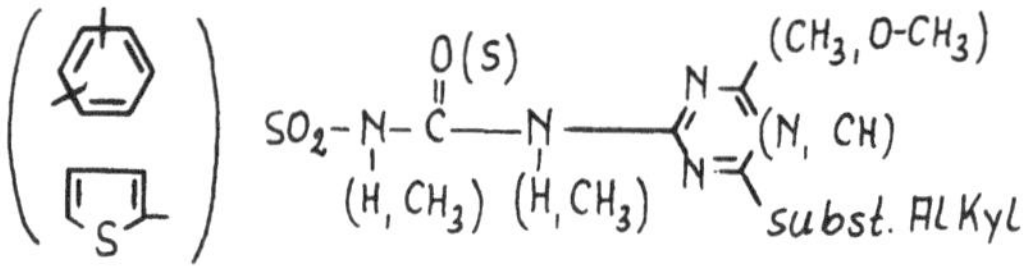

$$\left(\text{Phenyl / }\underset{S}{\square}\right)SO_2\text{-}\underset{(H,CH_3)}{\text{N-}}\overset{O(S)}{C}\text{-}\underset{(H,CH_3)}{\text{N}}\text{-Triazin}\begin{smallmatrix}(CH_3,O\text{-}CH_3)\\(N,CH)\\subst. Alkyl\end{smallmatrix}$$

EP   10 560   1.11.78/14. 5.80   Du Pont

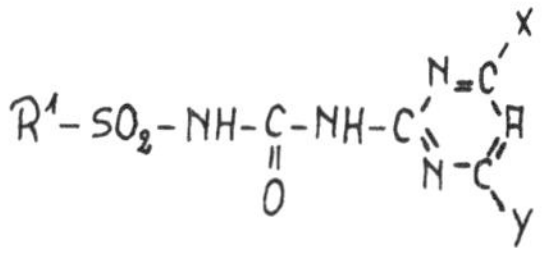

$$R^1\text{-}SO_2\text{-NH-}\overset{O}{C}\text{-NH-C}\cdots \text{Triazin}\begin{smallmatrix}X\\A\\Y\end{smallmatrix}$$

A = CH oder N

R^1 = subst. Phenyl u. $\underset{S}{\square}$

X = $N\begin{smallmatrix}H\\CH_3\end{smallmatrix}$, $N(CH_3)_2$

Y = CH_3, $O\text{-}CH_3$

Pre und post emergence in Sojabohnen, Weizen usw. wirksam mit o,1-2,0 kg/ha.

EP   13 480   3.12.79/23. 7.80   Du Pont
US Prior. 4.12.78 u. 22.10.79

R = (three heterocyclic ring structures: 1,3,5-triazinyl with X, Z, Y; pyrimidinyl with X¹, Y¹; pyrazolyl/triazolyl with X¹, Y¹)

Structure: pyridine ring substituted with R^1 and R^2, bearing $-SO_2-N(H, Alkyl)-C(=O(S))-N(H,CH_3)(R)$

R^1 = Halog., H, NO_2, COOR, Alkyl,
        (O) S-Alkyl usw.

X = CH_3, O-CH_3, O-C_2H_5
Z = -CH=, -N=
X^1 = H, CH_3, O-CH_3
Y^1 = H, CH_3, O-CH_3

Y = H, Cl, CH_3, CF_3, $N(H)(CH_3)$, $N(CH_3)_2$ usw.

z.B. (substituted pyridyl)-SO_2-NH-C(=O)-NH-(triazolyl with (O)CH_3, N=C-CH_3 substituents)

EP   15 683   22. 2.79/17. 9.80   Du Pont

$R-SO_2-NH-C(=O(S))-NH-R^1$ oder $R-SO_2-N=C(SR)-NH R^1$

R = subst. Phenyl, 3-Pyridyl
    (subst.)
Y = O, CH_2

R^1 = (triazinyl/pyrimidinyl ring with H, CH_3, O-CH_3, Cl usw. and $Y-(CH_2)_{1 u.2}$)

EP   17 473   2. 4.80/15.10.80   Du Pont
US Prior. 4. 4. 79

Structure: thiophene ring (R-substituted, S in ring)-SO_2-NH-C(=O)-NH-(pyrimidinyl with X, Z, Y)

R = Alkyl, NO_2, Cl, Br,

$SO_2 N<$

X = H, Cl, Br, CH_3, C_2H_5,
    O-Alkyl, CH_2-O-CH_3,
    S-Alkyl, CF_3 usw.
Y = CH_3 oder O-CH_3
Z = CH oder N

Wirksam gegen breitblättrige Unkräuter in Weizen und Gerste.

US   4 169 719   7. 4.76/ 2.10.79   Du Pont

$R^1-SO_2-NH-C(=O(S))-NH-$(pyrimidinyl with X, Z)

X = H, Cl, CH_3, CF_3, O-Alkyl
    usw.
Z = (O)-CH_3

R^1 = (subst. phenyl), (thiophene/furan ring with O, (S))

GB   1 560 918   7. 4.76/13. 2.80   Du Pont

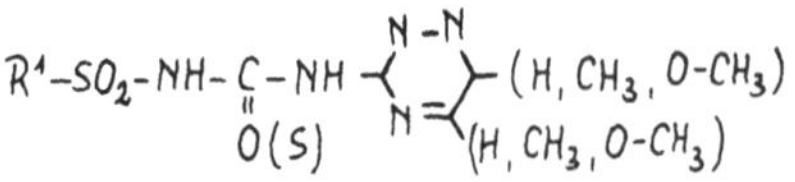

$$R^1-SO_2-NH-\underset{O(S)}{\overset{\|}{C}}-NH-\underset{N=\underset{(H,CH_3,O-CH_3)}{}}{\overset{N-N}{}}(H,CH_3,O-CH_3)$$

RD   181 049   20. 4.79/10. 5.79

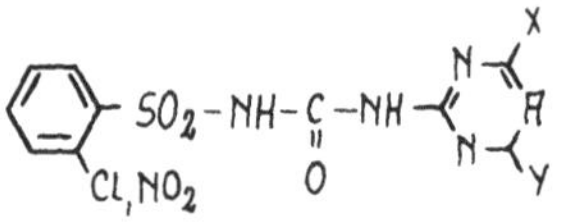

A = CH, N
X u. Y = CH$_3$, O-CH$_3$

Selektiv in Getreide

JA   54 044683   1. 8.77/ 9. 4.79   Du Pont

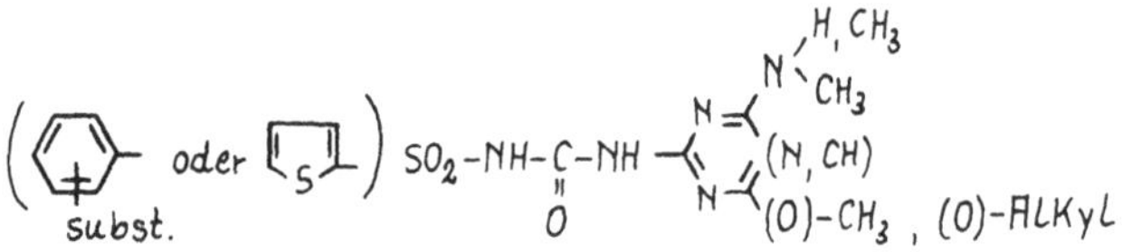

5.6 Sulfamide

Bd. 5/492

US   4 013 448   29.12.75/22. 3.77   Monsanto

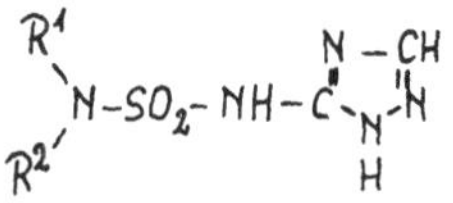

R^1 u. R^2 = H oder Alkyl

US   4 071 352   5. 9.72/31. 1.78   Stauffer

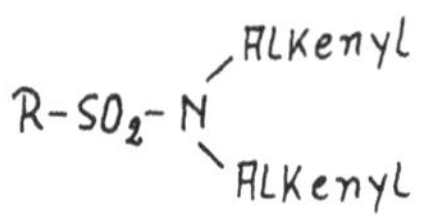

Antidot gegen Schäden durch
Thiocarbamat-Herbizide

218

US   4 071 353   5. 9.72/31. 1.78   Stauffer

ALKylen, Phenyl–NH–SO$_2$–N$\leqq$ Antidot für Thiocarbamat-
                                          Herbizide

5.6. 1 Aliphatische Nitro-Amine (neu)

SU   639 868   28.12.76/30.12.78   Marisk Univ.

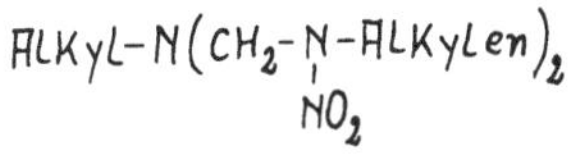

$$ALKyl-N\left(CH_2-\underset{\underset{NO_2}{|}}{N}-ALKylen\right)_2$$

5.7 Carbonsäuren, Nitrile und Aldehyde   Bd. 5/493

5.7.1 Aliphatische Carbonsäuren

a) Monocarbonsäuren, ihre Ester, Amide und Amidine
   (außer Anilide und Amidoheterocyclen) sowie Nitrile

DOS   2 620 101   6. 5.76/18.11.76   Prod. Chim. Ugine Kuhlmann
FR Prior.   7. 5.75

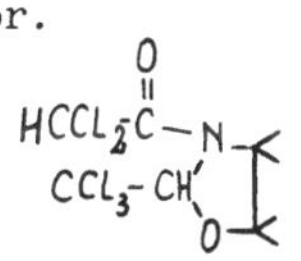

$$\underset{CCl_3-CH}{HCCl_2\overset{\overset{O}{\|}}{C}}-N\underset{O}{\overset{}{\langle}}$$

                                          Antidots für Herbizide zur
                                          Erhöhung der Selektivität von
                                          Thiolcarbamaten.

DOS   2 634 278   30. 7.76/24. 2.77   Stauffer
US Prior. 8.8.75 u. 10.6.76

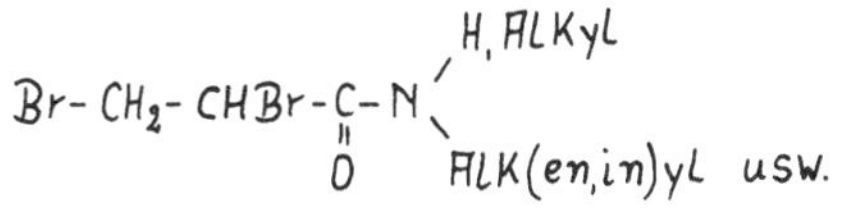

                                          Antidots gegen Herbizid-
                                          schäden durch Thiocarbamate.

$$Br-CH_2-CHBr-\overset{\overset{}{\underset{\underset{O}{\|}}{}}}{C}-N\overset{H, ALKyl}{\underset{ALK(en,in)yl\ usw.}{}}$$

DOS   2 637 580   20. 3.76/17. 3.77   Stauffer
US Prior.   28. 8.75

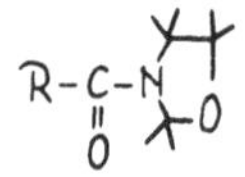

$$R-\underset{O}{\overset{\parallel}{C}}-N \overset{\diagup}{\underset{\diagdown}{}} \begin{array}{c} \\ O \end{array}$$

R = C_{1-10}-Halogen-Alkyl,
Alkyl-S-

Antidots für Herbizide und
z. T. selbst Herbizide.

DOS   2 729 672   30. 6.76/12. 1.78   Kao Soap

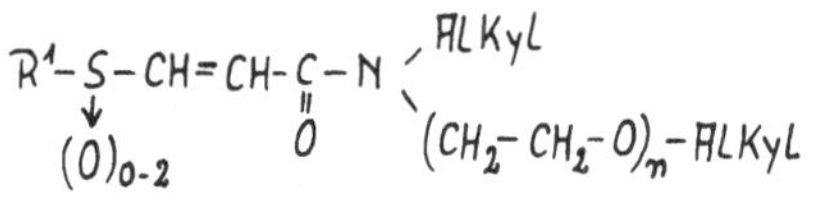

$$R^1-\underset{(O)_{0-2}}{S}-CH=CH-\underset{O}{\overset{\parallel}{C}}-N \overset{\diagup ALKyl}{\underset{\diagdown (CH_2-CH_2-O)_n-ALKyl}{}}$$

R^1 = Alkyl

Germicide Herbizide.

DOS   2 729 685   30. 6.76/ 5. 1.78   Kao Soap

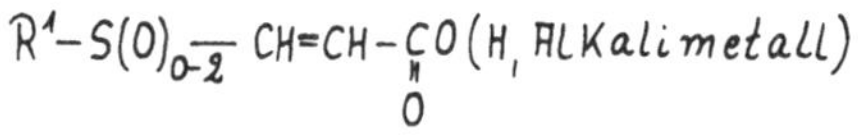

$$R^1-S(O)_{0-2}\, CH=CH-\underset{O}{\overset{\parallel}{C}}O\,(H, ALKalimetall)$$

R^1 = Alk(en)yl, C_1-C_{20}

Germicide Herbizide.

DOS   2 747 814   29.10.76/ 3. 5.78   Montedison
IT Prior.   29.10.76

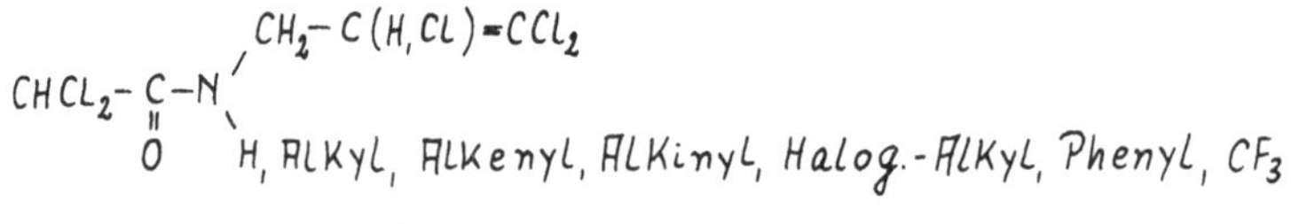

$$CHCL_2-\underset{O}{\overset{\parallel}{C}}-N \overset{\diagup CH_2-C(H,CL)=CCL_2}{\underset{\diagdown H, ALKyl, ALKenyl, ALKinyl, Halog.-ALKyl, Phenyl, CF_3}{}}$$

Antidot gegen "Glycin"-Schäden in Mais.

Glycin siehe DOS 2 311 897

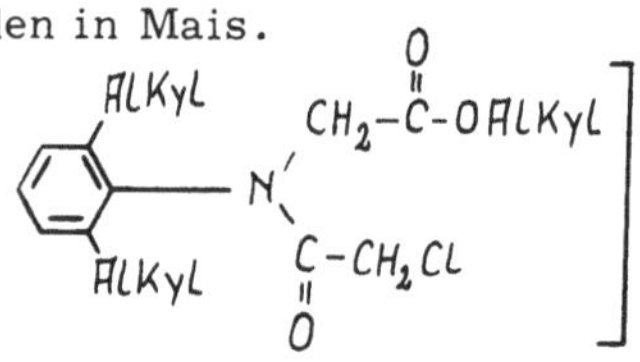

$$\left[\begin{array}{c} \underset{ALKyl}{\overset{ALKyl}{\bigcirc}} - N \overset{\diagup CH_2-\overset{O}{\overset{\parallel}{C}}-OALKyl}{\underset{\diagdown \underset{O}{\overset{\parallel}{C}}-CH_2CL}{}} \end{array} \right]$$

DOS   2 753 945   3.12.77/ 7. 6.79   Bayer

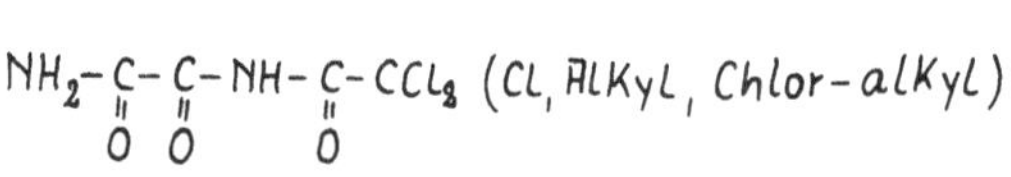

$$NH_2-\underset{O}{\overset{\parallel}{C}}-\underset{O}{\overset{\parallel}{C}}-NH-\underset{O}{\overset{\parallel}{C}}-CCL_3 \ (CL, ALKyl, Chlor-alkyl)$$

220

DOS   2 757 813   23.12.77/29. 6.78   Egypt Gyog.

Spez. Darstellungsverfahren.

$CL_2-\overset{H}{\underset{}{C}}-\overset{}{\underset{O}{C}}-N\big(\overset{ALKyl}{\underset{ALKyl}{\big<}}$, *Glieder eines Ringsystems*)

DOS   2 807 340   21. 2.78/23. 8.79   Egypt Gyog.

$CHCL_2-\underset{O}{C}-N(CH_2-CH=CH_2)$ *neues Darstellungsverfahren mittels* $CHCL_3-\overset{H}{\underset{O}{C}}$

DOS   2 828 265   28. 6.78/17. 1.80   Bayer

Einfache Säureamide als Safener für Acetanilid-Herbizide

der Formel   [Formel: Phenyl mit ALKyl-Substituenten, $-N\big<\,^{CH_2-Heterocycl.\ mit\ N}_{C-CH_2-Halog.}$]

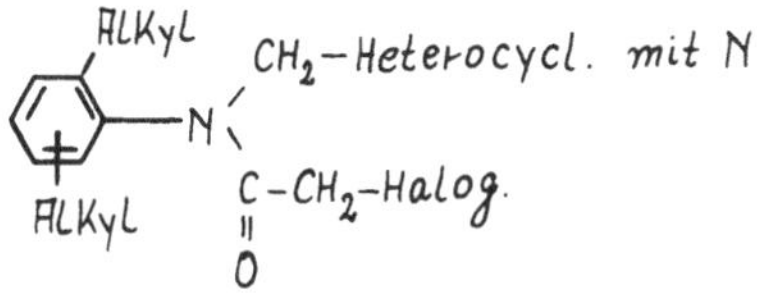

DOS   2 828 331   28. 6.78/10. 1.80   Bayer

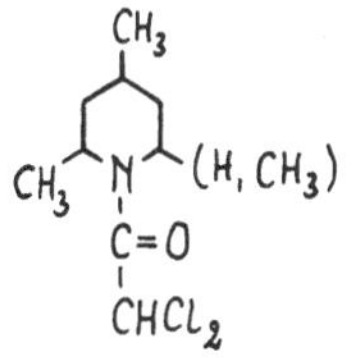

Safener für Herbizid-Schäden
Wirkstoffmischung

[Formel: Piperidinring mit CH_3, CH_3, (H, CH_3), $C=O$, $CHCL_2$]

DOS   2 831 654   19. 7.78/15. 2.79   Chinoin Gyogyszer

R = H, O-Alkyl, S-Alkyl,
    Alkyl, Aryl usw.

R^1 = H, Alkyl, Aryl, Aralkyl,
    $N\!<$

R^2 = H, Alkyl, Aryl, Aralkyl,
    $N\!<$

R^3 = H, Alkyl, Aralkyl, Heterocy.

DOS   2 832 950   27. 7.78/21. 2.80   BASF

$$CHCl_2-\underset{O}{\overset{\|}{C}}-N\begin{cases}\text{Alkyl, subst. Alkyl}\\ \text{Alkyl}\end{cases}$$

Safener für Herbizide der Klasse:

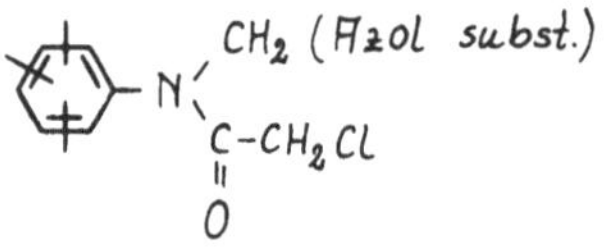

$$\text{Ring}-N\begin{cases}CH_2\ (Azol\ subst.)\\ \underset{O}{\overset{\|}{C}}-CH_2\,Cl\end{cases}$$

DOS   2 832 974   27. 7.78/14. 2.80   BASF

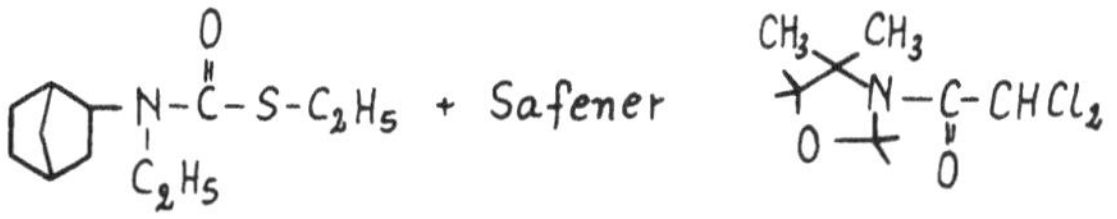

$$\text{Ring}-\underset{C_2H_5}{\overset{O\ \ \ \|}{N-C}}-S-C_2H_5\ +\ Safener\qquad \overset{CH_3\ \ CH_3}{\underset{O}{\underset{\|}{N}}-\underset{O}{\overset{\|}{C}}-CHCl_2}$$

Herbizid für Zuckerrüben

DOS   2 855 229   21.12.78/ 5. 7.79   Montedison
Ergänzung zu DOS 2 747 814
IT Prior.   23.12.77   Antidot für Herbizide in Mais-
                                                         kulturen

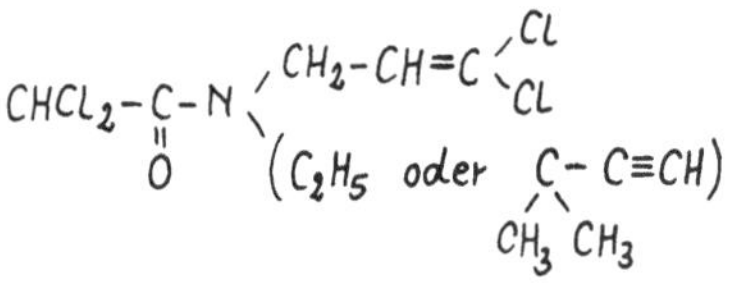

$$CHCl_2-\underset{O}{\overset{\|}{C}}-N\begin{cases}CH_2-CH=C\begin{cases}Cl\\Cl\end{cases}\\ \left(C_2H_5\ oder\ \underset{CH_3\ CH_3}{\overset{}{C}}-C\equiv CH\right)\end{cases}$$

Herbizid: N, N-disubstituiertes
Glycin

DOS   2 905 650   14. 2.79/21. 8.80   Nitro Kemia Ungarn

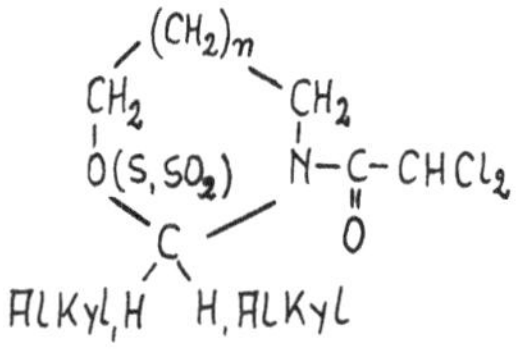

Antidot für Thiolcarbamate,
Triazin-Chloracetanilid- usw.
herbizide.

DOS   2 937 867   19. 9.78/27. 3.80   Sumitomo

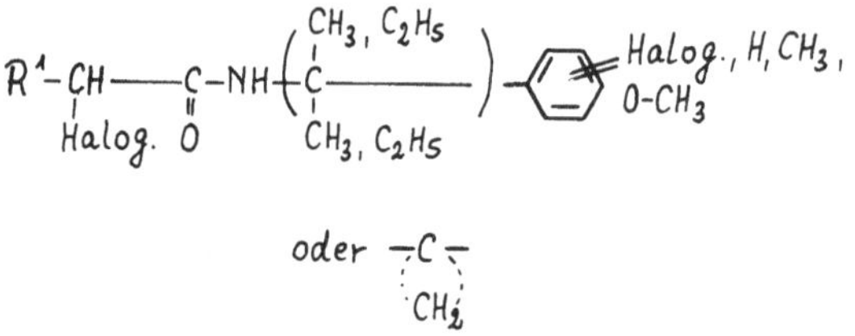

R^1 = tert. Alkyl

$$R^1-\underset{Halog.}{\overset{}{CH}}-\underset{O}{\overset{\|}{C}}-NH-\left(\underset{CH_3,\,C_2H_5}{\overset{CH_3,\,C_2H_5}{C}}\right)-\text{Ring}=\begin{cases}Halog., H, CH_3,\\ O-CH_3\end{cases}$$

oder $-C-$
      CH_2

DOS   2 952 483   27.12.79/31. 7.79   Hokko Chemic.

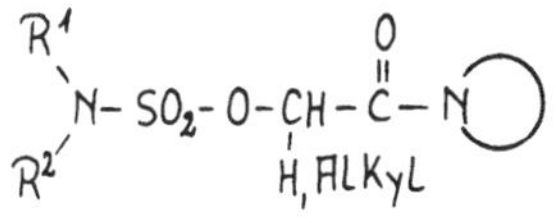

R¹ u. R² = Alkyl, Cycloalkyl, Alkoxy usw.

(Structure: R¹,R²–N–SO₂–O–CH(H,Alkyl)–C(=O)–N⟨ring⟩)

R^1 u. R^2 = Alkyl, Cycloalkyl, Alkoxy usw.

oder Glieder eines Ringes

EP   852   3. 8.77/21. 2.79   Ugine Kuhlmann

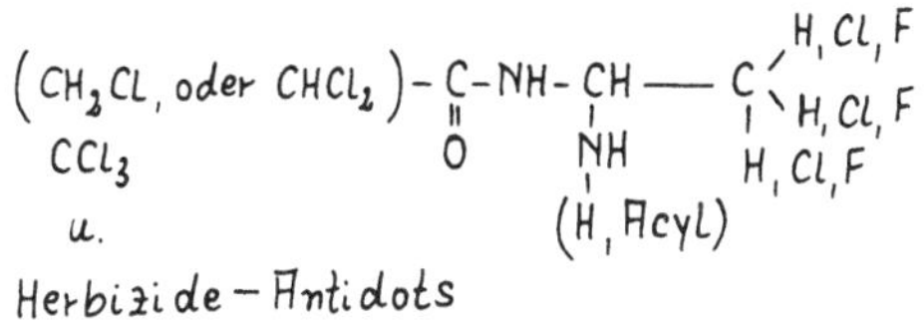

(Structure: (CH₂Cl, oder CHCl₂ / CCl₃)–C(=O)–NH–CH(NH(H,Acyl))—C⟨H,Cl,F ; H,Cl,F ; H,Cl,F⟩)

u.

Herbizide – Antidots

EP   5 321   4. 4.79/14.11.79   Monsanto
US Prior.   6. 4.78

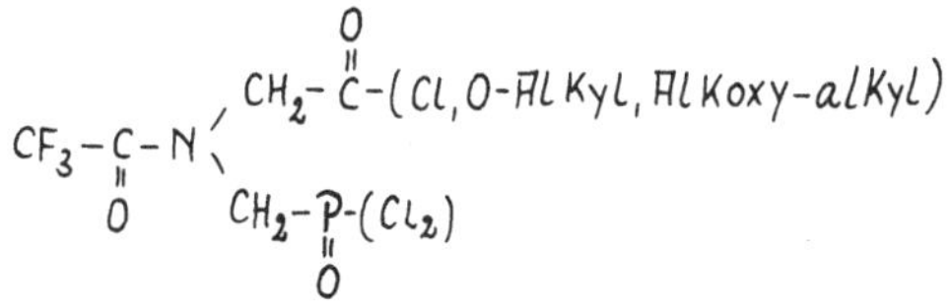

(Structure: CF₃–C(=O)–N⟨CH₂–C(=O)–(Cl,O-Alkyl,Alkoxy-alkyl) ; CH₂–P(=O)–(Cl₂)⟩)

post emergence Herbizide.

EP   5 341   24. 4.79/14.11.79   Amer. Cyanamid
US Prior.   5. 5.78

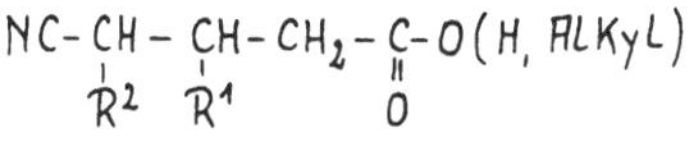

R^1 u. R^2 = Aryl, subst. Pyridyl, Thienyl

(Structure: NC–CH(R²)–CH(R¹)–CH₂–C(=O)–O(H,Alkyl))

EP   6 540   5. 6.79/ 9. 1.80   Bayer
US Prior.   28. 6.78

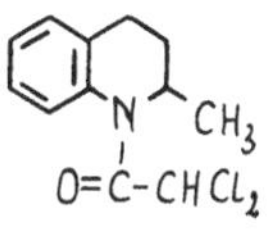

Safener gegen Herbizid-schäden

(Structure: 1,2,3,4-tetrahydroquinoline, N–CH₃, O=C–CHCl₂)

EP   6 541   15. 6.79/ 9. 1.80   Bayer
GE Prior.   28. 6.78
=DOS   2 828 303

$$CHCl_2-\overset{O}{\underset{\|}{C}}-N(CH_2-CH=CH_2)_2$$

Safener gegen Herbizidschäden durch Acetanilide der Formel

EP   6 542   28. 6.78/ 9. 1.80   Bayer

Safener gegen Herbizidschäden
durch Acetanilide.

EP   13 429   29.12.78/23. 7.80   Chem. Werke Hüls
DOS   2 940 231   4.10.79
DOS   2 856 651   29.12.78/10. 7.80

X = H, CH₃, Cl, Br

US   3 981 716   24. 5.73/21. 9.76   Stauffer

R = Naphthyl, Phenyl

A Schutzsubstanz für
Thiocarbamat-Herbizide

US   4 075 007   12. 3.74/21. 2.78   National Patent Dev.

224

US   4 087 278   21. 3.73/ 2. 5.78   Dow Chem.

$$CCl_3\text{-}CH \langle (O\underset{O}{\overset{\|}{C}}\text{-}CCl_2\text{-}CH_2Cl)_2$$

US   4 097 491   10. 6.77/28. 6.78   Velsicol Chem.

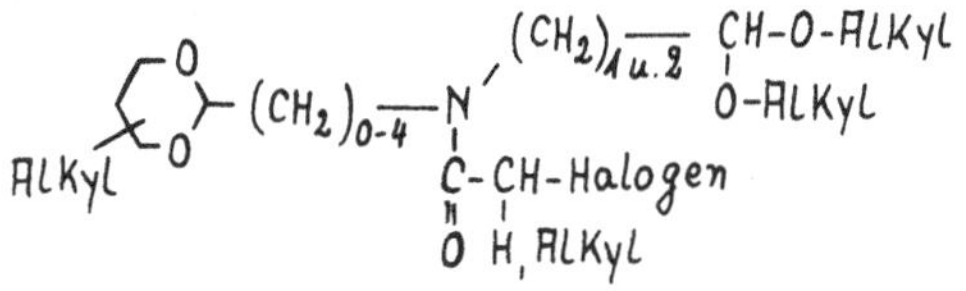

pre emergence Grass-Herbizide

US   4 110 101   6. 4.77/29. 8.78   Velsicol Chem.

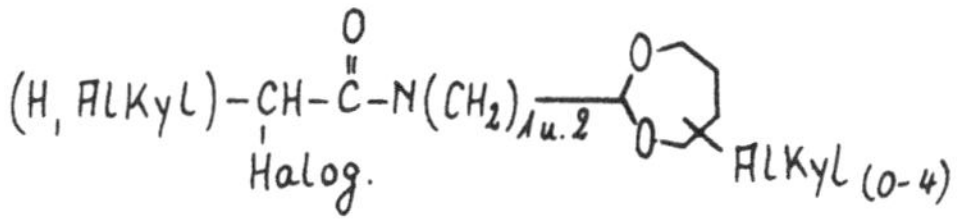

US   4 113 464   4. 4.77/12. 9.78   Velsicol Chem.

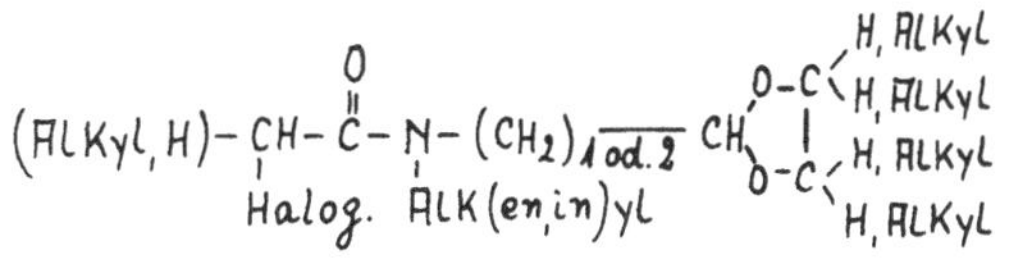

Gräserherbizid u.a. auch gegen Wildhafer und Acker-furchsschwanz

US   4 116 670   9. 4.77/26. 9.78   Velsicol Chem.

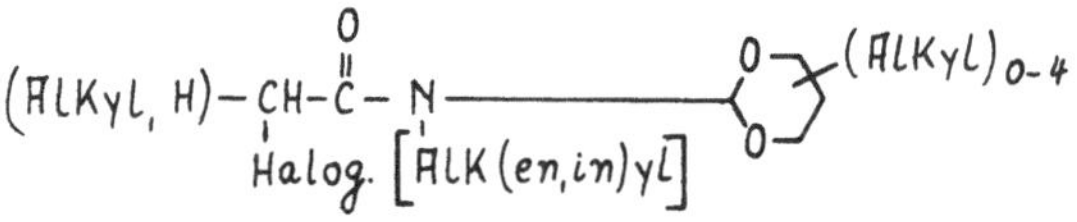

US   4 118 216   25. 4.77/ 3.10.78   Velsicol Chem.

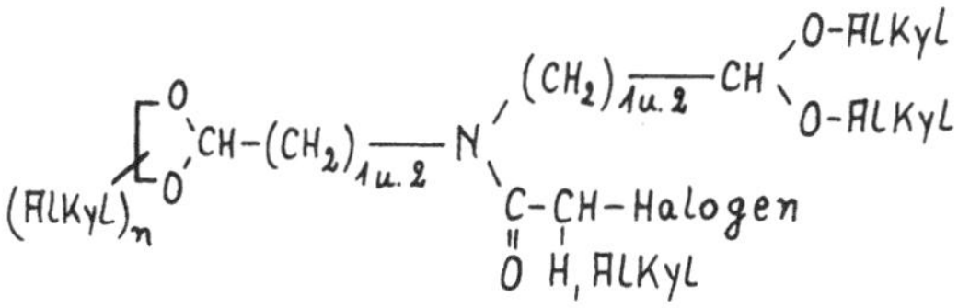

US   4 154 594   17.11.77/15. 5.79   Gaf Corp.

$$R-\underset{\underset{O}{\|}}{C}-NH-\underset{\underset{O}{\|}}{C}-R$$

R = Alkyl, Alkenyl, Halog.-
Alkyl

US   4 154 595   4. 4.78/15. 5.79   Stauffer

$$Cl-CH_2-\underset{\underset{O}{\|}}{C}-NH-CH_2-CH\overset{O}{\underset{O}{\big|}}R$$

R = Acetyl-methyl, Alkyl,
Benzyl, Phenoxy, Phenyl

US   4 155 744   17. 6.77/22. 5.79   Monsanto

Y = Cycloalken-1-yl

$$Cl-CH_2-\underset{\underset{O}{\|}}{C}-\underset{\underset{Y}{|}}{N}-\underset{\underset{H,ALK(en)yl}{|}}{\overset{H,ALK(en)yl}{C}}\text{---- Heterocycl.}$$

US   4 155 745   4. 4.78/22. 5.79   Stauffer

R u. R^1 = Alkyl, Halog.-Alkyl,
Phenyl oder H

$$Cl-CH_2-\underset{\underset{O}{\|}}{C}-NH-CH_2\underset{O(S)}{\overset{O(S)}{\big<}}\underset{R^1}{\overset{R}{\big>}}$$

US   4 158 560   17.11.77/19. 6.79   Gaf Corp.

R = CH_2-Cl, CHCl-CH_2Cl,
CHCl-CH_3

R^1 = C_1-C_3 mono-Halogen
z.B. CH_2Cl

$$R-\underset{\underset{O}{\|}}{C}-NH-\underset{\underset{O}{\|}}{C}-R^1$$

US   4 186 130   13.10.72/29. 1.80   Stauffer

R = $CHCl_2$, CBr_3, CCl_3,
CH_2Br usw.

R^1 = CH_3; R^2 = Alkyl

$$R-\underset{\underset{O}{\|}}{C}-N\overset{\big|}{\underset{R^1\quad R^2}{\big|}}O$$

US   4 208 203   17. 9.71/17. 6.80   Gulf Res.

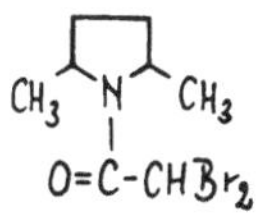

Antidot gegen Schäden im
Mais durch Thiocarbamat-
Herbizide.

BE   853 558   14. 4.76/13.10.77   Thomae

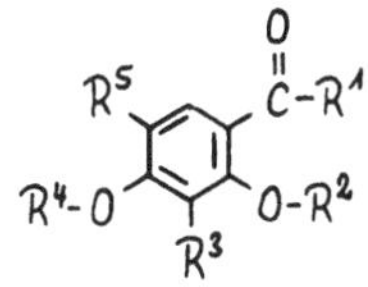

R^1 = Perfluoralkyl
R^2 u. R^4 = <u>Acylreste</u>
R^3 u. R^5 = <u>Alkyl, NO_2,</u>

    Halogen, Alkyl-SO_2-
    Cycloalkyl
unter vielen Verwendungs-
möglichkeiten auch Herbizide
genannt.

BE   868 189   17. 6.77/18.12.78   Monsanto Co.

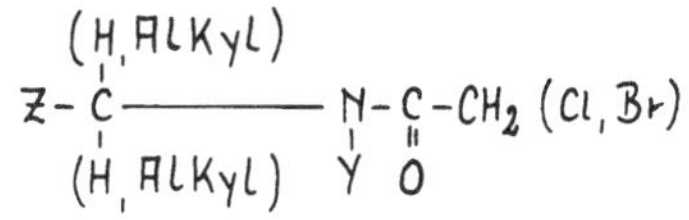

Y u. Z = Cyclo-1-enyl,
         Cycloalkadienyl
Z auch = C_1-C_2 Heterocyclus

    mit O, N, S

BE   874 361   22. 2.79/18. 6.79   Nitrokema

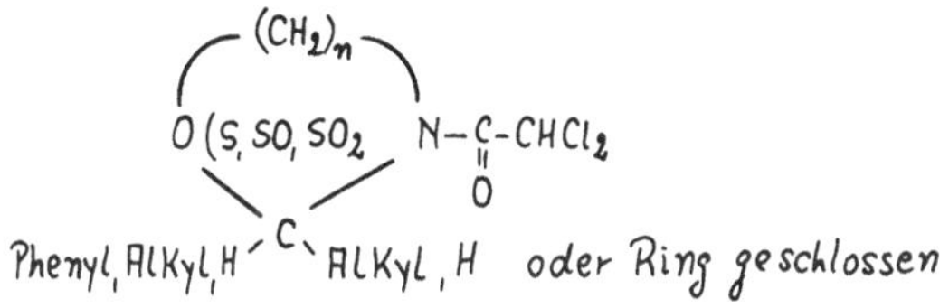

Antidot für Thiolcarbamate,
Triazine, Chloracetanilide,
Phenoxyessigsäuren

FR   2 300 084   10. 2.75/ 8.10.76   Stauffer

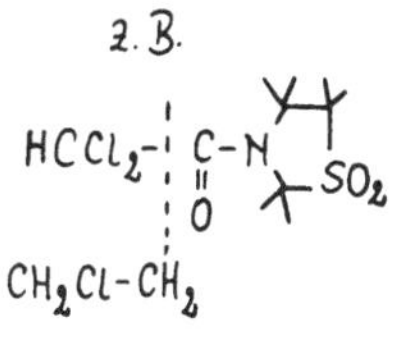

Antidots

GB   1 534 145   25.11.75/29.11.78   Shell Int. Res.

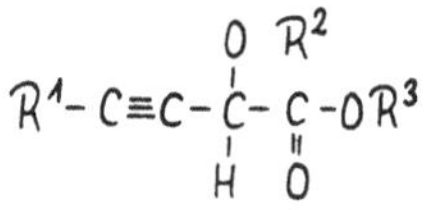

R^1 = H, Alkyl
R^2 = H, Acyl
R^3 = Alkyl, Aryl (subst.)

NE   7 909 358   28.12.78/ 1. 7.80   Hokko Chem.

R^1 u. R^2 = Alkyl, Cycloalkyl, O-Alkyl, S-Alkyl usw.

R^1 u. R^2 Glieder eines Ringsystems

$$\overset{R^1}{\underset{R^2}{\diagdown}}N-SO_2-O-\underset{CH_3}{\overset{|}{C}}H-\underset{O}{\overset{||}{C}}-N\quad\quad z.B.\ N$$

SU   503 849   4. 2.74/16. 3.76   As Belo Phys.

$$CCl_2=CCl-CH_2-C\overset{\diagup O}{\diagdown_{OH}}$$

JA   53 098298   31. 1.77/28. 8.77   Stauffer

Antidot für Herbizide; soll auch die Fischgiftigkeit herabsetzen (ev. zus. mit Ascorbinsäure)

$$Cl_2-CH-\underset{O}{\overset{||}{C}}-N\overset{\diagup}{\diagdown}O(S)$$

b) <u>Aliphatische Amino- und Hydroxylamino(ether)-Carbonsäuren</u>

DOS   2 640 484   8. 9.76/ 7. 4.77   Philagro
FR Prior. 11.9.75 u. 8.7.76

$$Ar-\underset{NH_2}{\overset{|}{C}}-N-O-\underset{(H,Alkyl)}{\overset{|}{C}}H-\overset{\overset{O}{||}}{C}-O\ (H,\ Alkyl,\ Kation)$$

DOS   2 831 983   20. 7.78/15. 2.79   Philagro
FR Prior.   18. 7.77

$$\underset{NH_2}{\overset{S}{\diagdown}}C=N-O-CH-\overset{(H,Alkyl)}{(CN,\ \underset{O}{\overset{||}{C}}-N\diagup)}$$

US   4 071 423   29. 4.77/31. 1.78   Stauffer

Substituenten: (Cyclo)Alkyl, Phenyl usw. u.a. auch Herbizide

$$NC-\overset{|}{\underset{|}{C}}-N=N-\overset{|}{\underset{|}{C}}-CN$$

228

GB   1 455 963   19.10.73/17.11.76   Ajinomoto

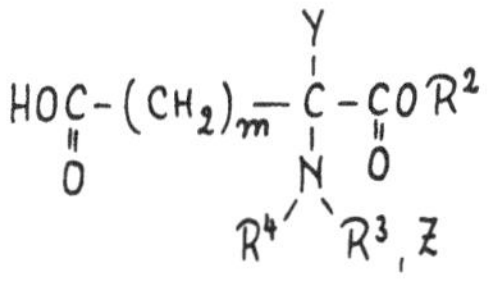

$$HO\underset{\parallel}{\underset{O}{C}}-(CH_2)_{\overline{m}}\underset{\underset{R^4'}{\underset{}{N}}\underset{R^3,Z}{}}{\overset{Y}{C}}-\underset{O}{CO}R^2$$

Y, R^3 u. R^4 = H, CH$_3$

R^2 = C$_2$-C$_{22}$-Alkyl

Z = Anorg. oder org. Säuregruppe

siehe JA 52 111577 (5.7.2. b Diaminodinitrile)

SU   519 885   22. 8.74/30. 9.79   AS UKR Org. Chem.

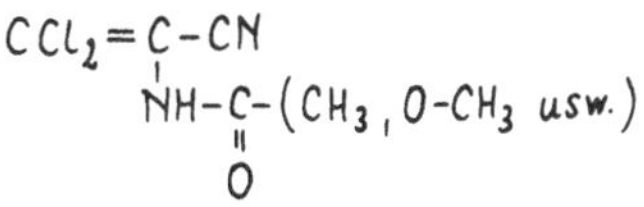

$$CCl_2=\underset{\underset{NH-\underset{\parallel}{\underset{O}{C}}-(CH_3,O-CH_3\ usw.)}{}}{C}-CN$$

Fungicid, bacterizid u. herbizid wirksam

JA   51 128433   30. 4.75/ 9.11.76   Ajionomoto

Höhere Alkyl-aminosäuren mit beliebiger Stellung

der Aminogruppe als Reisherbizid.

c) <u>Aliphatische Di-carbonsäuren, sowie cyclische Imide</u>

EP   10 396   24.10.78/30. 4.80   Fisons
GB Prior.   17. 8.79

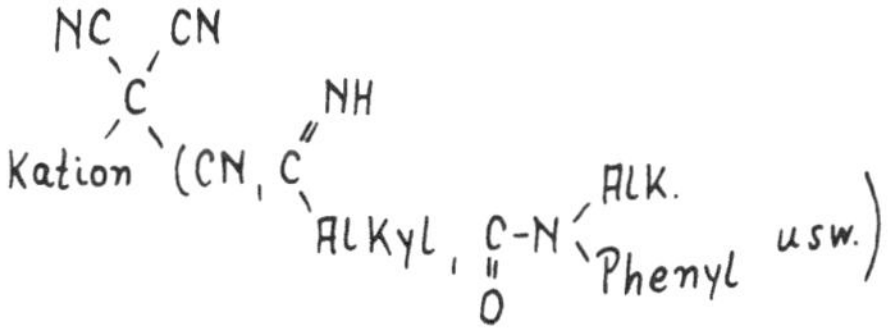

Fungizid + Herbizid

$$\underset{Kation}{}\underset{\ }{\overset{NC\diagdown\ \diagup CN}{\underset{\diagup\ \diagdown}{C}}}\ (CN,\ \underset{\underset{ALKyl,\ \underset{\parallel}{\underset{O}{C}}-N\diagup^{ALK.}_{\diagdown Phenyl}\ usw.}{}}{\overset{NH}{C}})$$

JA   52 111577   12. 3.76/19. 9.77   Kurarzy

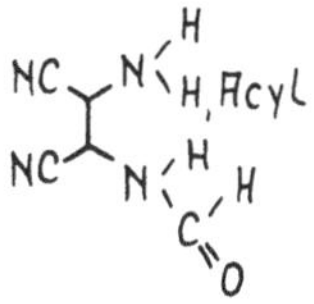

$$\underset{NC}{\overset{NC}{}}\diagup N\diagdown\ ...$$

JA   53 038622   14. 9.76/ 8. 4.78   Godo Shusci

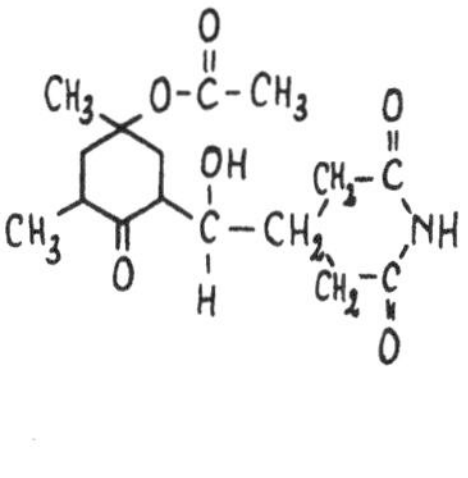

DOS   2 803 991   31. 1.78/10. 8.78   Fisons

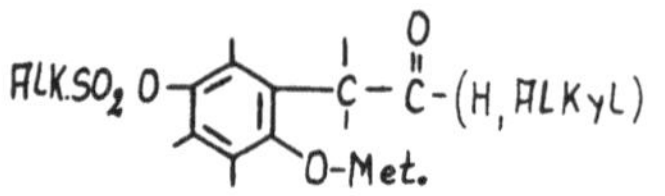

$ALK.SO_2-O-\langle\text{Ring}\rangle-\overset{|}{\underset{|}{C}}-\overset{O}{\overset{\|}{C}}-(H, ALKyl)$ (siehe 5.14.2 a)

(am Ring: O-Met.)

US   4 021 229   5. 9.72/ 3. 5.77   Stauffer

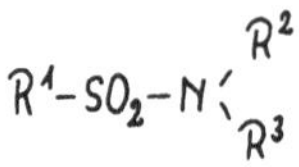

$R^1-SO_2-N\langle \begin{smallmatrix} R^2 \\ R^3 \end{smallmatrix}$

R^1 = Alkyl, Halog.-Alkyl
R^2 u. R^3 = Halog.-Alk(en)yl

Antidots gegen Thiocarbamat-
schäden

US   4 071 349   5. 9.72/31. 1.78   Stauffer

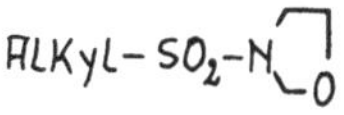

Antidot für Thiocarbamat-
Herbizide.

$ALKyl-SO_2-N\langle \rangle O$ (Morpholinring)

US   4 071 351   5. 9.72/31. 1.78   Stauffer

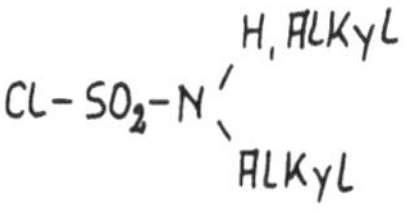

Antidot für Thiocarbamat-
Herbizide.

$Cl-SO_2-N\langle \begin{smallmatrix} H, ALKyl \\ ALKyl \end{smallmatrix}$

US   4 098 599   31. 8.70/ 4. 7.78   Stauffer

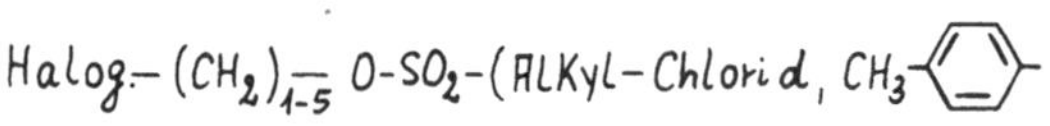

Antidots für Thiocarbamate
gegen Herbizidschäden in Mais.

$Halog.-(CH_2)_{1-5}-O-SO_2-(ALKyl-Chlorid, CH_3-\langle\text{Ring}\rangle-)$

US   4 115 099   31. 8.70/19. 9.78   Stauffer

Antidot gegen Herbizid-Schäden
durch eine Mischung von
Triazinherbiziden mit Thiolcarbamaten.

$Br-CH_2-CH_2-O-SO_2-CH_3$

US   4 125 397   3. 1.77/14.11.78   Chevron Res.

X, Y, Z = Cl, Br

$$\text{C}=\text{NO}-\text{SO}_2-(\text{Chlor})-\text{Alkyl}$$

(with aromatic ring bearing X, Y, Z substituents)

US   4 125 555   21. 5.76/14.11.78   Dow Chem.

$$(\text{CF}_3-\text{SO}_2-\text{O}-\text{CH}_2)_m - \text{C}(\text{CH}_2\,R^1)_{4-m}$$

$R^1 = Br,\ O-$ (phenyl)$-Br_{1-4}$

US   4 146 557   6.12.67/27. 3.79   Exon Res.

R^2-R^7 = H, Cl, F, Alkyl

$$\text{Alkyl}-\text{SO}_2-\underset{\underset{H,Cl}{|}}{N}-\underset{\underset{R^3}{|}}{\overset{\overset{R^2}{|}}{C}}-\overset{\overset{R^4}{|}}{C}=\overset{\overset{R^5}{|}}{C}-\underset{\underset{R^7}{|}}{\overset{\overset{R^6}{|}}{C}}-Cl$$

JA   54 063034   27.10.77/21. 5.79   Mitsubishi Chem.

$$(\text{CH}_3,\,\text{C}_2\text{H}_5)-\text{SO}_2-\underset{\underset{(\text{Alkyl},\ \text{Cyclohexyl})}{|}}{N}-\overset{\overset{O}{\|}}{C}-\text{CH}_2-O-\text{(C}_6\text{H}_3)(\text{Cl},\text{CH}_3)-\text{Cl}$$

5.7.2 <u>Cyclo-aliphatische Carbonsäuren</u>

a) <u>Monocarbonsäuren und Nitrile</u>   Bd.5/504

DOS   2 642 850   23. 9.7ö/ 7. 4.77   Ciba-Geigy
CH Prior. 26.9.75, 29.4.76, 29.7.76

R^1 u. R^2 = H oder zus.

C-C Bindung. Akarizide, Insektizide, Fungizide u.z.T. Herbizide

DOS 2 642 861 23. 9.76/ 7. 4.77 Ciba-Geigy
CH Prior. 26.9.75 u. 23.7.76

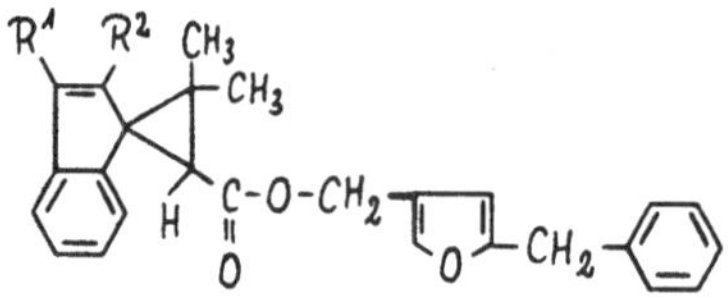

Verwendung wie zuvor

EP 11 670 30.11.78/11. 6.80 Bayer

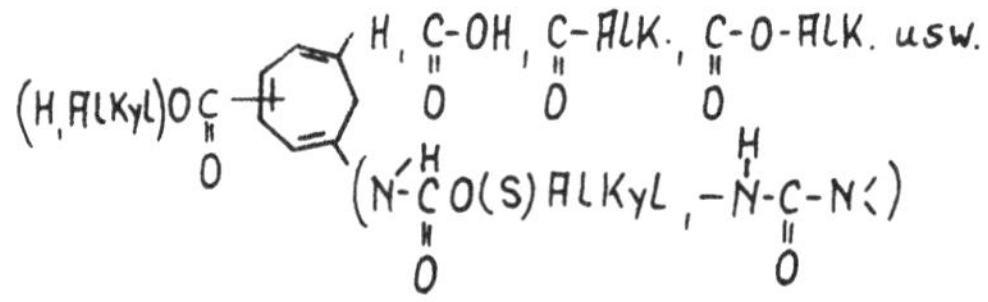

(siehe 5.7.2 a)

b) Dicarbonsäuren und Nitrile

DOS 2 733 115 22. 7.77/26. 1.78 Takeda Chem.
JA Prior. 23. 7.76

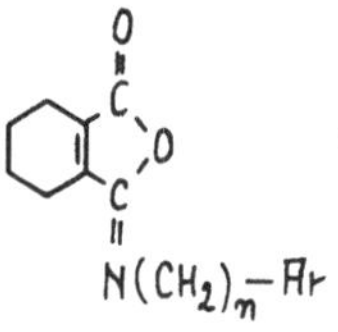

n = 0 u. 1 vorzugsweise

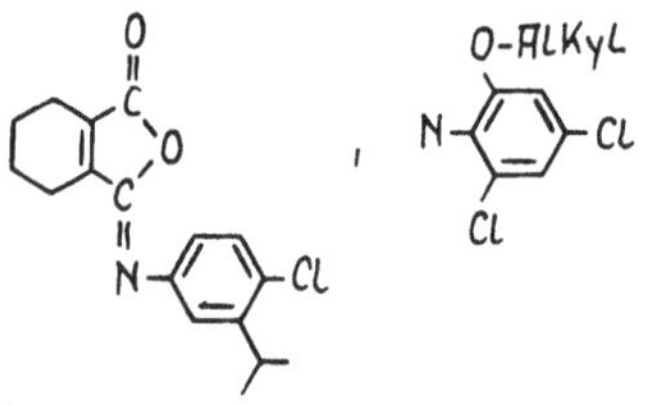

Gegen Unkräuter u. Ungräser in Reis.

JA 50 121428 12. 3.74/23. 9.75 Mitsubishi Chem.

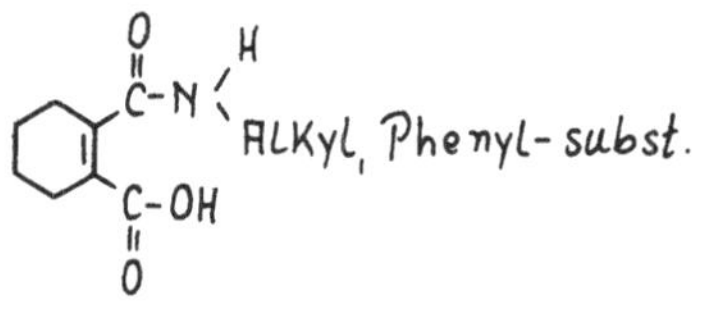

5.7.3 Aromatisch-aliphatische Carbonsäuren und Nitrile

a) Ar-aliphatische Carbonsäuren, Ester, Amide und Nitrile   Bd. 5/505

DOS   2 554 532   4.12.75/16. 6.77   Bayer

$$Cl-\langle\bigcirc\rangle-CH_2\text{-}CH\text{-}\underset{O}{\overset{}{C}}\text{-}O\text{-}CH_3 \;+\; C_{12}H_{25}\text{-}(O\text{-}CH\text{-}CH_2)_n\text{-}OH$$

(mit Cl am CH)

DOS   2 647 484   21.10.76/27. 4.78   Bayer

$$Cl-\langle\bigcirc\rangle-CH_2\text{-}CH\text{-}\underset{O}{\overset{}{C}}\text{-}O\text{-}CH_3 \;+\; \text{Paraffinöl}$$

(mit Cl am CH)

DOS   2 717 279   11. 4.77/26.10.78   Amer. Cyanamid
=US 4 164 415

$$\text{Naphthyl-}CH(\text{Alkyl})\text{-}\overset{O}{\overset{\|}{C}}\text{-}O\text{-}(CH(H,CN)\text{-}\langle\bigcirc\rangle\text{-subst.}, \; O\text{-}\langle\bigcirc\rangle\text{-subst.})$$

unter anderem auch Herbizide

DOS   2 741 171   18.10.76/20. 4.78   Kao Soap

$$\langle\bigcirc\rangle_{X_n}\text{-}S(O)_{1\,u.\,2}\text{-}CH=CH\text{-}\underset{O}{\overset{}{C}}\text{-}O(H, \text{Alkyl})$$

X = Halog., $S(O)_{1\,u.\,2}$-Alkyl

n = 0-3

DOS   2 946 024   14.11.79/29. 5.80   Ishihara Chem.

$$\left(\langle\bigcirc\rangle_{\text{subst.}} \; \text{oder} \; \langle\bigcirc_N\rangle_{\text{subst.}}\right) O\text{-}\langle\bigcirc\rangle\text{-}\underset{(H, \text{Alkyl})}{\overset{}{C}}\text{-}CH=CH\,(CH_2OH, \underset{O}{\overset{}{C}}O\text{Alkyl}, \underset{O}{\overset{}{C}}\text{-}N\langle \; \text{usw.})$$

EP   5 341   24. 4.79/14.11.79   Amer. Cyanamid
US Prior.   5. 8.78
                                                          pre und post emerg. Herbizide

statt Phenyl auch Pyridyl oder Thienyl

BE   845 827   4. 9.75/ 3. 3.77   Ciba-Geigy

BE   870 066   30. 8.78/28. 2.79   Ciba-Geigy

                                                          Herbizide-Antidot

BE   870 067   30. 8.78/28. 2.79   Ciba-Geigy

SU   535 285   7. 1.75/20.12.76   Ufa Plant Prot.

JA   50 135229   18. 4.74/27.10.75   Ajinomoto

                                                          Selektiv in Reis.

234

JA   53 053638   21.10.76/16. 5.78   Sumitomo Chem.

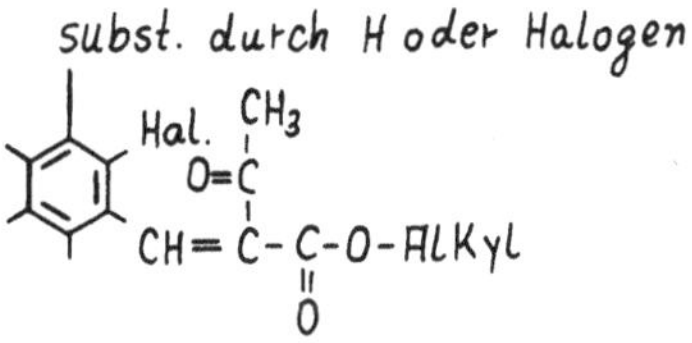

subst. durch H oder Halogen

$$CH=\overset{\overset{\displaystyle CH_3}{|}}{\underset{\underset{O}{\parallel}}{\overset{\overset{O=C}{|}}{C}}}-C-O-Alkyl \quad (Hal.)$$

JA   54 126 726   24. 3.78/ 2.10.79   Tanabe

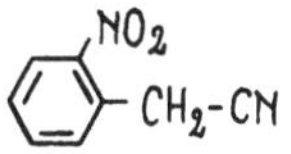

$$\text{o-}NO_2\text{-}C_6H_4\text{-}CH_2\text{-}CN$$

235

5.7.3.b.1 α-Phenoxy-essigsäuren, Phenthio-essigsäuren und
             ihre Ester
    b.2   -Amide
    b.3   Anilide u. Amide von heterocyclischen Aminen
    c α-Phenoxy-propionsäure-derivate u. höhergliedrige
             Phenoxy-alkan-Carbonsäuren,   Ester u. Amide nebst
             Aniliden u. Amiden von Amino-heterocyclen.

Die selteneren Phenthioalkansäuren werden ebenfalls hier registriert.

Die früher getrennten Gruppen werden hier wegen zu häufiger Überschneidungen zusammengefasst. Es werden auch umfasst Diphenylether-
oxyessigsäuren, Pyridyl-phenyl-ether-oxy-alkancarbonsäuren und andere
Heterocycl.-phenylether-oxy-alkancarbonsäuren-derivate.   Bd.5/508

DOS   2 535 832   12. 8.75/24. 2.77   Lentia

Reinigung von Phenoxy-alkan-carbonsäuren.

DOS   2 537 201   21. 8.75/ 3. 3.77   Bayer
= BE   845 382

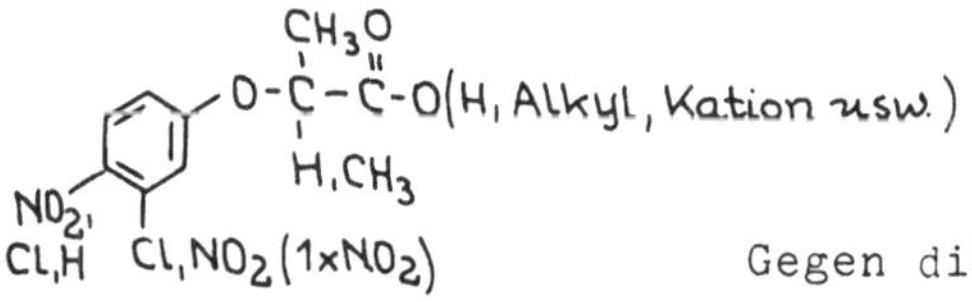

Gegen dikotyle Unkräuter.

DOS   2 601 376   15. 1.76/21. 7.77   Bayer
= BE   850 372

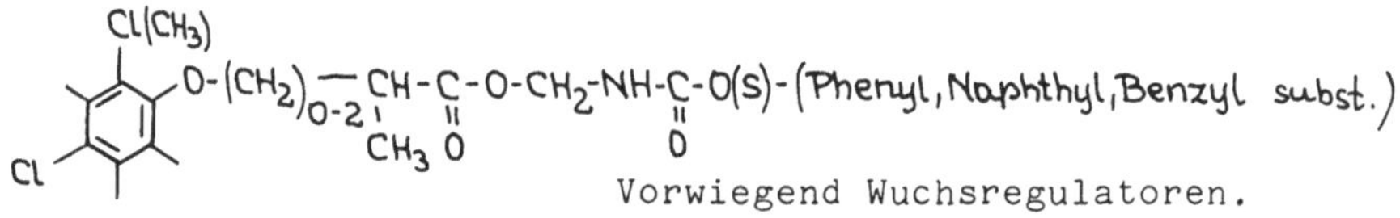

Vorwiegend Wuchsregulatoren.

DOS   2 601 548   16. 1.76/21. 7.77   Höchst
= BE   850 450
Siehe Grundpatent 2 223 894

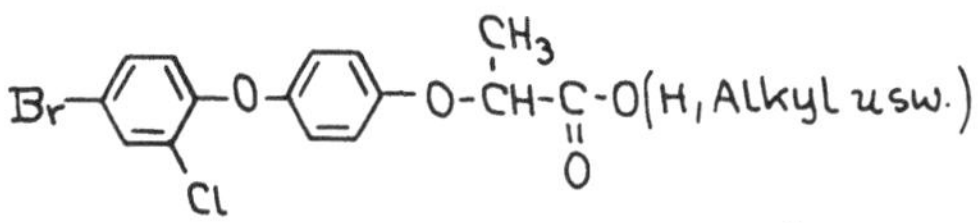

Überlegene herb. Wirkung.

DOS   2 609 461   8. 3.76/22. 9.77   Höchst
= BE   852 210

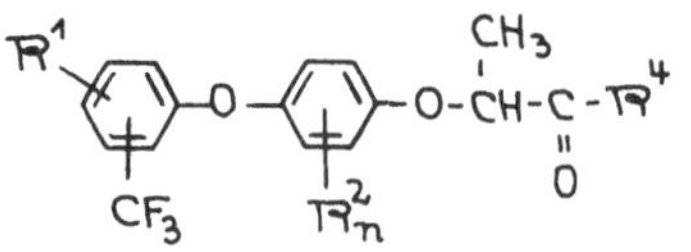

R^1 = H, Halogen, R^2 = Halog.,
    Alkyl., CF_3,
R^4 = HO-, O-Alkyl,
    $ArO(CH_2)_n$-O-, O-Phenyl,
    NH-Ar usw.

236

DOS   2 613 434   20. 3.76/13.10.77   Bayer

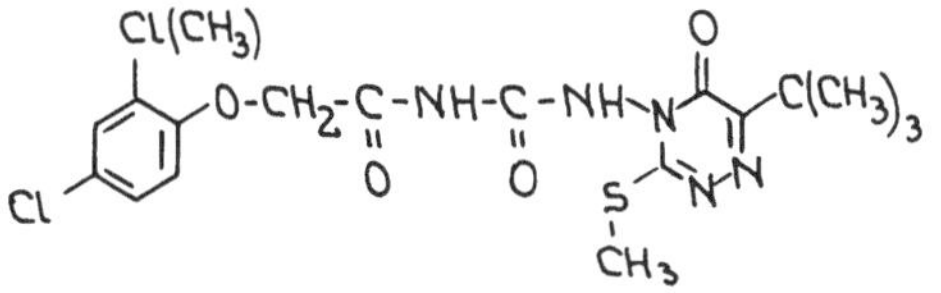

$Cl(CH_3)$

$O-CH_2-\overset{O}{\underset{O}{C}}-NH-\overset{O}{\underset{O}{C}}-NH-N$ ... $C(CH_3)_3$, $S-CH_3$

DOS   2 613 697   31. 3.76/13.10.77   Höchst

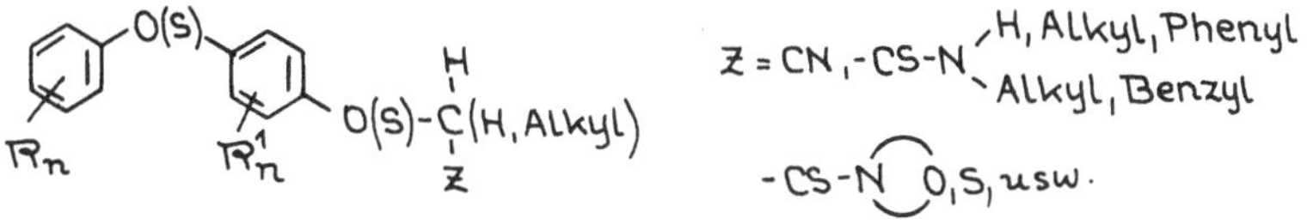

$Z = CN$, $-CS-N\underset{Alkyl, Benzyl}{\overset{H, Alkyl, Phenyl}{}}$

$-CS-N\,O,S,$ usw.

R_n ... R_n^1 ... $O(S)-\overset{H}{\underset{Z}{C}}(H, Alkyl)$

DOS   2 617 804   23. 4.76/ 3.11.76   Höchst

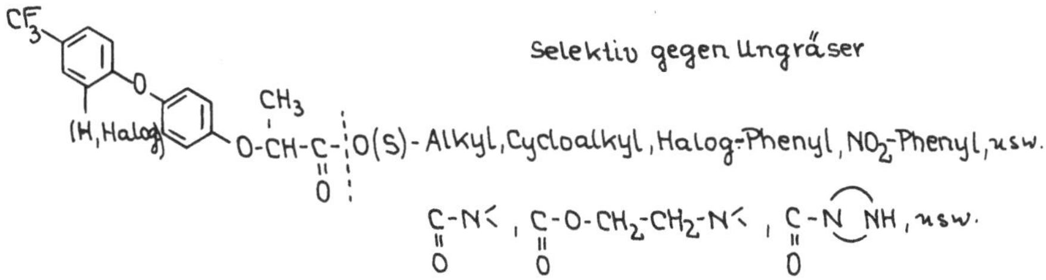

Selektiv gegen Ungräser

CF_3 ... (H, Halog) ... $O-\overset{CH_3}{\underset{O}{CH-C}}-O(S)-$ Alkyl, Cycloalkyl, Halog-Phenyl, NO_2-Phenyl, usw.

$\overset{}{\underset{O}{C}}-N<$, $\overset{}{\underset{O}{C}}-O-CH_2-CH_2-N<$, $\overset{}{\underset{O}{C}}-N\,NH$, usw.

DOS   2 623 558   26. 5.76/15.12.77   Höchst
= BE   855 094

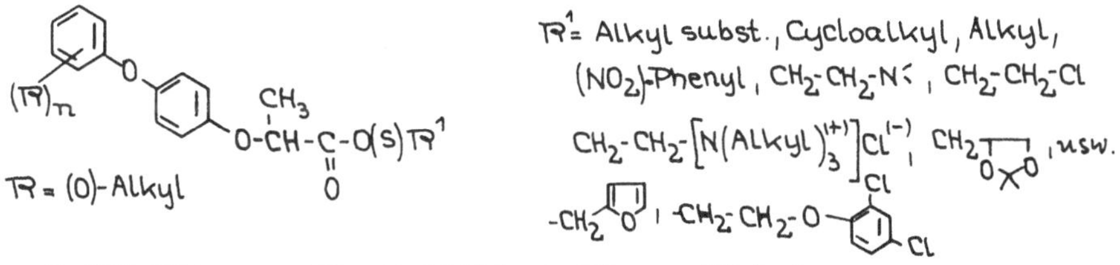

R^1 = Alkyl subst., Cycloalkyl, Alkyl,
(NO_2)-Phenyl, $CH_2-CH_2-N<$, CH_2-CH_2-Cl
$CH_2-CH_2-[N(Alkyl)_3^{(+)}]Cl^{(-)}$, CH_2 ... usw.
$-CH_2$ ... $-CH_2-CH_2-O$ ... Cl, Cl

$(R)_n$ ... $O-\overset{CH_3}{\underset{O}{CH-C}}-O(S)R^1$

$R = (O)$-Alkyl

DOS   2 632 581   20. 7.76/26. 1.78   Höchst
= BE   857 001

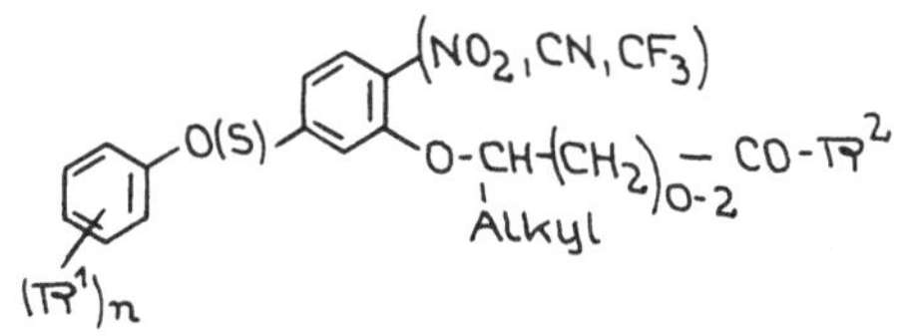

R^2 = OH, O-Alkyl(subst.)
$NH-NH_2$, $NH-C_6-H_5$
$O(S)$-Phenyl usw.

(NO_2, CN, CF_3) ... $O-\overset{}{\underset{Alkyl}{CH}}(CH_2)_{0-2}- CO-R^2$

$(R^1)_n$

Pre u. post emergence Wirkung gegen dikotyle Unkräuter.

DOS   2 635 099   4. 8.76/ 9. 2.78   Celamerk
= BE   857 444

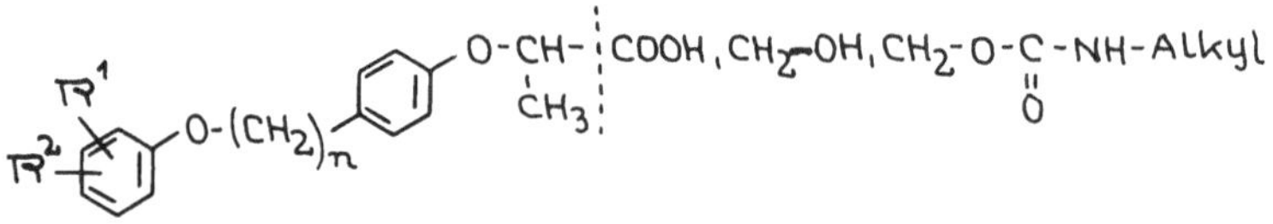

R^1 = Halog.; R^2 = H, Halogen

DOS   2 635 100   4. 8.76/ 9. 2.78   Celamerk

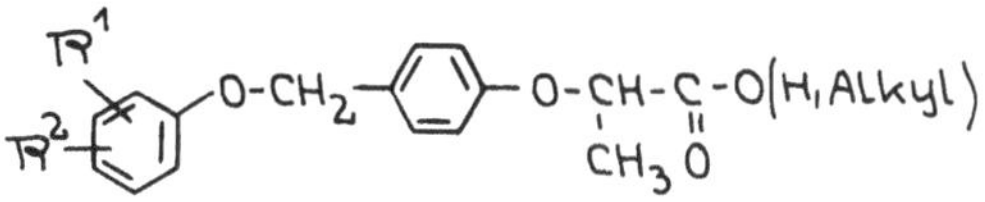

gegen Unkräuter.

R^1 = H, F, Cl, Br, I
R^2 = H, F, Cl, Br, I

DOS   2 639 796   3. 9.76/31. 3.77   Rohm u. Haas
                           US Prior. 3. 9.75; 30. 7.76

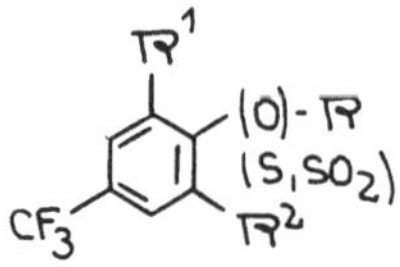

R^1 = H, Halog.; R^2 = H, Halog.,CN

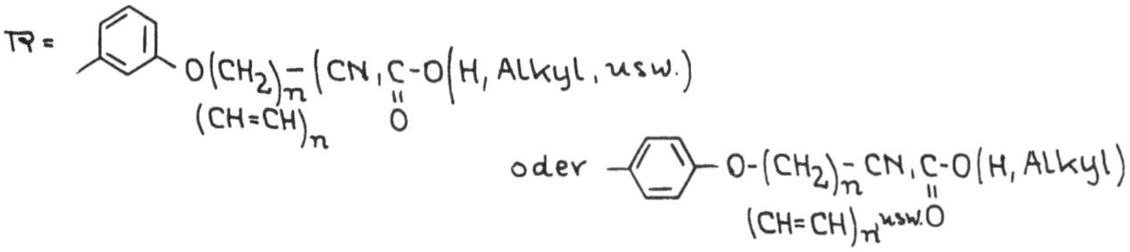

oder

DOS   2 640 730   10. 9.76/16. 3.78   Höchst

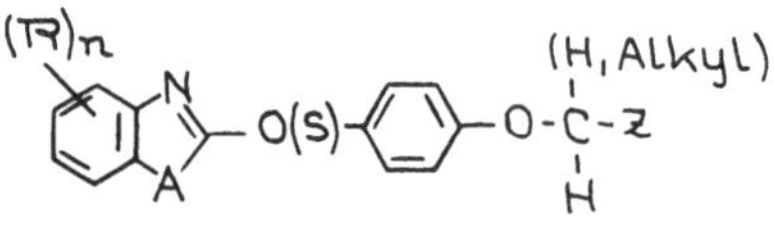

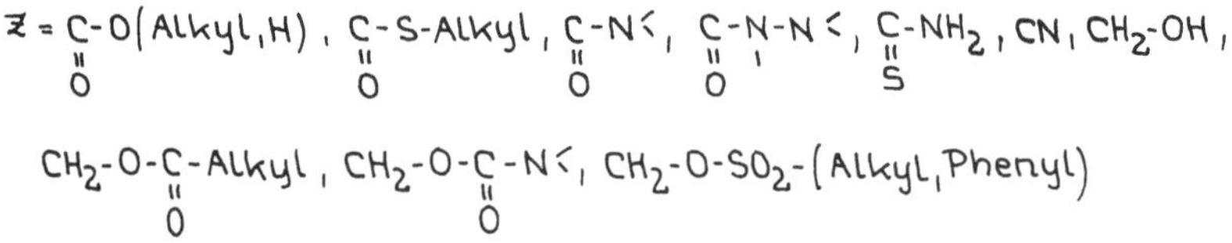

R = Halogen, NO_2,CN, NH_2, Alkyl, CF_3, O-Alkyl usw.
    n = 0 - 4
A = O, S, NH, N-Alkyl

238

DOS   2 643 438   27. 9.76/ 7. 4.77   Ishihara Sangyo

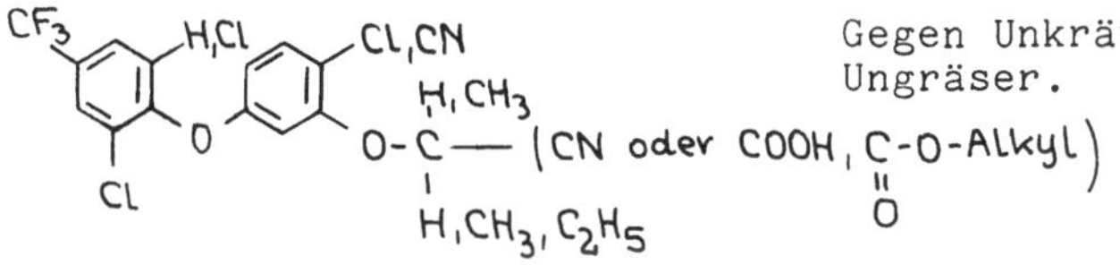

Gegen Unkräuter u. z. T. Ungräser.

DOS   2 646 124   13.10.76/27. 4.78   Höchst

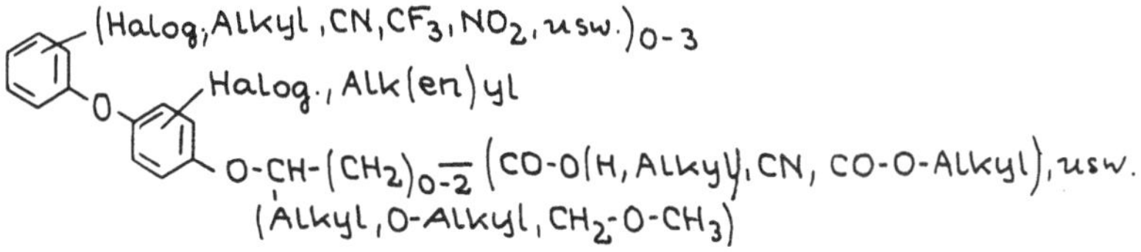

DOS   2 649 706   29.10.76/ 5. 5.77   Ishihara Sangyo

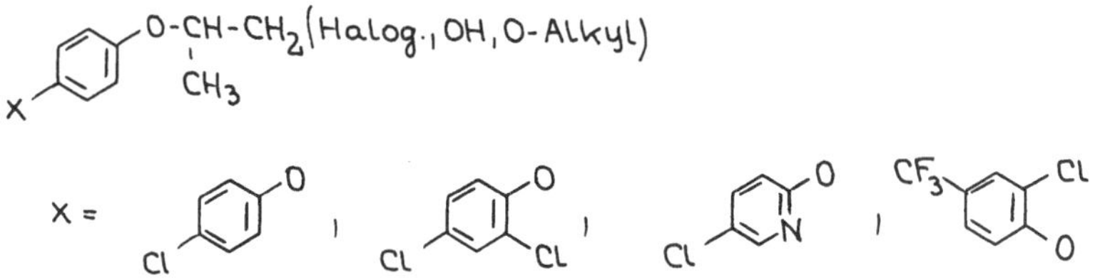

DOS   2 652 384   17.11.76/26. 5.77   Ciba Geigy
                                                  CH Prior. 20.11.75

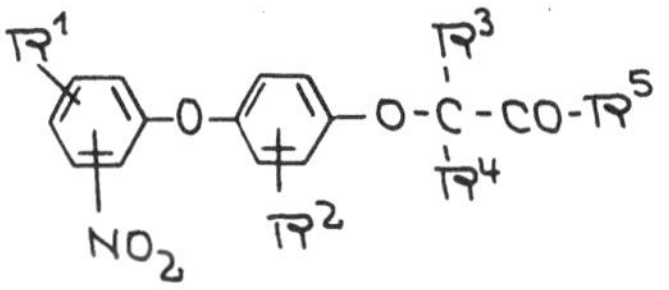

R^1 = Halog., CF_3, (O,S,)-Alkyl
       CN, SO_2-N$<$, CO-O(H,Alkyl,N$)$
R^2 = H, Halog., Alkyl

R^3 u. R^4 = H, Alkyl
R^5 = , OH, O-Kation- O-Alkyl, S-Alkyl, O-Alk(en,in)yl, O-CH_2-C_6-H_5,
       N(Alkyl)$_2$ usw.

DOS   2 701 108   12. 1.77/13. 7.78   Stauffer Chem.

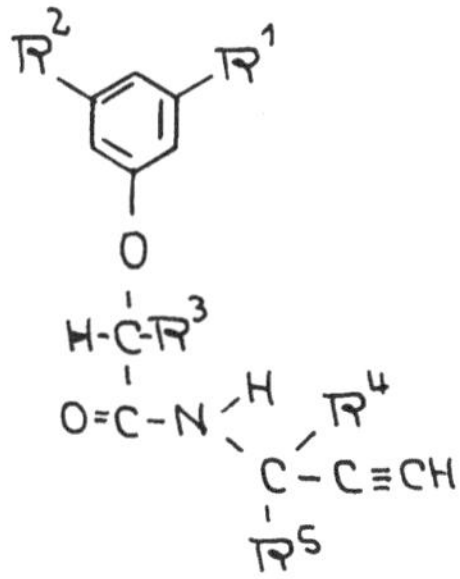

R^1 u. R^2 = Halogen, Alkyl
R^3 = Alkyl
R^4 = H, Alkyl
R^5 = H, Alkyl

DOS   2 709 032   2. 3.77/ 7. 9.78   Höchst

$$Y = \text{Halogen}, C_1-C_4, \text{Alkyl}, CF_3$$

$$(Y)_n\text{—}\langle\text{—}\rangle\text{—O—}CH_2\text{—}\langle\text{—}\rangle\text{—O—CH—}\Big(\underset{O}{\overset{||}{C}}\text{-OH}, \underset{O}{\overset{||}{C}}\text{-O-Alkyl}, CH_2\text{-O-H}, CH_2\text{-O-Alkyl},$$

$$n = 1 \text{ od. } 2, \qquad CH_3$$

Gräser herbizid.

$$CN, CO\text{-O-}CH_2\text{-}C_6H_5, \text{usw.}$$

DOS   2 714 662   8. 4.76/20.10.77   Ishihara Sangyo

$$CF_3\text{—}\langle\text{—}\rangle\text{—O—}\langle\text{—}\rangle\text{—O—}\underset{|}{\overset{CH_3}{C}}\text{-CO-O}(H, \text{Alkyl})$$

$$\left[Cl\text{—}\langle\overset{Cl}{\underset{N}{}}\rangle\text{—}\right] \qquad \underset{O}{\overset{||}{C}}\text{-O}(H, \text{Alkyl})$$

Gegen Gräser.

DOS   2 715 284   14. 4.76/27.10.77   Ishihara Sangyo

$$CF_3\text{—}\langle\text{—}\rangle\text{—O—}\langle\text{—}\rangle\text{—O—}\underset{CH_3}{CH}\text{-}CH_2\text{-}CH_2\text{-}\underset{O}{\overset{||}{C}}\text{-O}(H, \text{Alkyl}, \text{Benzyl})$$

$$H,Cl \quad H,Cl$$

$$Cl\text{—}\langle\overset{}{\underset{N}{}}\rangle\text{—}Cl$$

DOS   2 715 319   5. 4.77/ 3.11.77   Ishihara Sangyo
= BE   853 355   JA Prior. 13. 4.76   27. 7.76

$$X\text{—}\langle\underset{NO_2}{\text{—}}\rangle\text{—O—}\langle\text{—}\rangle\text{—O—}\underset{O}{\overset{CH_3}{CH}}\text{-}C\text{-O}(H, \text{Alkyl}, C_6H_5\text{-}CH_2\text{-})$$

$$O \quad \text{-N}\langle, \text{-N-}$$

$$\langle\text{—}\rangle\text{-}Cl$$

X = Halogen
Selektiv gegen Ungräser.

DOS   2 716 189   12. 4.77/27.10.77   Société de Rech. Ind.
= BE   845 308   20. 8.75/27. 2.77   GB Prior. 15. 4.76

$$\underset{X^3}{\overset{X^1}{X^2}}\text{—}\langle\text{—}\rangle\text{-Z-}\langle\text{—}\rangle\text{—O—}\underset{O}{\overset{CH_3}{CH}}\text{-}C\text{-O}(H, \text{Alkyl})$$

$$Z = O\text{-}CH_2, CH_2\text{-O}, CH=CH, CH_2\text{-}CH_2,$$

$$\underset{O}{\overset{||}{C}}\text{-}CH_2, CH_2\text{-}\underset{O}{\overset{||}{C}}, O, S, SO_2$$

u.a. auch herbizid wirksam.

DOS   2 718 310   25. 4.77/17.11.77   Philagro
                                                          FR Prior. 27. 4.76

$$\langle\overset{O\text{-}CH_2\text{-CO-N}\diagup H}{\underset{Cl}{\underset{CH_3}{}}}\rangle \diagdown CH_3$$

Selektiv in Weizen, Mais, Erdnüssen,
Zuckerrohr.

DOS   2 725 590   7. 6.77/21.12.78   Bayer
= EP   50

Gräser-Herbizid   R = H, Halog., Alkyl; n = 1 - 4

DOS   2 729 602   30. 6.77/11. 1.79   Celamerk

R^1 = H, Halogen
R^2 = H, Halogen   Selektiv in Mais.

DOS   2 730 591   9. 7.76/12. 1.78   Ciba Geigy

R^1 = H, Halog., CF_3, CN; R^2 = H, Halog.

Gegen Gräser in mono- oder dicotyledonen Kulturen wie Soja, Baumwolle,
Zuckerrüben, Getreide.

DOS   2 731 214   11. 7.77   1. 2.79 Celamerk

DOS   2 731 960   14. 7.77/ 2. 2.78   Stauffer Chem.
= BE   857 181   US Prior. 30. 7.76

R = 1 x CF_3, 3,5-Dimethyl
n = 1 u. 2; 3,5-Di-chlor

DOS   2 732 442   18. 1.77/26. 1.78   Ciba Geigy
                                                    CH Prior. 21. 7.76

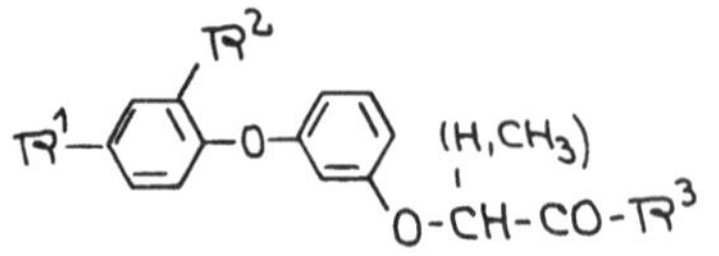

[Structure: R^1-/R^2-substituted phenyl-O-phenyl-O-CH(H,CH$_3$)-CO-R^3]

R^3 = OH, O-Kation$^+$, O-Alkyl, S-Alkyl,
        N<
        O(S)-Alkenyl, O-Cycloalkyl
        >C = NO-, O-(S)(CH$_2$)$_{0 \text{ u. } 1}$-Ar
        subst.

R^1 u. R^2 = 1 x CF$_3$, das andere H,
        Halogen oder NO$_2$ u. CN

Wirksam gegen dikolyle Pflanzen, wenig wirksam gegen Ungräser.
Selektiv in Reis.
Zu beachten ist die m-Stellung des Oxypropionsäure-Restes.

DOS   2 732 846   20. 7.77/26. 1.78   Ciba Geigy
                                                    CH Prior. 23. 7.76

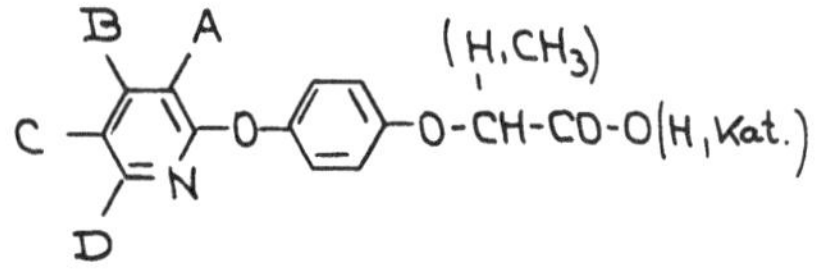

[Structure: pyridine ring with substituents A, B, C, D and N, -O-phenyl-O-CH(H,CH$_3$)-CO-O(H,Kat.)]

A = H, Halog.(O)-Alkyl CN, NO$_2$,
    CS-NH$_2$

B = H, Halog. CN, Alkyl O-Alkyl,
    $\overline{N}$< , CS-NH

C = H, Halog., CN, NO$_2$, CS-NH$_2$

D = H, Halog, N< , Alkyl

Gegen Mono u. Dikotyledone wirksam.
Auch Wuchsregulierende Eigenschaften.
Siehe BE 834 495 4-(Pyrid-2-yloxy)alkancarbonsäuren mit Gräserwirkung.

DOS   2 733 179   22. 7.77/18. 5.78   Höchst AG

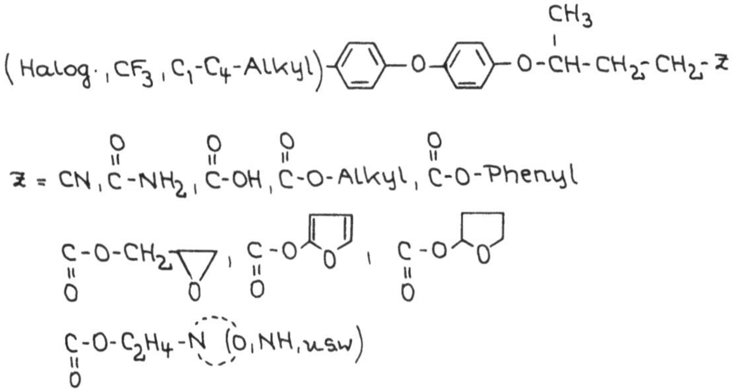

[Structure: (Halog., CF$_3$, C$_1$-C$_4$-Alkyl)-phenyl-O-phenyl-O-CH(CH$_3$)-CH$_2$-CH$_2$-Z]

Z = CN, C(=O)-NH$_2$, C(=O)-OH, C(=O)-O-Alkyl, C(=O)-O-Phenyl

C(=O)-O-CH$_2$-[epoxide] , C(=O)-O-[furyl] , C(=O)-O-[tetrahydrofuryl]

C(=O)-O-C$_2$H$_4$-N (O, NH, usw)

DOS   2 734 667   1. 8.77/ 9. 2.78   Shell Int. Res.
= BE   857 192   GB Prior. 2. 8.76

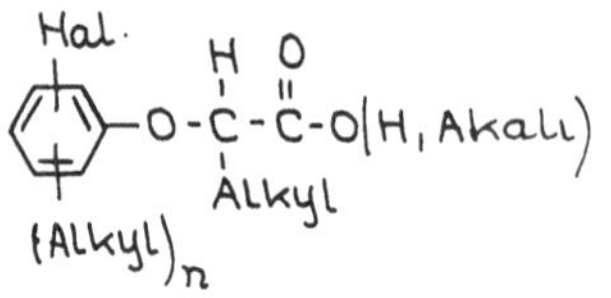

[Structure: Hal./(Alkyl)$_n$-substituted phenyl-O-CH(Alkyl)-CH(H)-C(=O)-O(H, Alkali)]

Darstellung optisch aktiver
Verbindungen.

242

DOS   2 739 280   31. 8.77/ 2. 3.78   Rohm u. Haas USA
                                                  US Prior. 31. 8.76

X = H, Halogen, Alkyl, CF_3, CN
Y = H, Halogen, CF_3
Z = C_1-C_4-Alkylen

DOS   2 741 171   18.10.76/20. 4.78   Kao Soap

X = Halog., $S(O)_{1u.2}$-Alkyl
n = 0 - 3

DOS   2 745 462   8.10.77/12. 4.79   Bayer

Selektiv gegen Gräser.

DOS   2 745 482   10.10.77/19. 4.79   Höchst

II dient als Synergist für das Gräserherbizid I̲.

DOS   2 745 869   12.10.77/19. 4.79   Höchst

Gräser-Herbizid

DOS   2 748 658   29.10.77/10. 5.79   Höchst

DOS   2 750 527   11.11.77/24. 4.78   Siegfried AG
                                                    CH Prior. 15.11.76

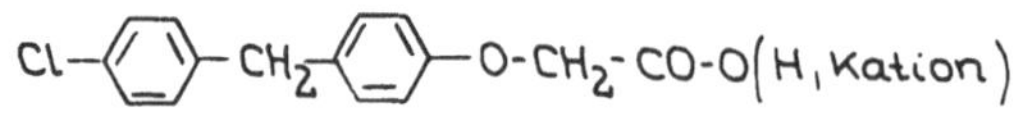

$$Cl-\bigcirc-CH_2-\bigcirc-O-CH_2-CO-O(H, Kation)$$

DOS   2 755 536   13.12.77/21. 6.79   Ciba Geigy

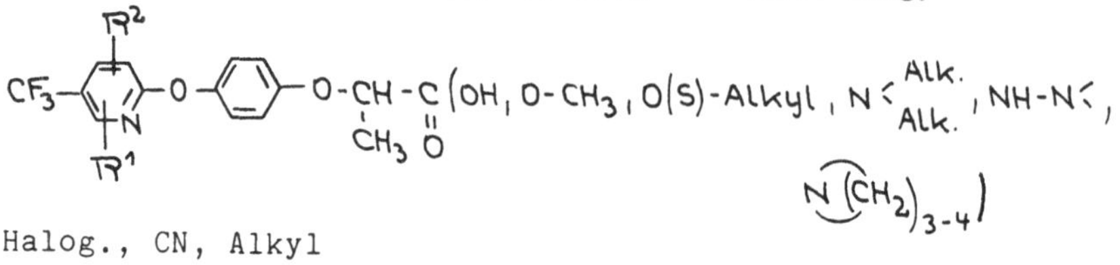

$$CF_3-\text{(Pyridin)}-O-\bigcirc-O-CH-C(OH, O-CH_3, O(S)-Alkyl, N{<}^{Alk.}_{Alk.}, NH-N{<}, N(CH_2)_{3-4})$$

mit Substituenten R^1, R^2 am Ring und CH_3, O an der Seitenkette.

R^1 = H, Halog., CN, Alkyl
R^2 = H, Halog., CN, Alkyl

DOS   2 758 002   24.12.77/ 5. 7.79   Höchst

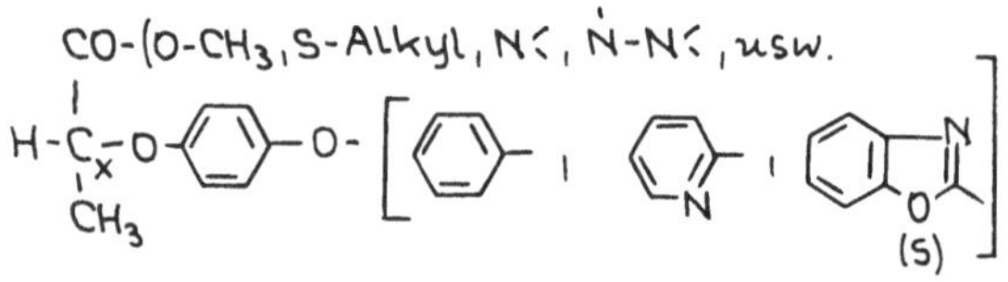

$$CO-(O-CH_3, S-Alkyl, N{<}, N-N{<}, usw.$$
$$H-C_x-O-\bigcirc-O-[\bigcirc-, \text{(Pyridin)}, \text{(Benzoxazol)}(S)]$$
mit CH_3.

Optisch aktive
Herbizide.

DOS   2 802 818   23. 1.78/26. 7.79   Celamerk

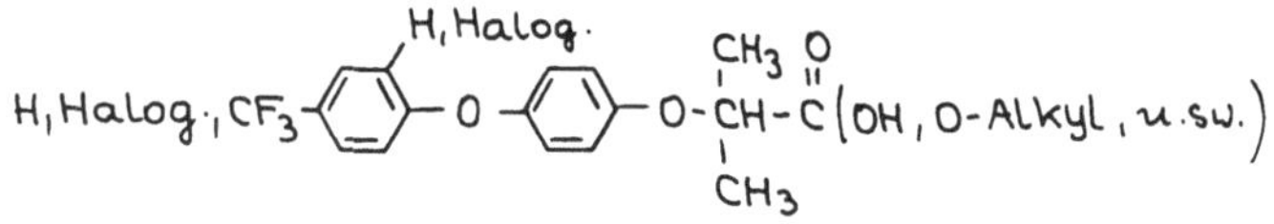

$$H, Halog.; CF_3-\text{(H, Halog.)}-O-\bigcirc-O-CH-C(\overset{O}{\overset{\|}{C}})(OH, O-Alkyl, u.sw.)$$
mit CH_3, CH_3 an der Seitenkette.

Zusammen mit Antidots zur Saatbehandlung.

Antidots: $CHCl_2-\overset{O}{\overset{\|}{C}}-N{<}^{Alk(en)yl}_{Alk(en)yl}$; $BrCH_2-CHBr-CO-N{<}$

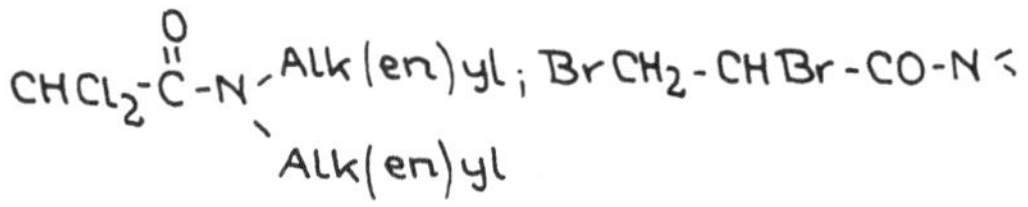

DOS   2 804 074   31. 1.78/10. 8.78   F. Hoffmann - La Roche
= BE   863 436   31. 1.77/31. 7.78   OE Prior 31. 1.77;
                                                                  31.1o.77

$$Cl-\text{(H, Cl, Alkyl)}-\bigcirc-O-\bigcirc-O-\overset{H}{\underset{Alkyl}{C}}-\overset{O}{\overset{\|}{C}}-O-Y$$

$$Y = -\overset{(H, Alkyl)}{\underset{(H, Alkyl)}{C}}-C{\equiv}CH, (CH_3, H)\text{(H, CH}_3)\text{(Lacton)}$$

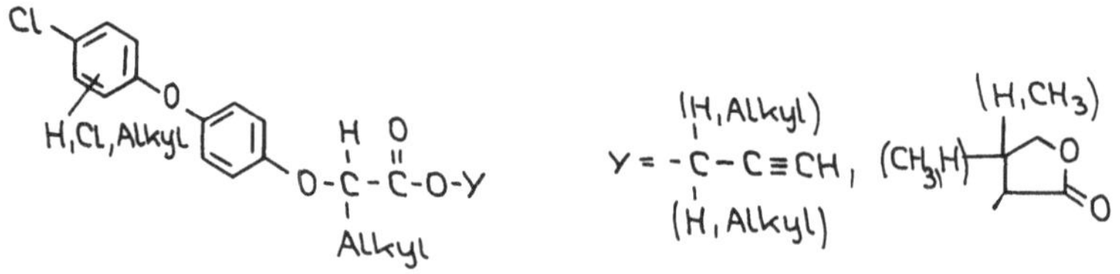

D-Isomere mit erhöhter herb. Wirksamkeit (2,4-6,2 fach)
Spaltung in opt. Isomere siehe z. B. Ind. and Engin. Chem. <u>60</u>(8)12 -28

DOS   2 805 981   13. 2.78/16. 8.79   Bayer

(H,Halog.)
CF_3—⬡—O—⬡
(H,Halog.)
CH_3 O
O—CH—C(OH, O-Alkyl, N<, S-Alkyl, N(CH$_2$)$_n$ usw.)

DOS   2 805 982   13. 2.78/16. 8.79   Bayer

H(Halog.)
CF_3—⬡—O—⬡
(H,Halog.)
O—CH_2—C(OH, O-Alkyl, O-Alk(en,in)yl), N(CH$_2$)$_{3-5}$, S-Alkyl, N<

DOS   2 809 541   6. 3.78/14. 9.78   Ciba Geigy
= BE   864 632   8. 3.77/ 7. 9.78   CH Prior. 8. 3.77

(Halog)$_{0-3}$
X—⬡—O—⬡—(Halog., CN)
Y   O—CH—A
                 H, Alkyl

A = CN, CO-B, OH, O-Kation
B = O-N = C< Alkyl / Alkyl, O-Alkyl,
    S-Alkyl, N<

DOS   2 810 527   10. 3.78/21. 9.78   Deutsche ITT Ind.
                                                     US Prior. 14. 3.77

Cl—⬡—O—⬡—O—CH—C—O—CH< CH_3 / C_4H_9 d-Form
(Cl) CH_3 O

Selektiv in Rasen
gegen Wiesenschwengel.

DOS   2 812 571   22. 3.78/ 1. 2.79   Nishiyama
                                                     JA Prior. 21. 7.77

H,Cl
CF_3(F,Cl)—⬡—O—⬡—O—CH< H,CH_3,C_2H_5 / (CH$_2$)$_{0-2}$—C—O—R
         N   O

DOS   2 815 287   8. 4.78/18.10.79   Höchst
Siehe Grundpatent DOS 2 640 730

Nur in Kombination mit
anderen Herbiziden.

Cl,Br
⬡—O—⬡—O—CH—C—O(H, Alkyl)
   O CH_3 O

DOS   2 820 032   8. 5.78/16.11.78   ICI Australia
                                                  US Prior. 6. 5.77

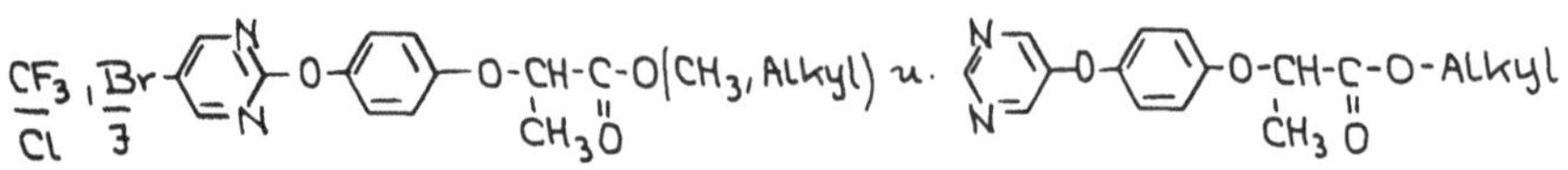

CF_3, Br pyrimidine-O-phenyl-$O-CH(CH_3)-C(=O)-O(CH_3, Alkyl)$ u. pyrimidine-O-phenyl-$O-CH(CH_3)-C(=O)-O-Alkyl$ (Cl, Cl, 3)

Selektiv in Getreide gegen Unkräuter (nicht Ungräser!)

DOS.   2 829 130   3. 7.78/ 3. 5.79   Kumiai Chem. Ind.
= BE   871 523   31.10.77/25. 4.79   Kumiai Chem. Ind.
                                                  JA Prior. 31.10.77

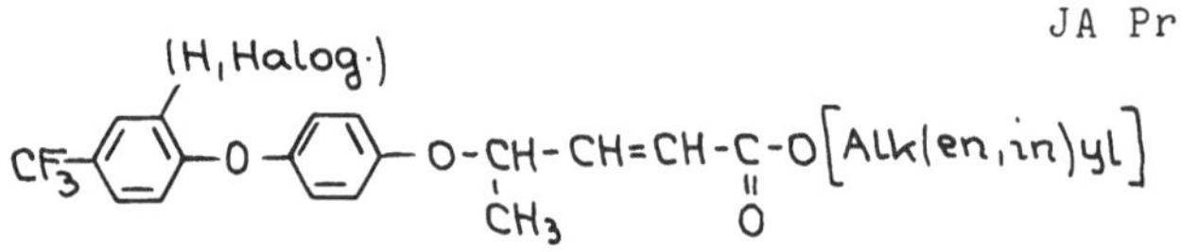

(H, Halog.) CF_3-phenyl-O-phenyl-$O-CH(CH_3)-CH=CH-C(=O)-O[Alk(en, in)yl]$

DOS   2 829 167   5. 7.77/18. 1.79   Ciba Geigy

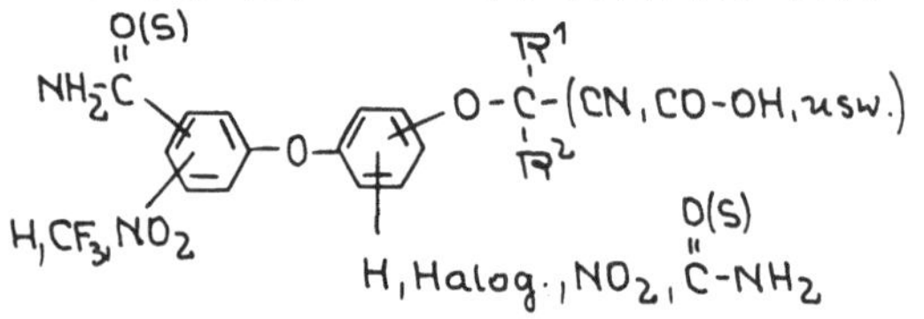

$O(S)$, NH_2-C, H, CF_3, NO_2-phenyl-O-phenyl-$O-C(R^1)(R^2)-(CN, CO-OH, usw.)$
$H, Halog., NO_2, C(=O(S))-NH_2$

R^1 = H, Alkyl,
       Alkoxy-Alkyl
R^2 = H, Alkyl, COO-Alkyl

DOS   2 829 169   3. 7.78/18. 1.79   Ciba Geigy

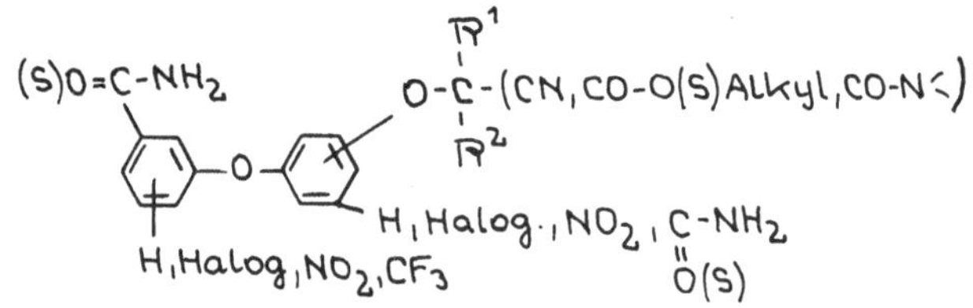

$(S)O=C-NH_2$, $H, Halog, NO_2, CF_3$-phenyl-O-phenyl-$O-C(R^1)(R^2)-(CN, CO-O(S)Alkyl, CO-N<)$
$H, Halog., NO_2, C(=O(S))-NH_2$

R^1 = H, CH_3
R^2 = H, Alkyl

Wachstumsregulatoren u. Herbizide.

DOS   2 830 066   8. 7.78/17. 1.80   Höchst

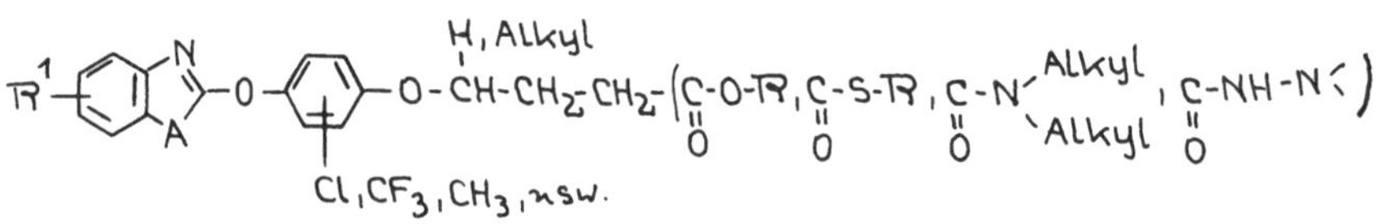

R^1-benzoxazol-O-phenyl-$O-CH(H, Alkyl)-CH_2-CH_2-(C(=O)-O-R, C(=O)-S-R, C(=O)-N<Alkyl/Alkyl, C(=O)-NH-N<)$
(Cl, CF_3, CH_3, usw.)

R^1 = Halog., CF_3, NO_2CN, (O,S)-Alkyl

DOS   2 830 141   8. 7.78/17. 1.80   Höchst

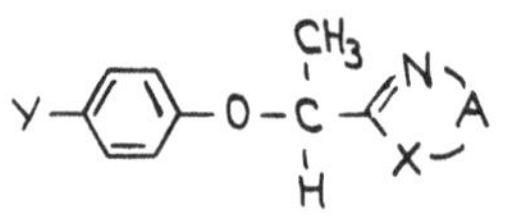

Y-phenyl-$O-C(CH_3)(H)$-(N, A, X ring)

X = O, S, NH, N-Alkyl
A = Alkylen, Benzolrest
Y = Alkyl, (O,S)-Alkyl <u>Phenyl-O</u>

246

DOS   2 832 435   27. 7.77/ 8. 2.79   Ciba Geigy
Siehe auch EP 3 295
Siehe ähnlich BE 856 975 21. 7.76/20. 1.78   Ciba Geigy
                                                  CH Prior. 27. 7.77

Z = CN, SCN, NO$_2$, Alk(en,in)yl
R^1 = H, Alkyl, CO-OAlkyl
R^2 = H, Alkyl, Alkoxy-Alkyl

DOS   2 834 744   8. 8.78/ 1. 3.79   Ciba Geigy

R^1 = H, Alkyl, Alkoxy-Alkyl; R^2 = H, Alkyl, CO-OAlkyl
C = CN, NO$_2$, Halog.; D = H, Halog
Herbizide bes. Wuchsregulator.

DOS   2 844 989   19.10.77/26. 4.79   Ciba Geigy

Spez. gegen Gräser.

DOS   2 854 492   16.12.78/19. 6.80   Höchst

X = Cl, Br, CF$_3$
Y = H, Cl

Darstellungsverfahren.

DOS   2 854 542   16.12.78/19. 6.80   Höchst

D-Form

DOS   2 904 563   7. 2.79/21. 8.80   Bayer
(Kombinationen von Prod. aus DOS 2 906 087)
DOS   2 906 087   17. 2.79/ 4. 9.80   Bayer
= EP   14 900

DOS   2 906 237   14. 2.79/28. 8.80   Bayer
= EP   14 880

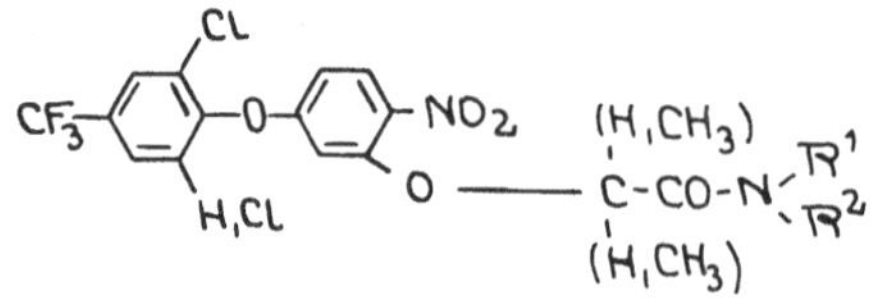

[Struktur: 2-Cl-4-CF$_3$-6-(H,Cl)-phenyl–O–(4-NO$_2$-phenyl)–O–C(H,CH$_3$)(H,CH$_3$)–CO–N$<$R^1/R^2]

R^1 u. R^2 = H, Alkyl, O-Alkyl,
                S-Alkyl
R^1 u. R^2 Glieder eines
              Heterocyclischen
              Ringes

DOS   2 907 089   23. 2.79/ 4. 9.80   Höchst
= EP   15 005

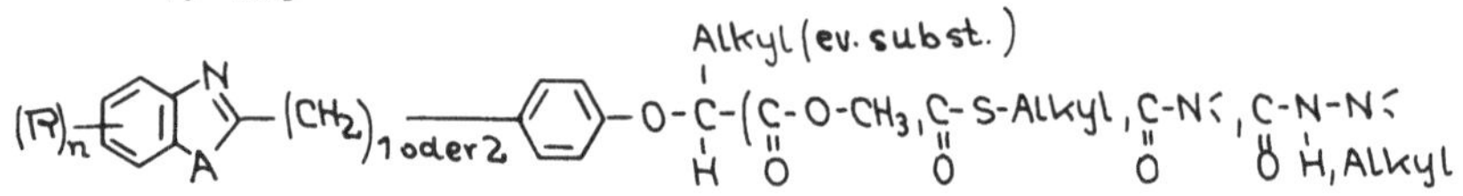

[Struktur: (R)$_n$-Benzazol(A)–(CH$_2$)$_{1 oder 2}$–phenyl–O–C(H)(Alkyl(ev. subst.))–(C-O-CH$_3$, C-S-Alkyl, C-N$<$, C-N-N$<$H,Alkyl)]

A = O, S, NH, N-Alkyl
R = Halog., NO$_2$, CN, NH$_2$, Alkyl O(S)Alkyl usw.; n = 0 - 4

DOS   2 908 937   7. 3.79/20. 9.79   Ishihara Sangyo

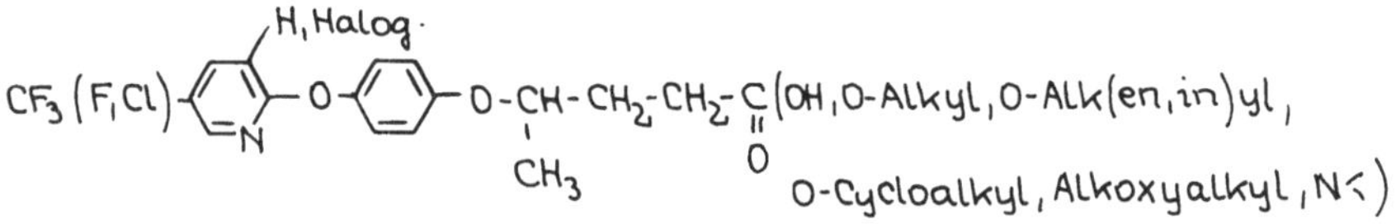

[Struktur: CF$_3$(F,Cl)-5-(3-H,Halog.)pyridin-2-yl–O–phenyl–O–CH(CH$_3$)–CH$_2$–CH$_2$–C(=O)(OH, O-Alkyl, O-Alk(en,in)yl, O-Cycloalkyl, Alkoxyalkyl, N$<$)]

DOS   2 909 816   13. 3.79/27. 9.79   Ciba Geigy

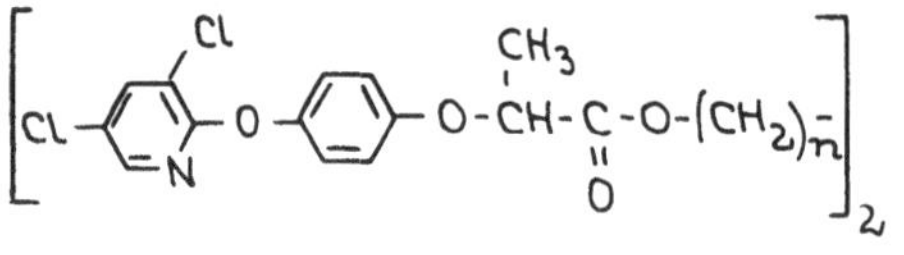

[Struktur: [Cl-5-(3-Cl)pyridin-2-yl–O–phenyl–O–CH(CH$_3$)–C(=O)–O–(CH$_2$)$_n$–]$_2$]

DOS   2 914 003   6. 4.79/16.10.80   Bayer

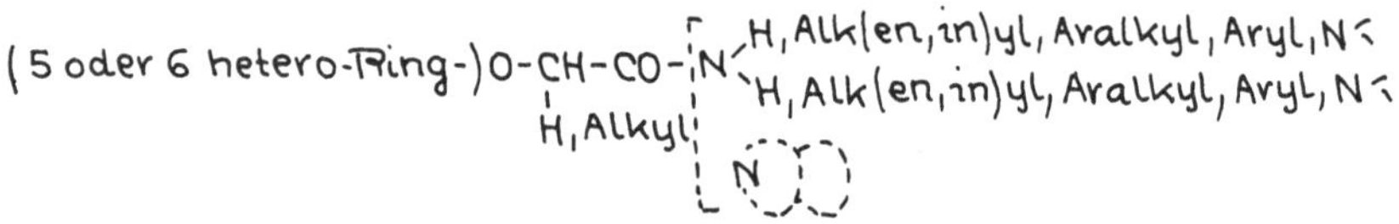

[Struktur: (5 oder 6 hetero-Ring-)–O–CH(H,Alkyl)–CO–N$<$ (H,Alk(en,in)yl, Aralkyl, Aryl, N$<$ / H,Alk(en,in)yl, Aralkyl, Aryl, N$<$); (N-Ring)]

DOS   2 914 300   9. 4.79/ 9.10.80   Höchst
= BE   882 705

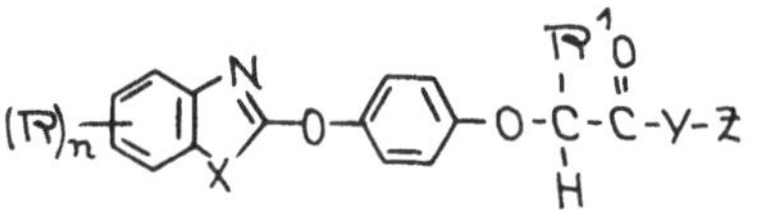

[Struktur: (R)$_n$-Benzazol(X)–O–phenyl–O–C(H)(R^1)–C(=O)–Y–Z]

R = Halog., CF$_3$, NO$_2$, CN,
    (O)-Alkyl
X = O, S, NH, N$<$ u. Alkyl
Y = O oder S;

Z = $-$C$-$(CH$_2$)$_{0-1}$$-$CN,   (CH$_2$)-COOAlkyl,

    CH$_2$-S-Alk(en)yl usw.; n = 0 - 2
         ↓
        (O)$_{0-2}$

248

DOS   2 918 541   8. 5.79/15.11.79   Nippon Soda
                                  JA Prior. 10. 5.78/19.12.78/13. 9.78

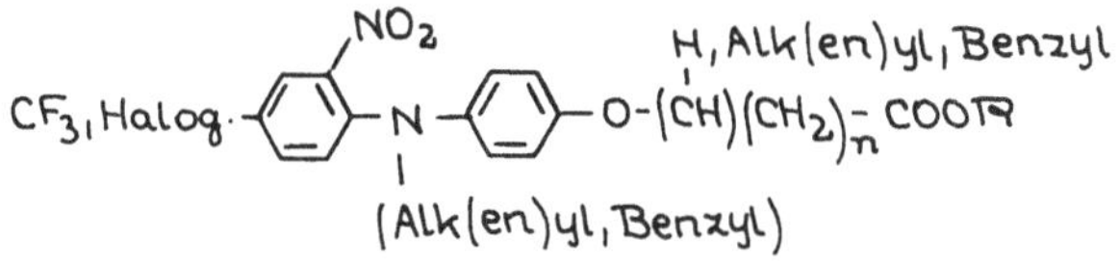

CF₃,Halog.⟨ring, NO₂⟩-N(Alk(en)yl,Benzyl)-⟨ring⟩-O-(CH)(CH₂)ₙ-COOR   [N-Substituent: H, Alk(en)yl, Benzyl]

DOS   2 934 300   24. 8.79/ 6. 3.80   P.P. Industr.

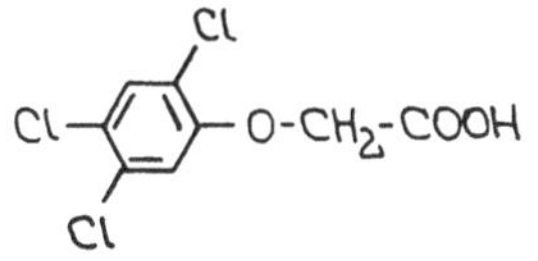

Cl,Cl,Cl⟨ring⟩-O-CH₂-COOH   Darstellung frei von Tetrachlor-
                                     dibenzdioxin

                                     durch chlorieren von 2,4-D

DOS   2 942 989   24.10.79/14. 5.80   Ishihara Chem. Ind.

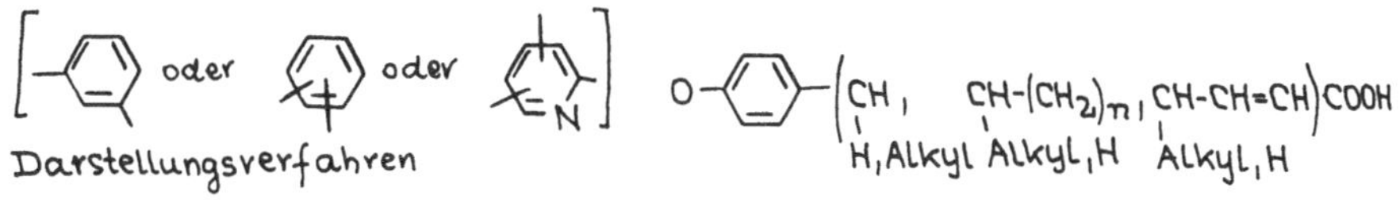

[⟨ring⟩ oder ⟨ring⟩ oder ⟨pyridyl N⟩]-O-⟨ring⟩-(CH, CH-(CH₂)ₙ, CH-CH=CH)COOH
Darstellungsverfahren   H,Alkyl Alkyl,H   Alkyl,H

DOS   2 946 652   19.11.79/29. 5.80   Shell Int. Res.
                                                   NE Prior. 21.11.78

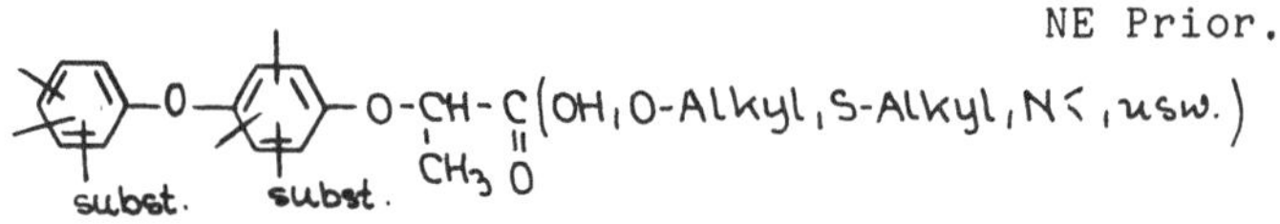

X⟨ring subst.⟩-O-⟨ring subst.⟩-O-CH(CH₃)-C(=O)(OH, O-Alkyl, S-Alkyl, N<, usw.)

Darstellung des R-Isomeren   Gräserherbizid in Getreide.

DOS   3 001 433   16. 1.80/17. 7.80   Comp. Francaise de
                                                   Produits Ind.
                                                   FR Prior. 16. 1.79

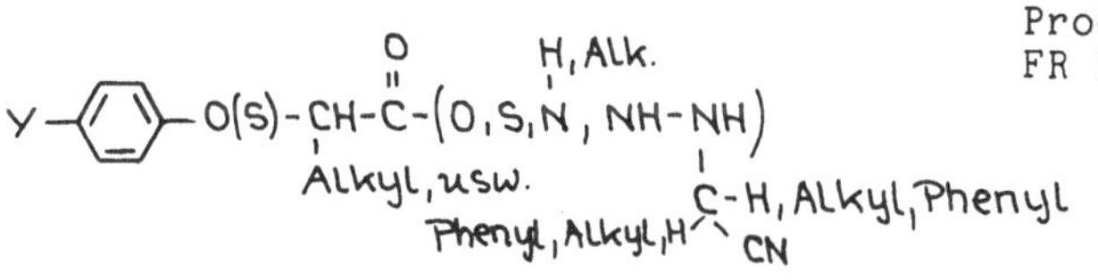

Y⟨ring⟩-O(S)-CH(Alkyl,usw.)-C(=O)-(O,S,N,NH-NH)(H,Alk.; Phenyl,Alkyl,H; C(-H,Alkyl,Phenyl)(CN))

Y = H, Halog., Alkyl oder Ar-O

DOS   3 004 770   8. 2.80/28. 8.80   Nissan Chem. Co.

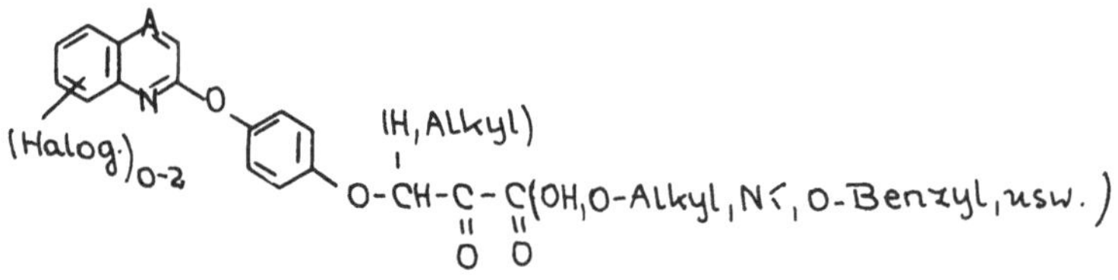

(Halog.)₀₋₂⟨quinoxaline, A⟩-O-⟨ring⟩-O-CH(H,Alkyl)-C(=O)-C(=O)(OH,O-Alkyl,N<,O-Benzyl,usw.)

A = CH oder N

EP Patentanmeldungen nach Nr. geordnet.

EP   50   7. 6.77/21.12.78   Bayer
= DOS   2 725 590   7. 6.77/21.12.78

R = H, Halog. Alkyl; n = 1 - 4   Gräserherbizid

EP   176   21. 6.78/10. 1.79   Ciba Geigy
                                                    CH Prior. 29. 6.77

                                                    A = CN, CO-OH,
                                                        2-Oxazolin

EP   351   25. 6.78/24. 1.79   Ciba Geigy
                                            CH Prior. 7. 7.77/27. 1.78

C = H, Halog., CN, Alkyl

D = H, Halog., CN, Alkyl, NO_2 CF_3, SO_2-Alkyl, SO_2-N<

E = H, Halog., CN, Alkyl, NO_2 CF_3, SO_2-Alkyl, SO_2-N<

EP   359   30. 6.78/24. 1.79   Ciba Geigy
                                                    CH Prior. 7. 7.77

C, D, E = H, Halog., CN, NO_2, CF_3, (O)-Alkyl, SO_2-Alkyl, SO_2N<

EP   474   30. 7.77/ 7. 2.79   Ciba Geigy

R^1 = H, Alkyl, Alkoxy-Alkyl
R^2 = H, Alkyl

250

EP   483   30. 6.78/ 7. 2.79   Dow Chem.
US Prior. 22. 7.77

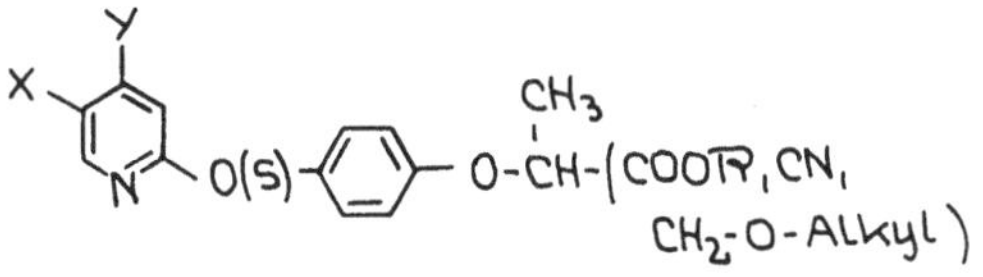

$X = Cl, Br, CF_3$
$Y = H, Cl, Br, CF_3$

EP   1 473   26. 7.78/18. 4.79   ICI
GB Prior. 12. 8., 26.10. u. 9. 2.78

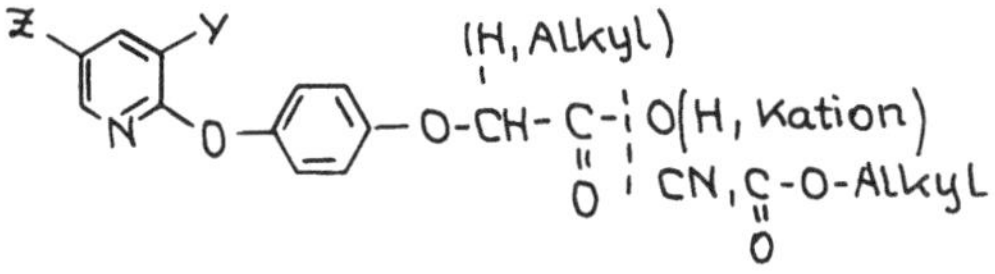

Z u. Y = Halogen, CF_3 $CFCl_2$ usw.

EP   1 641   24.10.78/ 5. 2.79   Ciba Geigy
CH Prior. 25.10.77

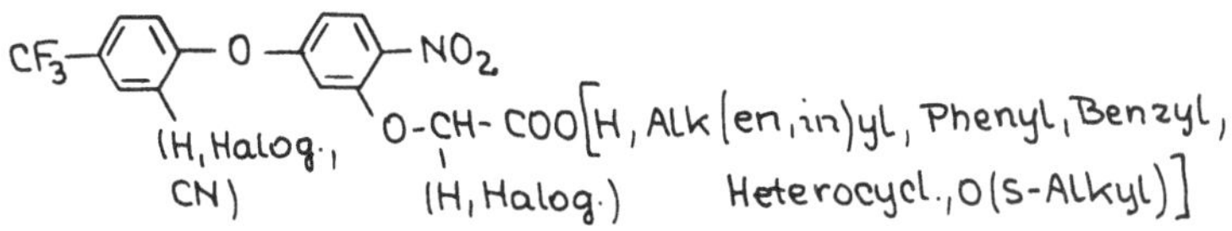

EP   2 246   28.11.79/13. 6.79   Hoffmann La Roche

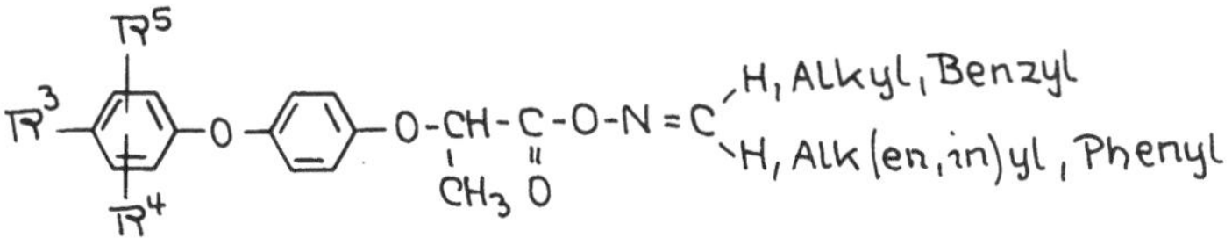

R^3 = Halog., CF_3, NO_2; R^4 u. R^5 = H, Cl

EP   2 757   28.12.77/12. 7.79   Ciba Geigy

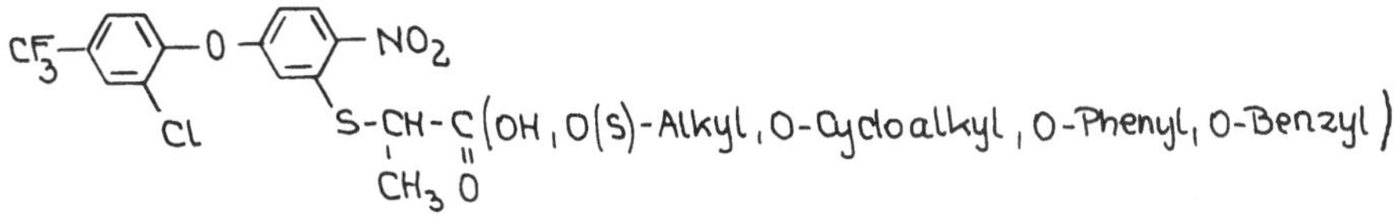

EP   2 800   20.12.78/11. 7.79   Höchst
= DOS 2 758 002

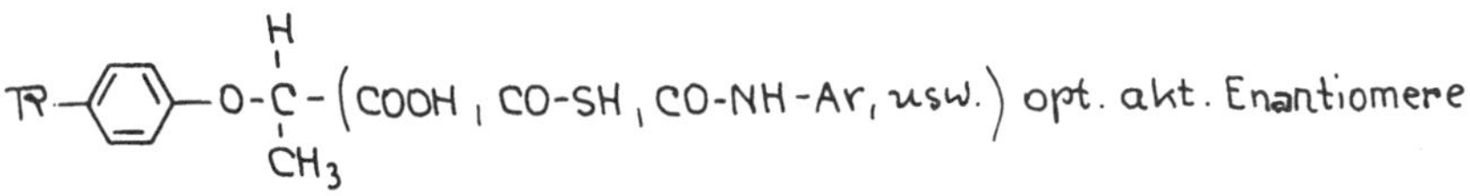

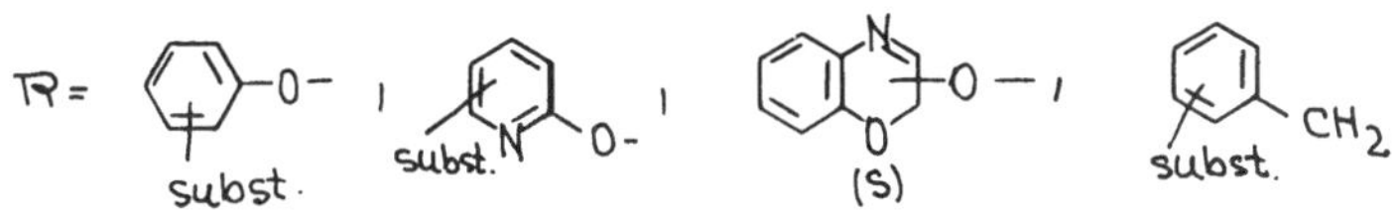

opt. akt. Enantiomere

EP   2 812   21.12.78/11. 7.79   Ciba Geigy
                                                    CH Prior. 2. 1.78

EP   2 925   18.12.78/11. 7.79   ICI
                                                    GB Prior. 23.12.77

EP   3 059   10. 1.78/25. 7.79   Ciba Geigy
                                                    CH Prior. 10. 1.78

EP   3 114   8. 1.79/25. 7.79   Ciba Geigy
                                                    CH Prior. 18. 1.78

EP   3 295   12. 1.79/ 8. 8.79   Ciba Geigy
                                                    CH Prior. 20. 1.78

X = H, Halog., CN, NO_2, CF_3; Y = H, Halog., CN, NO_2, CF_3

EP   3 297   12. 1.79/ 8. 8.79   Ciba Geigy
                                                    CH Prior. 25. 1.78

A = H, Halog., CN, NO_2, CO-N<
    CS-N<
Q = H, Halog., CN, NO_2, CO-N<
    CF_3

EP   3 313   17. 1.79/ 8. 8.79   Ciba Geigy
= US   4 213 774   CH Prior. 21. 1.78

X = Halog, CF$_3$; Y = H, Halog., CF$_3$

EP   3 517   22. 1.79/22. 8.79   Ciba Geigy
                                                       CH Prior. 3. 2.78

EP   3 584   5. 2.79/22. 8.79   Bayer
= DOS   2 805 981   DB Prior. 13. 2.78

EP   3 648   16. 1.79/22. 8.79   ICI
                                                       GB Prior. 15. 2.78

EP   3 890   5. 2.79/ 5. 9.79   ICI
                                                       GB Prior. 1. 3.78

Z = H, CF$_3$, CHF$_2$, CF$_2$Cl; Y = H, Cl, CF$_3$

EP   4 317   8. 3.79/ 3.10.79   Ciba Geigy
                                                       CH Prior. 17. 3.78

R^1 = Alkyl, Alkoxyalkyl, Phenyl, Benzyl, Alk(en,in)yl usw.

EP   3 586   5. 2.79/22. 8.79   Bayer
= DOS 2 805 982   DT Prior. 13. 2.78

EP   4 414   20. 2.79/ 3.10.79   ICI
                                          GB Prior. 2. 3.78

Z u. Y = mindestens 1 x CF_3, sonst H, Halog.

EP   4 433   13. 3.79/ 3.10.79   ICI
                                          GB Prior. 17. 3.78

EP   5 698   1. 6.78/12.12.79   Kalo Lab.

EP   5 709   5. 3.79/12.12.79   Ciba Geigy
                                          CH Prior. 15. 3.78

                                          Z = Halogen, OH, SH, O-Alkyl
                                              N< , S-Alkyl usw.

EP   6 608   26. 6.79/ 9. 1.80   Ciba Geigy
                                          CH Prior. 29. 6.78

     X = R-Isomeres

EP   7 573   18. 7.79/ 6. 2.80   Ciba Geigy
= CH 783 778   CH Prior. 20. 7.78

254

EP   8 624   5. 7.79/19. 3.80   Ciba Geigy
                                                    CH   3. 7.78

Cl—[pyridin ring: H,Halog.]—O—[phenyl]—O—CH—C-O(s)-Alkylen - C-O(s)Alkyl, N<Alk./Alk.
                                        | ‖   ‖
                                        CH₃ O   O
                                                    (subst. CH₃,CN,CO-CH₃ usw)

EP   9 285   19. 9.78/ 2. 4.80   AKZO Nv.
                                                      NE Prior. 19. 9.78

[phenyl: Hal.,CH₃]—O—CH<COOH / CH₃   Darstellung in opt.aktiv. Form

EP   9 796   4.10.78/16. 4.80   Höchst
= DOS   2 843 184

[phenyl: Rₙ]—(O.CH₂)—[phenyl]—O-CH—C-O-CH₃   spez. Darstellungsverfahren
                                   | ‖
                                   CH₃ O

R = Alkyl, Halog., CF₃, NO₂; n = 1 oder 2

EP   11 793   19.11.79/11. 6.80   Bayer
= DOS   2 850 902 u. 2 851 788   DB Prior. 30.11.78

[phenyl: (subst.)ₙ]—(O-CH₂-COOH)₁ oder 2   neues Darstellungsverfahren

EP   11 802   24.11.78/11. 6.80   Höchst
= DOS   2 850 902   24.11.78/12. 6.80   Höchst

[phenyl: (R¹)ₙ]—O—[phenyl]—NO₂
                        O-CH-C-N<H,Alkyl / Alk(en,in)yl,N<,CN,S-Alkyl,   ] zusammen
                          | ‖   Phenyl,Pyridyl   ] ein 4-7 glied-
                          CH₃ O   } riger
                                                                          ] Ring

R¹ = Halog., NO₂, CN, CF₃; n = 1 oder 2

EP   12 717   26.11.79/25. 6.80   Ciba Geigy
                                                       CH Prior. 1.12.78

[phenyl: Y—Q]—O—[phenyl: Z]   Y u. Z = Halog., Alkyl, CF₃ usw.
                O-CH-(CN,COOR, C-SR, C-N<)
                  | ‖   ‖
                  H,Alkyl   O   O
Q=CH od. N

Safener für 1,3,5-Triazin-Herbizide

EP   13 143   28. 3.79/29. 7.80   ICI Australia

R^1 = H, Alk(en,in)yl, COAlkyl usw.

A,B,D,E u. V = H, Halog., NO_2, CN, NH_2, N< usw.

pre u. post emergence gegen Gräser

EP   13 144   22.12.78/ 9. 7.80   ICI Australia
                                                   Al Prior. 8. 6.79

R^1 = H, Alk(en,in)yl, COAlkyl, HC = O, CH_2-Phenyl usw.

A,B,D,ET,U u. V = H, Halog., CN, CS, NH_2, N<, Alk(en)yl, O, S-Alkyl
Gräserherbizid   usw.

EP   13 145   22.12.78/ 9. 7.80   ICI Australia
                                                   Al Prior. 19. 3.79

Gräserherbizid

EP   13 542   4. 1.79/23. 7.80   Ciba Geigy
                                                   CH Prior. 4. 1.79

EP   13 660   4. 1.80/27. 7.80   Ciba Geigy
                                                   CH Prior. 9. 2.79

256

| EP | 14 880 | 4. 2.80/ 3. 9.80 | Bayer |
| = DOS | 2 906 237 | 17. 2.79 siehe dort | |

$$CF_3\text{-ring}(Cl)(H,Cl)\text{-O-ring-NO}_2,\ O\text{--}\overset{R^1\ R^2}{\underset{}{C}}\text{-CO-N}\overset{R^3}{\underset{R^4}{}}$$

| EP | 14 900 | 6. 2.80/ 3. 9.80 | Bayer |
| = DOS | 2 906 087 | 17. 2.79 siehe dort | |

$$CF_3\text{-ring}(Cl)(H,Cl)\text{-O-ring-NO}_2,\ O\text{-}\overset{R^1}{\underset{R^2}{C}}\text{-CO-O-}\overset{R^3}{\underset{R^4}{C}}\text{-}\underset{O}{C}(OH, N{\prec}, usw.)$$

| EP | 15 005 | 22. 2.80/ 3. 9.80 | Höchst AG |
| = DOS | 2 907 089 | siehe dort | |

$$(R)_n\text{-benzazol}(A)\text{-}(CH_2)_{1-2}\text{-ring-O-}\overset{R^1\ O}{\underset{H}{C}}\text{-}\underset{O}{C}(O\text{-Alkyl},\ \underset{S}{C}\text{-N}{\prec},\ C\text{-NH}_2,\ CN)$$

| US | 4 035 416 | 2.11.72/12. 7.77 | Dow Chem. |

$$\text{ring}(H,CH_3,Cl)\text{-O-}\underset{H,CH_3}{CH}\text{-COOH}$$

spez. Darstellungsverfahren

| US | 4 048 213 | 15.11.76/13. 9.77 | Dow Chem. |

$$Ar\text{-S-CH=C}\overset{CN}{\underset{CN}{}}$$

u.a. auch herbizid wirksam

| US | 4 049 423 | 30. 6.75/20. 9.77 | Stauffer Chem. |

$$CH_3\text{-ring-}CH_3\text{-O-}\underset{O\text{-}CH_3}{CH}\text{-}\underset{O}{C}\text{-NH-}\overset{CH_3}{\underset{CH_3}{C}}\text{-C}{\equiv}CH$$

| US | 4 049 424 | 30. 6.75/20. 9.77 | Stauffer Chem. |

$$R\text{-ring-}R\text{-O-}\underset{O\text{-Alkyl}}{CH}\text{-}\underset{O}{C}\text{-NH-}\overset{CH_3}{\underset{CH_3}{C}}\text{-CH}_3$$

R = Halogen, CH_3

US   4 050 923   30. 6.75/27. 9.77   Stauffer Chem.

            CH₃
            │
    (benzene ring) -O-CH-C(=O)-NH-CH-C₆H₅
            │   │   │
            CH₃   Alkyl   (H,Alkyl)

US   4 051 184   25. 6.71/27. 9.77   Stauffer Chem.

    Cl,CH₃   O   CH₃
            │   ‖   │
    (benzene ring) -O-CH-C-NH-C-C≡CH   allgem.Herbizid
            │   │   │
    Cl,CH₃   C₂H₅   CH₃

US   4 052 432   25. 6.71/ 9.10.77   Stauffer Chem.

    CH₃,Cl   O   CH₃
            │   ‖   │
    (benzene ring) -O-CH-C-NH-C-CN
            │   │   │
    CH₃,Cl   C₂H₅   CH₃

US   4 052 411   6.10.76/ 4.10.77   Shell Oil Co.

                                 Cl(CH₃)
              O   │
              ‖   Gegen Gräser!
    N   N-C-CH₂-O-(benzene ring)-Cl
    │ │   (Cl)
    Alkyl C-CO-O-Alkyl
         │
        NO₂

US   4 082 800   30. 6.75/ 4. 4.78   Stauffer Chem.

    R
     ╲   CO-NH-C(CH₃)₃
      (benzene ring)-O-CH   R = Halogen, C₁-C₄-Alkyl
     ╱   O-Alkyl
    R

US   4 083 867   30. 6.75/11. 4.78   Stauffer Chem.

    CH₃   CH₃
     │   │
    (benzene ring)-O-CH-CO-NH-C-C≡CH   (Acetal!)
     │   │   │
    CH₃   O-CH₃   CH₃

US   4 087 277   25. 6.71/ 2. 5.78   Stauffer Chem.
Teil aus 4 052 432

    R¹   R¹ u. R² = Cl, CH₃
     │   CH₃
    (benzene ring)-O-CH-CO-NH-C-CN
     │   │   │
    R²   C₂H₅   CH₃
        │   │
      (CH₃)   (CH₃)

258

US   4 116 677   30. 7.76/26. 9.78   Stauffer Chem.

R u. R^1 = CH_3, C_2H_5

US   4 119 433   5. 7.77/10.10.78   Stauffer Chem.

US   4 143 069   25. 6.71/ 6. 3.79   Stauffer Chem.

Gegen Gräser und
Unkräuter.

US   4 146 387   25.11.77/27. 3.79   Stauffer Chem.

R^2, R^3 u. R^4 = Alkyl,
R^5 = H, Alkyl

In Kombination mit einem Chloracetanilidderivat.

US   4 182 625   5. 7.77/ 8. 1.80   Stauffer Chem.

US   4 192 669   22.12.77/11. 3.80   Du Pont

BE   848 526   20.11.75/20. 5.77   Ciba Geigy

R = Halog., CF_3, Alkyl, O-Alkyl, S-Alkyl, CN, $SO_2N\!<$, usw.
R^1 = H, Halog., Alkyl; R^2 u. R^3 = H, Alkyl; R^4 = OH; O-Alkyl;
    O-Alkenyl usw.

BE   850 450   16. 1.76/18. 7.77   Höchst AG

Br-⟨phenyl⟩(Cl)-O-⟨phenyl⟩-O-CH(CH_3)-C(=O)-(OH, O-Alkyl, S-Alkyl, O-Cyclohexyl, O(S)-Phenyl, O-Benzyl) (S)

BE   855 094   26. 5.76/28.11.77   Höchst AG

⟨phenyl⟩R_n-O-⟨phenyl⟩-O-CH(CH_3)-C(=O)-O(S)-R^1

R^1 = isoAlkyl, Alkylen, Cyclohexyl, subst. Phenyl, N⟨, O-CH_2-CH_2-N⟨, [O-CH_2-CH_2-N⟨]$^{(+)}$ Anion, [-N⟨]$^{(+)}$ Anion, usw.

R = (O)-Alkyl, Halog.

BE   856 101   24. 6.76/27.12.77   Höchst AG

⟨phenyl⟩($R)_n$-X-⟨phenyl⟩($R^1)_n$-O-CH(CH_3)-C(=O)-(O,S,NH)-(CH_2-CH_2-CN, CH_2-CO-O-Alkyl)

X = O u. S

R u. R^1 = Halog., Halog.Alkyl, NO_2, O(S)-Alkyl; n = 0 - 3

Herbizide und Fungicide in Dicotyledonen.

BE   856 975   21. 7.76/20. 1.78   Ciba Geigy

R^1-⟨phenyl⟩(R^2)-O-CH(H, CH_3)-CO(OH, O-Alkyl, O-Alk(en,in)yl, O-N=C(CH_3)(Alkyl), N⟨, O-CH_2-C_6H_5, O(S)-C_6H_5 usw.)

Gegen breitblättrige Unkräuter in Reis.

BE   857 001   20. 7.76/20. 1.78   Höchst AG

R^1 = Halog., Alkyl, NO_2, CN

R^1-⟨phenyl⟩-O(S)-⟨phenyl⟩(NO_2, CN, CF_3)-O-CH(CH_3, Alkyl)-($CH_2)_{0-2}$-C(=O)-(OH, O-Alkyl, NH_2, NH-NH_2, O-Phenyl-subst.)

Pre u. post emergence Herbizid gegen breitblättrige Unkräuter in
Sorghum u. Mais

260

BE   858 104   25. 8.76/27. 2.78   Höchst

R^1 = Halogen, Alkyl, CF_3

BE   858 618   10. 9.76/13. 3.78   Höchst

(S)NH, N-Alkyl   Gegen Gräser.

BE   862 325   27.12.77/27. 6.78   Ciba Geigy

R^1 u. R^2 = H, CN, Alkyl

Z = COOH, CO-OAlkyl CO-S-Alkyl, CO-N<, CO-(Cycloalkyl, Alk(en,in)yl
    subst. Phenyl)

BE   864 632   8. 3.77/ 7. 9.78   Ciba Geigy
= DOS   2 834 744

A = CN, COON = C< Alkyl / Alkyl, CO-(O-Alkyl,
    S-Alkyl) CO-N<
X = H, CF_3; Y = H, CN, Alkyl
Z = Halog, CN

Man beachte die m-Stellung;   Wirksam gegen breitblättrige
                                     Unkräuter.

BE   866 696   6. 5.77/ 3.11.78   ICI Australia

A, B, D, E, V, = H, Halog., NO_2, CN, S-CN, N<
W = CN; $CS-NH_2$; COOH, CO-SH, CO-OAlkyl, CH_2Cl, CH_2OH, CH_2-O-Alkyl usw.

BE   868 875   21. 7.77/10. 1.79   Ishihara Sangyo

(F,Cl)—pyridine$(H,Cl)(CF_3)$—O—phenyl—O—CH(H,CH_3,C_2H_5)—(CH$_2$)$\overline{0-2}$—C(=O)—[OH, Alkyl, Phenyl, S-Alkyl, NH-C$_6$H$_5$, usw.]

BE   870 068   30. 8.78/28. 2.79   Ciba Geigy

phenyl$(H,Halog.)(H,Halog.,NO_2)$—O—phenyl—C(=N-O-(Alkyl, CO-O-Alkyl))—(CN, NO$_2$, Halog., CO-O-Alkyl, usw.)

BE   876 077   10. 5.78/ 3. 9.79   Nippon Soda

$(Hal., CF_3)$—phenyl(NO_2)—N$(H, Alk(en)yl, Benzyl)$—phenyl—O—CH$(Alkyl)$—(CH$_2$)$\overline{0-2}$—C(=O)—(O,S)-[H, Alkalimet., Alkyl]

GB   2 002 368   12. 8.77/21. 2.79   ICI

Z—pyridine(Y)—O—phenyl—O—CH$(H, Alkyl)$—(CN, CO-O-Alkyl, C(=O)-N<, C(=O)-S-Alkyl, usw.)

Z u. Y = Halog., CF$_3$, CHF$_2$, CCl$_2$F   Gegen Gräser wirksam.

GB   2 005 668   14. 9.77/25. 4.79   Shell Int. Res.

(CF_3)—phenyl—O—CH(CH_3)—[(CH$_2$)$_n$-OH, C(=O)-N<, C(=O)-O-Alkyl]

NE   7 704 533   27. 4.76/31.10.77   Philagro

$(Cl)(CH_3)$—phenyl—O—CH$_2$—C(=O)—N<$(H)(CH_3)$

RD   196 040   20. 7.80/10. 8.80   Anonym

$(Cl)(Cl)$—pyridine—O—phenyl—O—CH(CH_3)(CO-O(S)-CH$_2$-C≡CH)

262

| Z A | 7 801 076 | 9. 3.77/18.12.78 | ICI |

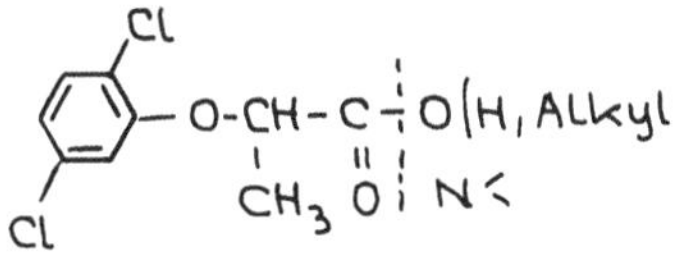

Desiccant für Baumwolle.

Structure: 2,5-dichlorophenyl–O–CH(CH$_3$)–C(=O)(=N<)–O[H, Alkyl]

| Z A | 7 803 204 | 5. 6.78/10. 5.79 | Nash |

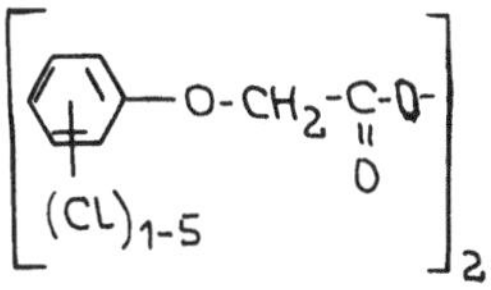

Structure: [phenyl(Cl)$_{1-5}$–O–CH$_2$–C(=O)–O–]$_2$

| J A | 50 135219 | 16. 4.74/27.10.75 | Hodogaya Chem. |

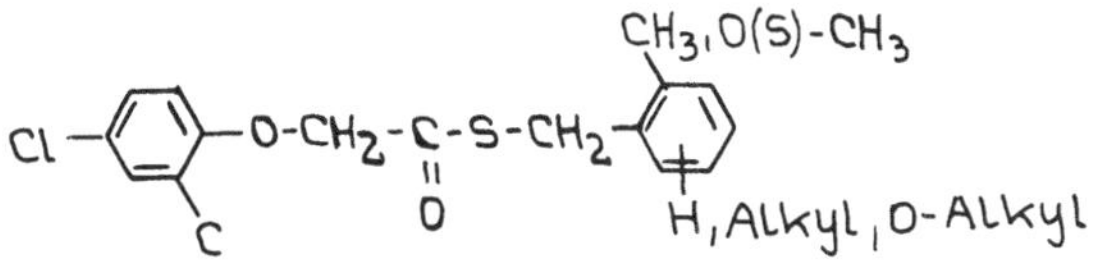

Structure: Cl–(phenyl, C)–O–CH$_2$–C(=O)–S–CH$_2$–(phenyl: CH$_3$, O(S)–CH$_3$, H, Alkyl, O-Alkyl)

| J A | 50 135220 | 7. 2.74/27.10.75 | Mitsui Toatsu Chem. |

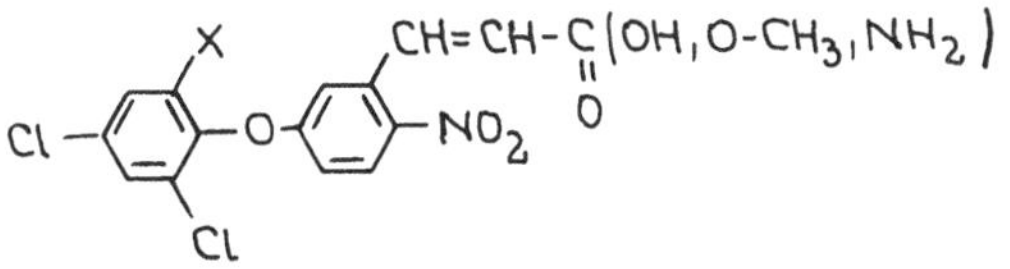

X = H, F, CH$_3$

Structure: Cl–(phenyl: X, Cl)–O–(phenyl: NO$_2$, CH=CH–C(=O)(OH, O–CH$_3$, NH$_2$))

| J A | 51 136823 | 16. 5.75/26.11.76 | Mitsui Toatsu Chem. |

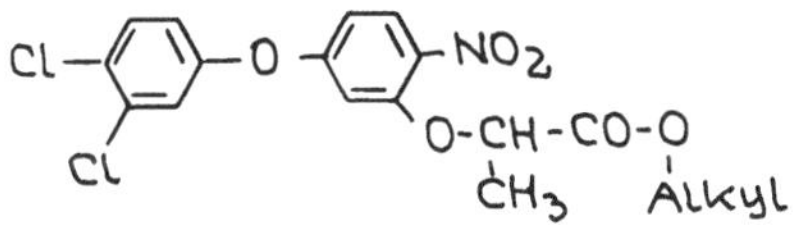

Structure: Cl–(phenyl: Cl)–O–(phenyl: NO$_2$, O–CH(CH$_3$)–CO–O–Alkyl)

Gegen Barnyard Gras in Reis.

| J A | 51 139622 | 28. 5.75/ 2.12.76 | Ishihara Sangyo |

Structure: X–(phenyl)–O–C(=CH$_2$)–COOH   X = CF$_3$–(phenyl)–O, bzw. 3,5-dichloro-2-pyridyl–O–

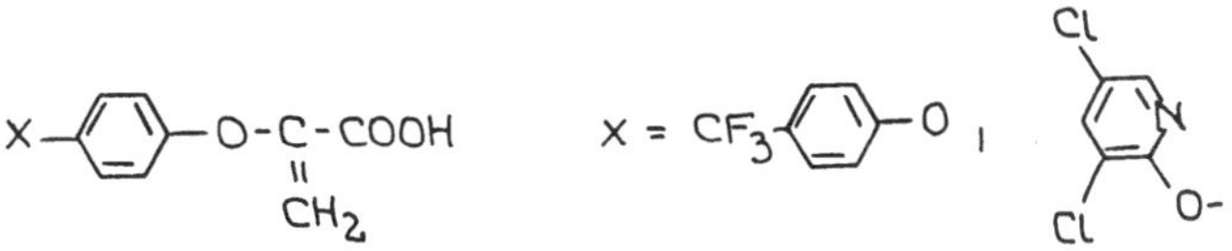

| J A | 51 139627 | 28. 5.75/ 2.12.76 | Ishihara Sangyo |

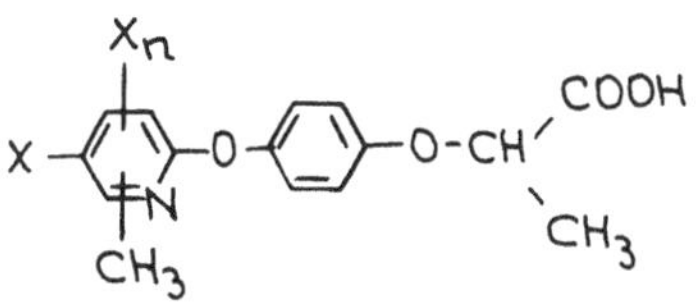

X = Halog.; n = 0 oder 1

Gegen Gräser.

Structure: X–(pyridine: X$_n$, CH$_3$, N$^+$)–O–(phenyl)–O–CH(CH$_3$)–COOH

JA   51 142534   30. 5.75/ 8.12.76   Ishihara Sangyo

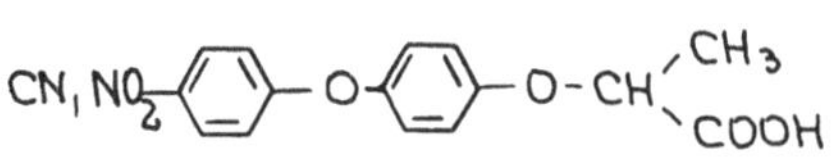

Selektiv gegen Gräser.

JA   51 142536   30. 5.75/ 8.12.76   Ishihara Sangyo

JA   51 142537   30. 5.75/ 8.12.76   Ishihara Sangyo

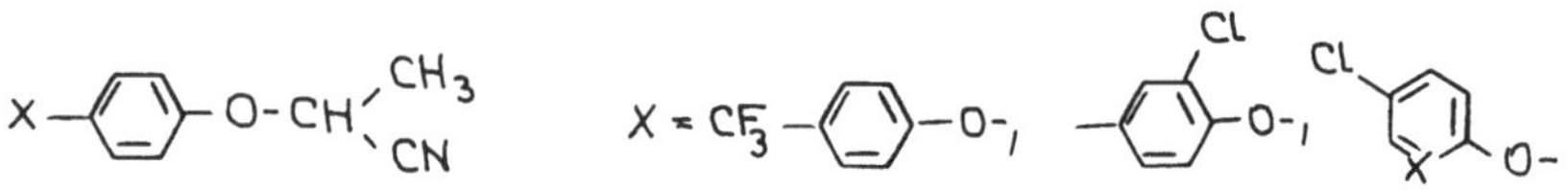

JA   52 015825   24. 7.75/ 5. 2.77   Ishihara Sangyo

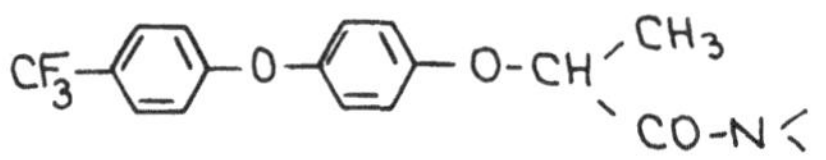

JA   52 041229   27. 9.75/30. 3.77   Ishihara Sangyo

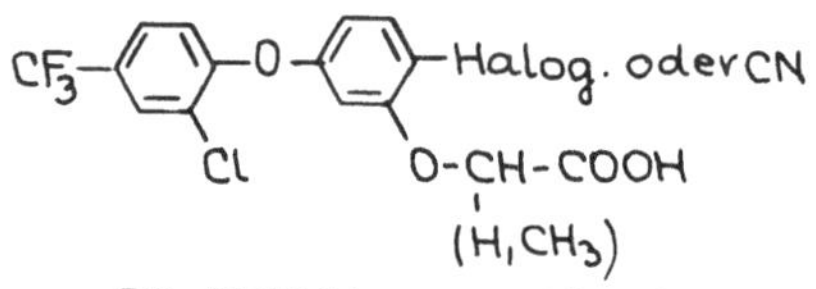

Gegen Unkräuter und z. T. Gräser.

JA   52 072814   12.12.75/17. 6.77   Mitsui Toatsu Chem.

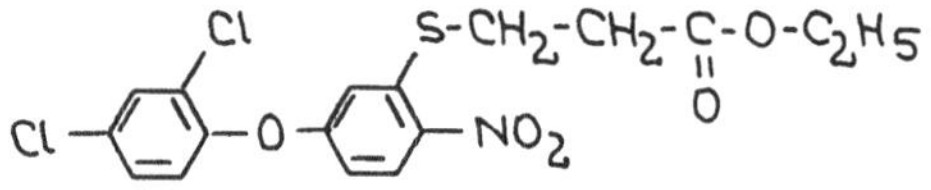

JA   52 072815   12.12.75/17. 6.77   Mitsui Toatsu Chem.

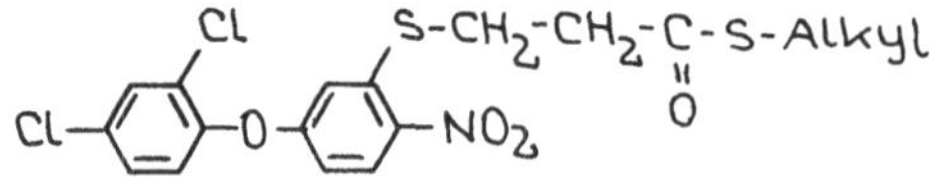

JA   52 087129   16. 1.76/20. 7.77   Ishihara Sangyo

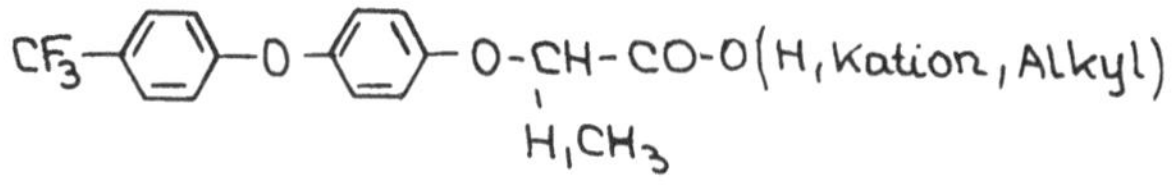

264

| JA | 52 087173 | 17. 7.76/20. 7.77 | Ishihara Sangyo |

R^1 = Halog., CH_3
R^2 = Halog., H, Alkyl

Spez. gegen Gräser.

| JA | 52 120123 | 2. 4.76/ 8.10.77 | Ube Industr. |

Gegen Unkräuter.

| JA | 52 125626 | 14. 4.76/21.10.77 | Ishihara Sangyo |

| JA | 52 128220 | 20. 4.76/27.10.77 | Ishihara Sangyo |

X = CF_3

Gegen Gräser in breitblättrigen Kulturen.

| JA | 52 128377 | 20. 4.76/27.10.77 | Ishihara Sangyo |

A = $COCH_3$; COC_2H_5

Gegen Gräser, auch als Wuchsstoff.

| JA | 52 130912 | 22. 4.76/ 2.11.77 | Ishihara Sangyo |

Gräserherbizid z. B.
in Sojabohnen oder
Rettichen.

Mitsui Toatsu

JA   52 131542   28. 4.76/ 4.11.77   Ishihara Sangyo

Cl, CF_3 — ⟨benzene⟩ — O — ⟨benzene⟩ — O — CH — C — O(H, Alkyl)
                                                    | ‖
                                                   CH_3 O

Cl Cl Cl $Cl, O-CH_3$
⟨pyridine⟩
  N O-

JA   52 133935   30. 4.76/ 9.11.77   Ishihara Sangyo

Halog. — ⟨benzene⟩ — O — ⟨benzene⟩ — O — CH⟨CN, CH_3⟩
                 |
              H, Halog.

JA   52 142034   19. 5.76/26.11.77   Mitsui Toatsu

                    $S(CH_2)_{\overline{n}} C — O$
⟨benzene⟩ — O — ⟨benzene⟩   ‖ |
   | NO_2 O   Alkyl
  X_n

                                X = Halogen, CF_3, Alkyl

JA   53 087330   12. 1.77/ 1. 8.78   Ihara Chem. KK

  X^1 N-O-Alkyl
⟨benzene⟩ — O(S)-$(CH_2)_{1 od. 2}$ ‖
  | C-S-⟨benzene⟩ X^3 $X^1 - X^4$ = H, Alkyl
  X^2 X^4 Halogen

JA   53 096323   2. 2.77/23. 8.78   Hodagyay Chem.

  CH_3 O
⟨benzene⟩ — O — CH — C — NH — ⟨benzene⟩   Reisherbizid
  | | ‖ |
  Cl (H, CH_3) (H, Halog.)

JA   53 098971   10. 2.77/29. 8.77   Ishihara Ind.

       X^1 COOH
X^2 ⟨pyridine⟩ — O — ⟨benzene⟩ — O — CH⟨COOH, CH_3⟩ X^1, X^2, X^3 = H u. bes. <u>Cl</u>
       N
       X^3

JA   53 109942   7. 3.77/26. 9.78   Nissan Chem. Ind.

        Br   Br
HO-C-CH_2-⟨benzene⟩-SO_2-⟨benzene⟩-O-CH_2-COOH
   ‖
   O   Br   Br

266

JA   53 023941   13. 8.76/ 6. 3.78   Ishihara Ind.

CF_3—[ring: H,Halog. / Cl]—O—[ring: (Cl,CN)]—O——C(CH$_3$)—CO-O-H-Alkyl, CO-O(H, Alkyl)

JA   53 005130   6. 7.76/18. 1.78   Ishihara Ind.

(Halog.,CF_3)—[ring: NO_2]—O—[ring]—O-CH(CN)(CH$_3$)

gegen Gräser, nicht wirksam
gegen breitblättrige Pflanzen

JA   53 130672   18. 4.77/14.11.78   Ishihara Ind.

Cl—[pyridine ring: (H,Cl), N]—O—[ring]—O-CH(CH$_3$)-CH$_2$-CH$_2$-C(=O)-(S-Alkyl, N$<$)

JA   54 023125   19. 7.77/21. 2.79   Ube Industr.

Y—[ring: X, CH$_3$]—O-CH(CH$_3$)-CH$_2$-CH$_2$-C(=O)-O(H, Alkyl, usw.)   X u. Y = H. CH$_3$

JA   54 032427   18. 8.77/ 9. 3.79   Mitsui Toatsu Chem.

[ring (Halog.,CF_3,Alkyl)$_{1-3}$]—O—[ring: NO_2]—O-CH(H,Alkyl)——C(O-Alkyl)(O-Alkyl)-CH$_2$-(H, Alkyl)

JA   54 032477   17. 8.77/ 9. 3.79   Ishihara Ind.

(Halog.,CF_3)—[pyridine ring: H,Cl, N]—O—[naphthalene ring]—O-CH(CH$_3$)-C(=O)-O-Alkyl

JA   54 039031   30. 8.77/24. 3.79   Ishihara Ind.

[CF_3—[ring]—O—, CF_3—[ring: Cl, Cl]—O—, Cl—[ring: Cl]—O—, Cl—[pyridine ring: N]—O—] —[ring]—O-CH(H,Alkyl)-C(=O)-O-CH$_2$—[epoxide O]

Gegen Gräser.

JA   54 039034   29. 8.77/24. 3.79   Mitsui Toatsu

JA   54 039035   1. 9.77/24. 3.79   Mitsui Toatsu

JA   54 046757   17. 9.77/12. 4.79   Ishihara Ind.

JA   54 055575   5. 10.77/ 2. 5.79   Ishihara Ind.

JA   54 063034   27.10.77/21. 5.79   Mitsubishi Chem.

JA   54 089022   26.12.77/14. 7.79   Ishihara Ind.

JA   54 100331   23. 1.78/ 8. 8.79   Nippon Soda

JA   54 117027   28. 2.78/11. 9.79   Nihon Noyaku

JA   54 122728   10. 3.78/22. 9.79   Ishihara Sangyo

$CF_2(F,Cl)$—pyridine(3-(H,Cl); 2-O)—phenyl—O-CH(CH_3)-CH_2-CH_2-C(=O)-(OH, O-Alkyl, N< usw.)

JA   54 135734   13. 4.78/21.10.79   Nihon Tokushu

R^1/R^2-phenyl—O-CH(H,Alkyl)-C(=O)-O-(CH_2)_{0-1}—phenyl—O—phenyl(subst.)

R^1 u. R^2 = Alkyl, Halog.

JA   54 144374   27. 4.78/10.11.79   Ishihara Sangyo

CF_3—pyridine(3-Halog.; 2-O)—phenyl—O-C(CH_3)(CO-O-R^1)(CO-O-R^2)

R^1 u. R^2 = H, Me^+

JA   55 017303   17. 7.78/ 6. 2.80   Ishihara Sangyo

(Halog., CF_3)—pyrimidine-2-O—phenyl—O-CH(CH_3)-C(=O)-O-(H, Na usw.)

JA   55 069566   17.11.78/26. 5.80   Ishihara Sangyo

CF_3—pyridine(3-X; 2-S(O)_n)—phenyl—O-CH(CH_3)-C(=O)-(OH, O-Kat., O-Alkyl usw.)

X = H, Halogen

JA   55 108837   10. 2.79/21. 8.80   Kumiai Chem.

(CF_3—phenyl oder Cl—pyridine(3-Cl))—O—phenyl—O-CH(CH_3)-CH=CH-C(=O)-O-R

R = -CH_2CH_2CN, CH_2CH_2-OC_2H_5, Benzyl, pyridyl-CH_2, Phenyl usw.

JA   55 111467   22. 2.79/28. 8.80   Ishihara Sangyo

CF_3—pyridine(3-X^1; 5-X^2; 2-O)—phenyl(Y_n)—O-CH(CH_3)-(CH_2)_n-C(=Z)(=O)

X^1 u. X^2 = H, Halog.
Y = Halog.
Z = OH, OAlkyl, S-Alkyl, O-CH_2-C_2H_5, O-C_2H_5

JA   55 118450   5. 3.79/11. 9.80   Kumiai Chem.

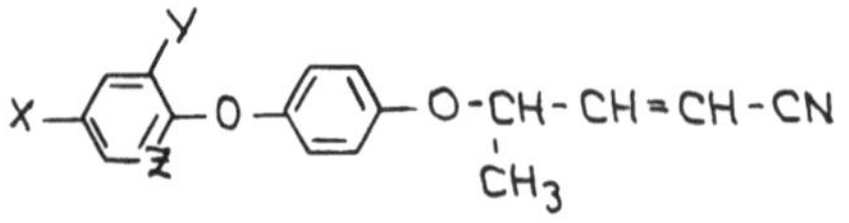

X = Halog., CF$_3$
Y = H, Halog; Z = = N- oder = CH-

JA   55 118475   8. 3.79/11. 9.80   Mitsui Toatsu

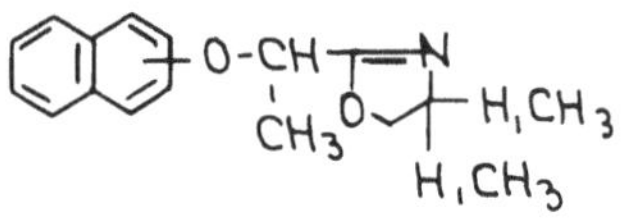

JA   55 120545   9. 3.79/17. 9.80   Nippon Soda KK

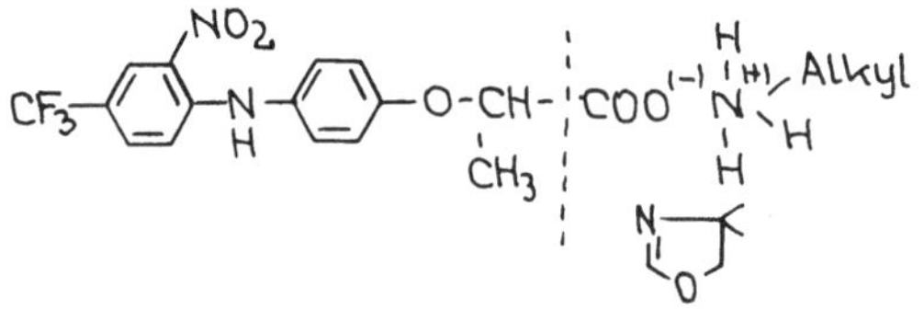

5.7.3.c2   <u>Phenoxy-acryl u. crotonsäure Derivate</u>

             z.T. unter 5.7.3.c mitbeansprucht

DOS   2 741 171   18.10.76/20. 4.80   Kao Soap

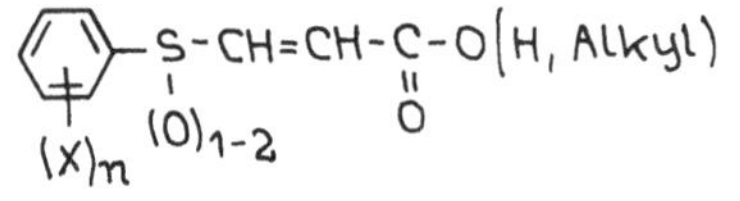

                                                          n = 0 - 3
                                                          X = Halog., S-Alkyl
                                                                       O(0 - 2)

DOS   2 829 130   3. 7.78/ 3. 5.79   Kumiai Chem. Ind.
= BE   871 523   31.10.77/25. 4.79   Kumiai Chem.
                                                          JA Prior. 31.10.77

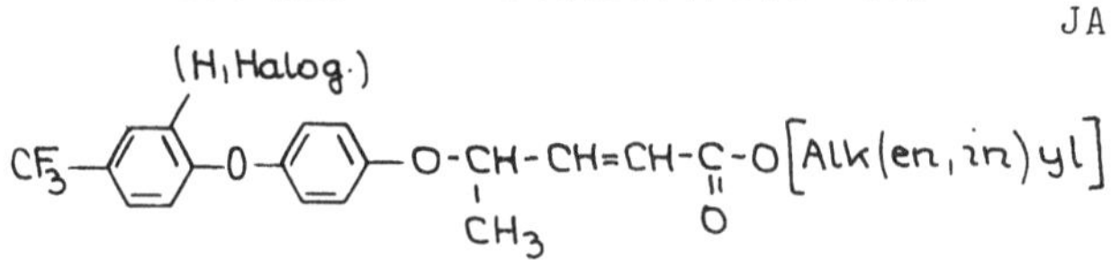

DOS   2 905 458   18. 2.78/30. 8.79   Kumiai Chem.

(Halog.,CF$_3$)—[ring, (H,Halog.)]—O—[ring]—O—CH—CH=CH$\left(\underset{O}{\overset{}{C}}\text{-O-Alkyl},\underset{O}{\overset{}{C}}\text{-S-Alkyl}\right)\underset{O}{\overset{}{C}}$-N$\langle$ usw.)
                                          |
                                          CH$_3$

DOS   2 915 686   18. 4.79/ 8.11.79   Ishihara Sangyo

CF$_3$—[pyridine ring, H,Alkyl,Halog.]—O—[ring]—O—CH—CH=CH—C—O—R
                                                         | ‖
                                                         CH$_3$ O

DOS   2 921 567   28. 5.79/24. 1.80   Kumiai Chem.
                        JA Prior. 14. u. 15. 7.78; 14.10.78

Cl—[pyridine ring, Cl]—O—[ring]—O—CH—CH=CH—Z   Z = CH$_2$OH, CO-OH(Alkyl)
                                    | CO-SAlkyl usw.
                                    CH$_3$

BE   875 880   27. 4.78/16. 8.79   Ishihara Sangyo

CF$_3$—[pyridine ring, (H,Halog.)]—O—[ring]—O—CH—CH=CH—C—O(H, Alkyl)
                                              | ‖
                                              CH$_3$ O

JA   52 094421   27. 1.76/ 9. 8.77   Tanabe Seiyaku

[ring R^1, R^2, R^3]—O—CH$_2$—CH=CH—C—O(H, Alkyl)   R^1 = H, Halog. (O)-Alkyl
                                        ‖   R^2 = H, Halog., Phenyl
                                        O   R^3 = H

                                                             Gegen Unkräuter in Reis.

JA   5 310132   17. 2.77/ 4. 9.78   Kumiai Chem.

Cl—[ring, Cl,CH$_3$]—O—CH$_2$—CH=CH—C—O(Alkyl, CH$_2$-CH=CH$_2$, Alkoxy-alkyl)
                                     ‖
                                     O   Gegen Gräser wirksam!

JA   53 101527   17. 2.77/ 5. 9.78   Kumiai Chem.

Cl—[ring, Cl,CH$_3$]—O—[ring]—O—CH$_2$—CH=CH—COOH

                                Gegen mehrjähr. Gräser in Reis.

JA   54 009234   20. 6.77/24. 1.79   Kumiai Chem.

$$Cl-\text{(Ring)}-O-CH_2-CH=CH-\overset{\overset{O}{\|}}{C}-NH-\text{(Ring)}$$

(erster Ring: Cl,CH_3; zweiter Ring: H,Cl,CH_3)

JA   54 109935   14. 2.78/29. 8.79   Kumiai Chem.

$$CF_3,(H,Halog.)\text{-(Ring)}-O-\text{(Ring)}-O-\underset{CH_3}{CH}-CH=CH-COOH$$

Gräser-Herbizid.

JA   54 135734   13. 4.78/22.10.79   Nihon Tukushu

$$R^1,R^2\text{-(Ring)}-O-\underset{(H,CH_3)}{CH}-\overset{\overset{O}{\|}}{C}-O(CH_2)_{0-1}-\underset{H,NO_2}{\text{(Ring)}}-O-\text{(Ring)}-Alkyl, S-Alkyl\ usw.$$

JA   54 154723   23. 5.78/ 6.12.79   Kumiai Chem.

$$Halog.\text{-(Ring, }H(Halog.))-O-\text{(Ring)}-O-\underset{CH_3}{CH}-CH=CH-\overset{\underset{O}{\|}}{C}-O-Kat.$$

JA   55 013248   15. 7.78/30. 1.80   Kumiai Chem.

$$(Halog., CF_3)\text{-(Ring, }(H,Halog.))-O-\text{(Ring)}-O-\underset{CH_3}{CH}-CH=CH-\overset{\underset{O}{\|}}{C}-O(Alkylen)-O-Alkyl$$

JA   55 015450   21. 7.78/ 2. 2.80   Mitsui Toatsu Chem.

$$(Halog., CF_3)\text{-(Ring, }Cl)-O-\text{(Ring, }NO_2)-O-\underset{(H,CH_3)}{CH}-C(=N-C(H,CH_3)_3)-O-H$$

(seitliche Substituenten: H,CH_3)

JA   55 020707   31. 7.78/14. 2.80   Mitsubishi

$$Z-O-\underset{R^2}{\overset{R^1}{\text{(Ring)}}}-\overset{\underset{O}{\|}}{C}-\text{(Ring)}-Y$$

R^1 u. R^2 = Halogen, NO_2

Z = CO(H, CH_3, O-Alkyl usw,

      CO-$(CH_2)_{0-2}$-CO-OAlkyl,

      SO_2-Alk(en,in)yl CH-CO-OH usw.

                                    |

                                 H, Alkyl

JA   55 098142   13. 1.79/25. 7.80   Kumiai Chem.

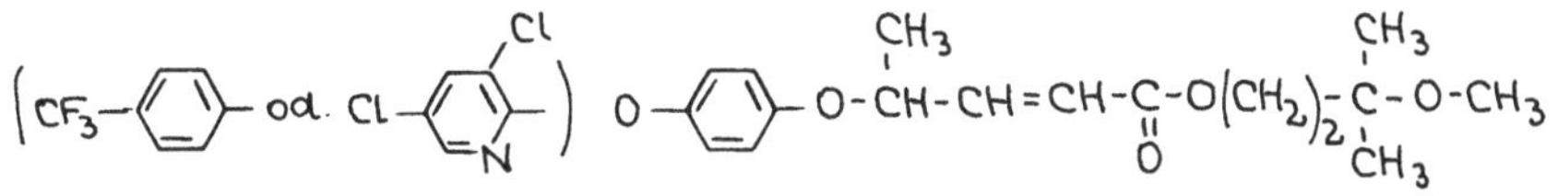

$\left(CF_3\text{—}\underset{}{\bigcirc}\text{—O— od. Cl—}\underset{N}{\bigcirc}^{Cl} \right)$ O—◯—O—CH—CH=CH—C—O(CH$_2$)$_2$—C—O—CH$_3$

(Kumiai Chem. structure with CH$_3$ substituents)

3.d   <u>Naphthoxy-essig- und -propionsäurederivate</u>   Bd.5/518

DOS   2 636 384   12. 8.76/24. 2.77   Stauffer
                                                     US Prior. 15. 8.75

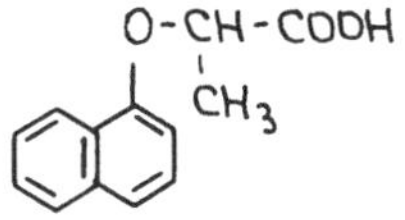

(Naphthalene with O-CH-COOH, CH$_3$ substituent)

                                    Verbessertes Darstellungsverfahren.

DOS   2 748 658   29.10.77/10. 5.79   Höchst AG

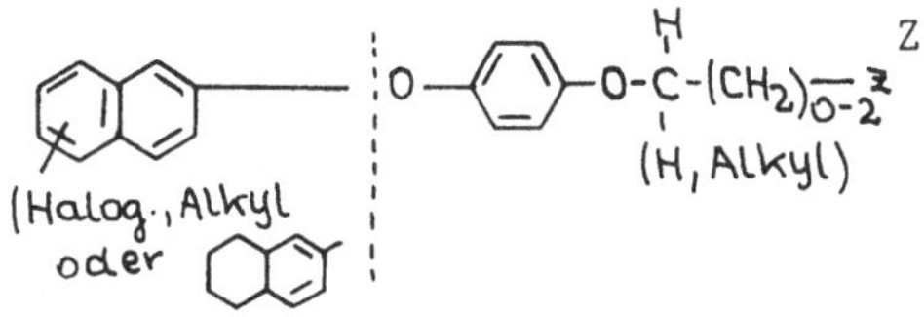

(Naphthalene structure) O—◯—O—C—(CH$_2$)$_{0-2}$—Z   Z = CO-OR, CO-N< , CO-S-R,
(Halog., Alkyl   CS-N< , CH$_2$OH, C$\underset{O}{\overset{H}{\lessgtr}}$
  oder)   (H, Alkyl)
                                                         CH$_2$-O-CO-N< usw.

US   3 928 880   15. 8.75/21.12.76   Stauffer
Siehe auch US 3 840 671

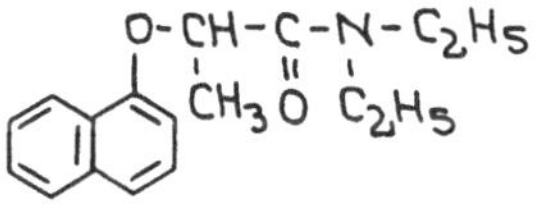

(Naphthalene with O-CH-C-N-C$_2$H$_5$, CH$_3$ O C$_2$H$_5$)

                              Verbesserte Darstellung.

JA   54 032622   11. 8.77/10. 3.77   Ube Ind.

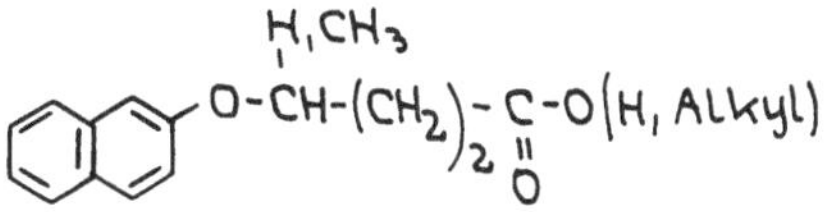

(Naphthalene with H,CH$_3$; O-CH-(CH$_2$)$_2$-C-O(H, Alkyl), O)

JA   55 118475   8. 3.79/12. 9.80   Mitsui Toatsu

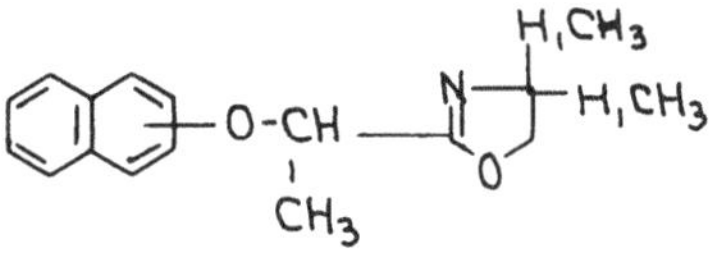

(Naphthalene with O-CH, oxazoline ring with H,CH$_3$ substituents, CH$_3$)

JA 54 088229 21.12.77/13. 7.79 Mitsui Toatsu Chem.

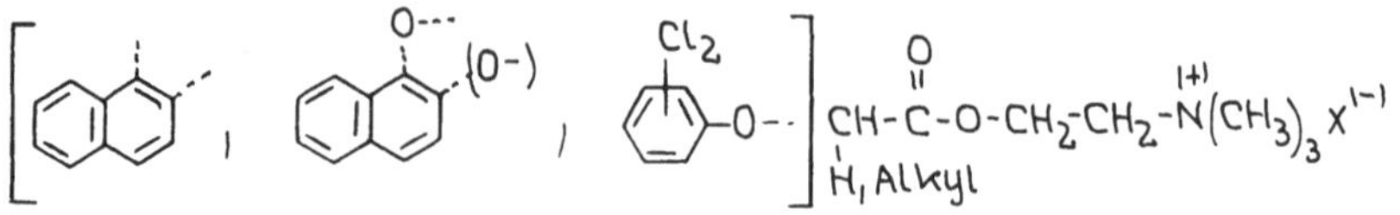

JA 54 143525 28. 4.78/ 8.11.79 Mitsui Toatsu Chem.

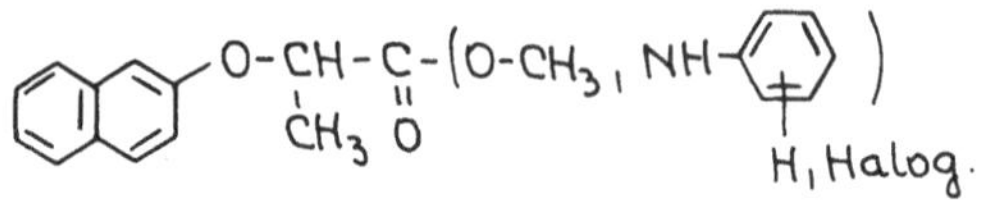

3.e **Aryl-amino-essig- und propionsäuren und ihre Acyl-verbindungen** Bd.5/519

Siehe auch bei Acylaniliden sec. Aniline 5.12.2

DOS 2 618 664 28. 4.76/11.11.76 Shell Int. Res.
= BE 841 109 30. 4.75/26.10.76 GB Prior. 30. 4.75

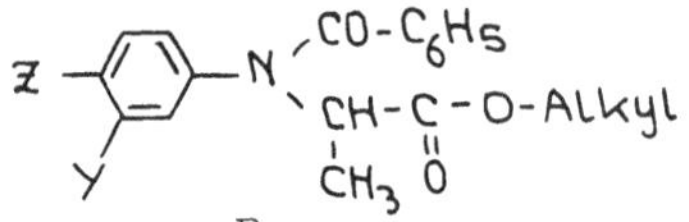

Z u. Y = Halogen

Gegen Flughafer.

+Zusatz+(Basagran[R] BASF)

DOS 2 624 094 29. 5.75/ 2.12.76 Sumitomo Chem.

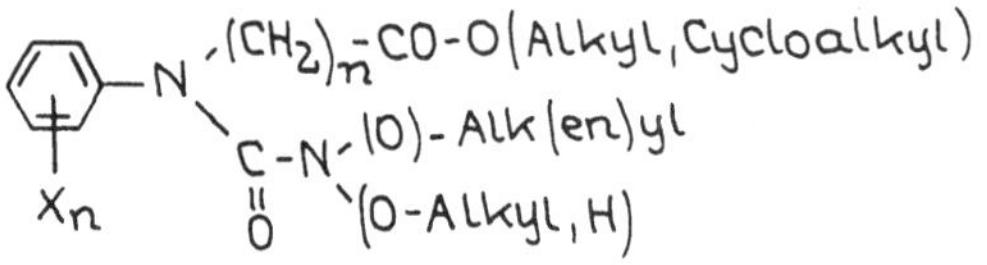

X = Halogen, Alkyl -O-Alkyl, CF_3

n = 0 - 2

DOS 2 628 901 28. 6.76/20. 1.77 Shell Int. Res.
= NE 7 607 051 30. 6.75/ 3. 1.77

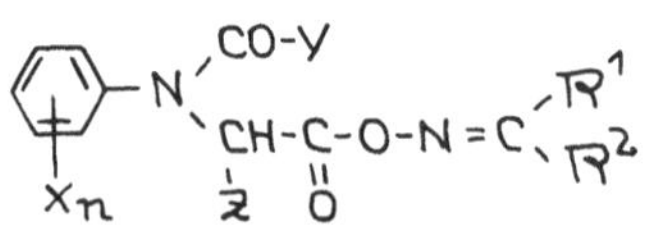

X = H, Halog., n = 1 - 5; Y = (subst.)-Aryl; Z = Alkyl;
R^1 = Acyl, COOH, CO-OAlkyl, CO-N< ; R^2 = Alk(en,in)yl, Aryl, Aralkyl (S)-O-Alkyl, Heterocycl.

DOS   2 633 729   29. 7.75/17. 2.77   Shell Int. Res.

X = + drehende opt.akt.Form
R = H, O-Alkyl, S-Alkyl, N<,
    ON = C<
X = F, Cl; Y = H, Cl, F

+ Zusatz

Gegen Unkräuter in Getreide.

DOS   2 650 434   3.11.76/18. 5.77   Shell Ind.
                                                    GB Prior. 5.11.75

X = Konfiguration Darstellung

Z u. Y = H, Cl, F; R = H, C_2H_5-CO(S)

DOS   2 706 823   17. 2.77/ 1. 9.77   Shell Int. Res.
                                                    GB Prior. 19. 2.76

Verbessertes Verfahren zur
Darstellung reiner Ester.

DOS   2 854 932   20.12.78/ 5. 7.79   Fisons Ltd.
= BE   872 919   GB Prior. 24.12.77/25. 7.78

A = CO-OH, CS-SH, CO-ON = O$<^{CH_3}_{CH_3}$, CN
R = Alkylen
R^1 = Alkyl, Aralkyl, Aryl, Heterocycl., N<
R^2 – R^6 = H, OH, O-Alkyl, NO_2, N<,
        SO_2N< usw.

Wuchsregulator   z.T. Herbizid.

EP   2 842   6.10.78/11. 7.79   Shell Int. Res.
                                                US Prior. 17.10.77

X = H, Halog., Alkyl, CF_3 usw.

EP   6 713   23. 6.78/ 9. 1.80   Janssen Pharm.

EP   10 396   9.10.70/30. 4.80   Fisons
                        GB Prior. 24.10.78; 9. 2.79 u. 17. 8.79

$$\text{Alk.}\;N\!\!\begin{array}{c}C\text{-}NH_2\\[2pt]CH\text{-}CN\\[2pt]CO\text{-}(N\!<,\,O\text{-Alkyl})\end{array}$$
subst.

US   3 981 717   24. 9.69/21. 9.76   Amer. Cyanamid

$$\begin{array}{c}H(Alkyl)\\N\text{---}C\!\!=\!\!C\!\!\begin{array}{c}R^1\\R^2\end{array}\\H(Alkyl)\end{array}$$ R^1 u. R^2 = CN, COOAlkyl

(O)Alkyl, Halog., NO_2, CF_3

US   4 193 789   19. 8.77/18. 3.80   Shell Oil Co.

$$X\text{---}\!\!\begin{array}{c}NO_2\\[2pt]NH\text{-}CH\text{-}\overset{O}{\overset{\|}{C}}\text{-}N\text{-}O(S)\text{-}(Alkyl, Phenyl, Benzyl-subst.)\\[2pt]CH_3\quad Alkyl\\[2pt]NO_2\end{array}$$

X = (Halog.,)-Alkyl, Halog., Alkyl-SO_2

BE   843 498   30. 6.75/28.12.76   Shell Int. Res.

$$Cl,F\text{---}\!\!\begin{array}{c}CO(S)\text{-}C_6H_5\\N\text{---}CH\text{-}C\text{-}O\text{-}R(H, Alkyl, ON\!\!=\!\!C\!<, S\text{-Alkyl}, N\!\!<\!\!\begin{array}{c}Alkyl\\Alkyl\end{array}\!usw.)+\\CH_3\;O\\Y\end{array}$$

X^1 = H, Halog., Alkyl, CN, CH_2-CN, CO-N< usw.
Y^1 = H, Halog., Alkyl   Gegen Flughafer.

BE   847 891   15.11.75/ 3. 5.77   Shell Int. Res.

$$Z\text{---}\!\!\begin{array}{c}\overset{x}{N\text{-}CH\text{-}COOH}\\R\;CH_3\\Y\end{array}$$

Y u. Z = H, Cl oder F

Darstellung der R-Verbin-
dung.

BE   851 460   19. 2.76/16. 8.77   Shell Int. Res.

$$\begin{array}{c}Y\\X\\CH_3\\N\text{-}CH\text{-}CO\text{-}O(H, Alkyl)\\CO\text{-}C_6H_5\end{array}$$

X u. Y = F oder Cl

Aufbereitungsverfahren.

GB   1 513 726   29. 7.74/ 7. 6.78   Shell Int. Res.

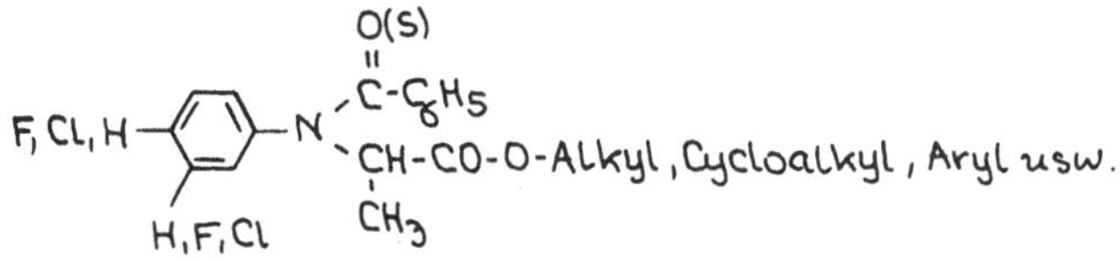

F,Cl,H—⟨benzene ring, CH₃ / H,F,Cl substituents⟩—N(—C(=O(S))—C₆H₅)—CH(CH₃)—CO-O-Alkyl, Cycloalkyl, Aryl usw.

FR   2 400 506   19. 8.77/20. 4.79   Shell Int. Res.

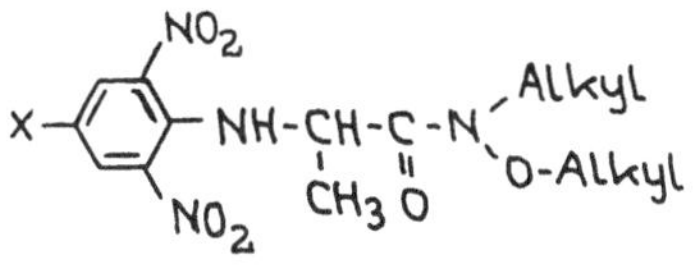

X—⟨benzene ring with NO₂, NO₂⟩—NH-CH(CH₃)-C(=O)-N(Alkyl)(O-Alkyl)

X = Halog., Halog.-Alkyl,
    Alkyl-SO₂-

3.f   <u>Heterocyclisch-aliphatische Carbonsäuren und</u>

       <u>heterocyclische Oxy-Essigsäuren</u>

3.f.1 Heterocyclische Alkancarbonsäuren   Bd.5/520

DOS   2 647 368   20.10.76/ 5. 5.77   Ciba Geigy
                               CH Prior. 23.10.75/30. 7.76 u. 5. 8.76

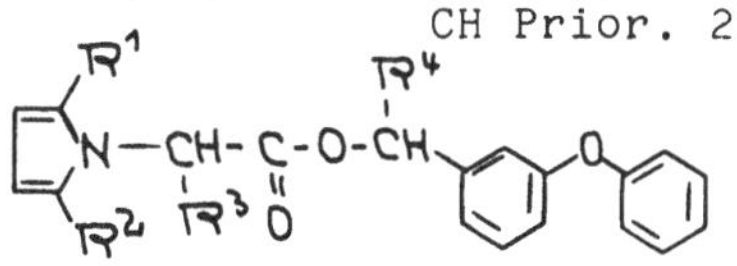

⟨pyrrole ring R¹, R²⟩—N—CH(R³)-C(=O)-O-CH(R⁴)—⟨phenyl⟩—O—⟨phenyl⟩

R^1 u. R^2 = H, CH_3
R^3 = nied. Alkyl u. Phenyl; R^4 = H, CN, CH_2-C≡CH

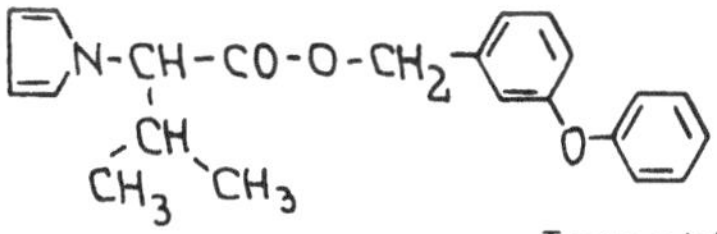

⟨pyrrole⟩N-CH(CH(CH₃)CH₃)-CO-O-CH₂—⟨phenyl⟩—O—⟨phenyl⟩

Insectizide z.T. auch Herbizide
Wirkung.

DOS   2 907 592   27. 2.79/ 6. 9.79   Nissan Chem. Ind.
                                                      JA Prior. 22. 2.78

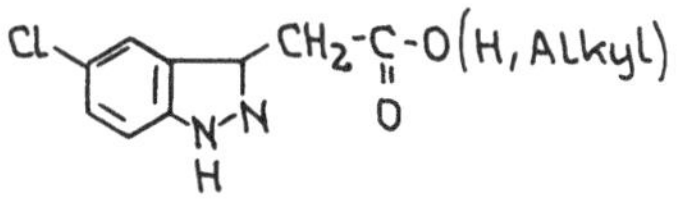

Cl—⟨indazoline ring, N-N-H⟩—CH₂-C(=O)-O(H,Alkyl)

Erhöhung des Zuckergehaltes
von Ananas oder Weinbeeren.

EP   1 349   19. 9.78/21. 9.77   Monsanto

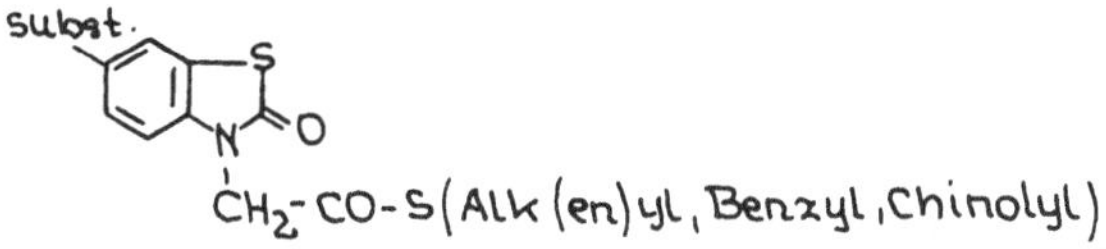

subst.—⟨benzothiazolone ring, S, N, =O⟩—CH₂-CO-S(Alk(en)yl, Benzyl, Chinolyl)

EP   7 772   18. 7.79/ 6. 2.80   Monsanto
= US   4 185 990

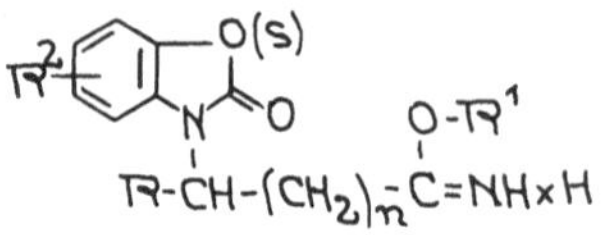

R = H, Alkyl; R^1 = Alkyl

R^2 = H, Alkyl, Halog., O-Alkyl

Herbizide u. Wuchsregulatoren.

EP   9 331   24. 8.79/ 2. 4.80   Monsanto
                                                   US Prior. 24. 8.78

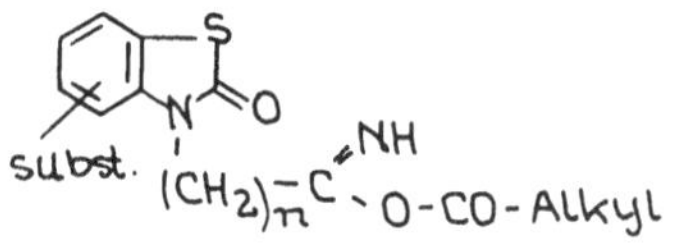

US   4 089 672   19. 2.70/16. 5.78   Upjohn Co.

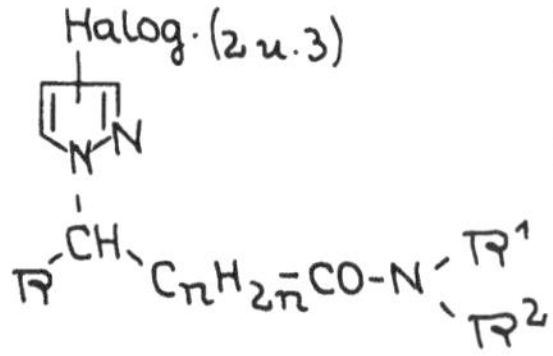

R = H, Alk(en)yl

R^1 u. R^2 = H. Alk(en,in)yl Aralkyl,
            Aryl, Cycloalkyl,

R^1 u. R^2 -Glieder eines
3 - 7 Ringes usw.

3.f.2 <u>Heterocyclische Oxy- und Thioessig- sowie</u>

<u>propionsäuren</u>   Bd.5/522

Ferner höhergliedrige Oxyalkancarbonsäuren
sowie Dicarbonsäuren.

DOS   2 546 845   18.10.75/28. 4.77   BASF

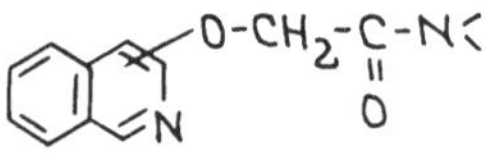

DOS   2 709 108   2. 3.77/ 8. 9.77   Ciba Geigy

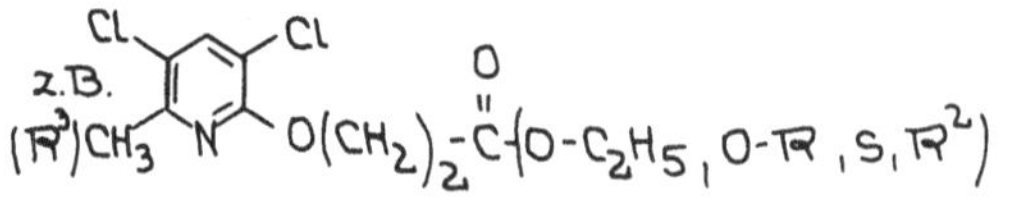

R^1 = H, Alkyl, er substit. durch Halog., O-Alkyl, S-Alkyl, Cycloalkyl
      Phenyl usw;
R^2 = Alkyl, Phenyl, Benzyl; R^3 = Alkyl, Phenyl, Halog.

DOS   2 732 924   21. 7.77/ 8. 2.79   Höchst

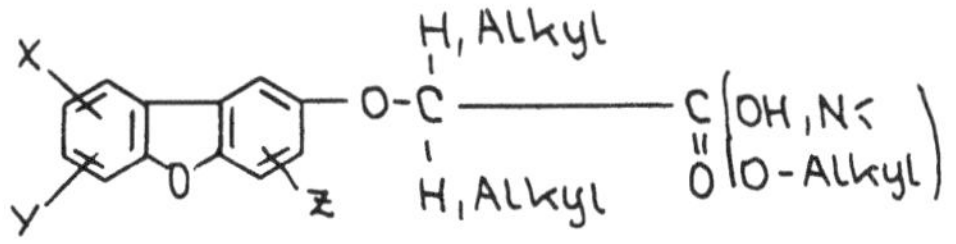

X, Y, Z = H, Alkyl,
          NO_2, O-Alkyl
          Halogen

DOS   2 823 424   29. 5.78/13.12.79   BASF

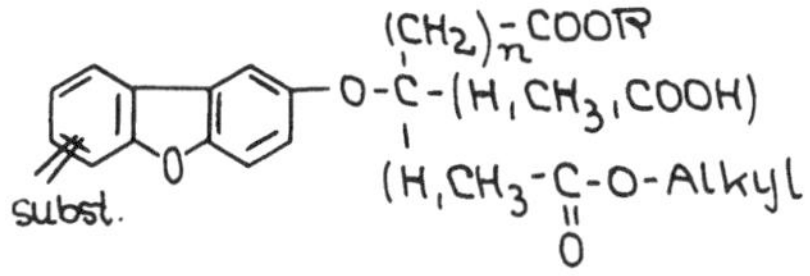

DOS   2 903 966   2. 2.79/ 7. 8.80   Bayer

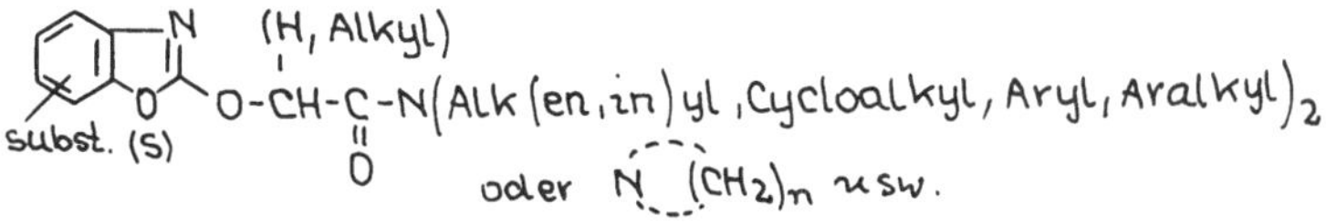

EP   5 501   2. 2.79/28.11.79   Bayer
= DOS   2 822 155   20. 5.78/22.11.79   Bayer

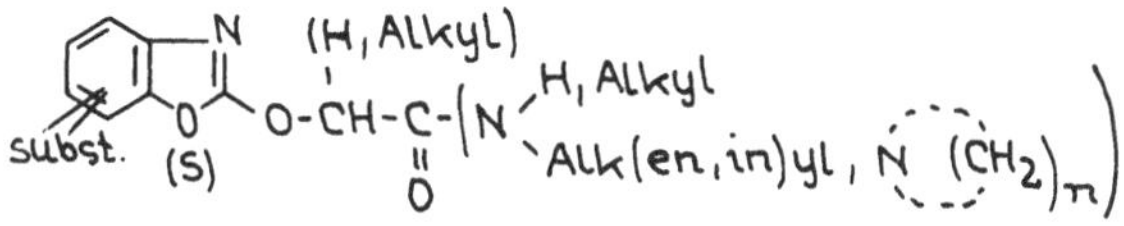

US   4 108 629   26. 7.71/22. 8.78   Dow Chem.

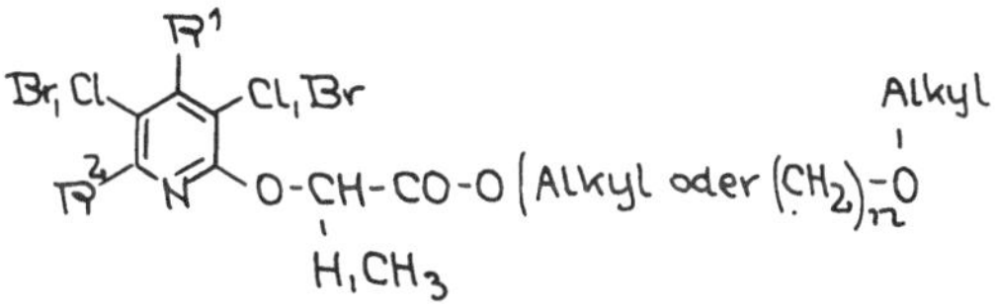

R^1 = H, NH_2, N(H, Alkyl)

R^2 = H, Alkyl, NH_2,
       N(H, Alkyl)

US   4 127 582   27.12.76/28.11.78   Dow Chem.

X u. Y = H, Halogen

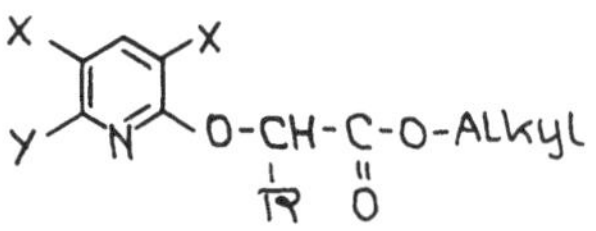

US   4 180 395   10. 7. 77/25.12.79   Dow Chem.

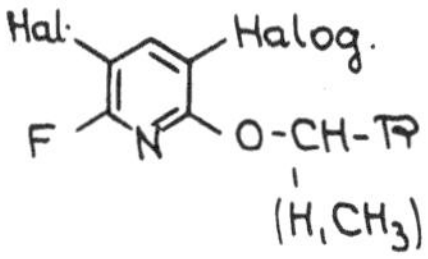

R = CH_2-OCO-Alkyl, CO-N<,
    CO-NH-CH_2-COOH usw.

BE   845 449   25. 8.75/24. 2.77   Ciba Geigy

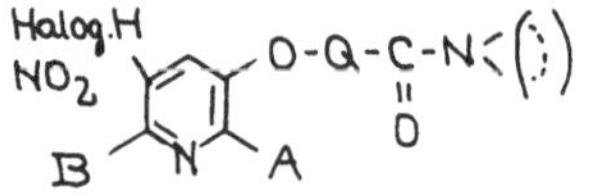

$$\text{Halog.H} \quad \text{NO}_2 \quad \text{B} \quad \text{N} \quad \text{A} \quad \text{O-Q-C-N}< \quad \text{O}$$

A u. B = H. Halog., Alkyl, CN,
                NO$_2$, O(S)-Alkyl, CO-N$<$

                Q = Alkylen, Alkenylen

Wuchsregulatoren und Antidots für Herbizide.

BE   852 028   5. 3.76/ 5. 9.77   Ciba Geigy

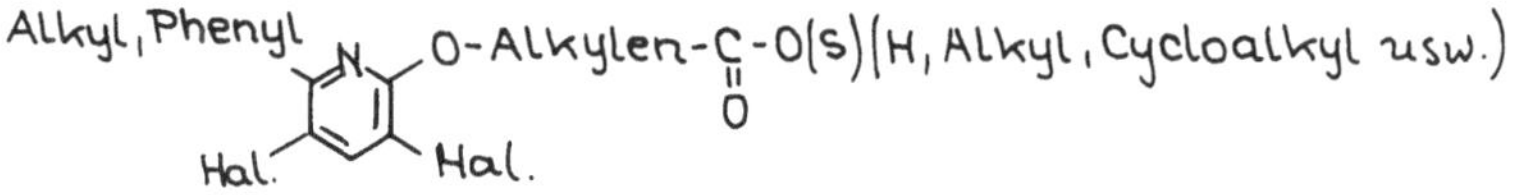

Alkyl, Phenyl   O-Alkylen-C-O(S)(H, Alkyl, Cycloalkyl usw.)

Hal.   Hal.

BE   862 081   27.12.76/20. 6.78   Dow Chem.

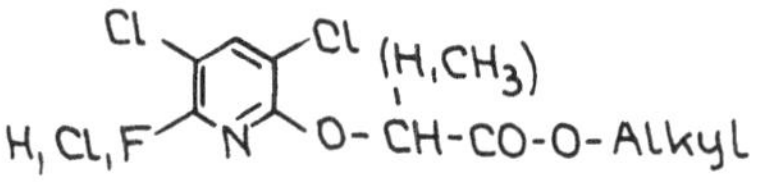

Cl   Cl (H,CH$_3$)

H, Cl, F   O-CH-CO-O-Alkyl

JA   53 124618   8. 4.77/31.10.78   Kao Soap

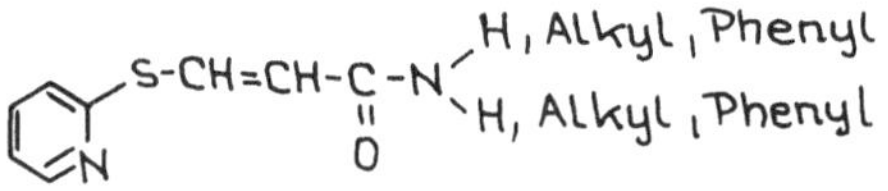

S-CH=CH-C-N   H, Alkyl, Phenyl
                             H, Alkyl, Phenyl

3.f.4   Heterocyclische-aliphatische Aminoalkylencarbonsäuren

US   4 207 090   25. 1.79/10. 6.80   Olin Corp.

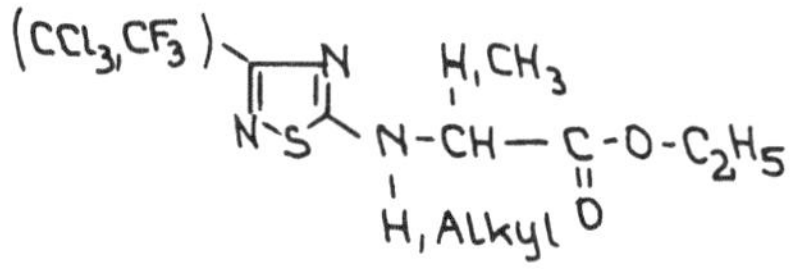

(CCl$_3$,CF$_3$)   H,CH$_3$
N-S   N-CH — C-O-C$_2$H$_5$
          H, Alkyl O

US   4 228 290   25. 1.79/14.10.80   Olin Corp.

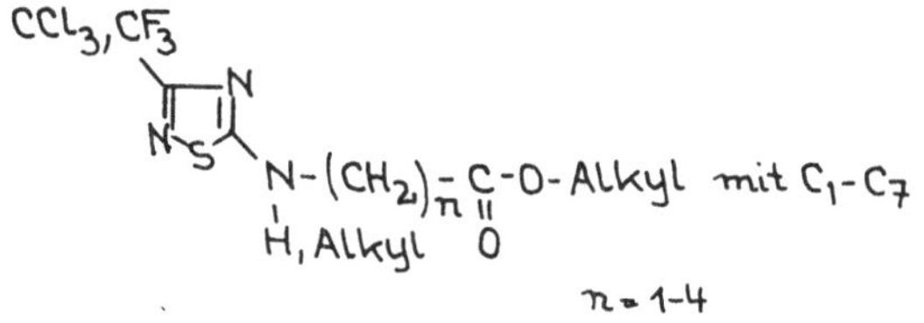

CCl$_3$,CF$_3$

N-S   N-(CH$_2$)$_n$-C-O-Alkyl mit C$_1$-C$_7$
          H, Alkyl O

                n = 1-4

280

5.7.4 **Aromatische Carbonsäuren**

4.a Mono-Carbonsäuren, Ester u. Amide Bd.5/524

DOS 2 605 609 12.2.76/26. 8.76 Hoffmann-La Roche

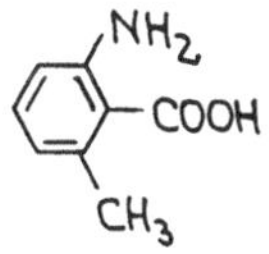

Wuchsregulator und Herbizid.

DOS 2 644 486 1.10.76/14. 4.77 Mobil Oil
US Prior. 1.10.75

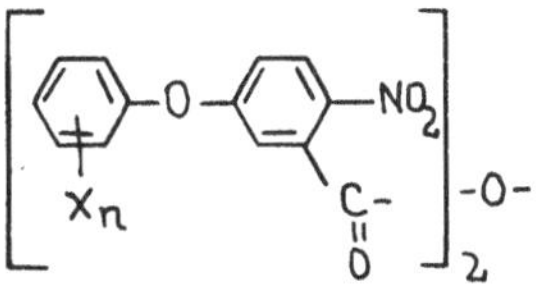

X = Halog., NO_2, Alkyl, CF_3, CN

Entsprechende Ester siehe
US 3 652 645

DOS 2 735 457 5. 8.77/ 9. 2.78 Hoffmann-La Roche
= BE 857 514 GB Prior. 6. 8.76 u. 25. 5.77

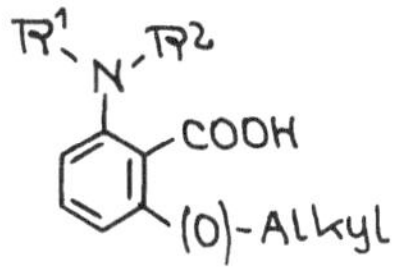

R^1 = H, Alkyl, Acyl

R^2 = CH_3, H

DOS 2 753 900 3.12.76/ 8. 6.78 Rohm und Haas
US Prior. 3.12.76

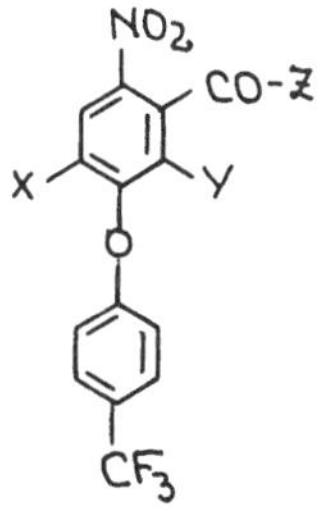

X = H, Halogen, $C(Hal.)_3$CN,Alkyl

Y = H, Halogen, $C(Hal.)_3$

Z = O-Alkylen, O-Alkinyl,
Aralkenyl-O-, O-Halog-Alkyl

N< H, Alkyl / Alk(en,in)yl, Aralkenyl

DOS 2 816 388 15. 4.78/25.10.79 Celamerk

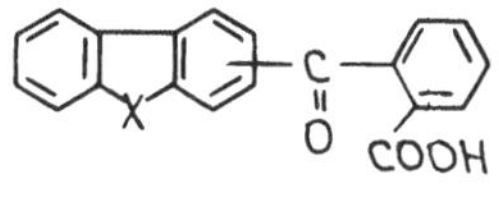

X = CH_2, O, S, CHOH usw.

DOS 2 833 274 28. 7.78/22. 2.79 Amer. Cyanamid
= US 4 188 487 8. 8.77/12. 2.80 Amer. Cyanamid

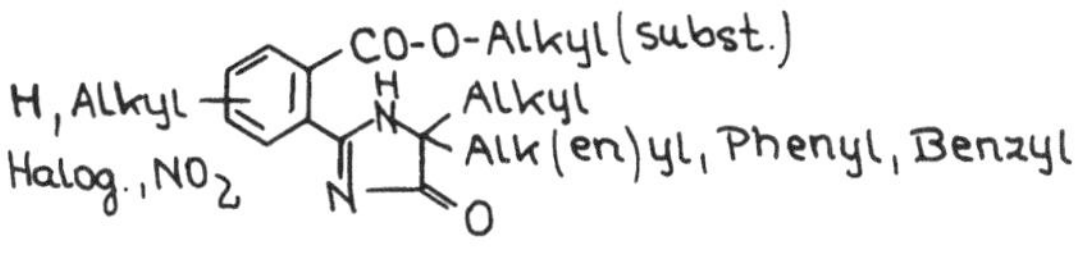

DOS   2 923 371   8. 6.79/13.12.79   Ishihara Sangyo

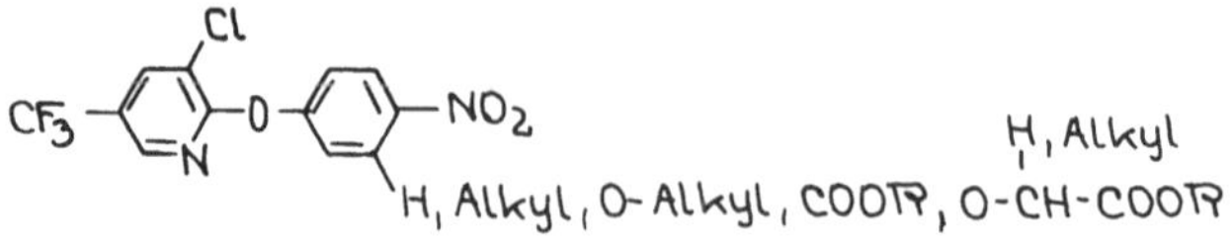

EP   13 660   4. 1.80/23. 7.80   Ciba Geigy
                                                        CH Prior. 8. 1.79

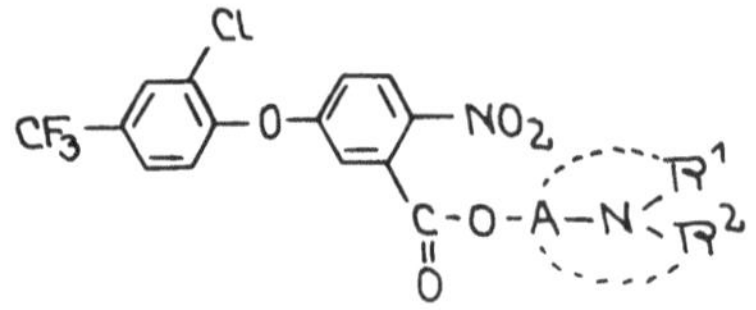

A = C_2-C_4Alkenylbrücke zu

R^1 oder R^2

oder direkte Bindung
nur zu -N⟨

EP   15 140   22. 2.79/ 3. 9.80   Monsanto
                                                         US Prior. 17. 5.79

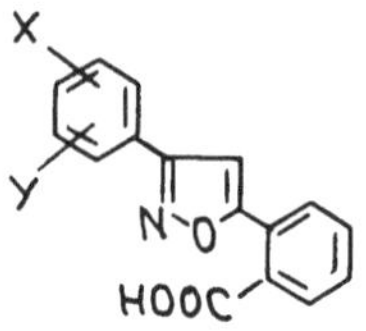

X u. Y = H, Halog., (O)-Alkyl,
         Phenoxy-phenyl

Herbizid und Wuchsregulator

US   3 983 168   25. 4.69/28. 9.76   Mobil Oil

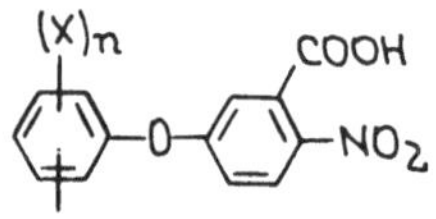

X = Halogen
n = 1 - 5

US   3 996 258   3. 9.75/ 7.12.76   Texaco Dev. Corp.

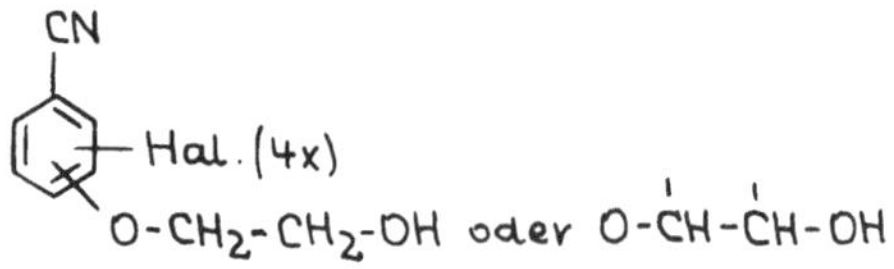

US   4 001 005   7. 6.74/ 4. 1.77   Mobil Oil
Siehe auch US 3 907 866

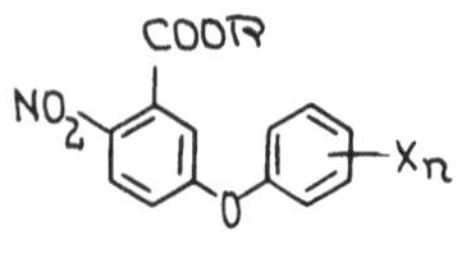

X = Halogen, n = 1 - 3
R = C_4-C_{12}Alkyl, Cycloalkyl usw.

US   4 022 610   15.10.73/10. 5.77   Velsicol Chem.
Siehe US   3 910 974

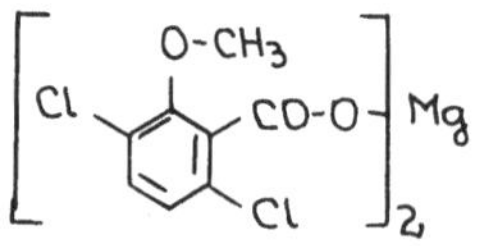

US   4 031 131   29. 9.75/22. 6.77   Rohm und Haas

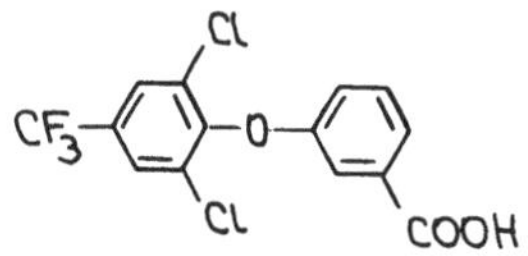

US   4 128 576   3. 1.78/ 5.12.78   Monsanto Corp.

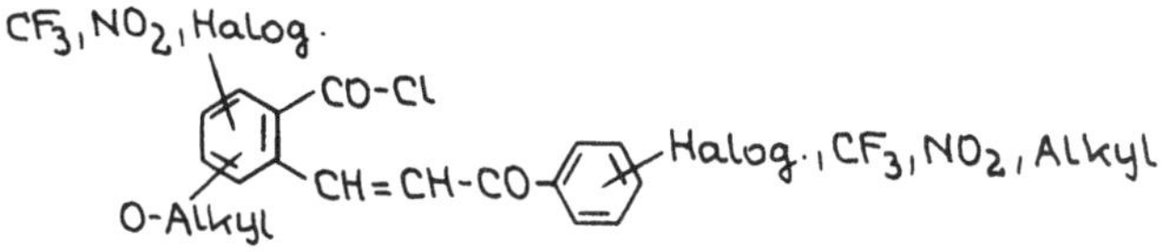

US   4 139 366   12. 5.77/13. 2.79   Monsanto Corp.

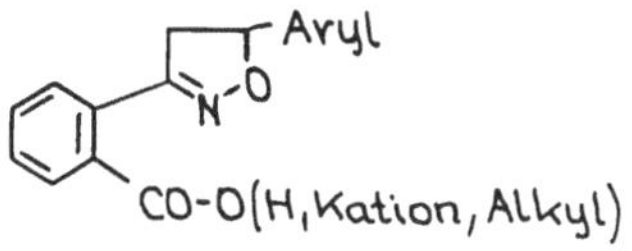

US   4 164 408   25. 4.69/14. 8.79   Mobil Oil
US   4 164 409   25. 4.69/14. 8.79   Mobil Oil
US   4 164 410   25. 4.69/14. 8.79   Mobil Oil

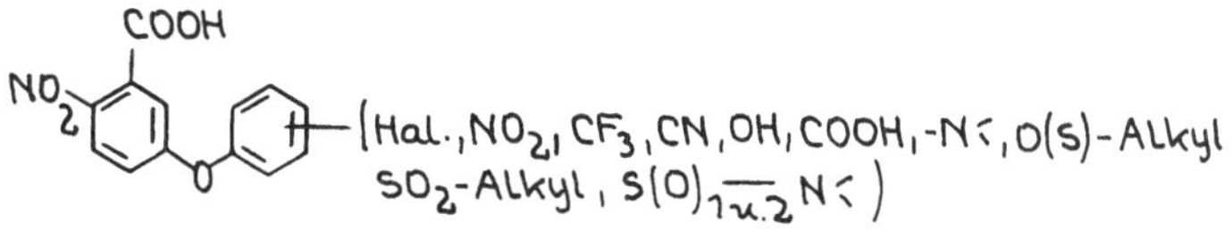

US   4 191 554   3. 10.77/ 4. 3.80   Du Pont

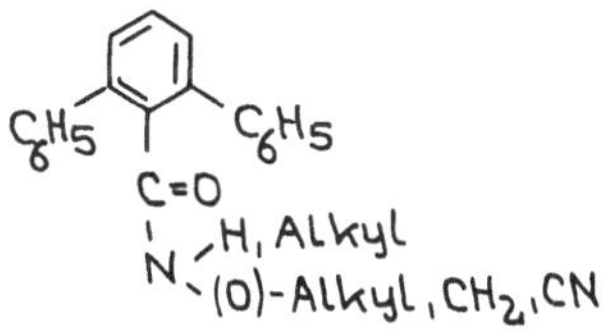

Selektiv in Reis, Weizen, Sojabohnen.

US   4 208 205   3. 10.77/17. 6.80   Du Pont

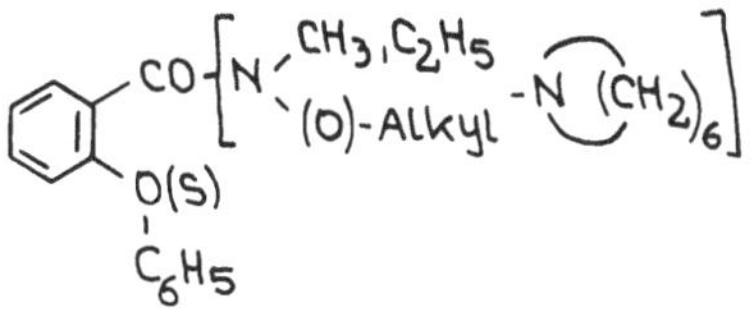

US   4 221 586   30. 1.79/ 9. 9.80   Amer. Cyanamid

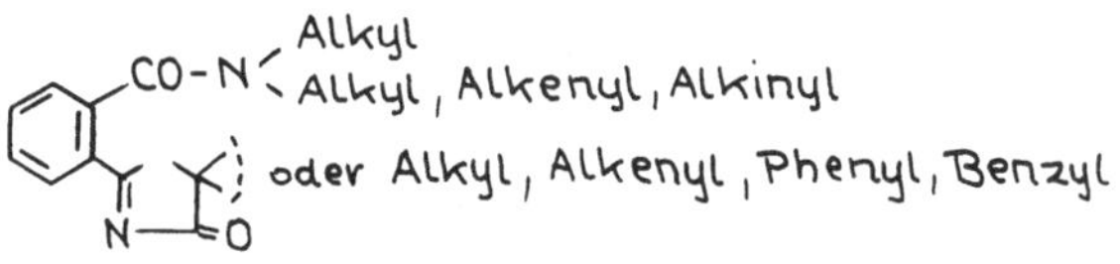

CA   1 002 336   9. 8.71/28.12.76   Uniroyal

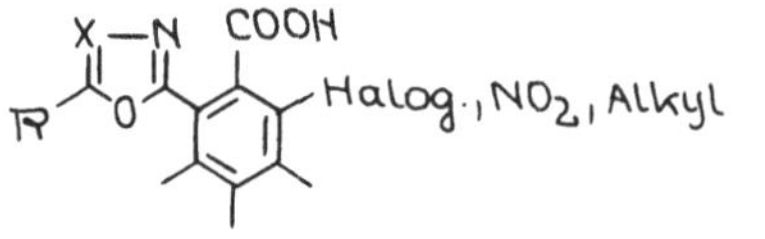

R= H, Alkyl, Heterocycl.
   Phenyl
X = -$\overset{|}{\underset{H}{C}}$=, CH_2

NE   7 705 037   7. 5.76/ 9.11.77   Sumitomo Chem.

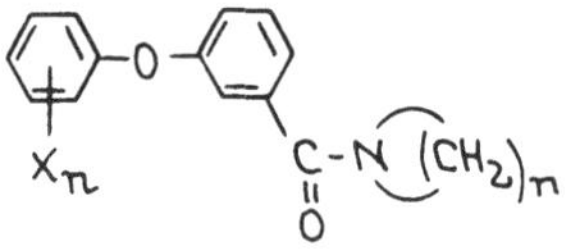

X = Halogen, (O-Alkyl)

JA   52 125141   13. 4.76/20.10.77   Chugai Pharm.

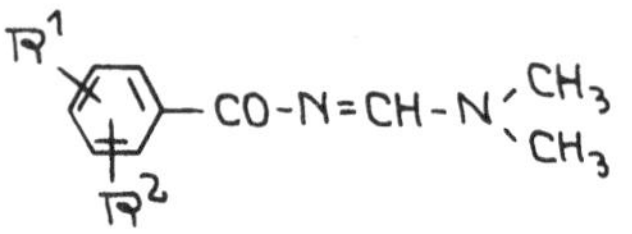

R^1 u. R^2 = H, Halogen, NO_2,OH,
              O-Alkyl, Alkyl usw.

JA   52 125144   13. 4.76/26.10.77   Chagat Pharm.

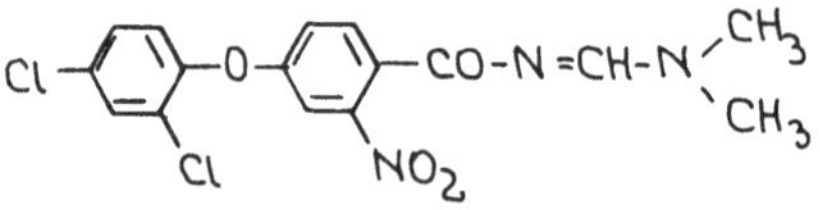

JA   53 034921   10. 9.76/31. 3.78   Nippon Kayaku

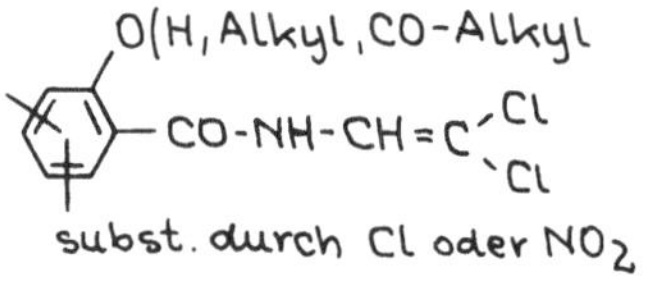

Selektiv gegen Gräser.

JA   53 063337   12.11.76/ 6. 6.78   Nippon Kayaku

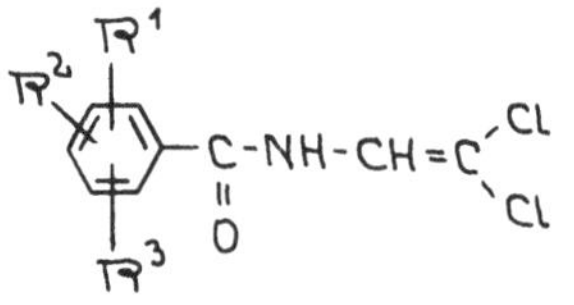

R^1= F, S-Alkyl, Alkyl,
      CF_3, Cl, CN
R^2 u. R^3 = H, Cl, CH_3

Selektiv in Reis.

JA   53 101528   14. 2.77/ 5. 9.78   Ihara Chem. KK

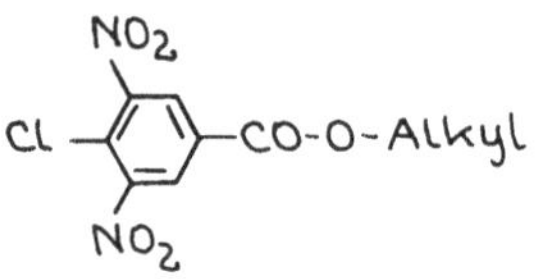

Gegen Gräser in Reis.

JA   55 035016   2. 9.78/11. 3.80   Nihon Noyaku

4.c   Aromatische Di- und Poly-carbonsäuren   Bd.5/531

JA   51 105070   12. 3.75/17. 9.76   Nippon Soda

R^1 u. R^2 = subst. Alkyl

R^3 = subst. Alk(en)yl

X = H, Halog.

Wirksam gegen Gräser.

5.7.5   Heterocyclische Mono- und Di.-Carbonsäuren   Bd.5/531

Carbamidsäureester und Amide; siehe auch 5.5.2.h
und 5.5.3.g

5.b   4- und 5-Ringe mit 1 x N   Bd.5/532

DOS   2 901 659   15. 1.79/24. 7.80   Schering

US   3 999 975   30. 1.70/28.12.76   Upjohn

Gegen breitblättrige Unkräuter.

5.c   5-Ringe mit 1 x N + 1 x O bzw. 1 x S   Bd.5/533

DOS   2 919 511   15. 5.79/29.11.79   Monsanto
= BE   876 245

R = $C(OAlkyl)_3$, CF_3 usw.

Antidot für Alachlor u. Butachlor.

US   4 187 099   16.11.77/ 5. 2.80   Monsanto

H,Alkyl,O-Alkyl,CF₃,Halog.,CN,NO₂

BE   26.12.78/24. 6.80   Monsanto

7.5.5.d.1)   <u>5-Ringe 2 x N</u>   Bd.5/533

DOS   2 732 531   19. 7.77/ 1. 2.79   Höchst

DOS   2 809 022   2. 3.78/ 7. 9.78   May u. Baker
= GB   2 028 821   GB Prior. 4. 3.78

R^1 = H, Alkyl, Alk(en,in)yl
R^2, R^3 u. R^4 = H, Halog., CF₃-O,
                   Alkyl, durch F
                   subst.-Alkyl

X asymetrisches C-Atom
stereoisomere Formen

DOS   2 829 289   4. 7.78/24. 1.80   BASF

R^1 = H, <u>CN</u> subst, Alkyl,
     Acetoacetyl, SO₂N‹ , ArSO₂,
     CO(S)-N
R^2 = O(S)Alkyl, Aryl, Heterocycl.
     usw.
R^3 = Halog., CN, NO₂, CO(S)-N‹ usw.
R^4 = H, Alkyl, O-alkyl, SAlkyl,
     Halog., COOAlkyl usw.

Selektiv in Zuckerrohr und Obst-
gärten usw.

286

d2)   <u>5-Ring 2 x S</u>

JA   50 123828   18. 3.74/29. 9.75   Nihon Nohyaku

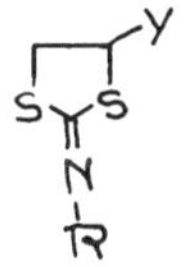

R = Octyl, Y = H, <u>COOH</u>, <u>CN</u>, C_6H_5

d3)   <u>5-Ring 2 x N + 1 x S, 2 x N + 1 x O, 3 x N</u>   Bd.5/333

DOS   2 617 339   21. 4.76/ 4.11.76   Roussel-Uclaf
= BE   840 996   22. 4.75/22. 10.76   FR Prior. 24. 4.75

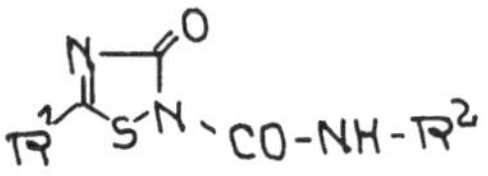

R^1 = O-Alkyl, S-Alkyl
       O(S)-Cycloalkyl
       O-Benzyl usw., N<
R^2 = Alkyl, Phenyl

DOS   2 728 523   23. 6.77/11. 1.79   Schering

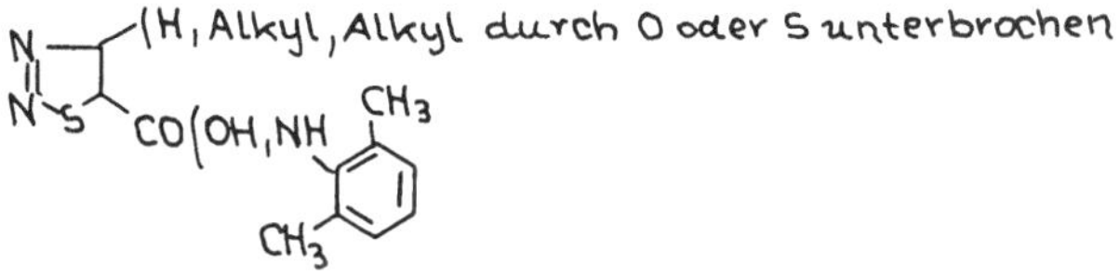

(H, Alkyl, Alkyl durch O oder S unterbrochen)

DOS   2 809 022   4. 3.77/ 7. 9.78   May und Baker

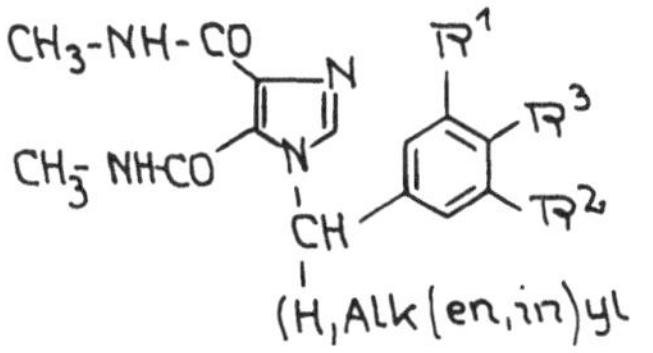

R^1, R^2 u. R^3 = H, Halog., CF_3O,
                CF_3, Alkyl, O-Alkyl
                usw.

Wirksam gegen Gräser und Unkräuter.

DOS   2 834 879   7. 8.78/21. 2.80   Schering

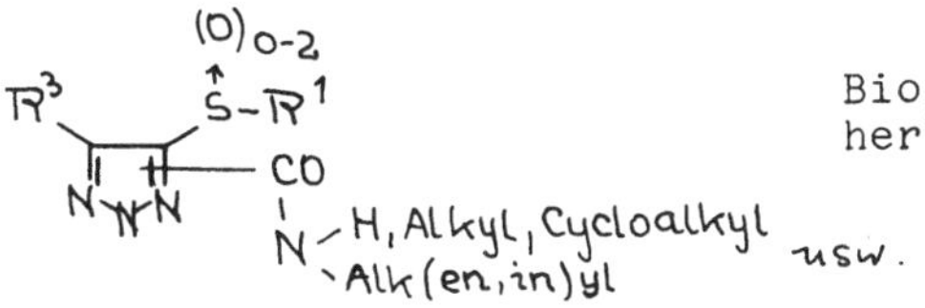

Biozid wirksam, auch
herbizid angeführt.

DOS   2 909 991   12. 3.79/ 2.10.80   Schering

          N—(H, Alkyl usw.)   Y = Halogen, N<Alk(en,in)yl
          N—S   H, Alk(en,in)yl
             C = X
             |
             Y   N( )   usw.

                        CH₃   X = N (Y)
    z.B.   N
            N—S
               C = N—cyclohexyl
               |
               Cl

US   4 115 095   23. 3.77/19. 9.78   Monsanto

    (H, Alkyl)O—C   N
               ‖   S
               O   —(H, Halog., CF₃, CN, CO-O-Alkyl)

Antidot in Reis, Sorghum u. Weizen gegen Herbizid-Schäden durch
Acetanilid-Herbizide.

BE   882 628   5. 4.79/ 3.10.80   Schering AG

          N   H, Alkyl oder S enthalt.   R¹ = H, Alkyl, Alk(en,in)yl,
          N—S   Aralkyl, Cycloalkyl
             CO—N—R²   R² = R¹ oder O-Alkyl,
                 | O-Alk(en,in)yl, N< usw.
                 R¹

GB   2 028 821   25. 8.78/12. 3.80   May und Baker

    CH₃-NH-CO   CO-NH-CH₃   R¹,R²,R³ = H, Halog., (O)-Alkyl
          N   N   CF₃ usw.
               | R¹
    CH₃O-CH   R²   Gegen breitblättrige Unkräuter.
                   R³

JA   52 083552   1. 1.76/12. 7.77   Mitsubishi Chem.

          N-CO-O-Alkyl
    (CH₂)ₘ |
          N-C-NH-Alk(en)yl, Cycloalkyl, Phenyl usw.
            ‖
            O(S)
    c)   3 x N

JA   52 125168   13. 4.76/20.10.77   Nippon Soda

          N—N-CO-N<R²   R¹ = H, Alkyl
                     R³
    R²-S   N O   R² = Alkyl, Cyclohexyl, Phenyl
               |
               NH₂   R³ = H, Alkyl

288

e)   6-Ring-Carbonsäuren, 1 x N

DOS   3 009 307   11. 3.80/ 2.10.80   Abott Lab.
                                                    US Prior. 12. 3.79

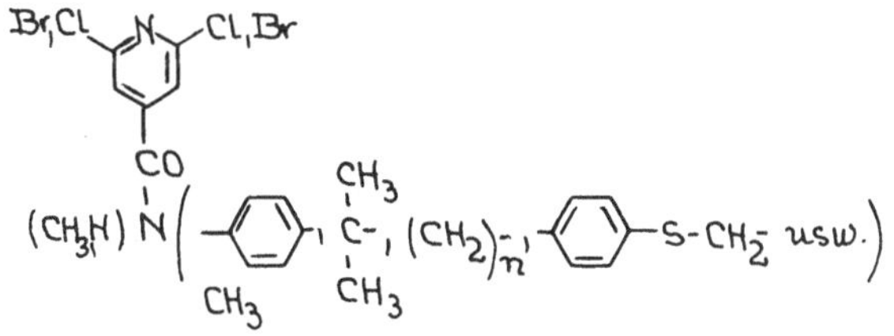

US   4 014 888   5.11.73/29. 3.77   Dow Chem.

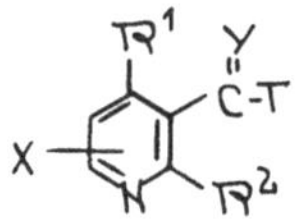

R^1 u. R^2 = 1 x NHSO$_2$N$<$ H, Alk(en,in)yl,
                                   Halog.-Alkyl
                                   Cycloalkyl
T = OH, Halog. O-Alkyl, Y = O
X = Alkyl, O-Alkyl, S-Alkyl, NO$_2$,
    N$<$, Cycloalkyl

Vorwiegend als Zwischenprodukt für Herbizide.

z.B.:

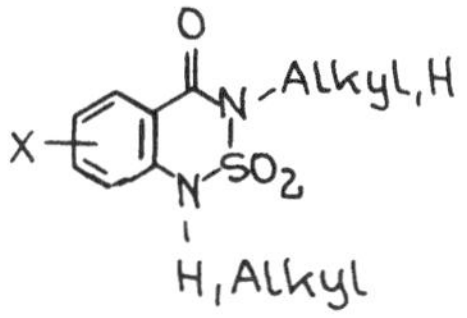

US   4 195 984   12. 3.79/ 1. 4.80   Abbot Lab.

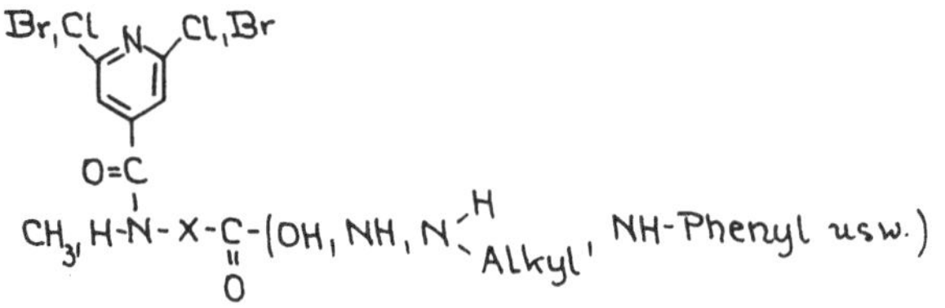

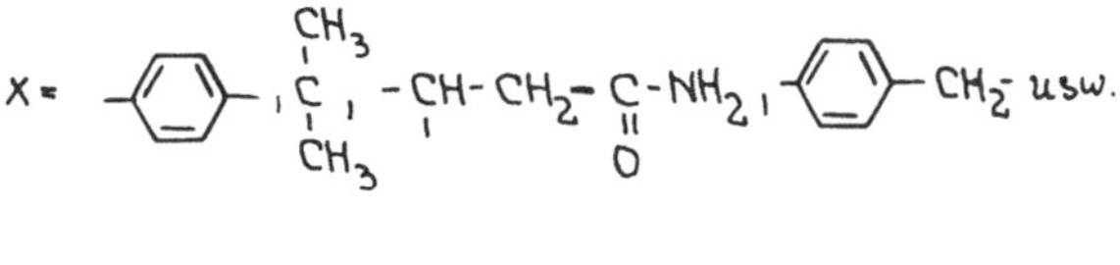

f) 6-Ringe mit 2 x N; 1 x O + 1 x S; 2 x S Bd.5/537

US 4 118 217 6. 1.77/ 3.10.78 Farmland Ind

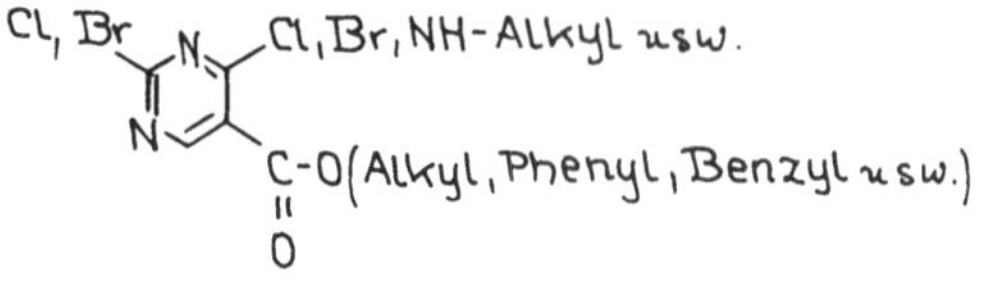

5.7.5.g) Heterocyclen mit mehr als 6-Gliedern. Mono u. Di-carbonsäuren (neu)

DOS 2 638 543 26. 8.76/ 3. 3.77 Ciba Geigy
CH Prior. 29. 8.75

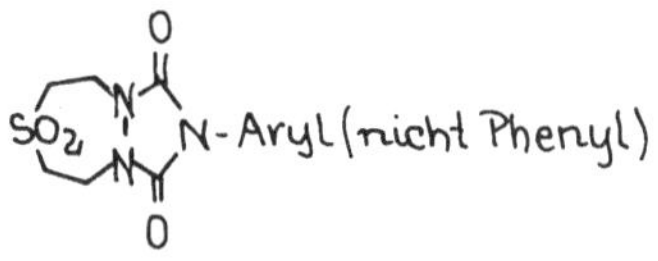

Herbizide auch Fungizide u. Insectizide.

5.8 Aromatische Aldehyde, Iminoverbindungen, Oxime und Hydrazone

Bd.5/538

Aromatische Dialdehyde
(Verbindungen Ar-CO-Alkyl siehe unter 5.3.2.)

5.8.1 Monoaldehyde

US 4 125 397 3. 1.77/14.11.78 Chevron Res.

X,Y,Z = Cl, Br

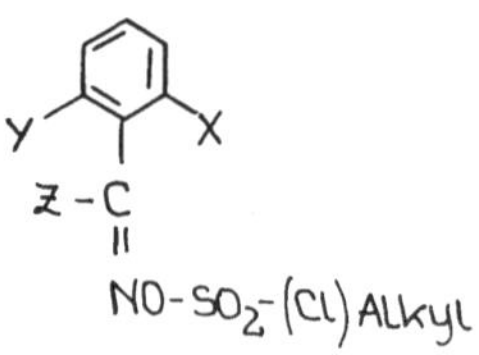

a)   <u>Aromatische Amine ohne NO_2 in 2 u. 6-Stellung</u>

DOS   2 605 609   12. 2.76/26. 8.76   F Hoffmann-La Roche

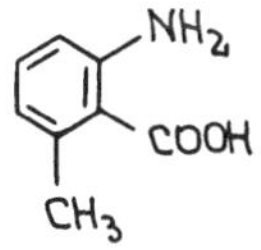

Wuchsregulator und Herbizid

DOS   2 735 457   5. 8.77/ 9. 2.78   F Hoffmann-La Roche
= BE   857 514   GB Prior. 6. 8.76   25. 5.77

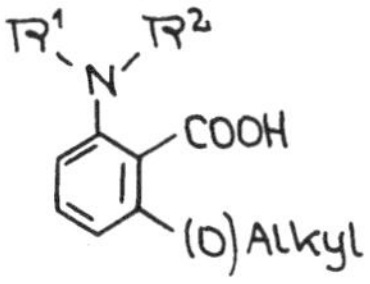

R^1 = H, Alkyl, Acyl

R^2 = CH_3, H

DOS   2 929 181   19. 7.79/14. 2.80   Duphar Int. Res.
                                                   NE Prior. 26. 7.78

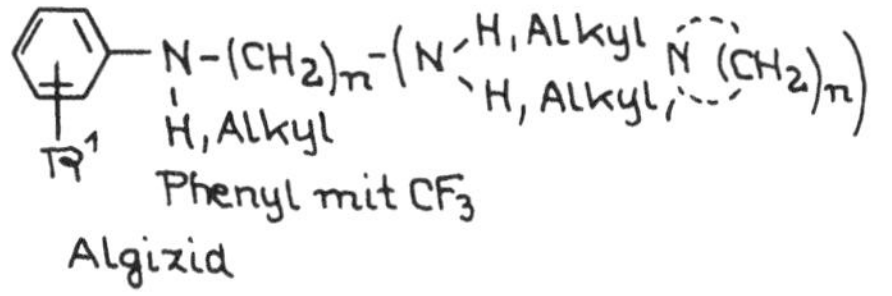

Algizid

R^1 = Halog., Halog.-Alkyl, O(S)-Alkyl, Cycloalkyl, Alkyl-SO_2-,
Phenoxy usw.

EP   5 288   21. 4.78/14.11.79   Chimac. Soc.

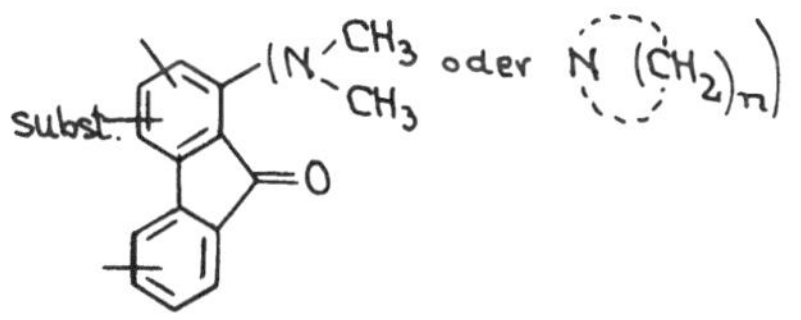

US   4 006 185   23. 7.73/ 1. 2.77   Olin Corp.

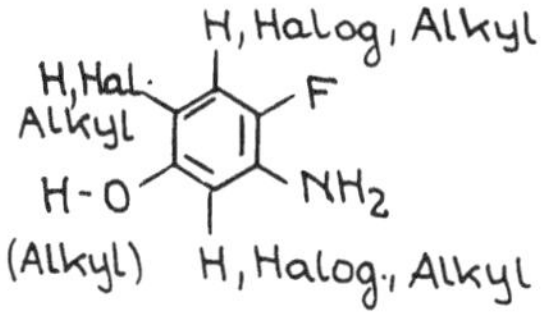

JA   54 126725   17. 3.78/ 2.10.79   Ube Industries

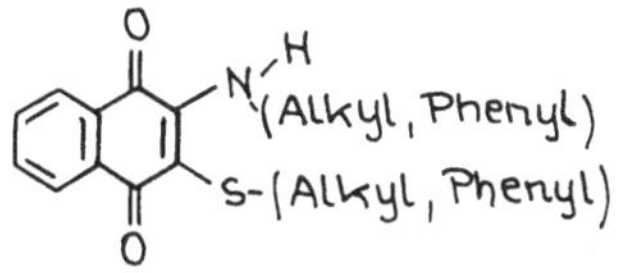

b) <u>Aromatische Amine mit 2 x NO_2 in 2 und 6ev.4-Stellung</u>

<u>oder anderen Elektronen anziehenden Gruppen in 2, 6</u>

<u>oder 4-Stellung</u>   Bd.5/544

DOS   2 627 349   18. 6.76/13. 1.77   Eli Lilly

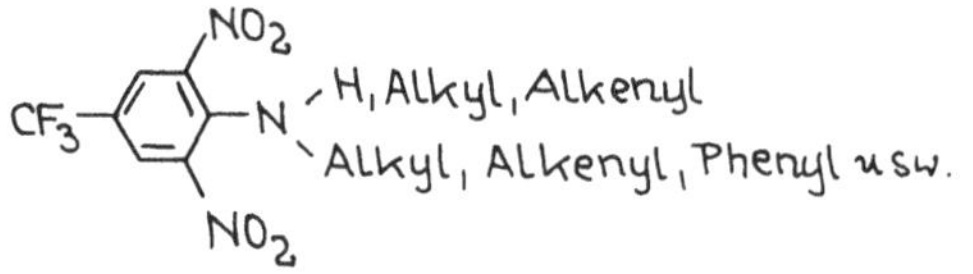

Fungizid-Beispiel für nahe Verwandschaft von Herbiziden mit Funziden.

DOS   2 627 351   18. 6.76/13. 1.77   Eli Lilly
Siehe auch US 3 867 452 und 3 891 706

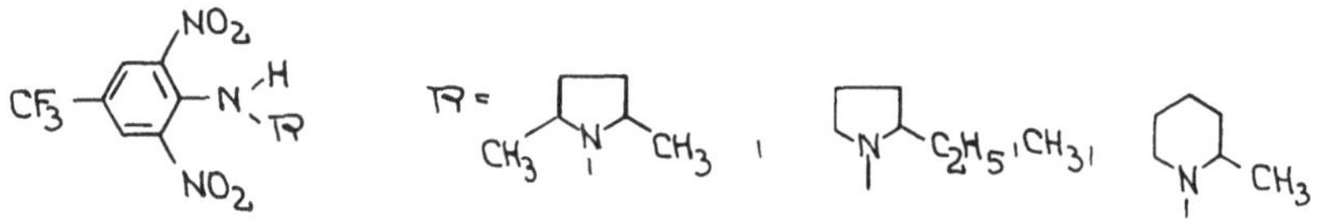

Herbizide und Fungizide.

DOS   2 657 327   17.12.76/ 7. 7.77   Amer. Cyanamid
                                        US Prior. 22.12.75   14. 6.76

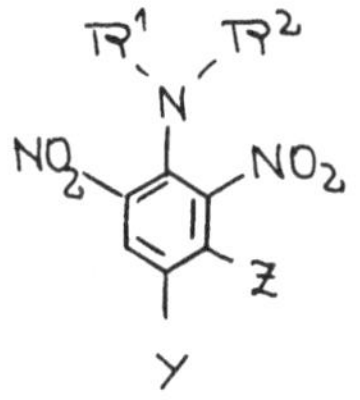

R^1 = H

R^2 = sec. Alkyl, Methoxy-sec. Alkyl

Z = $CH_3O-\overset{CH_3}{\underset{H}{C}}-$; Y = Cl, CH_3, C_2H_5,
                                n und iso.Propyl,
                                sec. Butyl

DOS   2 738 584   26. 8.77/ 2. 3.78   Eli Lilly
                                        US Prior. 31. 8.76

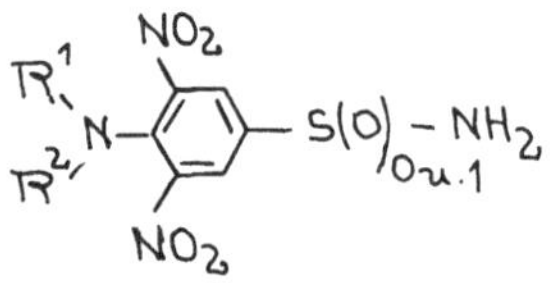

R^1 = H, Alk(en)yl; R^2 = C_3-C_7-Alkyl

DOS   2 708 440   26. 2.77/31. 8.77   Bayer

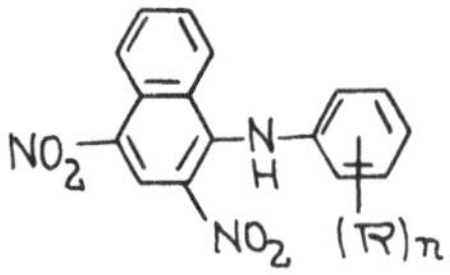

u.a. auch herbizid wirksam.
R = H, Alkyl, Alk(en,in)yl,O(S)-Alkyl
S-Alkyl, CN, NH-CO-Alkyl usw.
n = 1 - 5

292

DOS	2 754 053	5.12.77/15. 6.78	Eli Lilly
= BE	861 475		US Prior. 9.12.76
= NE	7 713 453	9.12.76/13. 6.78	Eli Lilly

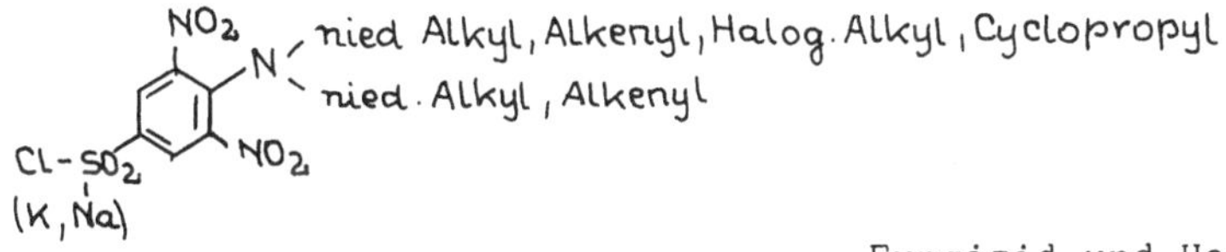

NO_2
N — nied. Alkyl, Alkenyl, Halog. Alkyl, Cyclopropyl
— nied. Alkyl, Alkenyl
$Cl-SO_2$
NO_2
(K, Na)

Fungizid und Herbizid.

| DOS | 2 805 251 | 8. 2.78/10. 8.78 | Lilly Ind. |
| | | | GB Prior. 9. 2.77 |

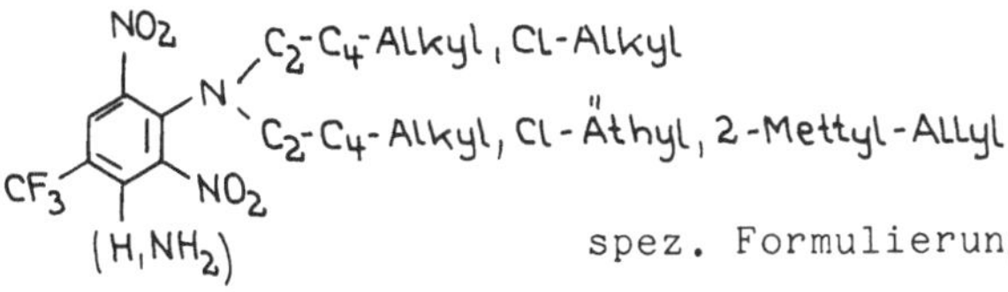

NO_2
N — C_2-C_4-Alkyl, Cl-Alkyl
— C_2-C_4-Alkyl, Cl-Äthyl, 2-Methyl-Allyl
CF_3
NO_2
(H, NH_2)

spez. Formulierung

| DOS | 2 808 628 | 28. 2.78/28. 9.78 | Amer. Cyanamid |
| = US | 4 082 537 | 23. 3.77/ 4. 4.78 | Amer. Cyanamid |

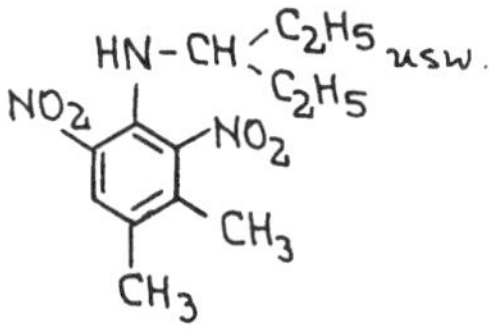

$HN-CH$ — C_2H_5 usw.
— C_2H_5
NO_2
NO_2
CH_3
CH_3

1 - 2% Na-di(C_6-C_8)alkylsulfo-
succinat als Stabilisator der
Suspensionen

| DOS | 2 827 451 | 22. 6.78/11. 1.79 | Eli Lilly |
| | | | US Prior. 27. 6.77 |

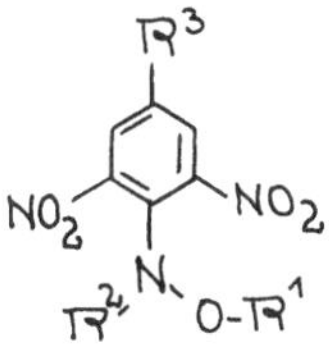

R^3
NO_2
NO_2
R^2 N $O-R^1$

R^1 = CH_3, C_2H_5
R^2 = H, Alk(en)yl
R^3 = CH_3, C_2H_5, CF_3

| DOS | 2 831 119 | 14. 7.78/ 8. 2.79 | Eli Lilly |
| | | | US Prior. 18. 7.77 |

Reinigung von
NO_2
NO_2
N
Alkyl Alkyl

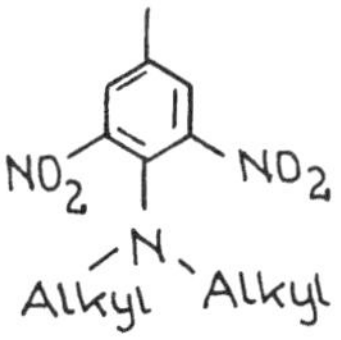

Spaltung von Nitrosaminen
durch Erhitzen mit HCl

EP   4 642   20. 1.79/17.10.79   Bayer
= DOS   2 805 757   DB Prior. 8. 4.78

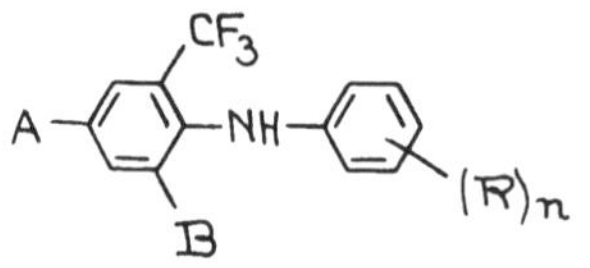

A u. B = CF_3, das andere NO_2
R = H, CN, NO_2, Alkyl usw.

u.a. auch Herbizide

EP   5 288   14. 2.79/14.11.79   Chimia SA
                                                  DB Prior. 21. 4.78

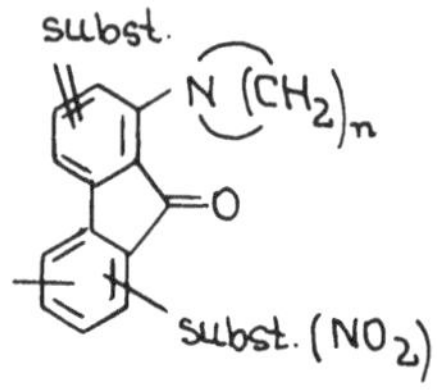

pre und post emergence
Herbizid

US   3 979 453   23. 6.75/ 7. 9.76   Eli Lilly

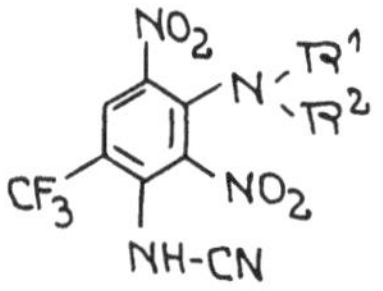

R^1 u. R^2 = Alk(en)yl, Cyclopropyl,
            Chlor-Alkyl usw.

US   3 987 076   23. 6.75/19.10.76   Eli Lilly
= NE   7 606 743

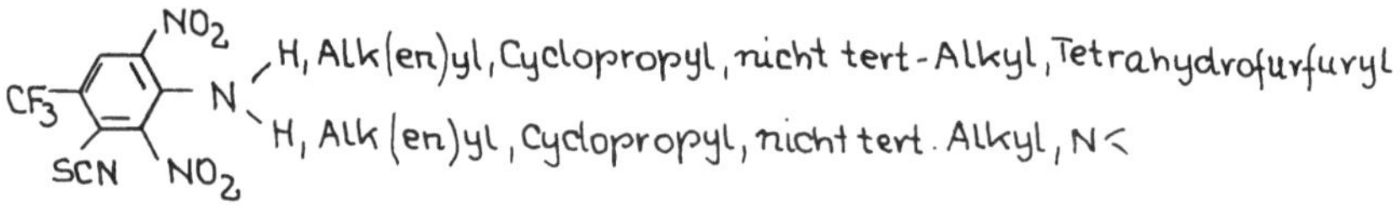

u.a. auch herbizid wirksam.

US   4 025 538   25. 8.71/24. 5.77   Amer. Cyanamid
Siehe US   3.23 000; 3 920 742

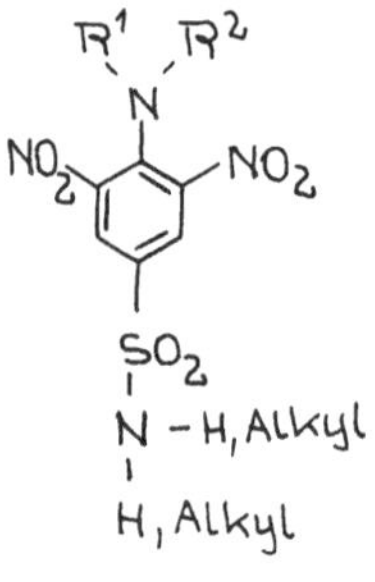

R^1 = H, Alk(en)yl, Alkinyl,
       Cycloalkyl

US   4 046 758   19.11.69/ 6. 9.77   US Borax

R^1 u. R^2 = Alkylen

US   4 046 809   16.12.74/ 6. 9.77   Wilcox

US   4 054 603   21. 8.76/18.10.77   Eli Lilly

R^1 = H, Alk(en)yl
R^2 = sec. Alkyl; n = 1 - 2

Herbizid und Fungizid

US   4 066 441   25. 8.71/ 3. 1.78   Amer. Cyanamid

Gegen Gräser in Sojabohnen.

US   4 087 270   3. 3.77/ 2. 5.78   Gaf. Corp.

US   4 087 460   27. 6.77/ 2. 5.78   Eli Lilly

R = CH_3, C_2H_5, CF_3

US   4 098 812   25. 8.71/ 4. 7.78   Amer. Cyanamid

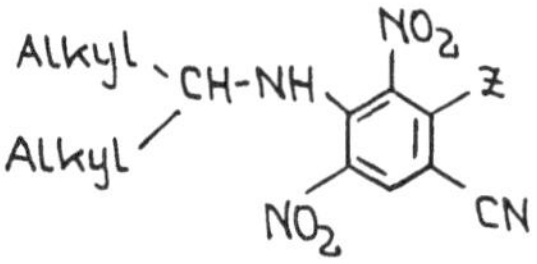

$Z = CH_3$, $CH_2-O-(CH_3, C_2H_5)$

US   4 101 582   25. 8.71/18. 7.78   Amer. Cyanamid

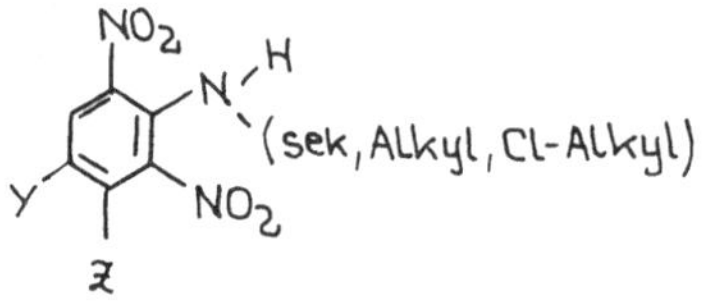

$Y = Alkyl$, iso-Alkyl, Cl, CF_3
$Z = CH_3$, CH_3-O-CH_2-,
$\qquad CH_3-O-\underset{\underset{CH_3}{|}}{CH}-$

US   4 124 639   25. 8.71/ 7.11.78   Amer. Cyanamid

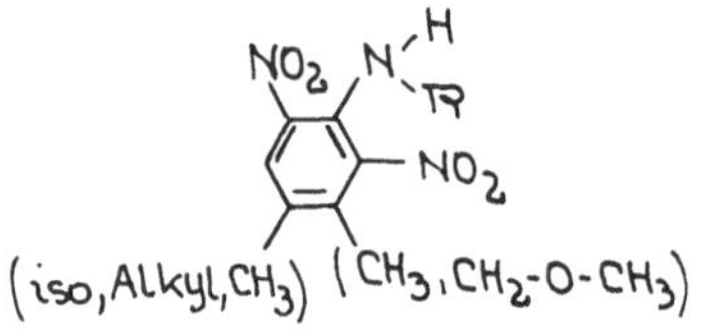

$R = CH_3-O-CH_2-\underset{\underset{(CH_3, C_2H_5)}{|}}{CH}-$

US   4 131 449   17.10.77/26.12.78   Shell Oil Co.

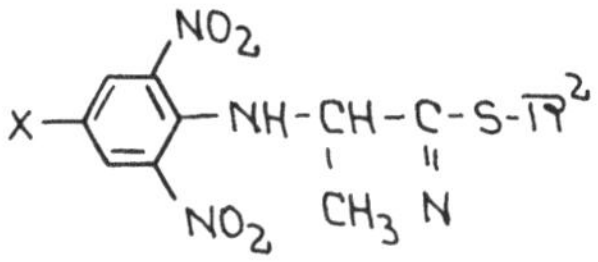

$X = H$, Halog., Alkyl
R^1 u. $R^2 = Alkyl$

US   4 136 117   25. 6.73/23. 1.79   Amer. Cyanamid

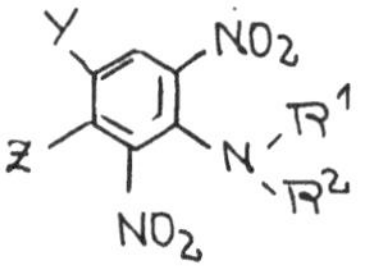

$Y = C_1-C_9$-Alkyl, Halog., CF_3
$Z = H$, Halog., (O)-Alkyl
$R^1 = H$ oder u. R^2
$R^2 = Alkyl$, Cycloalkyl, Halogen-Alkyl
usw.

US   4 138 244   4.12.74/ 6. 2.79   US Borax u. Chem. Corp.

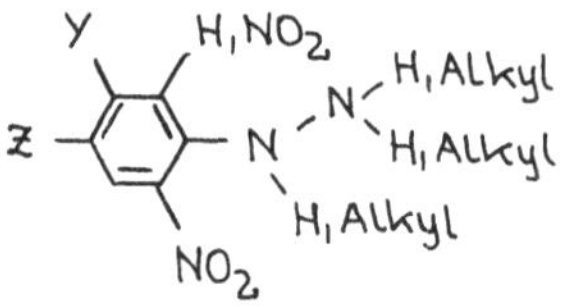

$Y = O$-Alkyl, NH_2, $N(Alkyl)_2$,
$\qquad NH-N<$
$Z = Halog.$, CF_3

Selektiv in Sojabohnen, Reis,
Baumwolle.

US   4 143 073   27. 6.77/ 6. 3.79   Eli Lilly

US   4 165 231   28. 8.71/21. 8.79   Amer. Cyanamid

US   4 169 721   16.12.74/ 2.10.79   Ciba Geigy

US   4 201 724   26. 7.73/ 6. 5.80   US Borax u. Chem. Co.

BE   844 580   30. 7.75/28. 1.77   Shell Int. Res.

BE   849 681   22.12.75/21. 6.75   Amer. Cyanamid

R¹ = H
R² = CH(C₂H₅), verzweigte Alkyle,
     z.T. durch Halog. substit.
X = C₂H₅, n-u. iso-Propyl
          n-u. iso-Buthyl,
    CH(CH₂)ₙ-OCH₃ usw.
    CH₃

BE   874 897   17. 3.78/16. 7.79   Penwalt Corp.

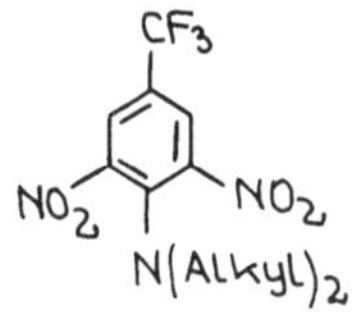

spez. Zurichtung

FR   2 295 011   16.12.74/20. 8.76   Ciba Geigy

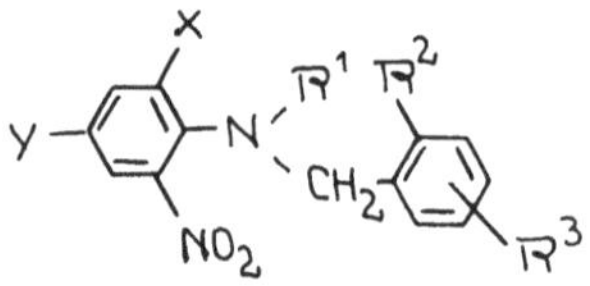

R^1 = H, Alkyl, Allyl
X u. Y = NO_2 u. CF_3, abwechselnd
R^2 = (O)-Alkyl, Halog.
R^3 = H, OCH_3, Alkyl

Wachstumsregulatoren gegen Austriebe beim Tabak.
In höheren Konzentrationen Herbizide.

GB   1 544 078   6. 3.75/11. 4.79   ICI

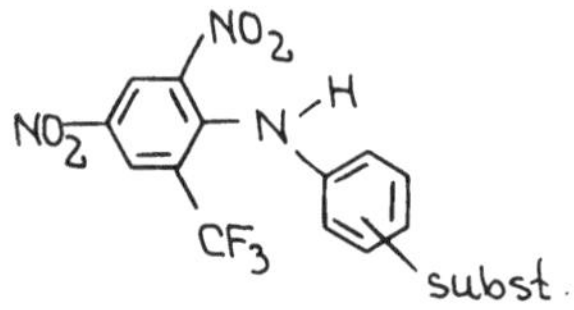

subst.

NE   7 606 742   23. 6.75/27.12.76   Eli Lilly

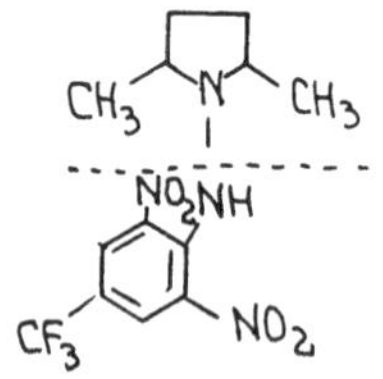

Gegen Gräser in Mais.

CH   585 706   25. 9.73/15. 3.77   Eli Lilly

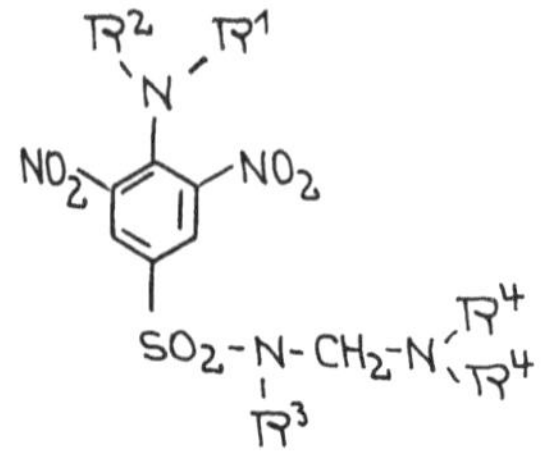

SO_2-N-CH_2-N   Darst. Verfahren

R^1 u. R^2 = Alk(en,in)yl
R^3 = H, CH_3
R^4 = C_3H_7, N<, usw.

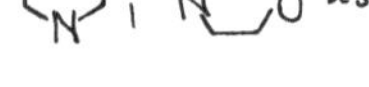

JA   55 120545   9. 3.79/17. 9.80   Nippon Soda

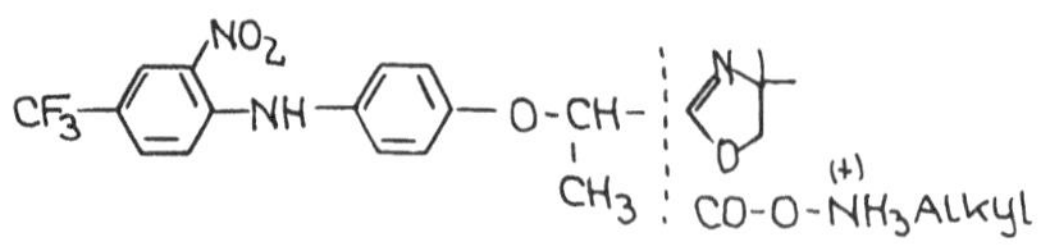

5.11.2.c   Phenyl-hydrazin-Derivate   neu

DOS   2 805 929   13. 2.78/17. 8.78   Monsanto
                                                 US Prior. 14. 2.77

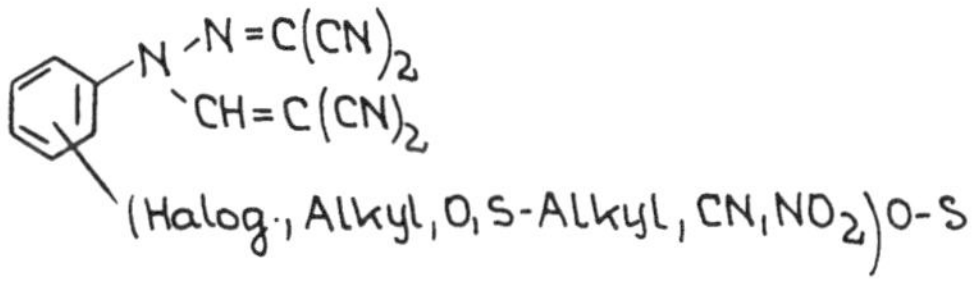

$$\text{(Halog., Alkyl, O,S-Alkyl, CN, NO}_2\text{)}\,\text{O-S}$$

US   4 160 090   4.12.74/ 3. 7.79   US Borax

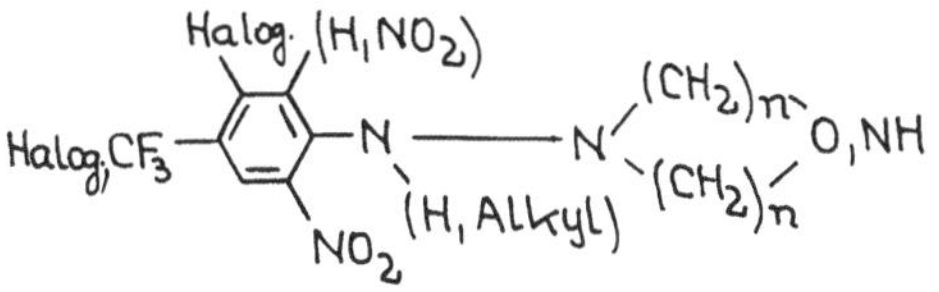

US   4 202 839   4.12.74/13. 5.80   US Borax

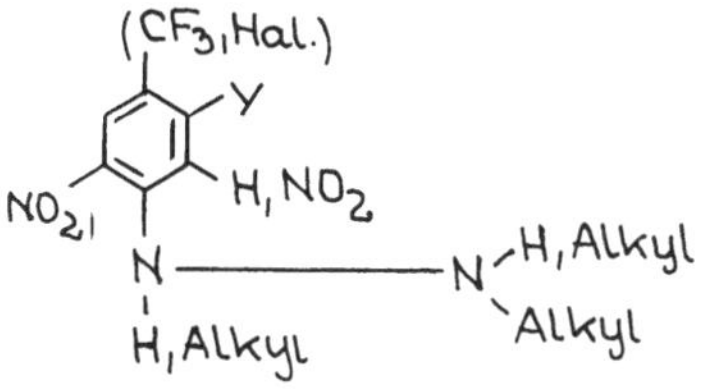

                                                 Y = O-Alkyl, NH$_2$,
                                                     N(Alkyl)$_2$, N-N$\langle$ R
                                                 usw.

5.11.2.d   Aromatische Diazoverbindungen   neu

US   4 006 132   28. 4.71/ 1. 2.77   Upjohn

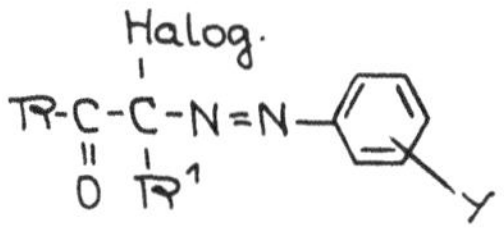

                                    R = (O)-Alkyl, Cycloalkyl, Y = Alkyl
                                    R^1 = Alkyl, Cycloalkyl, Phenyl,
                                          Halog.

US   4 008 217   28. 4.71/15. 2.77   Upjohn Co.

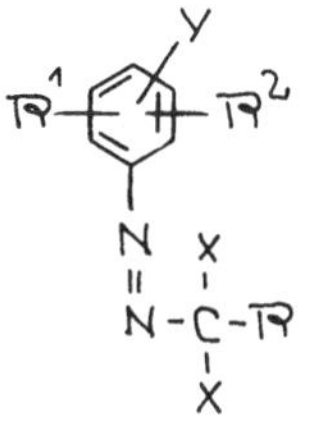

                                    R = H, Alkyl, Cycloalkyl, Phenyl
                                    X = F, Cl, Br
                                    R^1 u. R^2 = Halog., Alkyl, O-Alkyl
                                    Y = Alkyl

z.B. Benzylamine   Bd.5/554

DOS   2 738 873   31. 8.76/ 2. 3.78   Eli Lilly

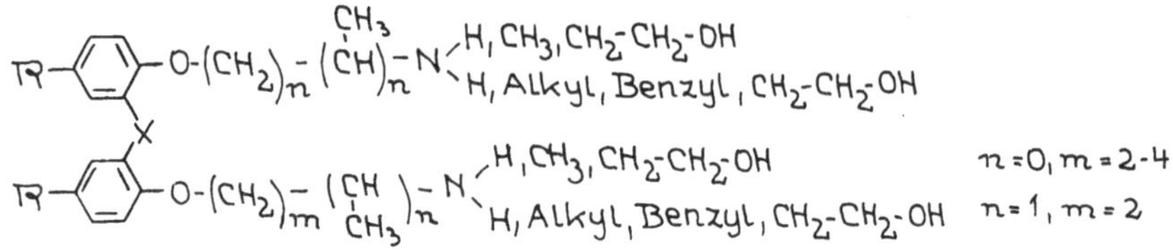

$$R\text{-}\!\!\bigcirc\!\!\text{-O-}(CH_2)_n\text{-}(\overset{CH_3}{CH})_m\text{-N}\!\!\begin{array}{l}H,CH_3,CH_2\text{-}CH_2\text{-OH}\\ H,Alkyl,Benzyl,CH_2\text{-}CH_2\text{-OH}\end{array}$$

$$R\text{-}\!\!\bigcirc\!\!\text{-O-}(CH_2)_m\text{-}(\overset{CH}{\underset{CH_3}{}})_n\text{-N}\!\!\begin{array}{l}H,CH_3,CH_2\text{-}CH_2\text{-OH}\\ H,Alkyl,Benzyl,CH_2\text{-}CH_2\text{-OH}\end{array}$$

$n = 0, m = 2\text{-}4$

$n = 1, m = 2$

R = H, Halogen, NO_2, (O)-Alkyl

X = CH_2, CH_2-CH_2, -CH-, S   Herbizid und Algizid
                          |
                         CH_3

DOS   2 738 902   29. 8.77/ 2. 3.78   Eli Lilly
                                            US Prior. 31. 8.76 u. 31.3.77

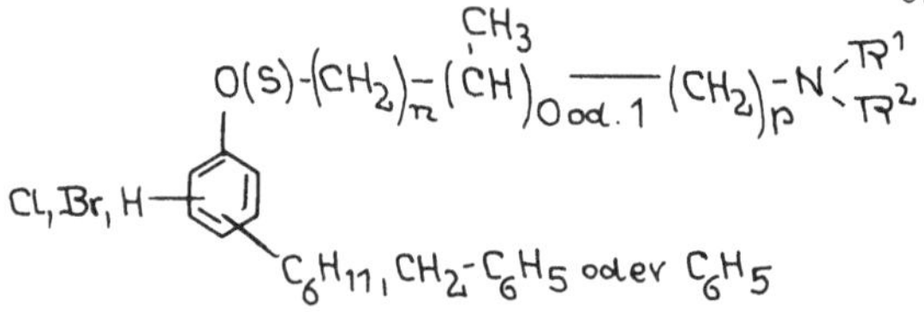

$$Cl,Br,H\text{-}\!\!\bigcirc\!\!\begin{array}{l}O(S)\text{-}(CH_2)_n\text{-}(\overset{CH_3}{CH})_{O\ od.\ 1}\text{-}(CH_2)_p\text{-N}\!\!<\!\!\begin{array}{l}R^1\\R^2\end{array}\\ C_6H_{11},CH_2\text{-}C_6H_5\ oder\ C_6H_5\end{array}$$

n + p = 2 - 10   Herbizid und Algizid

R^1 = H, Alkyl, CH_2-CH_2-OH

R^2 = H, Alkyl, Alkenyl, Cycloalkyl, CH_2-CH_2-OH, CH_2-C_6H_5, Phenyl,
     1-Adamantyl

US   4 201 569   3.10.77/ 6. 5.80   Gulf Oil

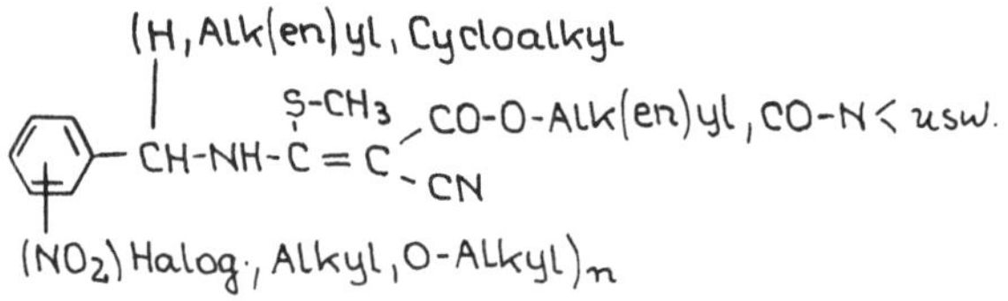

$$\bigcirc\!\!\text{-}\underset{(NO_2)Halog.,Alkyl,O\text{-}Alkyl)_n}{\overset{(H,Alk(en)yl,Cycloalkyl}{\underset{|}{CH}}}\text{-NH-C}=\text{C}\!\!\begin{array}{l}S\text{-}CH_3\\CO\text{-}O\text{-}Alk(en)yl,CO\text{-}N<usw.\\CN\end{array}$$

BE   850 154   7. 1.77/ 2. 5.77   Stauffer

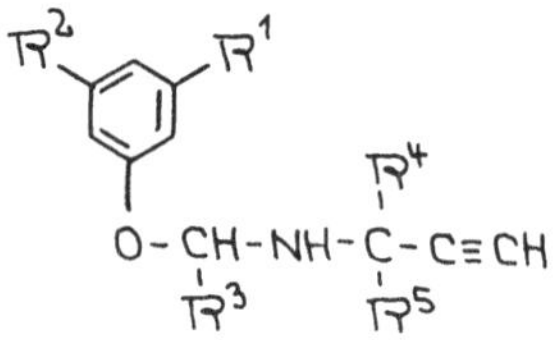

$$\overset{R^2}{}\!\!\bigcirc\!\!\overset{R^1}{}$$
$$\underset{R^3\quad R^5}{O\text{-}CH\text{-}NH\text{-}\overset{R^4}{C}\text{-}C\equiv CH}$$

R^1 u. R^2 = C_1-C_4-Alkyl

R^3 = C_1-C_4 Alkyl

R^4 u. R^5 = H oder C_1-C_3-Alkyl

Totalherbizide

DDR   123 049   23.12.75/20.11.76   Grimmecke

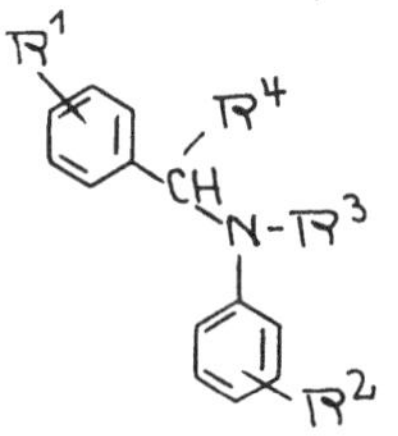

R^1 = H, Alkyl, Aryl, OH, O-Alkyl,
      COOH, Cl, NO_2, NH_2
R^2 = H, Alkyl, Aryl, OH, O-Alkyl,
      COOH, Cl, NO_2, NH_2

CH_2-NH-

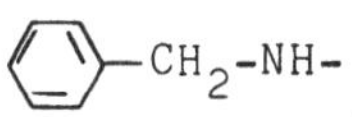

R^3 = H, $CO-CH_3$; R^4 = H, CH_2NO_2, CH_2,
      CH_2NH_2, CCl_3

JA   52 090630   21. 1.76/30. 7.77   Sumitomo

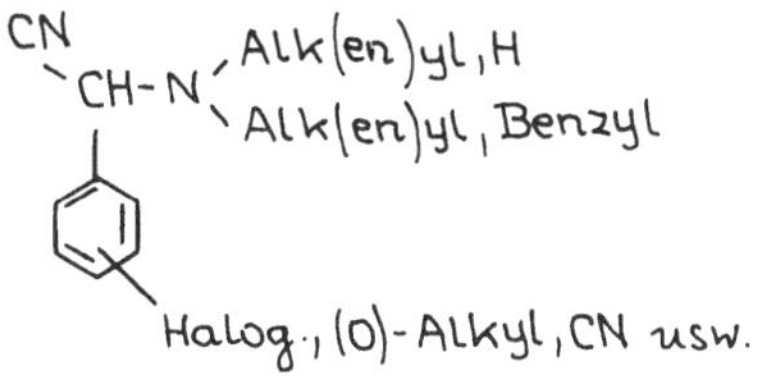

gegen Algen

5.12 Carbonsäureamide aromatischer und heterocyclischer Amine

Bd.5/555

5.12.1.a   <u>Acylderivate von primären Anilinen mit aliphatischen,</u>

<u>cycloaliphatischen, araliphatischen Carbonsäuren</u>

<u>sowie aromatische Carbonsäuren</u>

Umfasst z. T. auch sec. Acylanilide

DOS   2 604 224   4. 2.76/11. 8.77   Höchst
= BE   851 126   4. 2.76/ 4. 8.77   Höchst

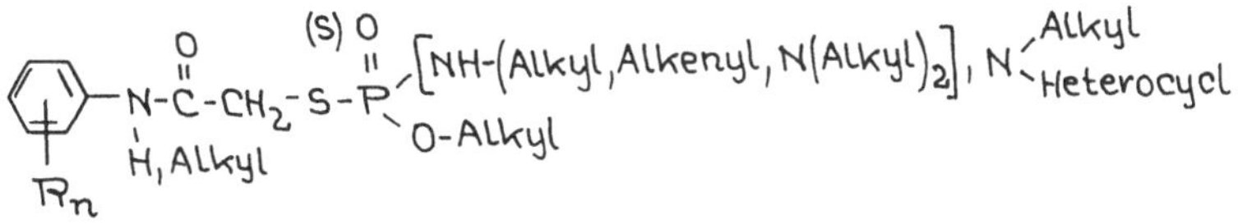

R = Halog.; Alkyl, O-Alkyl, Halog.-Alkyl; NO_2, SO_2-CH_3; SO_2NH_2,
    $COOCH_3$, CN   n = 1 - 3; Siehe ähnliche Verbindungen als
                                Fungizide DOS 2 604 225 u. a. auch
                                Herbizide.

DOS   2 606 858   20. 2.76/25. 8.77   Bayer
= BE   851 533

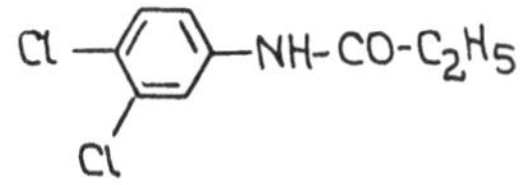

Cl—⟨ ⟩—NH-CO-C$_2$H$_5$ neues Darstellungsverfahren
   |
   Cl

Cl—⟨ ⟩—N=C + C$_2$H$_5$-COOH
   | ‖
   Cl   O

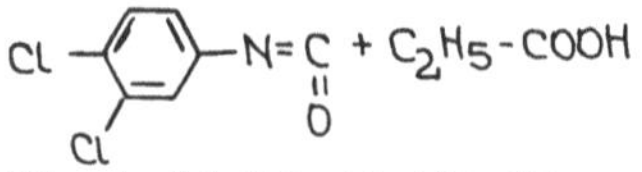

DOS   2 616 757   15. 4.76/28.10.76 (Zusatzpatent)
                                          Roussel-Uclaf
=. FR   2 320 936   16. 4.75
= BE   840 720

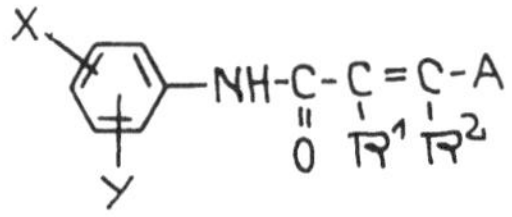

X—⟨ ⟩—NH-C-C=C-A   X u. Y = H, Halog., Alkyl, S-Alkyl
   | ‖ | |
   Y   O R^1 R^2 R^1 = H, Cl, CO-OAlkyl, NO$_2$,S-Alkyl
                                                              |
                              R^2 = Alkyl   (O)n

                              A = O(S)Alkyl

z.B. ⟨ ⟩—NH-C-CH=CH-O-CH$_3$
       | ‖
       CH$_3$ O

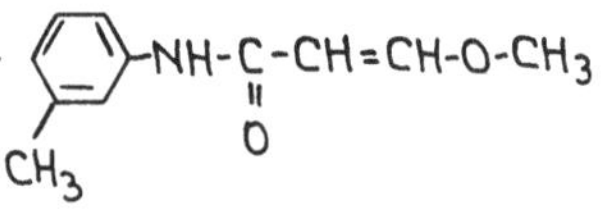

DOS   2 633 159   23. 7.76/26. 1.78   Höchst

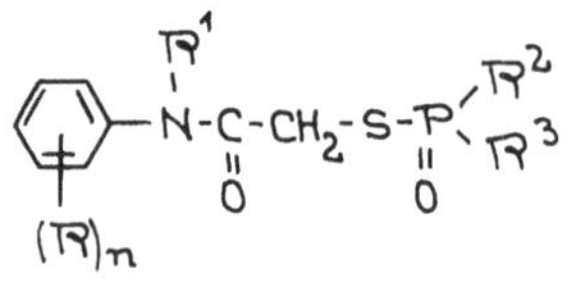

        R^1
        |
⟨ ⟩—N-C-CH$_2$-S-P$^{R^2}_{R^3}$ R = Halogen, Alkyl, O-Alkyl, NO$_2$,
   | ‖   ‖   CN, S(O$_2$)-Alkyl usw.
  (R)$_n$ O   O
                                         R^1 = H, Alkyl; R^2 = Alkyl-N<
                                         R^3 = Alkyl

Gegen Gräser und einige breitblättrige Unkräuter.

DOS   2 657 728   20.12.76/14. 7.77   Ciba Geigy
= BE   849 723   23.12.75/22. 6.77   Ciba Geigy

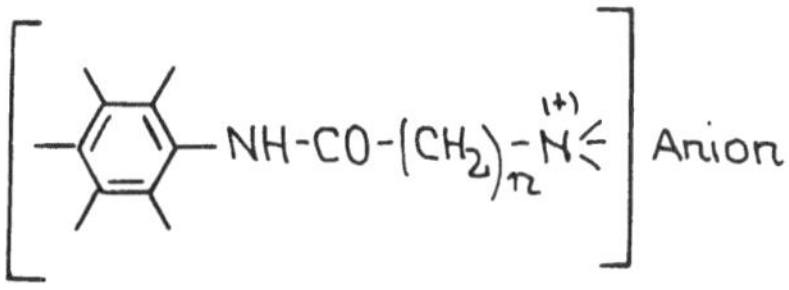

[—⟨ ⟩—NH-CO-(CH$_2$)$_n$-N$^{(+)}$<] Anion

DOS   2 802 282   19. 1.78/27. 7.78   Shell Int. Res.
                                                     US Prior. 21. 1.77
= US   4 168 153   Shell Oil Co.

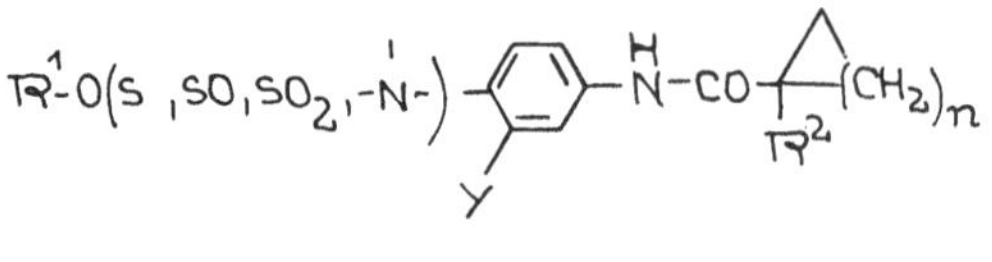

                                          Y = Halog., CN,NO$_2$,
                                              CF$_3$ usw.
R^1-O(S,SO,SO$_2$,-N-)—⟨ ⟩—N-CO-⟨▷⟩(CH$_2$)$_n$ R^1 = Alk(en)yl
                        | |
                        Y   R^2 R^2 = Alkyl

                                          Selektiv in Soja.

302

DOS 2 830 351 11. 7.78/24. 1.80 Cons. f. elektrochem. Ind.

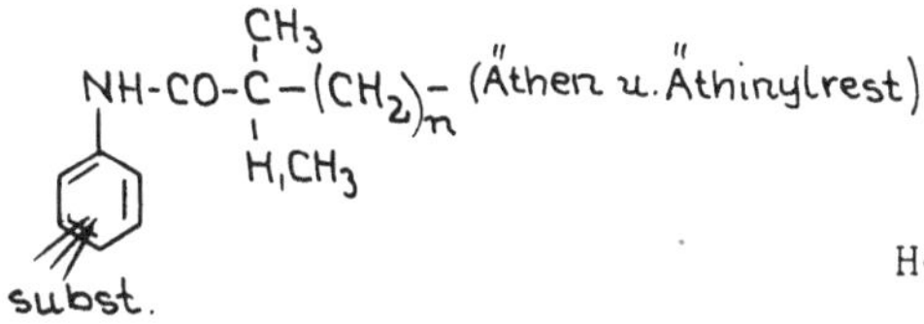
CH3
NH-CO-C-(CH2)n- (Äthen u. Äthinylrest)
H,CH3
subst.

Herbizide und Fungizide.

DOS 2 846 437 25.10.78/ 8. 5.80 Bayer
= EP 12 172 12.10.79/25. 6.80

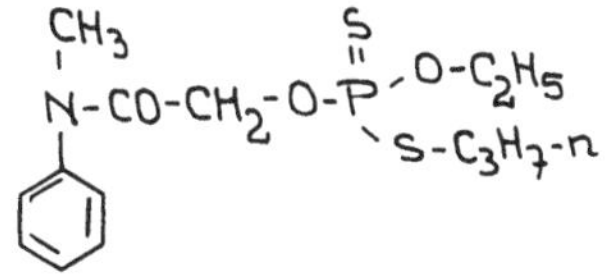
CH3
S
N-CO-CH2-O-P-O-C2H5
S-C3H7-n

DOS 2 855 699 22.12.78/28. 6.79 Sumitomo

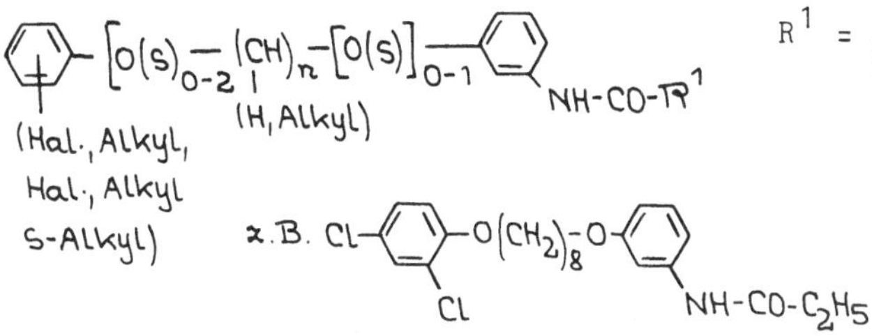
[O(S)]0-2-(CH)n-[O(S)]0-1
(H,Alkyl)
NH-CO-R1
(Hal., Alkyl,
Hal., Alkyl
S-Alkyl)
z.B. Cl
O(CH2)8-O
Cl
NH-CO-C2H5

R^1 = Alk(en,in)yl, Halog.-Alkyl, CN-Alkyl, Cycloalkyl, (S)O-Alkyl

US 3 993 677 3.11.75/23.11.76 Amer. Cynamid

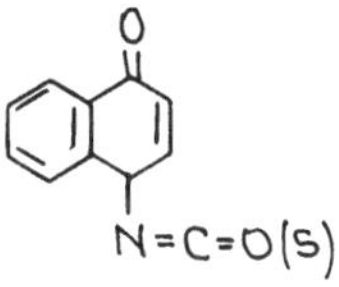
O
N=C=O(S)

US 4 123 251 2. 6.77/31.10.78 US Borax u. Chem. Corp.

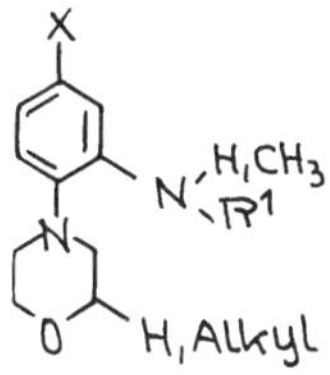
X
N
H,CH3
R1
N
O
H,Alkyl

X = Halog., Alkyl, Alkyl-SO_2, Halog.-Alkyl-(SO_2)
R^1 = CO-CH_3, O-Alkyl usw.

US 4 165 976 21. 4.78/28. 8.79 Stauffer Chem.

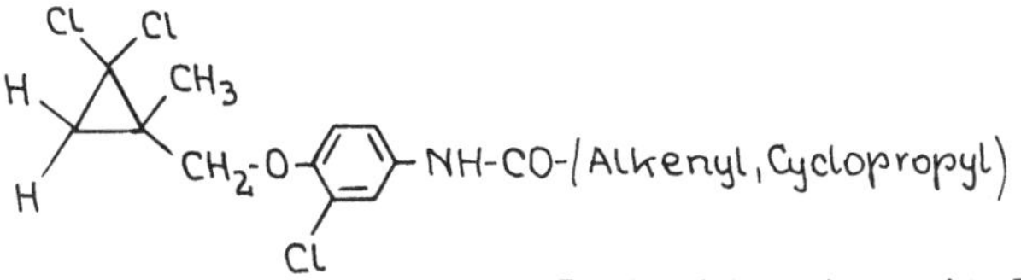
Cl
Cl
H
CH3
H
CH2-O
NH-CO-(Alkenyl, Cyclopropyl)
Cl

In Kombination mit Thiolcarbamaten.

US   4 166 735   21. 1.77/ 4. 9.79   Shell Oil

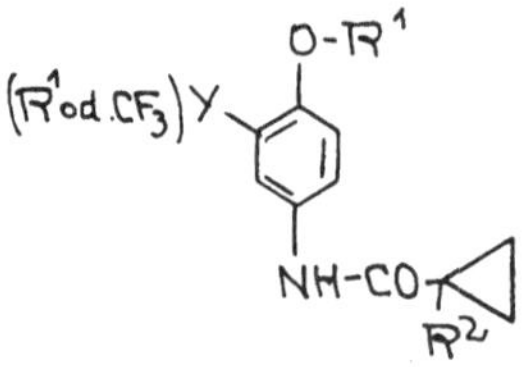

R^1 = CH_3, Y = CF_3, Cl
R^2 = CH_3, C_2H_5,

Y = CF_3, Cl

US   4 168 151   2. 6.77/18. 9.79   US Borax u. Chem. Corp.

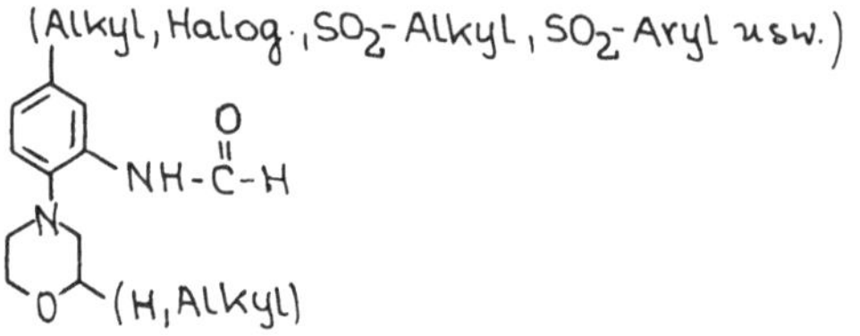

US   4 174 958   9. 1.78/20.11.79   Shell Oil

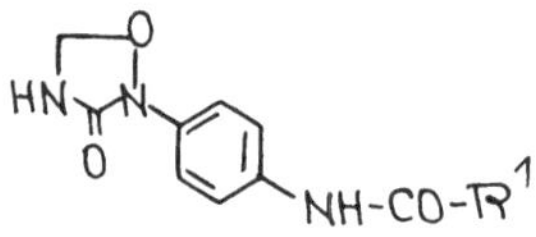

R^1 = H, Alkyl, Cycloprophyl

US   4 193 787   31. 8.78/18. 3.80   Stauffer Chem.

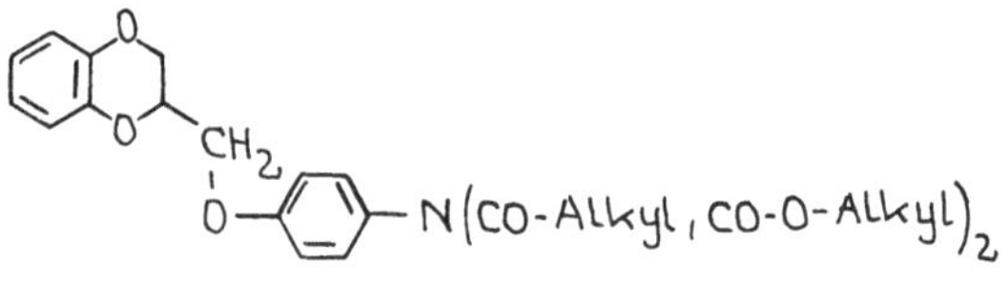

US   4 225 335   21. 6.79/30. 9.80   Velsicol

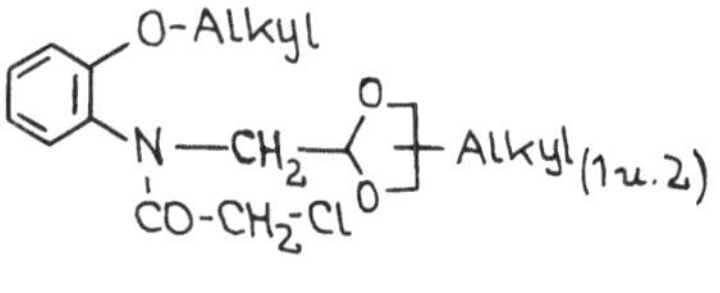

BE   849 723   23.12.75/22. 6.77   Ciba Geigy

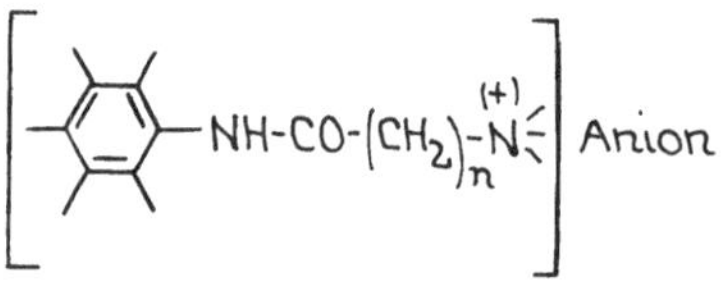

statt $(CH_2)_n$ auch isoAlkylen
u. ungesättigte Alkylidene

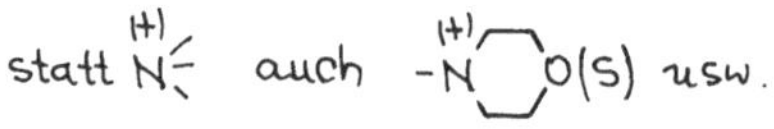

BE   863 074   21. 1.77/19. 7.78   Shell Int. Res.

R^1 = Alk(en)yl, Aryl(subst.)
       Allkynyl, O-Alkyl,
       Arylkal, Cycloalkyl usw.
R^2 = Alkyl, O(S)-Alkyl
Y = Cl, Br, F, CN, NO$_2$,
    verzw. Alkyl

BE   882 415   25. 3.80/25. 9.80   Monsanto

R u. R^1 = H, Alkyl, Alkoxy-methyl
R^2 u. R^3 = H, Alkyl, O-Alkyl
R^4 = H, Alkyl, Alkenyl, Phenyl(subst.)

GB   2 016 013   10. 2.78/19. 9.79   Shell Int. Res.

Y = CF$_3$, Cl; Z = H, Halogen

GB   2 018 754   17. 4.78/24.10.79   Sumitomo Chem.

JA   51 098327   25. 2.75/30. 8.76   Mitsui Toatsu Chem.

JA   51 142535   30. 5.75/ 8.12.76   Nihon Noyaku

JA   55 076851   5.12.78/10. 6.80   Hodogaya Chem. Ind.

DOS   2 728 523   23. 6.77/11. 1.79   Schering

DOS   2 909 991   12. 3.79/ 2.10.80   Schering

ähnlich wie zuvor (Imidchlorid)

DOS   3 009 307   11. 3.80/ 2.10.80   Aboti Lab.

US   4 097 265   30. 7.73/28. 6.78   Chevron Res. Co.

JA   53 040785   27. 9.76/13. 4.78   Mitsubishi Chem.

5.12.1.c   <u>Offene und cyclische Anilide von Dicarbonsäuren</u> Bd.5/567

DOS   2 733 115   22. 7.77/26. 1.78   Takeda Chem.

DOS   2 851 379   28.11.78/31. 5.79   Takeda Chem. Ind.

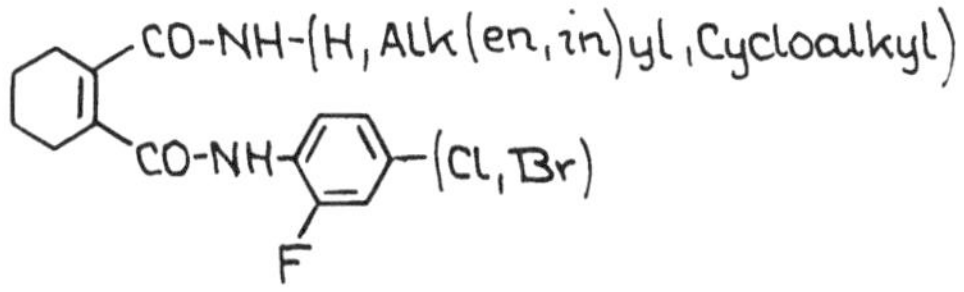

CO-NH-(H, Alk(en,in)yl, Cycloalkyl)

CO-NH—(Cl,Br)

F

DOS   2 921 002   25. 5.78/29.11.79   Mitsubishi Chem.
= JA   54 125640

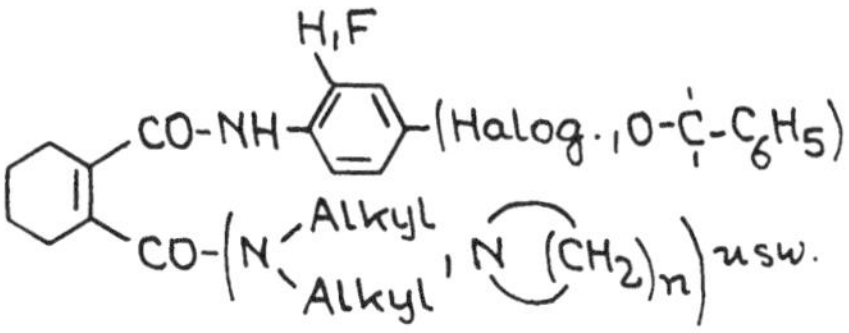

H,F

CO-NH—(Halog.,O-C-C₆H₅)

CO-(N〈Alkyl / Alkyl , N⌒(CH₂)ₙ) usw.

DOS   3 013 162   3. 4.80/30.10.80   Mitsubishi Chem.
= NE   8 002 020

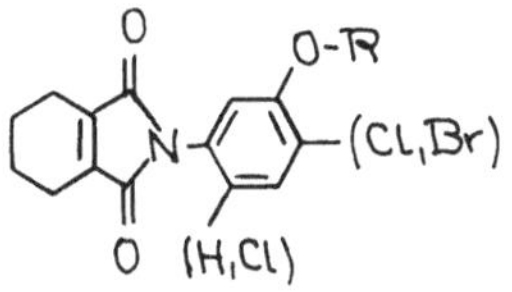

                                        R = Alk(en,in)yl, Aralkyl,
                                            Aryloxyalkyl, Cycloalkyl
                                            usw.

O-R
N—(Cl,Br)
O (H,Cl)

US   3 987 057   30. 1.75/19.10.76   Du Pont

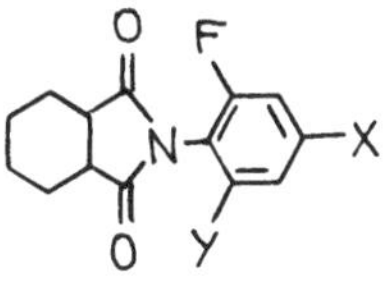

                                        X = Cl, Br, F
                                        Y = H, F

                                        wenn Y = F auch X = F

F
N—X
Y

US   4 003 926   30. 1.75/18. 1.77   Du Pont

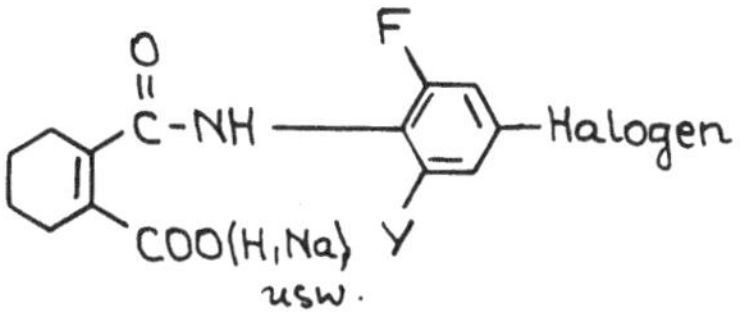

                                        Y = H, F
F
C-NH——Halogen
COO(H,Na) Y
  usw.

US   4 032 326   3. 9.74/28. 6.77   Du Pont

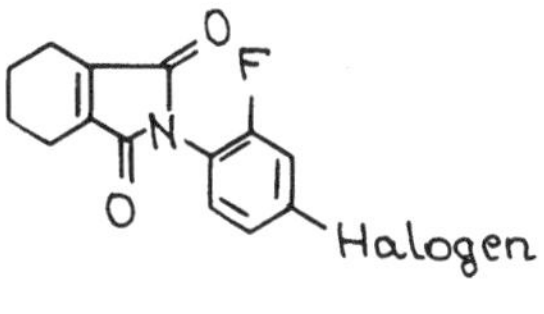

O F
N—
O   Halogen

US   4 045 208   23. 7.75/30. 8.77   Stauffer Chem.

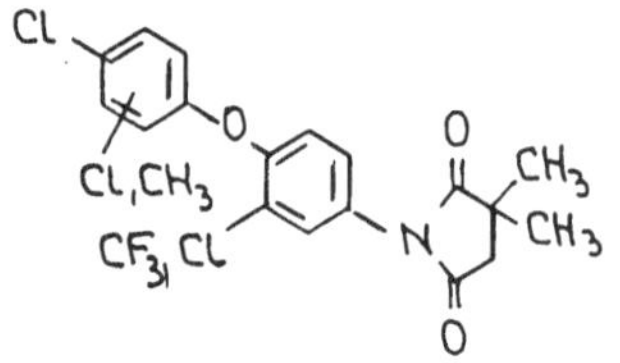

post emerg. gegen Unkräuter

US   4 138 242   3. 2.75/ 6. 2.79   Du Pont

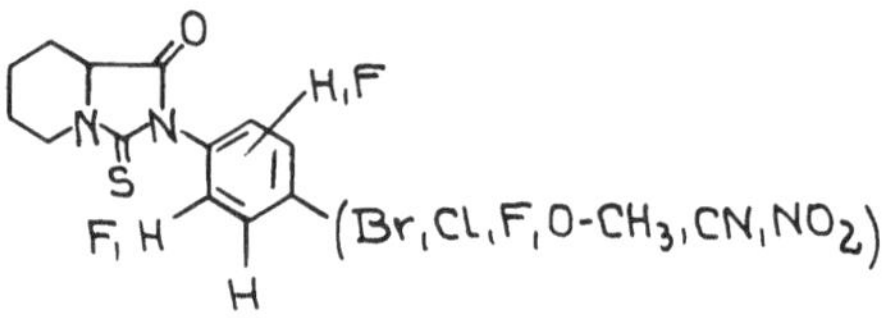

JA   50 121428   12. 3.74/23. 9.75   Mitsubishi Chem.

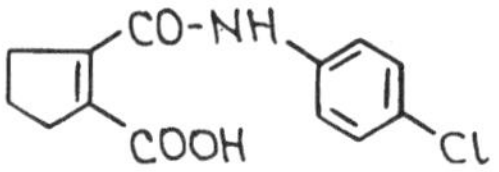

JA   50 148529   21. 5.74/28.11.75   Mitsubishi Chem.

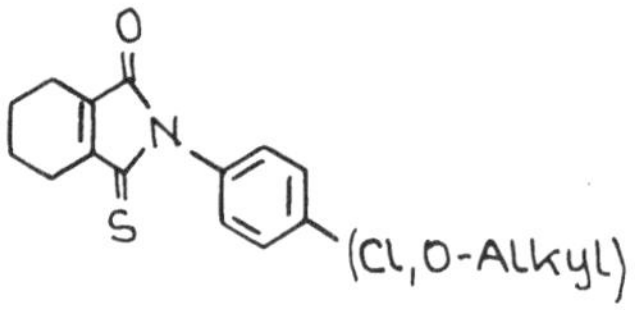

Gegen Gräser in Reis.

JA   52 083544   1. 1.76/12. 7.77   Mitsubishi Chem.

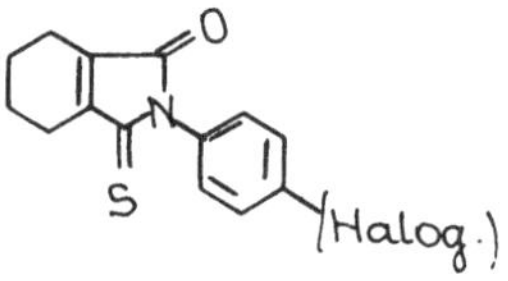

JA   52 087228   9. 1.76/20. 7.77   Nihon Noyaku

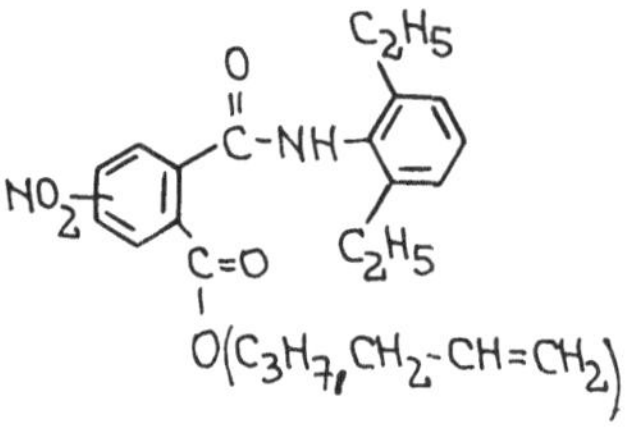

308

JA   53 018   3. 8.76/20. 2.78   Mitsubishi Chem.

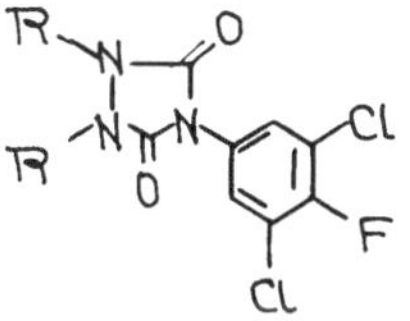

                                                    R = H, Alkyl
                                                    R u. R = $(CH_2)_{1-4}$

JA   53 044587   28. 9.76/21. 4.78   Mitsubishi Chem.

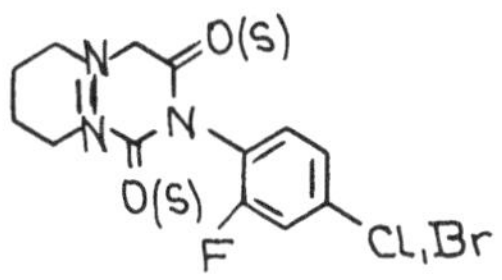

JA   53 053694   27.10.76/16. 5.78   Mitsubishi Chem.

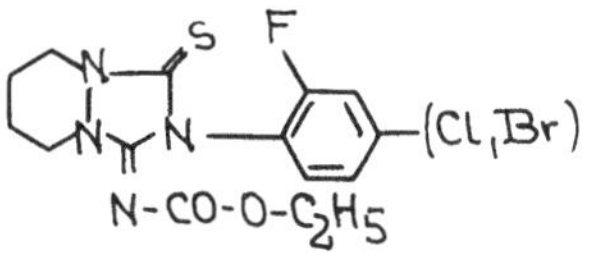

JA   53 073557   13.12.76/30. 6.78   Mitsubishi Chem.

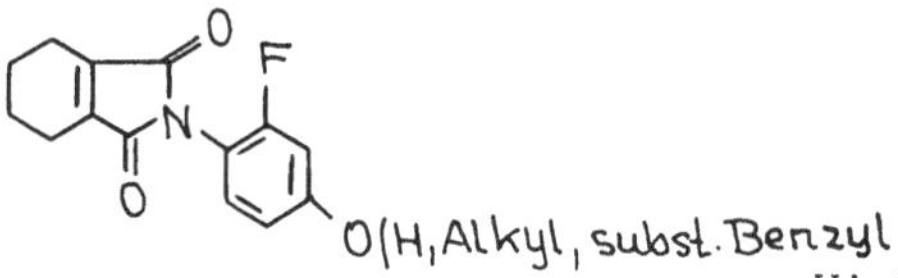

                                        Wirksam gegen Gräser u.Unkräuter.

JA   53 095961   28. 1.77/22. 8.78   Nihon Noyaku

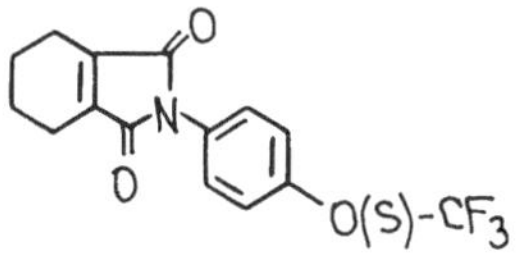

JA   54 125652   22. 3.78/29. 9.79   Mitsubishi Chem.

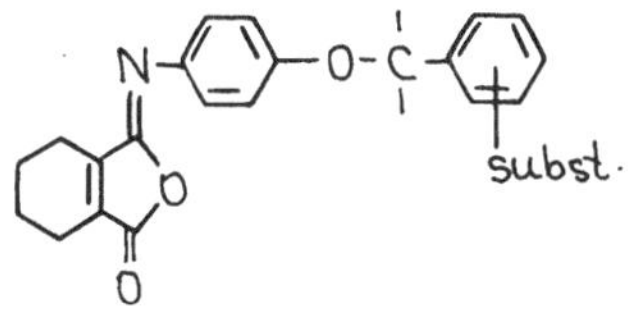

DOS   2 531 597   15. 7.75/ 3. 2.77   Eli Lilly

R^0 = -CO-CF$_2$(H, Cl, F, CF$_2$H, CnF$_{2n+1}$ usw.)

R^1 = H, CO-O(Alkyl, Phenyl)

Sehr breiter Anspruch.

US   4 011 341   3. 3.69/ 8. 3.77   Eli Lilly
= NE   7 508 283

R^0 = -CO-CF$_2$(H, F, Cl, CHF$_2$)
R^1 = H, C$_6$H$_5$-CO-, Furoyl usw.

u.a. auch herbizid wirksam

1.e   Acylverbindungen des Phenyl-hydroxylamins u.

-hydrazins   Bd.5/563

US   3 982 922   28.12.67/28. 9.76   Velsicol Chem. Co.

X = Halog., Alk(en)yl, NO$_2$,
CN, O-Alkyl

Z = Halog., (O)-Alkyl, NO$_2$

5.12.2. u. 2.a   Acylverbindungen von sec. Anilinen u.
           sec. 2,6-Diakyl-anilinen   Bd.5/563 u.5/566

Acylverbindungen von sec. Anilinen überschneiden sich oft mit Acyl-
verbindungen primärer Aniline siehe daher auch bei 5.12.1.a.
Die früher gesondert abgehandelten Acylverbindungen von 2,6-Dial-
kylanilinen (Bd.5.12.a) wurden wegen Überschneidung mit anderen
Aniliden hier miterfasst.

Die Europ. Patente von 5.12.1.a und 5.12.2. und 2a werden zusammengefasst und hier behandelt.

Einige Acylanilide sec. Amine wie
DOS 2 846 437 (Bayer), US 4 225 335 (Velsicol) siehe 5.12.1.a

| DOS | 2 624 094 | 29. 5.75/ 2.12.76 | Sumitomo Chem. |

$$\underset{X_n}{\boxed{}}-N\underset{\substack{|\\ C-N \\ \| \\ O}}{\overset{(CH_2)_n-CO-O[(Cyclo)-Alkyl]}{}}\begin{array}{l}(O)Alk(en)yl\\ (O) Alkyl, H\end{array}$$

X = Halog., Alkyl
    O-Alkyl, CF_3

n = 0 - 2

| DOS | 2 630 373 | 6. 7.76/27. 1.77 | Ciba Geigy |

R = $-CH(CH_3)_2$,
    CH_2-OCH_3

$\triangleleft\!\!-(CH_3)$

Keine Herbizide sondern Wuchsregulatoren.

| DOS | 2 632 437 | 19. 7.76/ 3. 2.77 | Amer. Cyanamid |

Selektives pre Emergence-
Herbizid gegen Gräser

| DOS | 2 632 619 | 20. 7.76/24. 2.77 | Stauffer |

US Prior. 23. 7.75 u. 23.7.75

X = Cl, CH_3, n = 0 - 2
Y = Cl, CF_3
R^1 = H, (O)-Alkyl, S-Alkyl
    Alkenyl
R^2 = H, O-Alkyl usw.

| DOS | 2 648 008 | 23.10.76/ 3. 5.78 | BASF |

subst. durch H, (O)-Alkyl, CF_3

A = Azol über Ringstickstoff
    an CH_2 gebunden, z.B.
    Pyrazol, 1,2,4-Triazol

DOS   2 704 281   2. 2.77/ 3. 8.77   Bayer

CH_2-(subst.Heterocycl)
Y, X, N, C-CH_2-Halog., O

X u. Y = Alkyl

DOS   2 711 928   18. 3.77/ 6.10.77   Monsanto
= BE   852 621   US Prior. 19. 3.76

R, C-CH_2-Halog., N, CH_2-O-R^1, Halog.

R = Alkyl $< C_{10}$

R^1 = subst. Alkyl;
Alk(en,in)yl

selektiv in Zuckerrüben

DOS   2 726 253   10. 6.77/21.12.78   Bayer
= EP   51   5. 6.78

R^1, R^2, C—CO-R^3, N, C-CH_2-Cl, O, $(R)_n$

R = Alkyl, Halog., S-Alkyl, SO_2-Alkyl,
CN, NO_2, SO_2-N$<$
R^1 u. R^2 = H, Alkyl, Phenyl
R^3 = Alkyl, Phenyl(subst.)

Gräser-Herbizid

DOS   2 742 583   22. 9.77/ 5. 4.79   Bayer

Alkyl, CH_2-Heterocycl., N, C-CH_2Cl, O, Alkyl

z.B. $-N$

Selektiv gegen Gräser.

DOS   2 744 396   3.10.77/12. 4.79   BASF

subst., C-CH_2(Cl,Br), N, O, C, CH_2-N, O, Halog, Alkyl, N, Halog.

DOS   2 758 418   28.12.77/13. 7.78   Monsanto
                                    US Prior. 29.12.77 u. 26.10.77

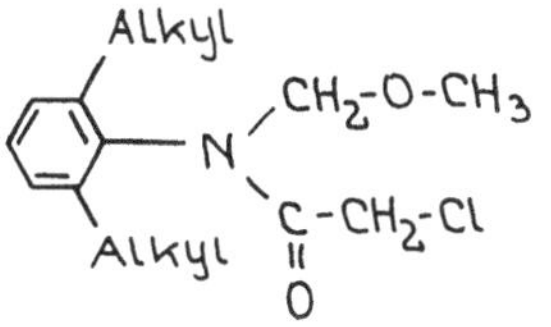

neues Darstellungsverfahren

DOS   2 803 662   25. 1.78/26. 7.79   Schering

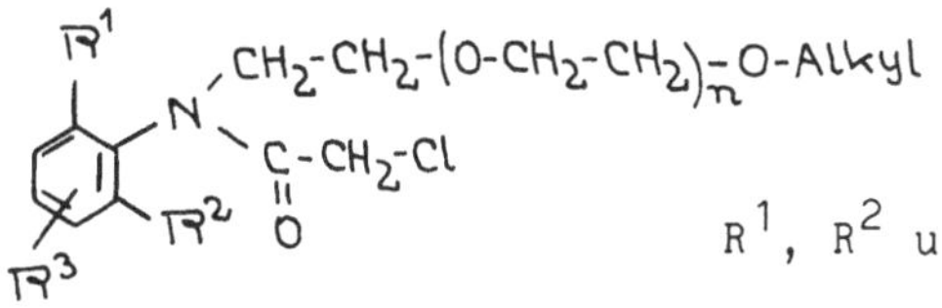

R^1, R^2 u. R^3 = H, Alkyl, Halog.,CF_3

DOS   2 805 525   9. 2.78/11. 8.78   Montedison
                                    IT Prior. 10. 2.77;20.5.77 u. 20.10.77

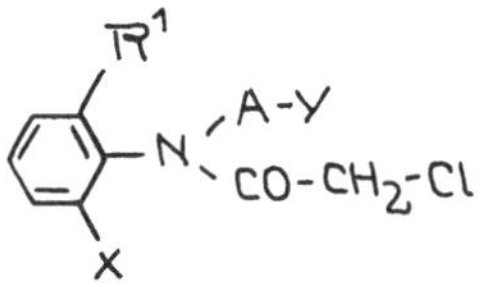

X = Alkenyl; R^1 = H, C_1-C_5-Alkyl

A = Alkylen

Y = H, Alkyl, Alk(en,in)yl, Phenyl,
    Halogen, heterocyclische Grup=
    pierung, O-Alkyl, S-Alkyl, N< ,
    CN usw.

Selektiv in Mais.

DOS   2 805 756   10. 2.78/16. 8.79   Bayer
 Vorprodukte für nachf. Patent
DOS   2 805 757   10. 2.78/16. 8.79   Bayer

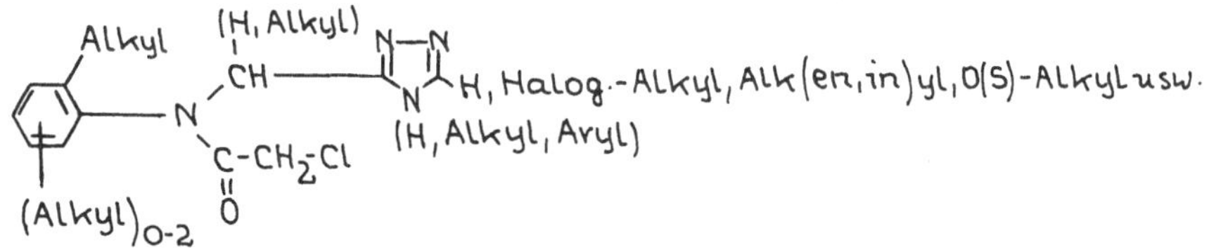

DOS   2 809 217   3. 3.78/13. 9.79   Bayer

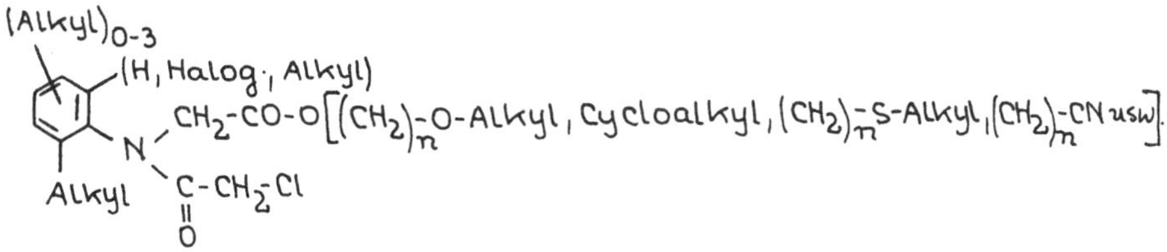

DOS   2 814 024   31. 3.78/11.10.79   Amer. Cyanamid

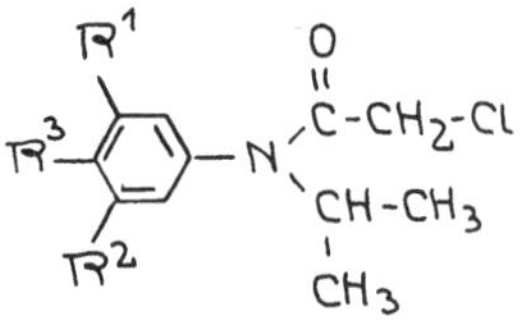

R^1 u. R^2 = niedr. Alkyl
R^3 = niedr. Alkyl u. H

DOS   2 817 023   19. 4.78/31.10.79   Höchst

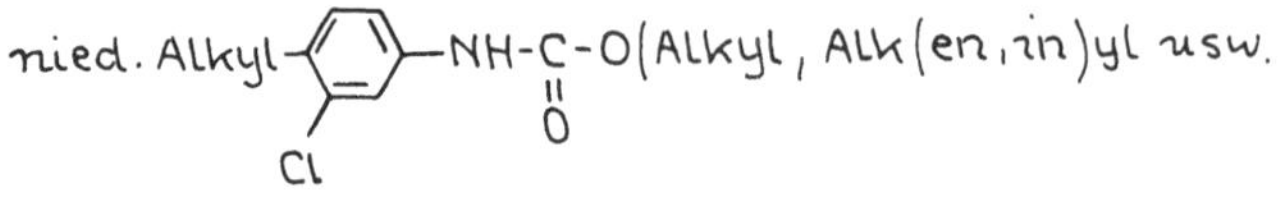

nied. Alkyl—⟨ ⟩—NH-C-O(Alkyl, Alk(en,in)yl usw.

Anwendung zur Gräserbekämpfung in Reiskulturen.

DOS   2 825 543   10. 6.78/13.12.79   Bayer

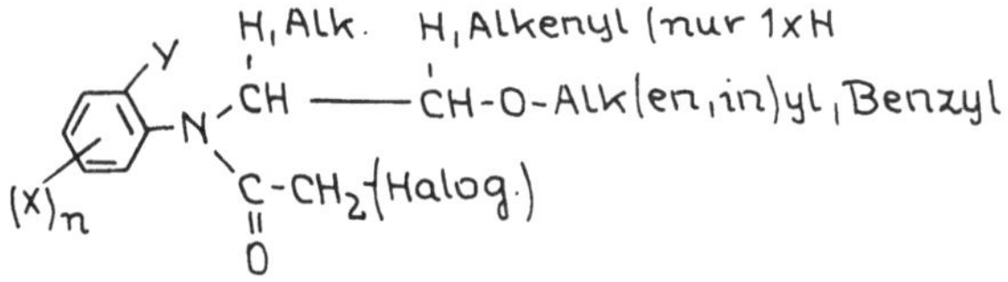

H,Alk. H,Alkenyl (nur 1xH
CH——CH-O-Alk(en,in)yl,Benzyl
C-CH₂(Halog.)

X u. Y = Alkyl
n = 0 - 2

DOS   2 832 890   27. 7.78/14. 2.80   BASF

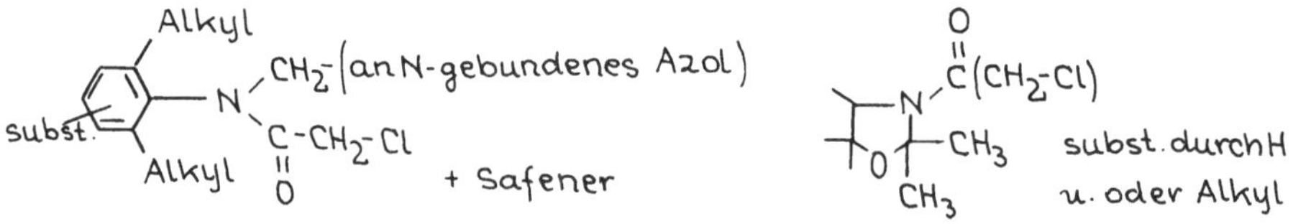

Alkyl
subst. durch H u. oder Alkyl
+ Safener

DOS   2 832 950   27. 7.78/21. 2.80   BASF

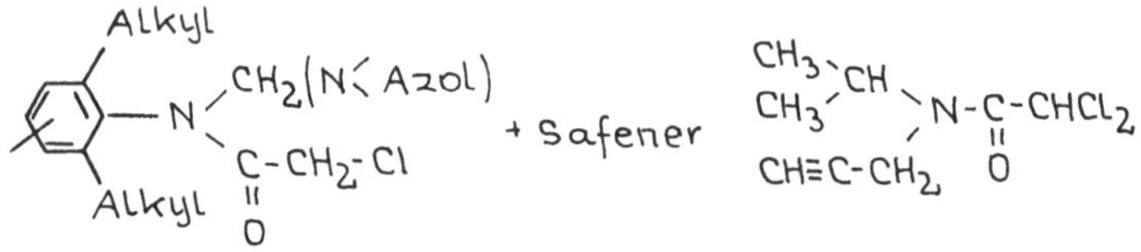

Alkyl
+ Safener

DOS   2 835 156   10. 8.78/14. 2.80   Bayer

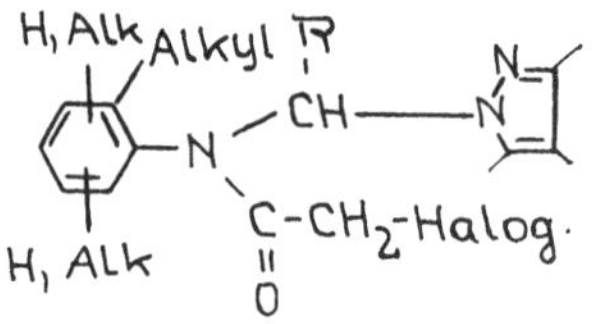

subst. durch H, Alkyl,
O-Alkyl, Halog.
R = Alkyl, Cycloalkyl,
    Halog.-Alkyl,
    (O)-Alk(en,in)yl,Phenyl

314

DOS   2 842 284   28. 9.78/17. 4.80   Bayer

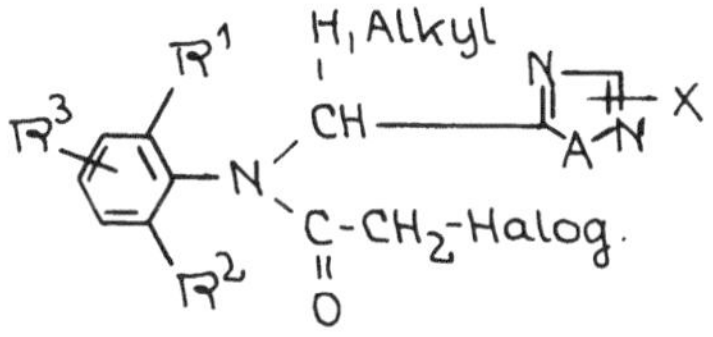

A = O, S, N(H, Alkyl)
R^1 = H, Alkyl, O-Alkyl
R^2 u. R^3 = H, Alkyl, O-Alkyl,
        Halog.
X = H, Alkyl, Halog.,
    O(S)-Alkyl, Alk(en,in)yl,
    CN, N< , O(S-)Phenyl,
    Phenyl usw.

DOS   2 842 315   28. 9.78/17. 4.80   Bayer

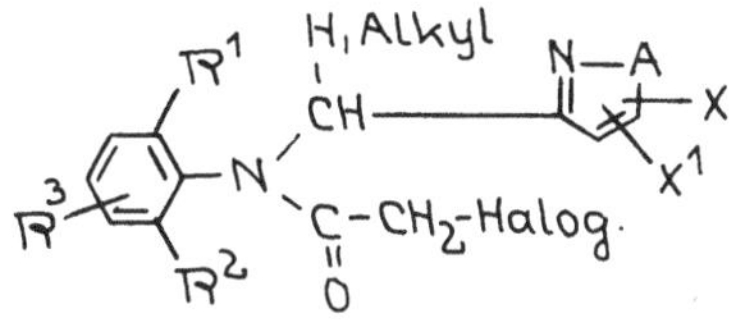

A, R^1, R^2 u. R^3 = siehe
        Patent zuvor
X^1 = X siehe zuvor

DOS   2 843 869   7.10.78/24. 4.80   BASF

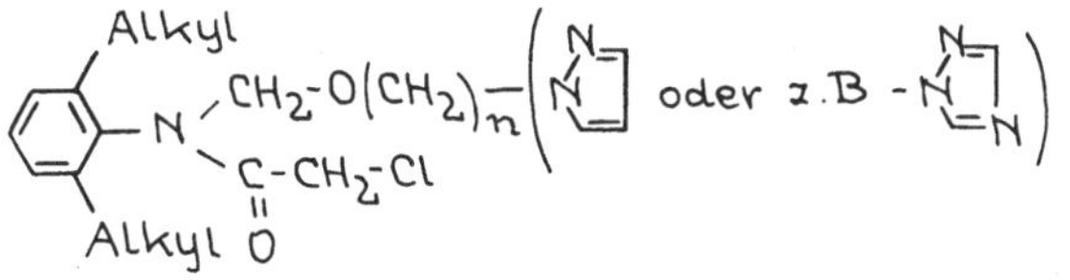

DOS   2 846 437   25.10.78/ 8. 5.80   Bayer

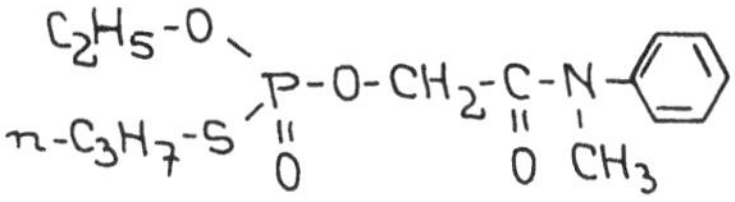

DOS   2 847 827   3.11.78/14. 5.80   Bayer

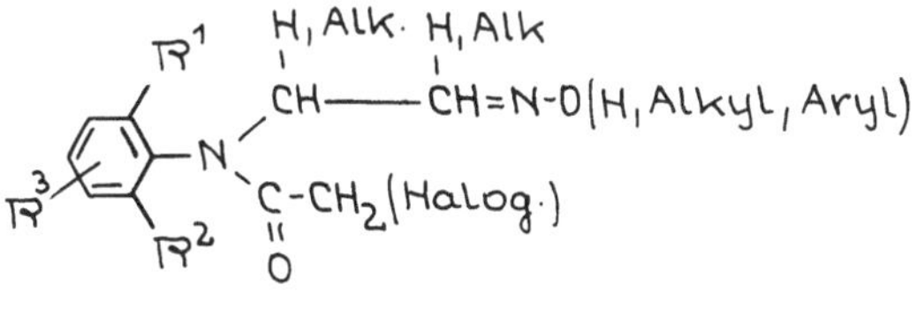

R^1 = H, (O)-Alkyl
R^2 = H, (O)-Alkyl,
        Halog.-Alkyl
R^3 = H, (O)-Alkyl,
        Halogen

DOS   2 849 442   15.11.78/29. 5.80   BASF

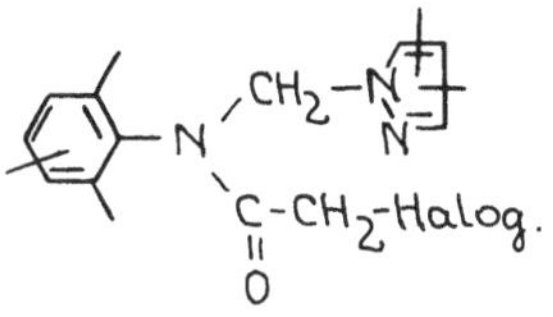

Verfahren zur Herstellung
weitgehend reiner Pyrazol-
verbindungen.

DOS   2 854 599   18.12.78/26. 6.80   BASF

A = Darst.Verfahren

Heterocycl., CH$_2$-S-CH$_3$, CH$_2$-O-Ar

DOS   2 901 593   17. 1.79/ 7. 8.80   BASF
= EP   13 873

Alkyl, Alkoxyalkyl, Alkinyl

C-N-Azolylmethyl

z.B.

DOS   2 901 659   15. 1.79/24. 7.80   Schering

oder Alk(en,in)yl

DOS   2 911 865   31. 3.78/11.10.79   Chevron Res.

C-CH$_2$-Halog., auch CCl$_3$, CHCl$_2$ R^1 u. R^2 = H, Alkyl, Phenyl
                                                             Benzyl subst. usw.
Ar-N   Ar = Phenyl subst. durch
     C-(CH$_2$)$_n$[CO-(H,Alkyl,Benzyl,Phenyl)]   Halog., Alkyl, O-Alkyl,
                                                      NO$_2$ usw.

     -C-(H,Alkyl,Benzyl,Phenyl
      N-O(H,Alkyl)

DOS   2 918 297   11. 5.78/15.11.79   Chevron Res.

(H,Alkyl)   X = O, S, N-(Alkyl, H)
(CH$_2$)$_n$ R^2 = H, Alk(en,in)yl,
Ar-N   Cycloalkyl, O(S)-Alkyl,
     C-CH$_2$-Cl   Phenyl(subst.)
     O

316

DOS   3 002 595   25. 1.80/21. 8.80   Chevron Res. Co.
                                                        US Prior. 9. 2.79

W u. Z = Heteroatome in einem aromatischen 5gliedrigem Ring

R^1 = (O)-Alkyl, Alkenyl usw.

R^2, R^3 u. R^4 = H, Alkyl, O-Alkyl, Alkenyl usw.

R^5 u. R^6 = H, Alkyl, O-Alkyl, Alkenyl, Phenyl usw.

Acylverb. der Phenylamino-alkancarbonsäuren

(hier nur DOS), siehe Näheres US, BE usw. bei 5.7.3e

Acylanilide der Formel

DOS   2 618 664   28. 4.76/11.11.76   Shell Int. Res.
= BE   841 109

                                                        siehe 5.7.3.e

DOS   2 628 901   28. 6.76/20. 1.77   Shell Int. Res.
= NE   7 607 052

                                                        siehe 5.7.3.e

DOS   2 650 434   3.11.76/18. 5.77   Shell Int. Res.

                                                        siehe 5.7.3.e

R - Konfiguration

317

DOS   2 706 823   17. 2.77/ 1. 9.77   Shell Int. Res.

siehe bei 5.7.3.e

C_6H_5-CO-N-CH-COOR   (Y, X substituted phenyl ring with CH_3)

DOS   2 908 739   6. 3.79/18. 9.80   Höchst

N-CH-CO-O-Alkyl   (H, Alkyl)
CO
CH_2-S-P<Alkyl, O-Alkyl   O(S) $(R)_n$

5.12.1.a u. 5.12.2   Umfasst Acylanilide primärer u.
Europ. Patent   sec. Anilide   Bd.5/555
                                                    bis Bd.5/563

EP   51   5. 6.78/20.12.78   Bayer
= DOS   2 726 253   DB Prior. 10. 6.77

$(R)_n$

CH_2CO Cl   N R^1 C-CO-R^3 R^2

R = Alkyl, Halog., S-Alkyl,
      SO_2-Alkyl, CN, NO_2, SO_2-N<
R^1 u. R^2 = H, Alkyl, Phenyl
R^3 = Alkyl, Phenyl-subst.

Gräserherbizid.

EP   324   1. 6.78/24. 1.79   Stauffer Chem.
Siehe auch US 4 165 976   21. 4.78/28. 8.79   US Prior. 21. 6.77

Cl Cl
O-CH_2-C<CH_3
Cl
N-C-R^1 R^2 (O)

R^1 = Alk(en)yl, O(S)-Alkyl,
       Cycloalkyl
R^2 = H, O-Alkyl, CO-Alkyl

EP   1 556   16. 5.78/ 2. 5.79   Stauffer Chem.

O
(CH_3,H) (CH_3,H) O(S)   Cl(0-1)

NH-CO-(Alkyl, Alkenyl, N<,) Cycloalkyl usw.

318

EP   3 539   21. 2.79/ 5. 9.79   Bayer
= DOS   2 805 757   DB prior. 10. 2.78

A = O, S, N(H, Alkyl,Aryl)
R^1 = H, Alkyl, Halog.-Alkyl
       Alk(en,in)yl, Aryl,
       Aralkyl subst.
X = Alkyl; Y = Alkyl,Halog.,
n = 0 - 2

EP   3 619   2. 1.79/ 5. 9.79   Shell Int. Res.
                                                  US Prior. 9. 1.78

A = Heterocycl. 5-Ringe
    ev. NO_2 subst.
R = Halog., NO_2, CN, CF_3,
    O(S)-Alkyl usw.

EP   5 066   21. 4.78/31.10.79   Stauffer Chem.
= US   4 165 976   21. 4.78/28. 8.79

In Kombination mit anderen Herbiziden.

EP   6 153   30. 6.78/ 9. 1.80   Bayer

EP   7 080   9. 7.79/23. 1.80   BASF
= DOS   2 830 764   DOS Prior. 13. 7.78

Darstellungsverfahren

Siehe auch DOS 2 648 008 BASF u. 2 704 281 Bayer
           DOS 2 744 396 BASF

EP   7 089   11. 7.78/23. 1.80   Cons. f. Elektrochem.
= DOS   2 830 351   Ind.

$$\text{subst.} \quad \text{Ar}-NH-\underset{O}{\underset{\|}{C}}-\overset{H,CH_3}{\underset{H,CH_3,C_2H_5}{C}}-(CH_2)_{\overline{n}}(CH_2-CH=CH_2 \text{ oder } CH_2-C\equiv CH$$

durch Cl, Alkyl, O-Alkyl, NO_2 usw.

EP   7 575   18. 7.79/ 6. 2.80   BASF
= DOS   2 832 046   DB Prior. 21. 7.78

$$\text{subst.} \quad Ar-N\underset{C-CH_2(Halog.)}{\overset{CH_2-(N \text{ Azol})}{\diagup}}$$

                                    Zusammen mit anderen Acetaniliden.

EP   8 057   30. 7.79/20. 2.80   Bayer
= DOS   2 835 157   10. 8.78/21. 2.80   Bayer
           u. 2 835 158

$$Ar-N\underset{CO-CH_2Cl}{\overset{Alkyl}{\diagdown}} CH_2-(N \text{ Azol}),$$ Furyl, Thiophenyl, Phenyl, Phenoxyl, Alk(en)yl usw.

                           neues Darstellungsverfahren

DOS   2 835 157

Zwischenprodukte $Cl-\underset{Alkyl,H}{\overset{}{CH}}-N$ Azol   für vorangehend angemeldetes Darstellungsverfahren

EP   8 091   3. 8.79/20. 2.80   Bayer
= DOS   2 835 156   DOS Prior. 10. 8.78

$$Ar-N\underset{CO-CH_2-Hal}{\overset{(H,Alkyl)}{\underset{\diagdown}{\overset{CH}{\diagup}}}} N \overset{subst.}{\underset{subst.}{}}$$

Alkyl usw.

EP   8 649   18. 7.79/19. 3.80   BASF
= DOS   2 832 950   DB Prior. 27. 7.78

$$\text{subst.} \quad Ar-N\underset{\underset{O}{\underset{\|}{C}}-CH_2-Hal}{\overset{CH_2-Azol}{\diagup}}$$ + $CHCl_2-CO-N(Alkyl)_2$ z.B Allyl

                                    als Safener

EP   9 091   18. 7.79/ 2. 4.80   BASF
= DOS   2 832 940   DB Prior. 27. 7.78

Inhalt wie zuvor aber als Safener

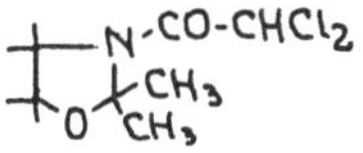

Siehe auch EP 10 159   14. 9.79/30. 4.80   Bayer

Inhalt ähnlich wie zuvor

EP   9 693   17. 9.79/16. 4.80   Bayer
= DOS   2 842 280   DB 28. 9.78

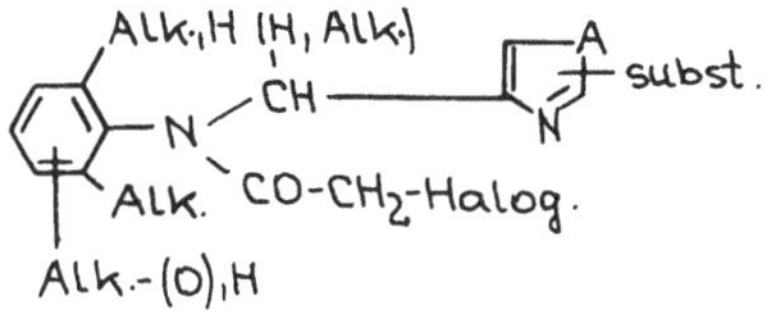

A = O,S,N(H, Alkyl)

EP   10 163   17. 9.79/30. 4.80   Bayer
= DOS   2 842 284   DB Prior. 28. 9.79

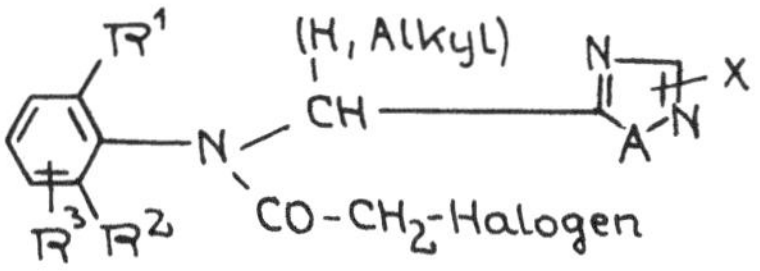

A = O, S, NR
R^1 = H, Alkyl
R^2 = H, Alkyl, O-Alkyl
R^3 = H, Alkyl, O-Alkyl,
        Halog.

X = H, Alkyl, Alk(en,in)yl,
    N<, CN usw.

EP   10 166   17. 9.79/30. 8.80   Bayer
= DOS   2 842 315   DB Prior. 28. 9.79

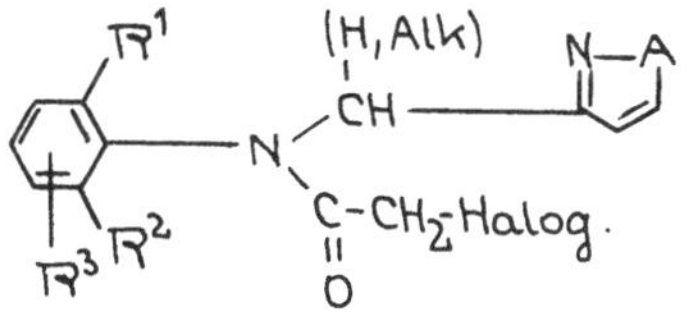

R^1, R^2, R^3 u. X siehe Patent zuvor

EP   10 715   20. 7.79/14. 5.80   Bayer
= DOS   2 847 827   DB Prior. 3.11.78 u. 31. 7.79

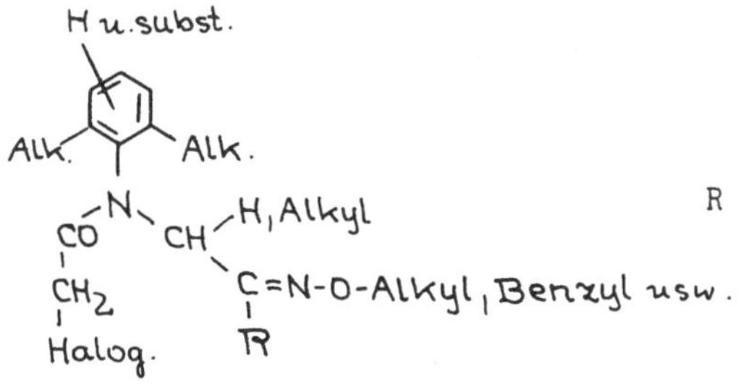

R = H, Alkyl, subst. Phenyl

EP   12 172   12.10.79/25. 6.80   Bayer
= DOS   2 846 437   DB Prior. 25.10.78

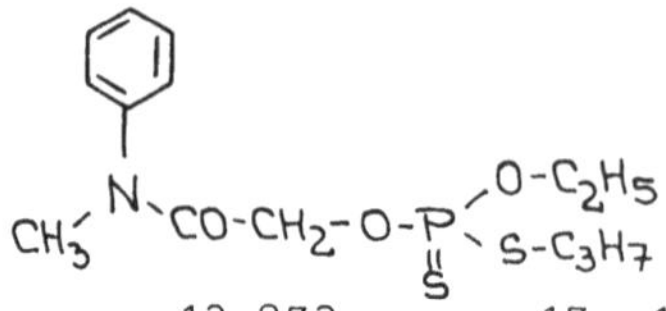

EP   13 873   17. 1.79/ 6. 8.80   BASF
= DOS   2 901 593

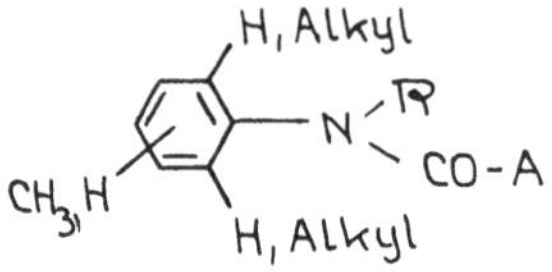

R = Alkyl, Alkoxy-Alkyl,
    Alkenyl oder A
A = N-Azolyl, (methyl-subst.)

EP   14 223   9. 2.79/20. 8.80   Ciba Geigy
Siehe DOS 2 657 728

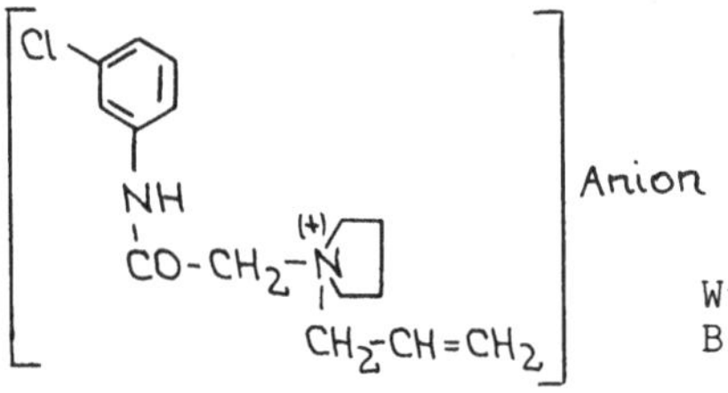

Wuchsregulator für Gras und zum
Blattabfall.

US   3 997 326   17.10.74/14.12.76   Chevron Res.

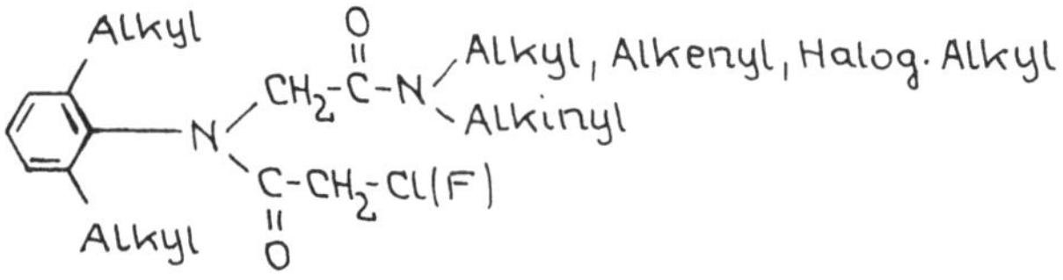

US   4 012 222   13.11.74/15. 3.77   Velsicol Chem. Corp.

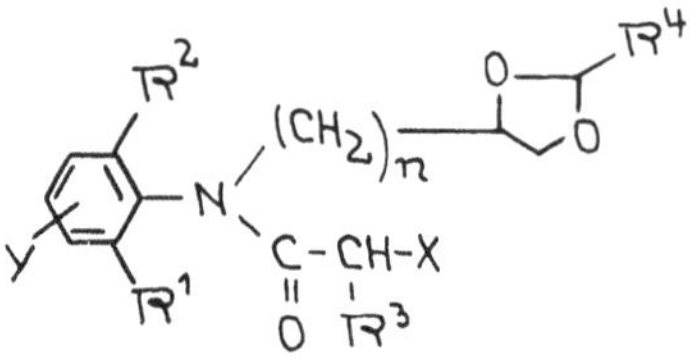

Y = H, nild. Alkyl, Halog.
R^1 = H, (O)-Alkyl
R^2 = Alkyl; R^3 u. R^4 = H,
                           Alkyl
X = Halog. n 1 oder 2

US   4 022 608   1. 8.74/10. 5.77   Velsicol Chem. Corp.
Siehe US   3 946 044

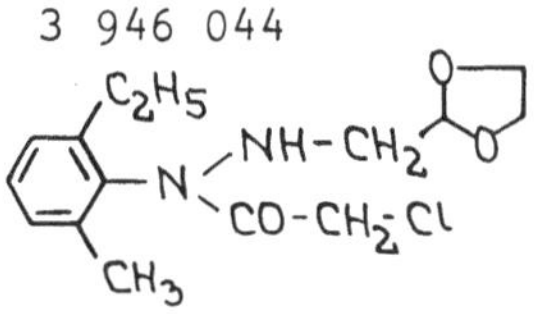

322

US   4 003 735   4.11.75/18. 1.77   Monsanto

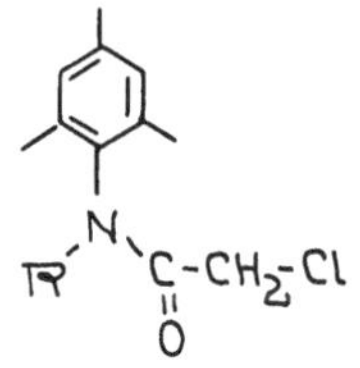

Schäden durch derartige Herbizide

vermindert durch $>N-\underset{\underset{S}{\|}}{C}-S-Alk(en)yl$, Phenyl, Benzyl

US   4 039 314   26. 7.74/ 2. 8.77   Velsicol Chem. Corp.
Siehe auch US 3 946 043

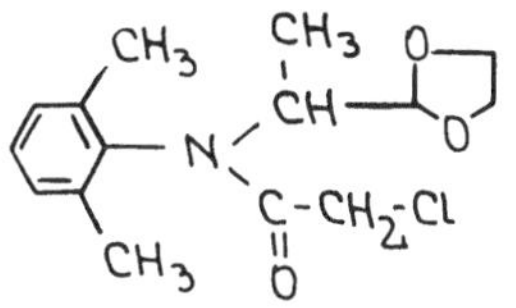

US   4 046 554   19.12.74/ 6. 9.77   Velsicol Chem. Corp.

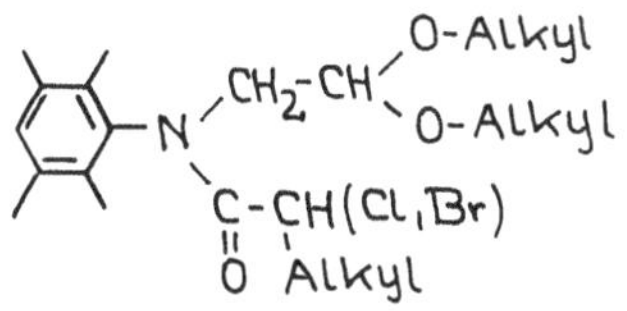

US   4 053 297   21. 4.75/11.10.77   Velsicol Chem. Corp.

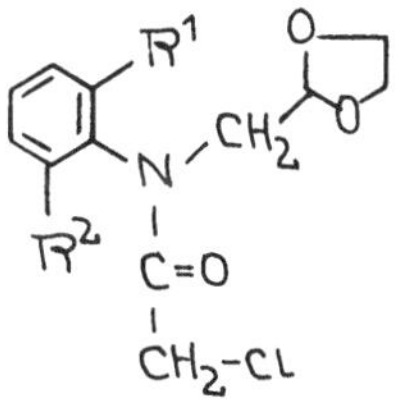

R^1 u. R^2 = Alkyl

Antidotzusatz $Cl_2CH-CO-N(Alkylen)_2$

US   4 055 410   15. 3.76/25.10.77   Du Pont

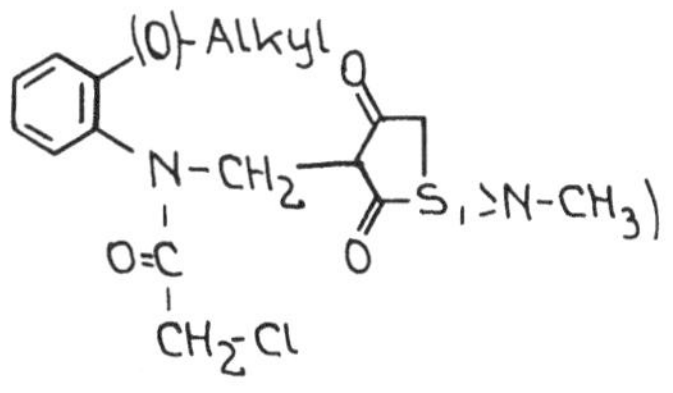

US   4 055 414   11. 3.76/25.10.77   Monsanto

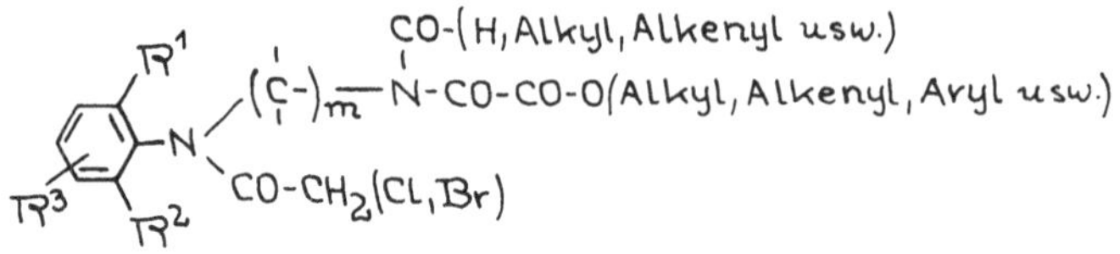

Selective pre emergence Herbicide in Zuckerrüben, Sojabohnen, Weizen
und Sorghum.

US   4 088 687   14. 5.73/ 9. 5.78   Stauffer Chem. Co.

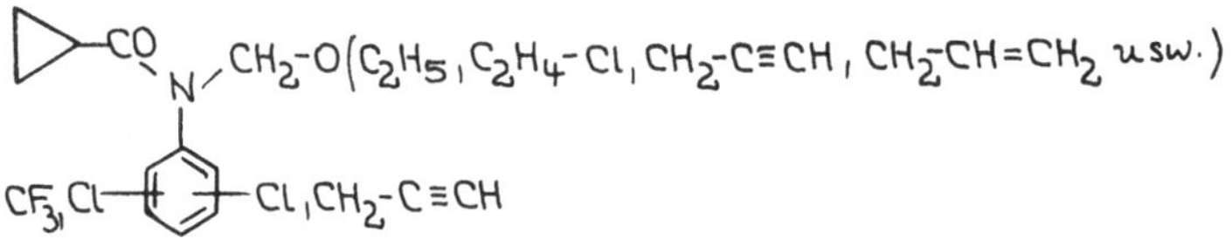

US   4 096 167   24. 9.73/20. 6.78   Calgon Corp.

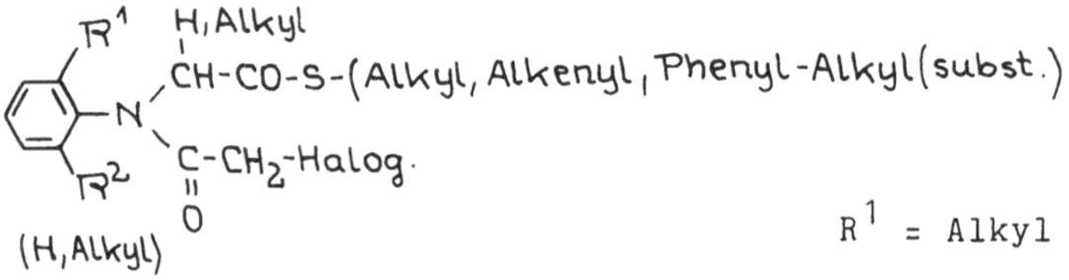

R^1 = Alkyl

US   4 097 262   22. 4.77/28. 6.78   Du Pont

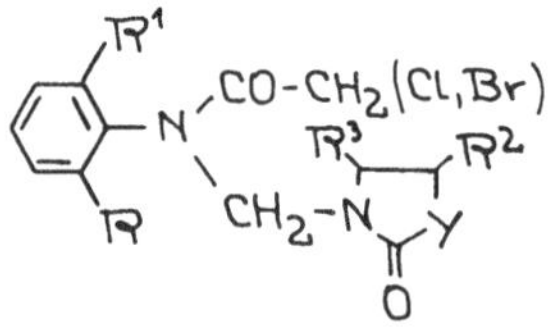

R = (O)-Alkyl
R^1 = H, Alkyl
R^2 u. R^3 = H, CH_3
Y = O, S

US   4 153 446   15. 9.78/ 8. 5.79   Gaf Corp.

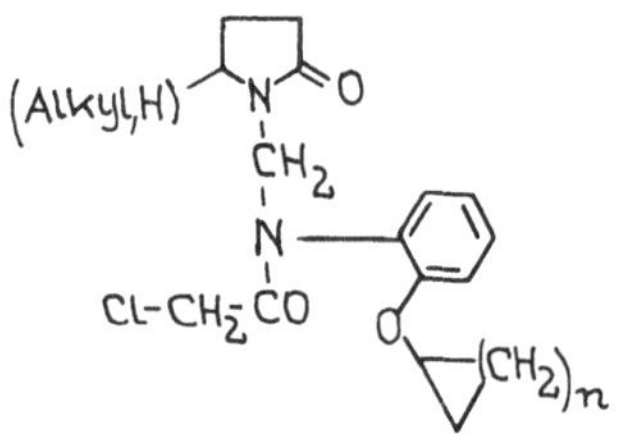

gegen Gräser

US   4 160 660   31. 3.73/10. 7.79   Ciba Geigy

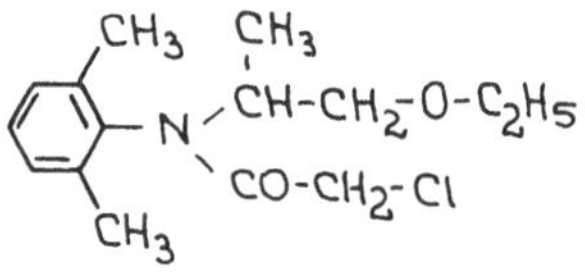

324

US   4 170 463   7. 4.78/ 9.10.79   Stauffer Chem. Co.

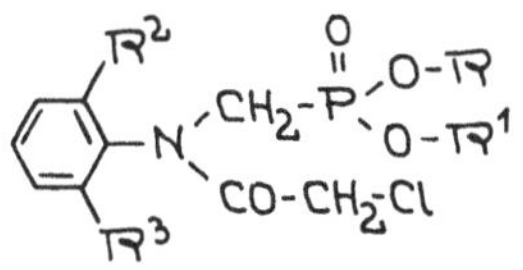

R u. R^1 = CH_3, C_2H_5

R^2 u. R^3 = H, CH_3, $CH<^{CH_3}_{CH_3}$

US   4 174 350   14. 5.73/13.11.79   Stauffer Chem. Co.

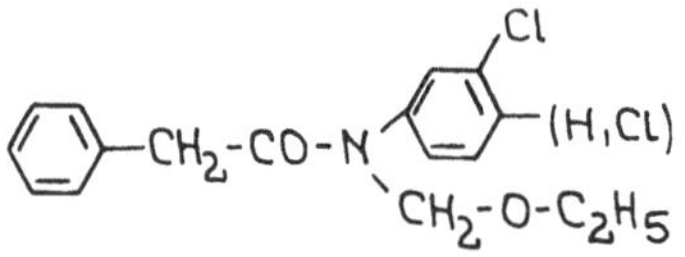

US   4 178 167   16. 6.77/11.12.79   Gaf Corp.

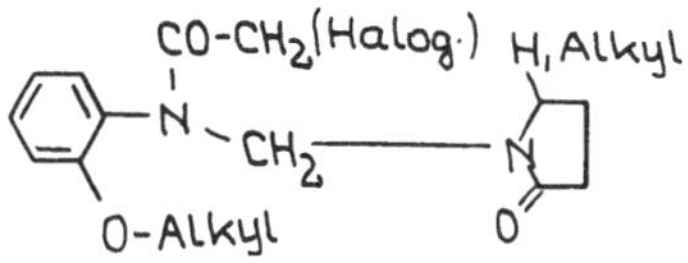

US   4 178 168   27. 7.78/11.12.79   Gaf Corp.

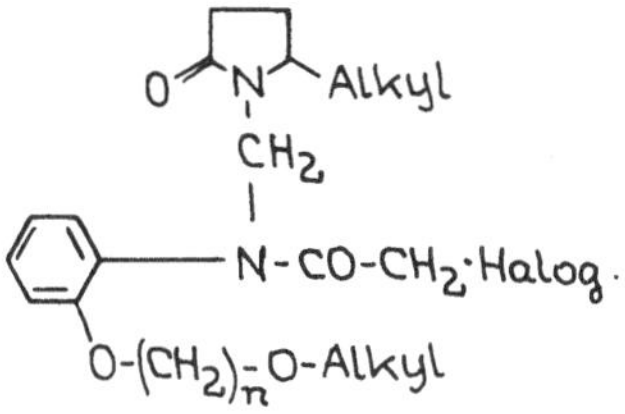

US   4 196 142   4. 3.77/ 1. 4.80   Amer. Cyanamid

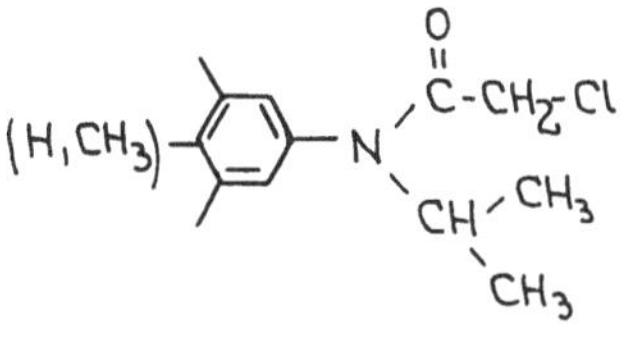

Gräserherbizid

US   4 205 168   24. 9.73/27. 5.80   Chevron Res.

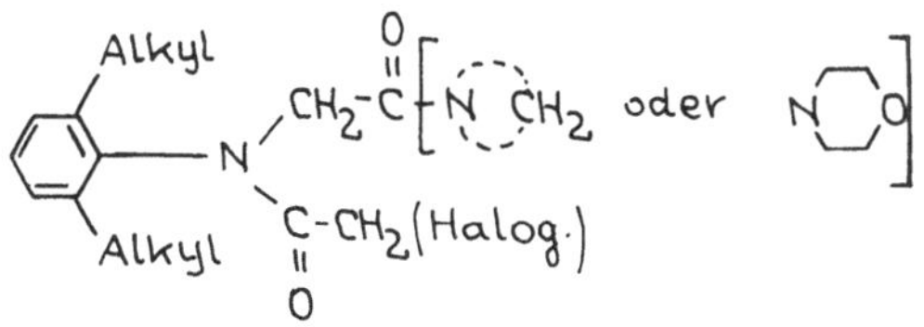

| BE | 844 501 | 25. 7.75/24. 1.77 | Amer. Cyanamid |

gegen Gräser

| BE | 860 043 | 23.10.76/24. 4.78 | BASF |

Belieb. substituiert z.B. durch H, Alkyl, O-Alkyl zwei benachbarte
Substituenten eine C_1-C_6-Alkylenkette bildend.
A = Azole z.B. Pyrazol subst. durch Cl,
   Phenyl, Bindung vom A an N.

| BE | 863 565 | 2. 2.77/ 2. 8.77 | Bayer |

R = N-Heterocycl.
X u. Y = Alkyl; Y bes. in
6-Stellung

| BE | 865 792 | 4. 3.77/ 9.10.78 | Amer. Cyanamid |

R^1 u. R^2 = CH_3, C_2H_5; R^3 = H, CH_3, C_2H_5

| BE | 880 509 | 24. 4.79/10. 6.80 | Mta. Mezogazd. HU |

spez. Darstellungsverfahren

| FR | 2 323 681 | 15. 3.72/13. 3.77 | Sumitomo Chem. Co. |

SU   426 638   27.10.72/15. 1.75   Dustmukhemedov

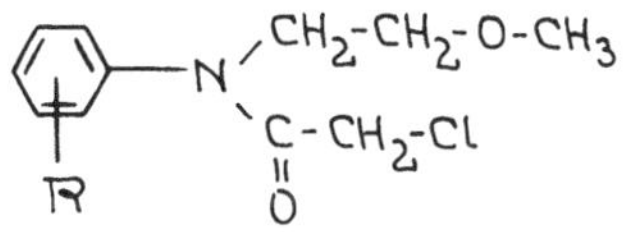

Gegen breitblättrige Un-
kräuter u. Ungräser in
Baumwolle.

R = H, CH_3, OCH_3, (O, mp)
    $CF_{3(p)}$

RD   186 063   20. 9.79/10.10.79

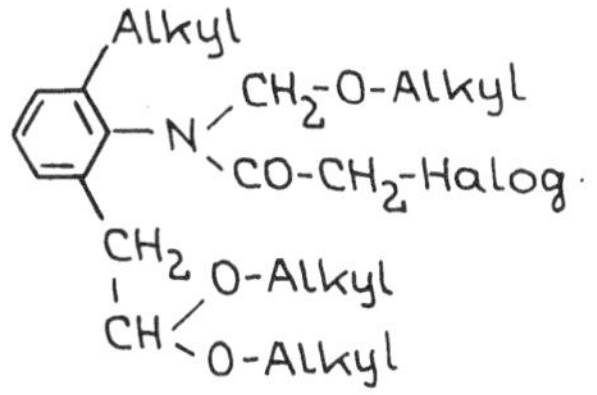

WP   8 000 701   7.10.78/17. 4.80   BASF

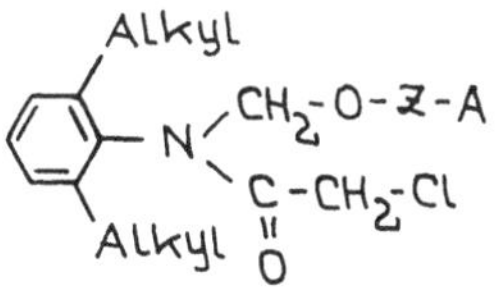

Z = CH_2, CN = CH

A = Pyrazol, Triazole,
    Imidazole alle am N
    angebundenen CH_2 usw.

RD   177 022   20.12.78/10. 1.79   Anonym

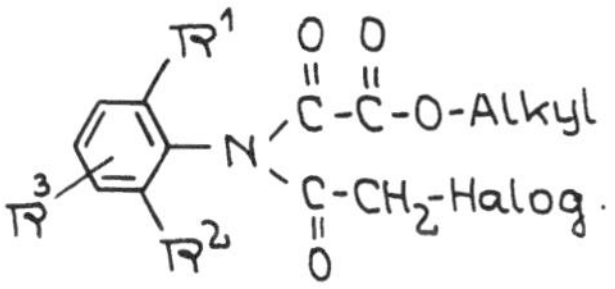

R^1 – R^3 = H, Alkyl, O-Alkyl,
         Halog,
         Alkoxy-Alkyl

JA   53 018540   2. 8.76/20. 2.78   Nihon Noyaku

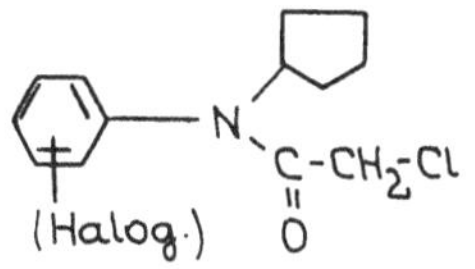

JA   76 046813   4. 7.68/11.12.76   Kaken Kayaku

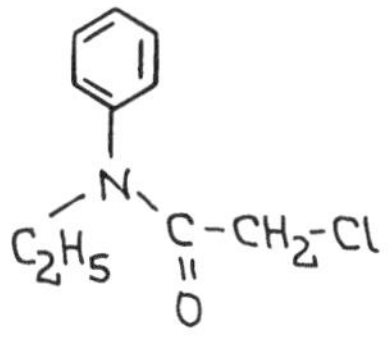

<u>Acylanilide zusätzlich eine herbizide Wirkungs-</u>
<u>gruppe enthaltend</u>   Bd.5/573
siehe auch unter N-Phenylcarbaminsäureestern
5.5.2.g

3.a   Zweite wirksame Gruppierung am Aryl (meist in
        3-Stellung)

| DOS | 2 844 806 | 11.10.78/24. 4.80 | Schering |
| = NE | 7 906 019 | 12.10.78/15. 4.80 | Schering |

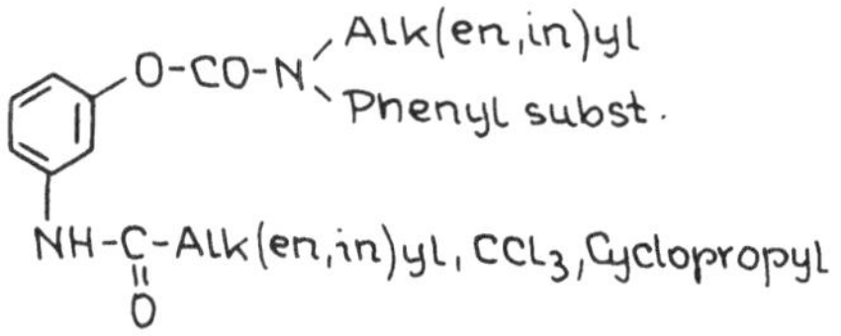

Gegen Mono- und Dicotyle
Unkräuter.

| DOS | 2 844 811 | 12.10.78/ 4. 9.80 | Schering |

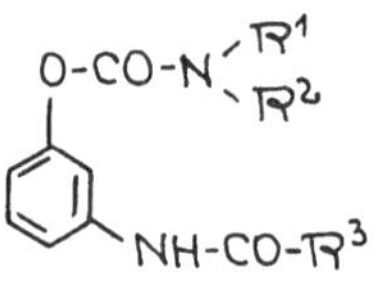

R^1 = Alkoxy-Alkyl, CN(Cl)-Alkyl,
       1,3-Dioxolan-2-yl-methyl
R^2 = subst. Phenyl, α-Cyanbenzyl
R^3 = Alkyl, Alkenyl, CCl_3,
       Cycloprophyl usw.

| EP | 10 692 | 18.10.79/10. 5.80 | BASF |
| = DOS | 2 846 625 | 26.10.78/ 8. 5.80 | BASF |

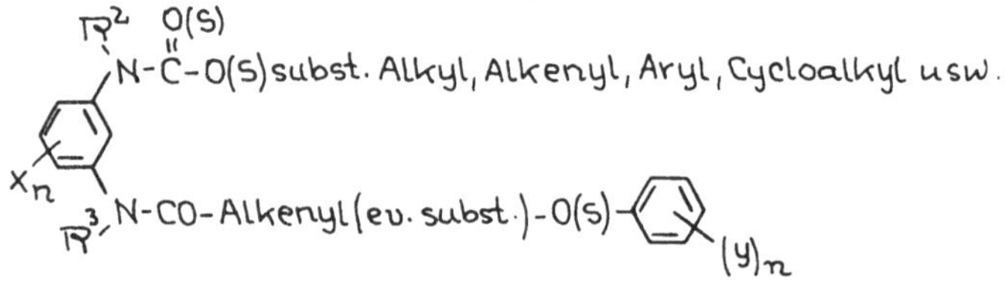

X = H, Alkyl, NO_2, NH_2, O-Alkyl usw.
Y = H, Alkyl, Halog.-Alkyl, Cycloalkyl, Aralkyl usw.
R^2 u. R^3 = H, Alkyl, O-Alkyl

| US | 3 979 202 | 12. 6.67/ 7. 5.76 | Monsanto |

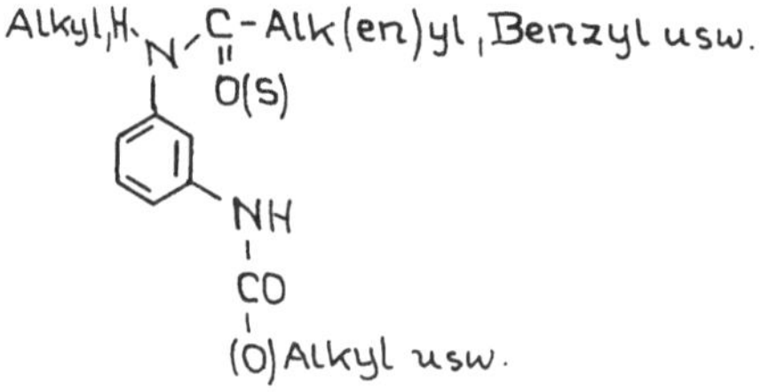

Sehr umfassendes Patent.

328

5.12.3.c   <u>Acylverbindungen von Benzylaminen</u>

DOS   2 937 867   19. 8.78/27. 3.80   Sumitomo Chem. Co.

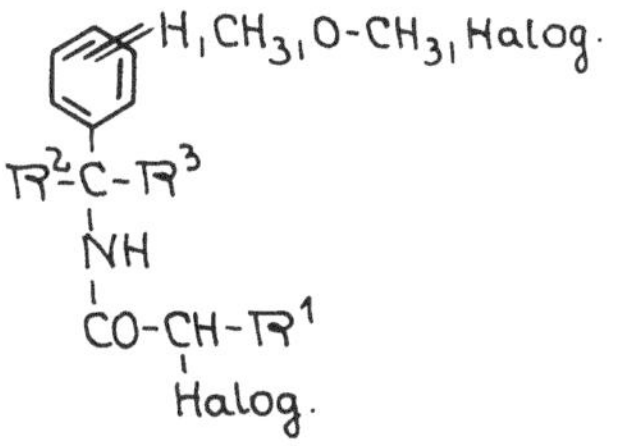

R^2 u. R^3 = CH_3, C_2H_5 oder zusammen
C_2-C_5-Alkylen

DOS   2 945 388   9.11.78/22. 5.80   Sumitomo

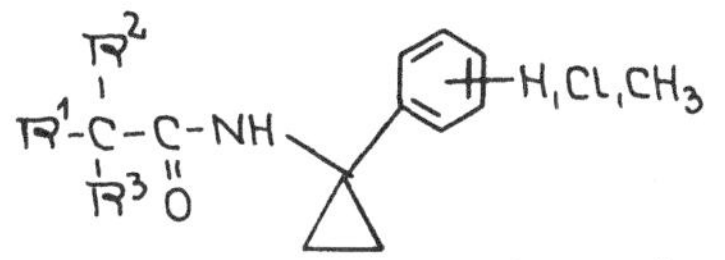

$\left.\begin{array}{l}R^1\\R^2\end{array}\right\}$ Halog., CH_2Cl, Alkyl

R^3 = H, Halog., CH_3

statt Cyclopropyl auch höhergliedrige Ringe

DOS   2 945 837   13.11.79/ 4. 6.80   Sumitomo
= NE   7 908 294   JA Prior. 13.11.78 u. 15.11.78

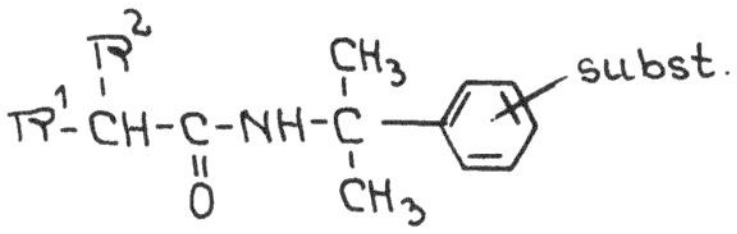

R^1 = Alkyl, Cycloalkyl
R^2 = H, Halog., O-Alkyl, Alkyl
R^1 u. R^2 = Alkylenrest eines Ringes

US   4 098 600   24. 5.76/ 4. 7.78   Monsanto

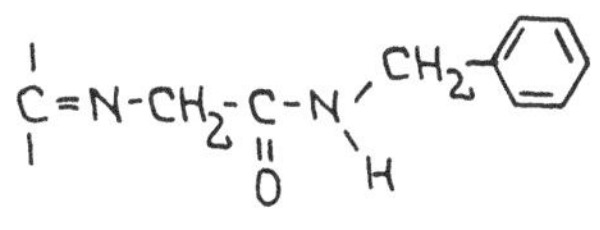

US   4 154 599   3.10.77/15. 5.79   Gulf Oil Corp.

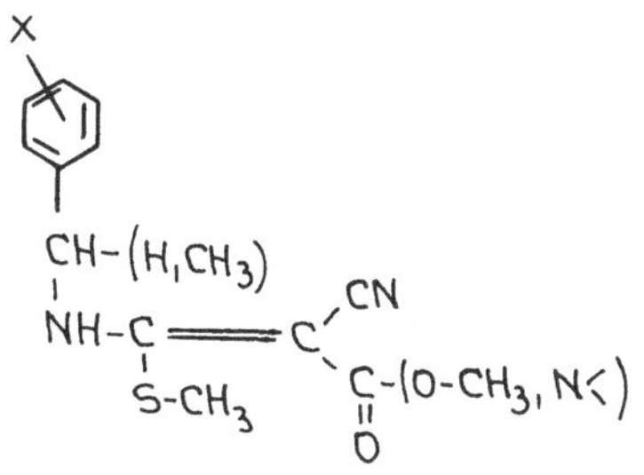

5.12.4 Acylverbindungen von Amino-Heterocyclen Bd.5/576

4.a Acylamide von 5-Ringen 1 x O oder 1 x S

DOS 2 627 935 22. 6.76/20. 1.77 W. R. Grece
US Prior. 23. 6.75

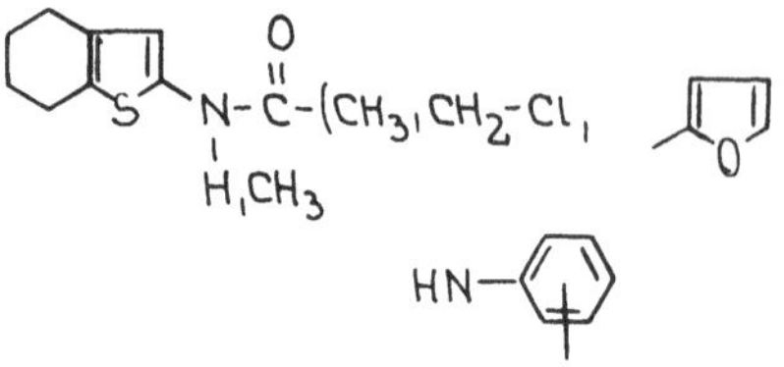

Reifemittel für Zuckerrohr.

4.c Acylamide von 5-Ringen 1 x N + 1 x O oder S

US 3 980 464 5. 5.75/14. 9.76 PPG. Industr.

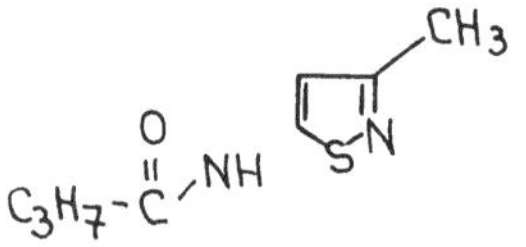

spez. gegen Unkräuter

US 4 013 675 5. 5.75/22. 3.75 PPG. Industr.

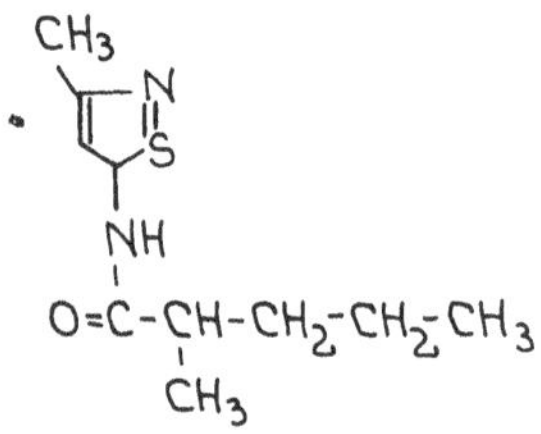

4.e Acylamide von 5-Ringen 1 x N + 2 x S, 2 x N + 1 x O bzw. S Bd.5/578

US 4 092 148 17. 2.70/30. 5.78 Air Prod.

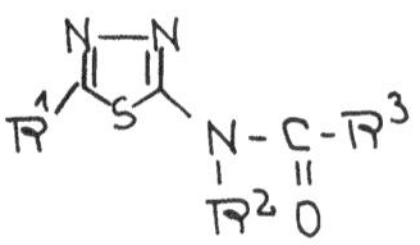

R^1 = CF_3, CF_2Cl, C_2F_5
R^2 = H, Alkyl; R^3 = H, ◁
(Chlor)-Alkyl

US   4 097 263   22.11.76/28. 6.78   Gulf Oil

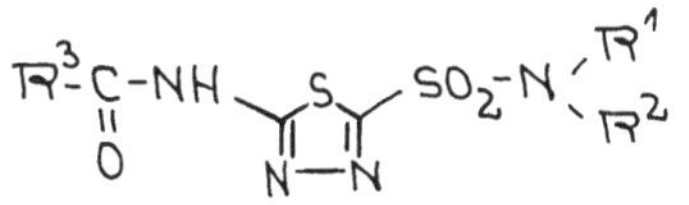

R^1 u. R^2 = H, Alk(en,in)yl,
O-Alkyl zus.
einen Ring bild.
R^3 = C_2H_5, C_3H_7, CH_3-O-,

tert. C_4H_9,

Cyclopropyl

US   4 187 098   24.11.78/ 5. 2.80   Gulf Oil

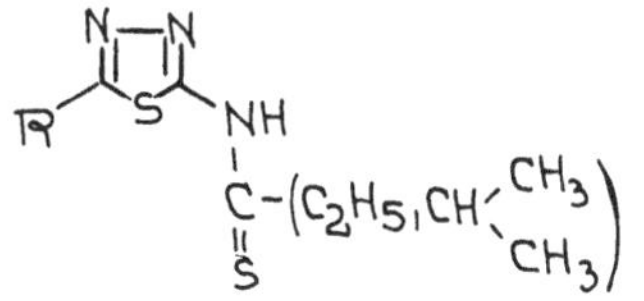

R = CF_3, $(CH_3)_2$N-SO_2

JA   50 117934   1. 3.74/16. 5.75   Sumitomo Chem. Co.

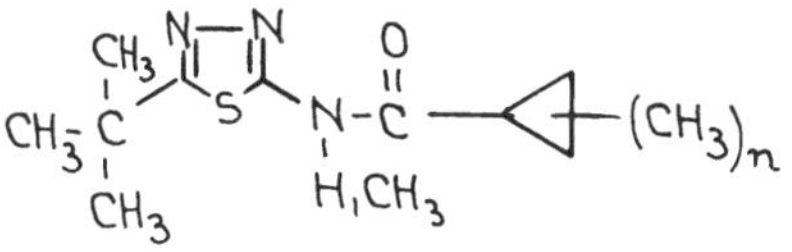

5.12.5   <u>Aromatisch- und heterocyclisch-aliphatische</u>

<u>Amidine und Hydrazidine</u>   Bd.5/581

DOS   2 558 839   27.12.75/22. 7.76   Sandoz

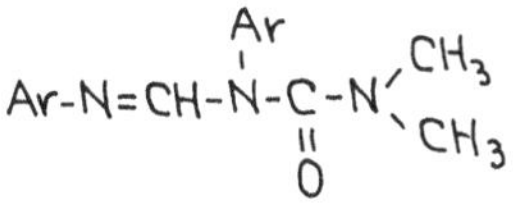

Ar = subst. Phenyl

Selektiv in Kartoffelkulturen, Mais usw. gegen Unkräuter und einige
Ungräser.

US   4 012 224   2.12.75/15. 3.77   Eli Lilly

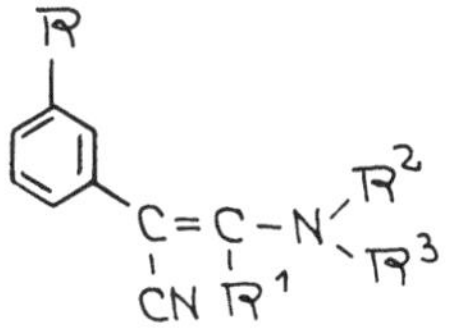

R = H, Cl, F, Br, CF_3
R^1 - R^3 = H, CH_3, C_2H_5

DOS   2 845 996   23.10.78/30. 4.80   Bayer
= EP   10 249   8.10.79/30. 4.80   Bayer
Siehe auch EP   10 250 Bayer

R^1 = O-Alkyl, O(S)-Phenyl; R^2 = Halog., Alkyl, CF_3, CH_3S, $SO_2-C_6H_5$
R^3 = H, Halog.   (O)

DOS   2 850 821   23.11.78/31. 5.79   Ishihara Sangyo
= BE   872 281   29.11.77/25. 5.79   Ishihara Sangyo

R^1 = Alkyl; R^3 = H, SO_2-Alkyl
                          Acyl, Alkyl
R^2 = H, Alkyl, NO_2, CF_3
R^3 = H, SO_2-Alkyl, Alkyl,
        SO_2-Alkyl
R^4 = H, Halog.

X^1 = H, Halog., Alkyl, CF_3; X^2 = H, Halog.

DOS   2 854 932   20.12.78/ 5. 7.79   Fisons Ldt.
= BE   872 904

$CH_3-SO_2-N-CH_2-COOR$ als Beispiel   Näheres siehe 5.7.3.e
                                                  Araliph. Aminosäuren

DOS   2 900 685   10. 1.79/24. 7.80   BASF
= EP   13 742

R^1 = Alk(en,in)yl, Cycloalk(en)yl,
       subst. Aralkyl
R^2 u. R^3 = H, Halog., CN, SCN, NO_2,
        C(O,S)-Alkyl, CO-Alkyl,
        SO_2Alkyl
R^4 = CO-OAlkyl, CO-Alkyl, SO_2-Alkyl,
        SO_2-Phenyl
R^5 = Halogen-Alkyl

332

DOS   2 916 714   25. 4.79/ 8.11.79   Sandoz
                                                US Prior. 1. 5.78

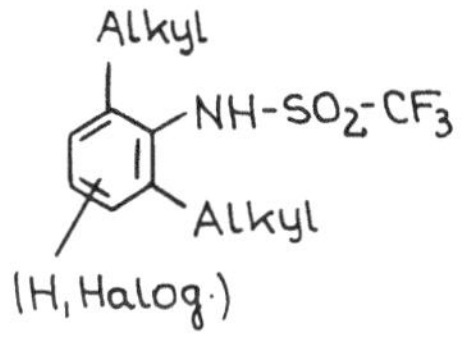

DOS   3 002 880   28. 1.80/ 7. 8.80   MMM
                                                US Prior. 29. 1.79

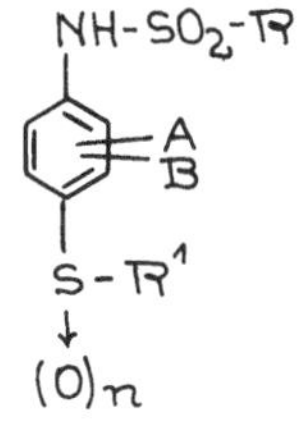

R = Alkyl, CH_2Cl
R^1 = Alkyl
A = Halog., CF_3
B = H, Halog.; n = 0 - 2

Wuchsregulatoren und Herbizide.

DOS   3 002 895   28. 1.80/ 7. 8.80   MMM
                                                US Prior. 29. 1.79

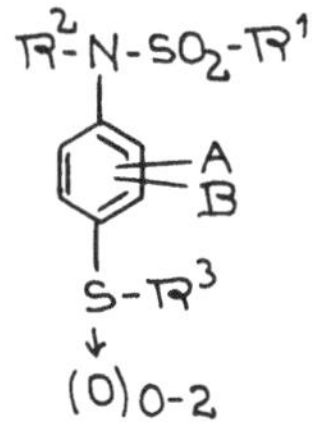

R^1 = Alkyl, CH_2Cl
R^2 = Alkyl, OAlkyl, Alkyl,
    $CH_2-C \equiv CH$ usw.
R^3 = Alkyl, Phenyl
A = Halog., CF_3; B = H, Halog.

DOS   3 002 905   28. 1.80/31. 7.80   MMM
                                                US Prior. 29. 1.79

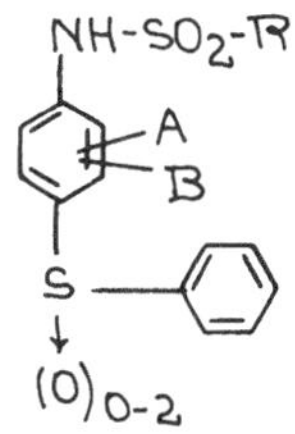

R = Halog.-Alkyl, Alkyl
A = Halog., CF_3(CF_3 dann nicht A = CH_3)
B = H, Halogen

DOS   3 002 906   28. 1.80/31. 7.80   MMM
                                                US 29. 1.79

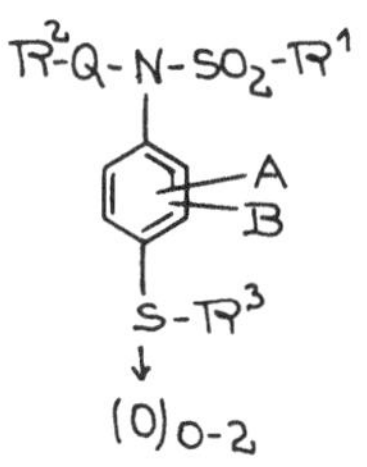

R^1 u. R^2 = Alkyl, Halog.-Alkyl
R^3 = Alkyl, Phenyl
Q = SO_2, O-CO-; A = Halog., CF_3
B = Halog., H

EP   11 693   1.10.79/11. 6.80   Ciba Geigy
                                              CH Prior.   3.10.78

A Glieder eines Ringsystems ev.
unterbrochen durch O, S, N oder P

subst. (Halog.)-Alkyl, CN, NO$_2$, $\overset{\text{S}}{\overset{\|}{\text{C}}}$-NH$_2$, S(O)$_n$-Alkyl, SO$_2N\langle$ usw.

US   3 981 914   3. 8.71/21. 9.76   Minnesota Mining Co.

Y = Halog., NO$_2$, CN, (O,S)-Alkyl usw.
R = Alkyl, R^1 = Perfluor-Alkyl

US   4 005 141   13. 4.70/25. 1.77   Minnesota Mining Co.

R = H, Alkyl-SO$_2$, CO-Alkyl; Y = Halog, (O)-Alkyl, NO$_2$, NH$_2$
                                  S(O)$_{1-2}$- Alkyl   (Alkyl)$_2$-N-SO$_2$ usw.
Y^1 = Y (nicht Cl u. Br)

US   4 013 448   29.12.75/22. 3.77   Monsanto Co.

                                        R^1 u. R^2 = H, Alkyl

US   4 071 349   5. 9.72/31. 1.78   Stauffer

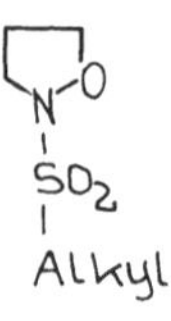

Antidot für Thiocarbamat - Herbizide

334

US   4 071 352   5. 9.72/31. 7.78   Stauffer

$$R-SO_2-N\underset{\diagdown Alkenyl,\ z.B.\ Allyl}{\overset{\diagup Alkenyl}{}}$$

Antidot für Thiocarbamat-Herbizide

$$R = -N=C=O,\quad HN-\text{(2-Cl, 4-Cl/Alkyl, 5-Cl am Phenylring)}$$

US   4 086 255   13. 4.70/25. 4.78   Minnesota Mining Co.

$$\text{Perfluor-Alkyl-SO}_2-\underset{(CN,\ Alkyl-O-CO-)}{N}-\text{(Phenyl, subst.)}-\overset{(O)_{0-2}}{S}-\text{(Phenyl, bel. subst.)}$$

US   4 163 659   21.10.66/ 7. 8.79   Minnesota Mining Co.

$$CF_3-SO_2-NH-\text{(Phenyl, CH}_3\text{)}-NH-\overset{O}{\underset{\|}{C}}-CH_3$$

vorwiegend wuchshemmend

US   4 164 412   13. 4.70/14. 8.79   Minnesota Mining Co.

$$C_nF_{2n+1}-SO_2-\underset{(H,CN,Alkyl,\ Alkyl-SO_2\ usw.)}{N}-\text{(Phenyl, subst.)}-S(O)-\text{(Phenyl, subst.)}$$

BE   875 968   1. 5.78/30.10.79   Sandoz

$$\text{Hal.}-\text{(Phenyl, 2-Alkyl, 6-Alkyl)}-NH-SO_2-CF_3$$

GB   1 453 565   17. 9.74/21.10.76   ICI

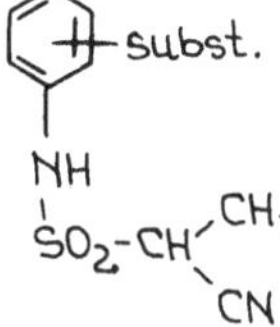

verhindert Chlorophyllbildung

5.12.6.b   <u>Benzol- und Heterocycl.-sulfonsäure-amide,</u>

<u>anilide und amidoheterocyclen</u>   Bd.5/586

DOS   2 636 452   13. 8.76/16. 2.78   Amer. Cyanamid

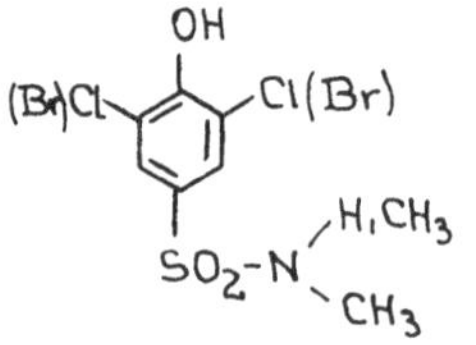

Gegen Wildhafer in Getreide,
Zuckerrüben, Flachs oder Raps.

DOS   2 644 504   1.10.76/ 2. 6.77   Stauffer
Siehe 5.5.2.a

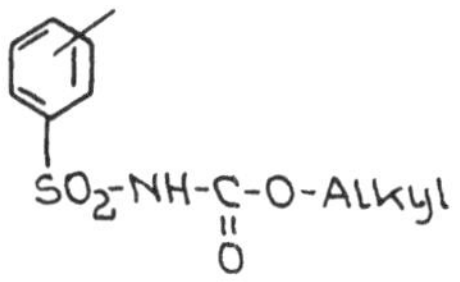

Antidot gegen Thiocarbamat-Schäden.

DOS   2 715 786   7. 4.77/13.10.77   Du Pont
Weitere ähnliche Patente siehe 5.5.3.k.4   US Prior. 7. 4.76/23. 2.77

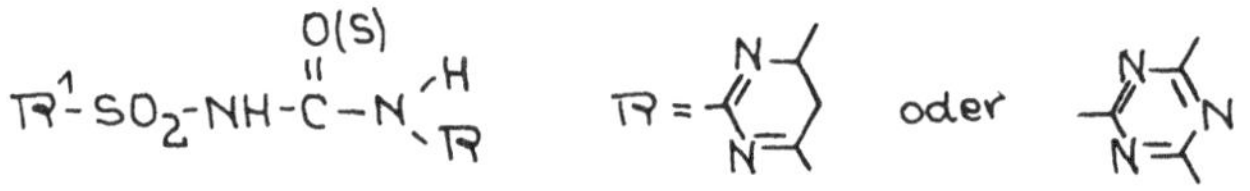

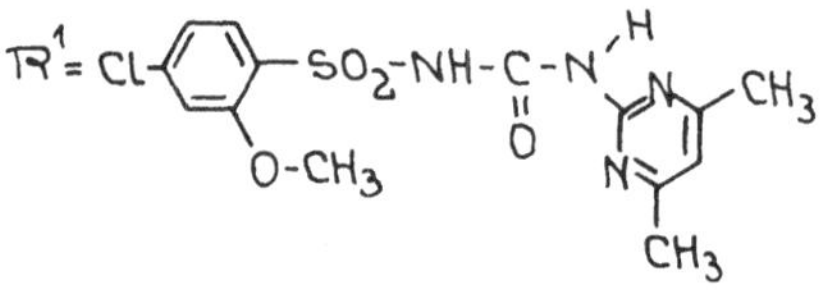

Sehr umfassendes
Patent.
247 Seiten.

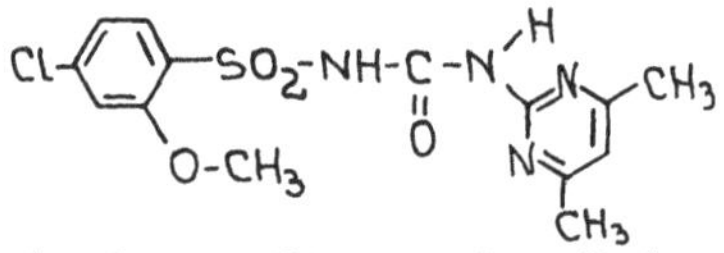

DOS   2 744 137   30. 9.77/ 6. 4.78   Utsunomiya Univ.
                           JA Prior. 1.10.76/20. 5.77, 2. 8.77

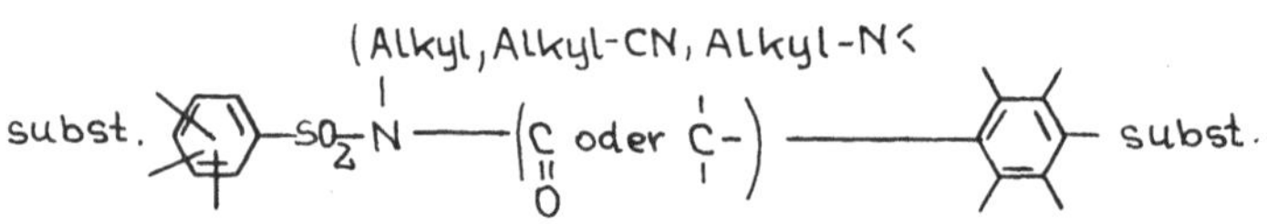

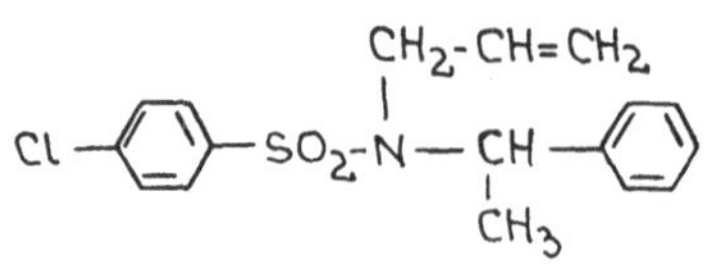

DOS   2 845 996   23.10.78/30. 4.80   Bayer

CF$_3$, NH-SO$_2$, Halog., NO$_2$ substituted diphenylsulfone structure

US   3 979 203   19. 1.72/ 7. 9.76   Amer. Cyanamid

R-O, SO$_2$-N(Alk(en)yl / Alk(in)yl) substituted benzene with X, Y

X = H, Cl, Br
Y = NO$_2$, Cl, Br
Z = H, CO-Alkyl, $>$N-CO-

US   4 071 350   5. 9.72/31. 1.78   Stauffer

SO$_2$-N(H,Alk(en)yl / Alkyl usw.) structure

Antidot für Thiocarbamatherbizide.

US   4 074 059   19. 1.72/14. 2.78   Amer. Cyanamid

[H,CH$_3$ N-C-O]/H / Cl,Br substituted phenyl SO$_2$-N(H,CH$_3$ / CH$_3$) with Cl,Br

Gegen Flughafer in Getreide.

BE   846 894   2.10.75/ 1. 4.77   Stauffer Chem.

SO$_2$-NH-C(=O)-O(Halog.-Alkyl, Alk(en, in)yl, bel. subst. Alkyl)

(H,Br,Cl(O)-CH$_3$, CF$_3$)$_{1-3}$

BE   846 895   2.10.75/ 1. 4.77   Stauffer Chem.

(H,(O)CH$_3$,Cl,Br)$_{1-3}$

SO$_2$-N(H,CH$_3$)—C(=O)-S(Alkyl, Cl-phenyl-SO$_2$-NH-C(=O)-S-CH$_2$, Benzyl,Phenyl)

Safener für Thiocarbamat-Herbizid

337

BE   845 158   13. 8.76/14. 2.77   Amer. Cyanamid

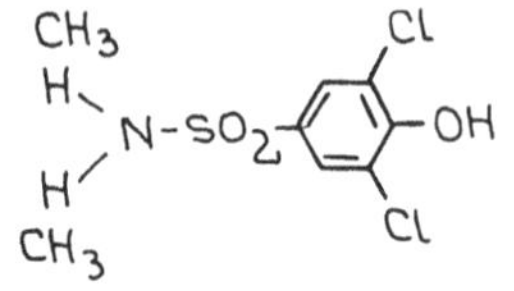

Selektive Anwendung gegen
Flughafer in Getreide.

JA   54 046743   22. 9.77/12. 4.79   Utsunomiya

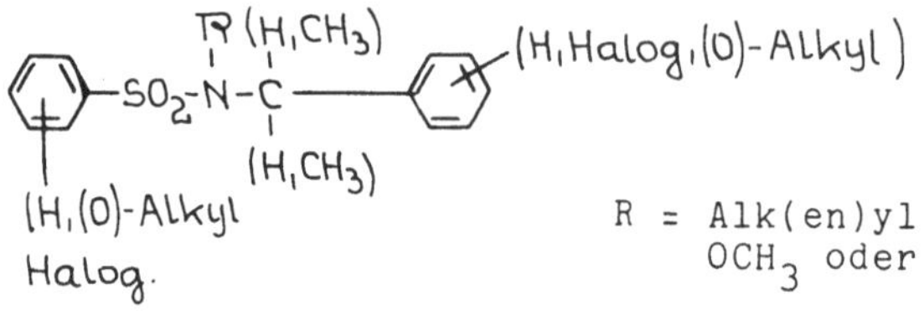

R = Alk(en)yl subst. durch, CN,
    OCH_3 oder $N(CH_3)_2$

JA   55 064567   10.11.78/15. 5.80   Ube Ind.

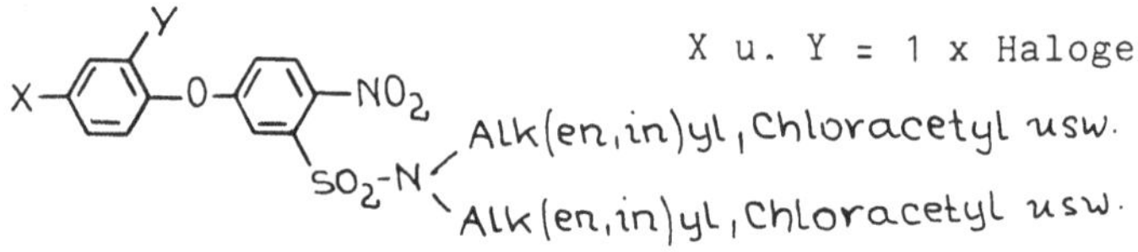

X u. Y = 1 x Halogen u. 1 x CF_3

5.13 Quartäre Ammonium- und Hydrazonium-Verbindungen, ferner Sulfonium-Verbindungen

<u>5.13.1   Aliphatische Vertreter</u>   Bd.5/588

<u>1.a   Quartäre Ammoniumverbindungen</u>

DOS   2 557 791   22.12.75/30. 6.77   Höchst

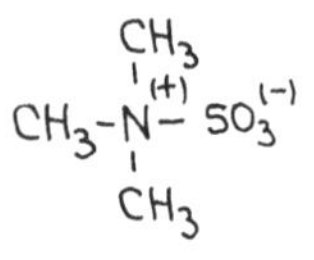

DOS   2 600 466   8. 1.76/10. 2.77   Buckmann Labs.
= BE   843 220   21. 7.75/18.10.76

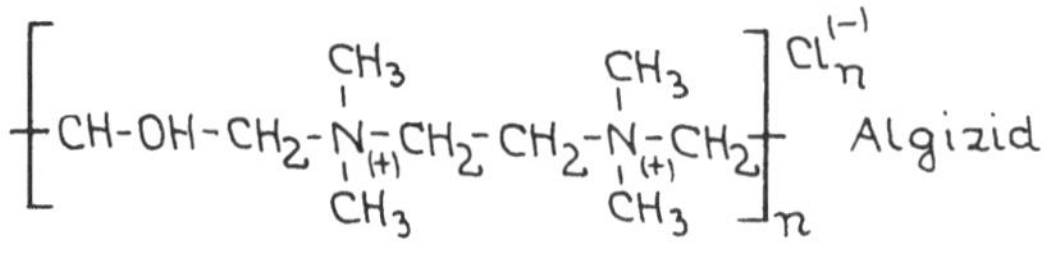

338

DOS   2 657 728   20.12.76/14. 7.77   Ciba Geigy
= BE   849 723   23.12.75/22. 6.77

$$\left[\text{Ar}-NH-\underset{\underset{O}{\|}}{C}-(CH_2)_m-\overset{(+)}{N}\underset{R^3}{\overset{R^1}{\underset{\diagdown}{\diagup}}}R^2 \right] \text{Anion}^{(-)}$$

bes. 1,4,5, substituiert durch Cl, CH_3, CF_3, Halog. CH_3, OCH_3 bzw. H, Cl, CF_3

spez. als Wuchshemmstoffe z. B. für Gräser und als Desiccant

EP   14 223   9. 2.79/20. 8.80   Ciba Geigy

$$\left[\text{(3-Cl-C}_6\text{H}_4\text{)}-NH-\underset{\underset{O}{\|}}{C}-CH_2-\overset{(+)}{N}\underset{CH_2-CH=CH_2}{<} \right] \text{Anion}^{(-)}$$

5.13.2   Heterocyclische Ammonium- und Sulfonium-Verbindungen

Bd.5/591

2.a   <u>5 u. 6-Ringe 1 x O, 1 x S, 1 x N</u>

DOS   2 701 681   17. 1.77/21. 7.77   ICI
                                            GB Prior. 20. 1.76

[2,2'-Bipyridinium] 2 x Anion   oder, und $[CH_3-\overset{(+)}{N}(C_5H_4)-(C_5H_4)\overset{(+)}{N}-CH_3]$ 2 x Anion

+ Mißbrauch verhindernde Duftmittel wie z.B. Buttersäure.

DOS   2 709 307   3. 3.77/ 3.11.77   ICI
= BE   852 862   GB Prior. 15. 4. u. 19. 8.76

[Diquat] 2 x Anion   oder $[CH_3-\overset{(+)}{N}(C_5H_4)-(C_5H_4)\overset{(+)}{N}-CH_3]$ 2 x Anion

unter Zusatz von [Heterocyclus]$-N<$ als Vomizid gegen Vergiftungen

DOS   2 712 032   18. 3.77/29. 9.77   ICI
                                                    GB Prior. 19. 3.76

Bispyridiliumverbindungen mit Geruchabschreckungsmitteln

DOS   2 801 152   12. 1.78/13. 7.78   Montedison
                                                    IT Prior. 12. 1.77

$$\left[CH_3\overset{(+)}{N} \bigcirc \hspace{-2pt} \bigcirc \overset{(+)}{N}-CH_3 \right] 2 \times Cl^{(-)}$$

                                                neues Darstellungsverfahren

DOS   2 905 843   15. 2.79/21. 8.80   Cheng Mong Chem.
                                                    Taiwan

R = C_1-C_4 Alkyl, O-Alkyl, Halog., CF_3

EP   6 359   18. 6.79/ 9. 1.80   Eli Lilly
                                                    US Prior. 19. 6.78

R = C_1-C_4 Alkyl, O-Alkyl, Halog., CF_3
R^1 = Halog., O(S)-Alkyl, S-Benzyl, $N(CH_3)_2$
R^2 = H, O(S)-Phenyl, O(S)-Alkyl, Alkyl, Phenyl usw.

US   4 116 675   4. 5.76/26. 9.78   Ciba Geigy

R^1 = Alkyl, O-Alkyl, Halog.-Alkyl, CN, Phenyl usw.
R^2 u. R^3 = H, Halog.-Alkyl, Alkyl

US   4 174 209   19. 6.78/13.11.79   Eli Lilly

R^1 = Halog., O-CH_3, $N{<}^H_{CH_3}$
R^2 = H, O-Phenyl, (O)-Alkyl, subst. Phenyl

340

BE   841 122   24. 4.75/26.10.76   Höchst

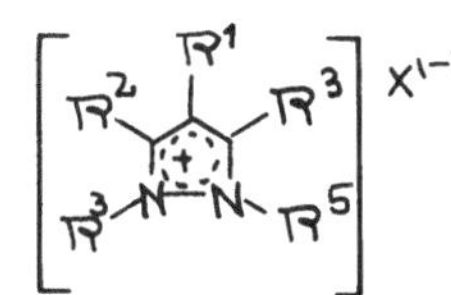

R^1 = (Cyclo)Alkyl, Naphthensäure-
           derivat
R^2 = CH$_3$, C$_2$H$_5$, C$_6$H$_5$CH$_2$
X = Cl, Br, CH$_3$SO$_2$ usw.

Algizid

FR   2 309 140   28. 4.75/31.12.76   ICI

$$\left[\text{Alkyl}^1-\text{N}^{(+)}\text{—}\text{—N}^{(+)}\text{—Alkyl}^2 \right] 2\times \text{Anion}$$

Anwendung gegen Flughafer in Getreide.

JA   52 007434   4. 7.75/20. 1.77   Nihon Noyaku

5.13.2.c   <u>Quartäre heterocyclische Hydrazonium-Verbindungen</u>

              <u>5-Ringsysteme mit 2 oder 3 N</u>   Bd.5/594

DOS   2 523 143   24. 5.75/ 9.12.76   Bayer

R = H, Halog., Alkyl
R^2 u. R^3 = Alkyl, subst. Phenyl
R^4 u. R^5 = Alkyl; X = Anion

DOS   2 523 144   24. 5.75/ 9.12.76   Bayer
= BE   842 146   24. 5.75/24.11.76   Bayer

Ähnliche Verbindungen als Wuchsregulatoren.

DOS   2 609 961   10. 3.76/30. 9.76   Amer. Cyanamid

R^1 u. R^2 = Alkyl
R^3 = H, Alkyl, Benzyl usw.
R^4 = H, (S)-Alkyl
R^5 = subst. Phenyl; X = Anion

Fungizide und Herbizide

DOS   2 618 412   27. 4.76/18.11.76   Amer. Cyanamid

R^1 u. R^2 = Alkyl (C$_2$-C$_4$)
              Cycloalkyl (C$_3$-C$_7$)
              Phenyl, Benzyl
R^3 = Alk(en,in)yl, Benzyl usw.

Darstellung: R^1-C-CH-C-R^4 + Alkyl-NH-NH$_2$ ⟶

DOS   2 653 447   25.11.76/ 8. 6.77   Fisons
= BE   848 615   26.11.75/23. 5.77
                            GB Prior. 26.11.75; 19. 6.76; 23. 9.76

oder

R^1 = Alk(en,in)yl, Cycloalkyl, Aryl, Aralkyl, Heterocyclyl
R^2 u. R^4 = H, S-(Alkyl, CH$_2$Aryl, Aryl) oder = R^1
R^3 = R^1 oder subst. N< ;   R^{11} = SO$_2$-Alkyl usw.

Herbizid und Fungizid

342

DOS   2 720 416   6. 5.77/ 1.12.77   Fisons
= BE   854 689   21. 5.76/16.11.77   Fisons
                                  GB Prior. 21. 5. u. 26.11.76

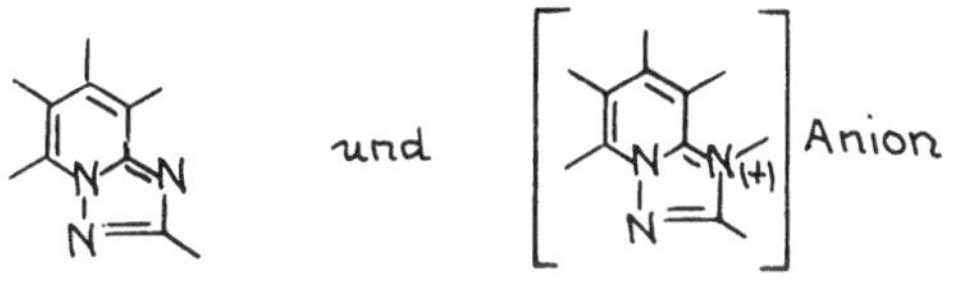

und ... Anion

subst. durch H, Halog. Alkyl usw.

DOS   2 822 014   19. 5.78/30.11.78   Fisons
              GB Prior. 21. 5.77; 24. 1.78; 28.3.,13.4., u. 21. 4.78

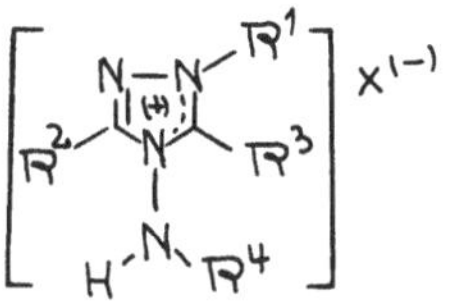

R^1 = Alk(en)yl
R^2 = Alkyl, Phenyl
R^3 = Alkyl Phenyl
R^4 = CH_3

z.B.

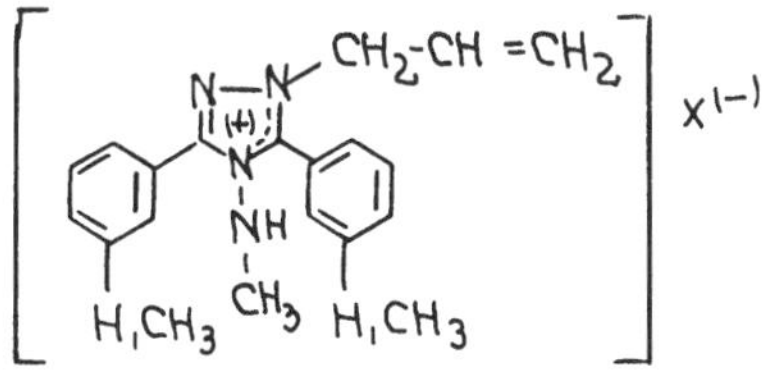

Herbizid und Fungizid

US   4 002 636   11. 8.71/11. 1.77   Höchst

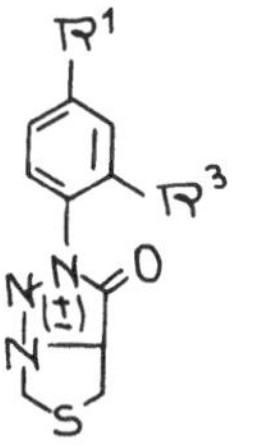

R^1 = Halog., R^3 u. R^4 = H,
                                     Halog.

Mesoionische Oxo-thiazolo-
(3,4-)-triazoliniumsalze

wirksam gegen Gräser

US   4 218 238   19. 6.78/19. 8.80   Eli Lilly

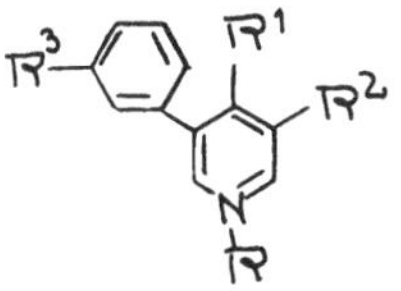

$R = CH_3, C_2H_5$
$R^1 = S$-Alkyl, C_2H_5-CH_2-S
$R^2 = H, O$-⟨⟩, S-⟨⟩, Alkyl, S-Alkyl
     Phenyl
$R^3 = $ Halog., CF_3, (O)-Alkyl

US   4 003 732   1. 7.74/18. 1.77   Amer. Cyanamid
= BE   841 231   2. 5.73/28.10.76   Amer. Cyanamid

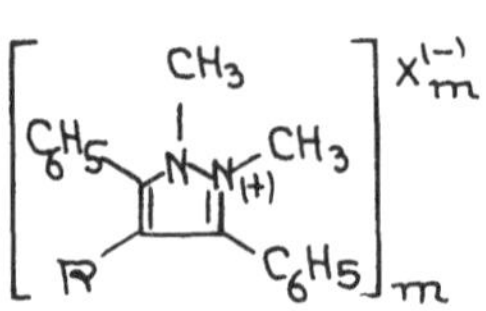

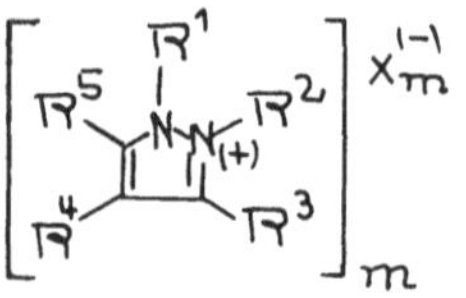

R = Alkyl, Allyl, Propargyl, Benzyl, $COOC_2H_5$
X = Anion; m = 1 - 3
R^1 u. R^2 = Alkyl
R^3 = C_2-C_4-Alkyl, C_3-C_7-Cycloalkyl, Phenyl, Benzyl
R^4 = C_2-C_4-Alkyl, C_3-C_7-Cycloalkyl, Phenyl

US   4 126 443   21.11.77/21.11.78   Amer. Cyanamid

Verhinderung der Kristallbildung von 1,2-Dimethyl-
3,5-diphenylpyrazolium-Methylsulfat

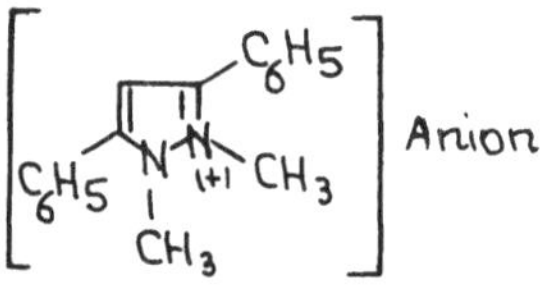

BE   839 377   10. 3.75/10. 9.76   Amer. Cyanamid

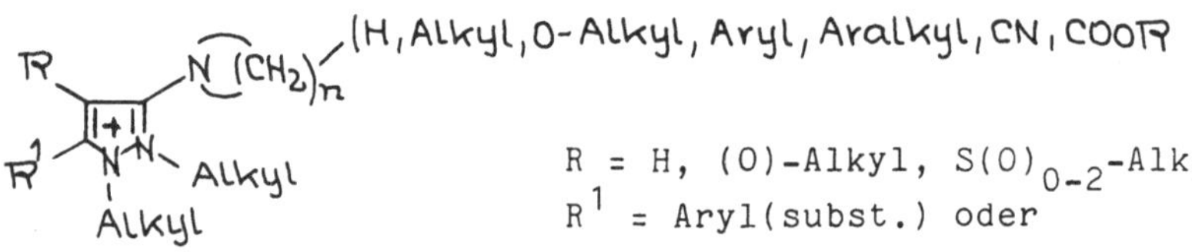

R = H, (O)-Alkyl, $S(O)_{0-2}$-Alkyl
R^1 = Aryl(subst.) oder
     -N<$(CH_2)_n$-Alkyl, CN usw.
X = Anion

selektiv in Weizen gegen Flughafer

SU   552 942   20. 6.75/23. 5.77   Melnikov

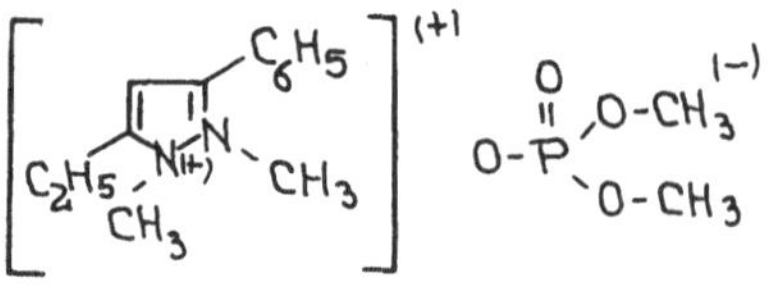

gegen Flughafer

344

5.14 Heterocyclische Verbindungen

5.14.1   <u>3 u. 4-Ring-Verbindungen</u>

DOS	2 520 165	6. 5.75/11.11.76	Dow Chemical
= US	4 018 801	6. 8.70/19. 4.77	

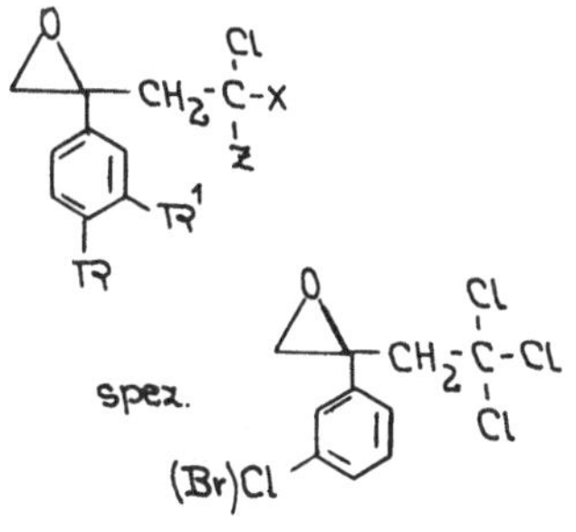

X = Halogen

Z = H, Halogen, CN, C_1-C_4-Alkyl

R u. R^1 = H, Halog.,CN, NO_2, O-Alkyl, Alkyl, CF_3, $O-CH_2-C_2H_5$

spez.

(Br)Cl

Besonders gegen Gräser.

US	4 075 000	27. 5.75/21. 2.78	Eli Lilly

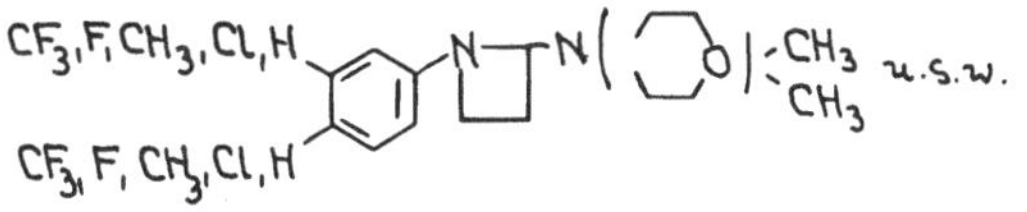

CF_3,F,CH_3,Cl,H ... CF_3,F,CH_3,Cl,H   u.s.w.

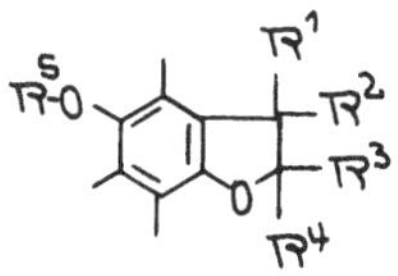

R^1-R^4 = H, Alkyl, Alkylenkette O-Alkyl, S-Alkyl, O-Aryl(Subst.)

5.14.2   <u>5-Ring-Verbindungen</u>

a)   <u>5 Ringe mit 1 x O</u>

DOS	2 651 046	9.11.76/18. 5.77	Fisons
			GB Prior. 12.11.75
= BE	848 115	12.11.75/ 9. 5.79	

siehe Grundpatent GB 1 271 659

Spez. Formulierung

DOS	2 733 115	23. 7.76/26. 1.78	Takeda Chem.

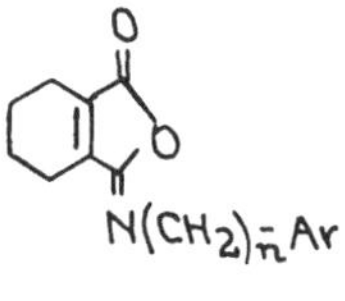

$N(CH_2)_{\overline{n}}Ar$

n = 0 und 1

DOS   2 749 974   8.11.77/11. 5.78   Shell Int. Res.
                                            GB Prior. 10.11.76

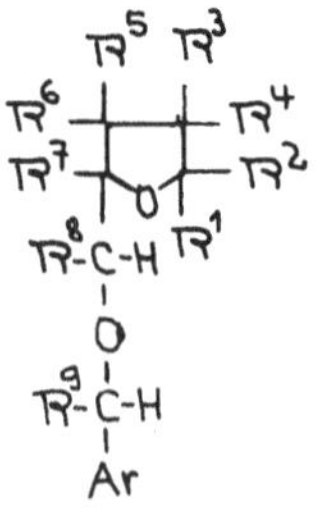

R^1 u. R^2 = H, Halog. Alkyl,
                Cycloalkyl, Aryl

R^3, R^4, R^5 u. R^6 = H, Halog.,
                (O,S)-Alkyl,
                Aryl

R^3 u. R^4, R^5 u. R^6 = C-C-Bindung
                oder $-CH_2-O-$

R^7 = H, Alkyl

Selektiv in Getreide gegen Unkräuter.

DOS   2 803 991   31. 1.78/10. 8.78   Fisons
= BE   863 471   GB Prior.5.2. u.5..8.77

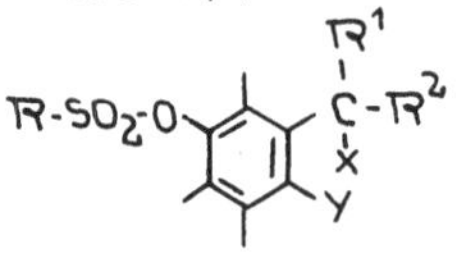

$X = -CH-O-R^4$
        R^3

$Y = -O-R^5$

X u. Y zusammen eine
        Gruppierung!

        $-CH-$ usw.
        R^3

R^1, R^2, R^3 = H, Alkyl

R^2 u. R^3 zusammen einen Ring bildend   Beispiel:

R = Alkyl, Phenyl, Aralkyl usw.

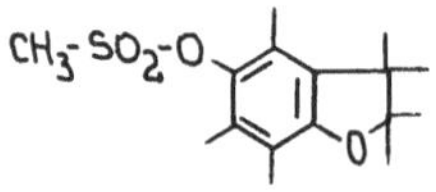

EP   5 917   9. 5.79/12.12.79   Fisons
                    (GB 30.5.78 u. 3.8.78)   GB Prior. 3. 8.78

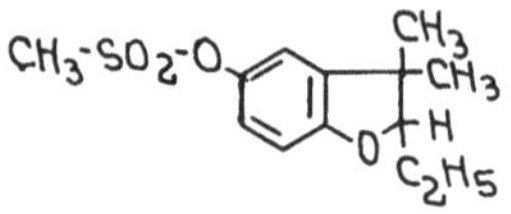

Anwendung in Kombination mit anderen Herbiziden oder allein   und
nachfolgend andere Herbizide.

EP   7 719   3. 7.79/ 6. 2.80   Fisons
= GB   3 164 678   GB Prior. 29.7.78 u. 24.10.78

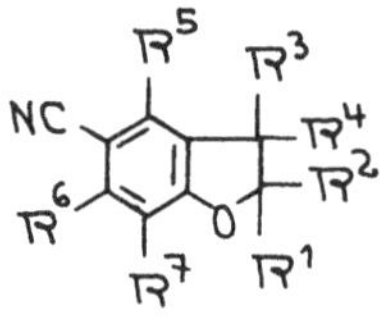

$R^1 + R^2$ = O oder R^1 = H, R^2 = H,OH,
                O-Alkyl, O-CO-Alkyl,
                Heterocycl. usw.

$R^3 + R^4$ = Alkylen, R^3 = H, Alkyl,
                R^4 = H, Alkyl

R^5, R^6, R^7 = H, Halog.,CN,
                O-Alkyl usw.

346

EP   13 581   8. 1.79/23. 7.80   Shell Int. Res.
                                                  GB Prior. 8. 1.79

                                            Het. = 5,6 Glieder, N haltiger
                                                  aromat. Heterocyclus

Siehe ähnlich DOS 2 749 974 zuvor

US   3 990 880   6.11.75/ 9.11.76   Du Pont

                                    X = H, F
                                    Y = H, F

US   3 992 415   18. 4.66/16.11.76   Air Prod.
siehe auch 3 862 220

BE   858 646   13. 9.76/13. 3.78   Int. Dev. Res. Cent.

                                            A = H
                                            B = H oder zusammen
                                                  -CH$_2$-
                                            X = CH$_2$

JA   2 083 734   01.01.76/12. 7.77   Kuraray
siehe auch J 5 083 457; 2 083 458; 2 083 459

                            R^1-R^3 = H, Alkyl, Phenyl usw.
                            A = -CH$_2$-CX$_3$; -CH=CX$_2$; X = Halogen

                            Herbizide und Fungizide.

347

DOS   2 612 731   25. 3.76/ 7.10.76   Stauffer Chem.
                                           US Prior. 28.3.75 u. 9.1.7.76

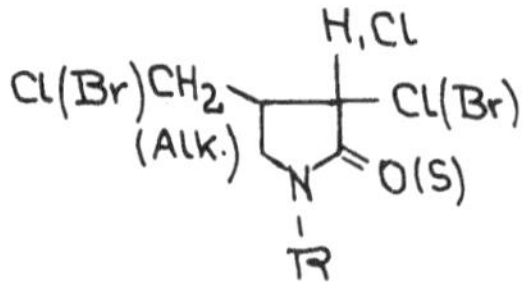

R = Alk(en)yl, Cyclo-
    alkyl, (Chlor)-
    Benzyl, Phenyl-
    (subst.)

Herbizide

DOS   2 735 841   9. 8.77/16. 2.78   Ciba Geigy
= BE   857 684   SZ Prior. 12. 8.76

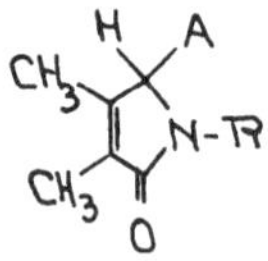

R = Aryl, Aralkyl,
    $(CH_2)_n$ -Heterocyclus
    z.B. 2-Pyridyl
A = Halogen, OH, O-Acyl

Fungizide und Herbizide

DOS   2 831 654   19. 7.78/15. 2.79   Chinoin Gyogyszer
= BE   869 236   25. 7.77/16.11.78   Chinoin Gyogyszer

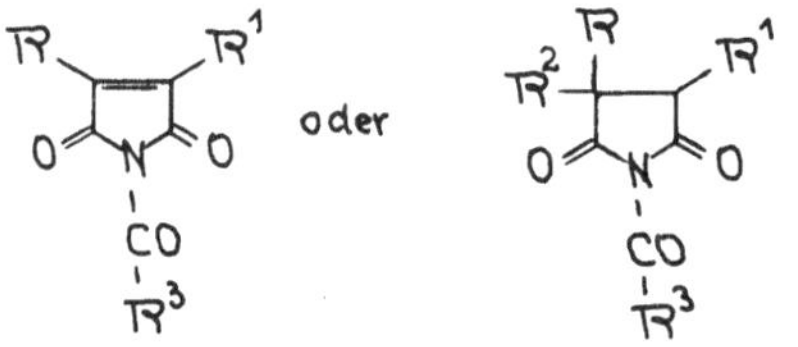

R = H, Halog., O(S)H
    O(S)-Alkyl
(Cyclo)Alkyl, Aryl,
Aralkyl,Heterocycl.

R^1 = H, Alkyl, Aryl,
    Aralkyl, N< ,
    Heterocycl.
R^2 = H, Alkyl, Aryl,
    Aralkyl, N< ,
    Heterocycl.,Halog.
R^3 = H, Alkyl, Aryl,
    Aralkyl, Heterocycl.

oder

DOS   2 831 770   19. 1.78/ 1. 2.79   Ciba Geigy
                                                 SZ Prior. 22. 7.77

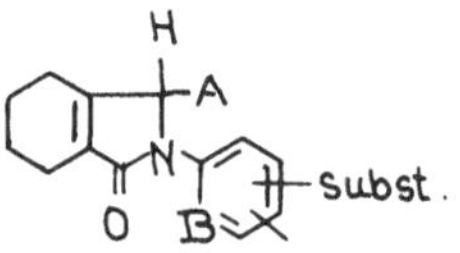

A = H, Halog.,
    $O(S)CO(S)-(O)_{0-1}$-Alkyl

B = -CH=,-N=

348

DOS   2 907 688   27. 2.79/ 6. 9.79   Shell Int. Res.
= BE   874 470   1. 3.78/27. 8.79   GB Prior. 1.3. u. 2o.4.78

substituiert durch H, Halog.,
                          O-Alkyl

(H,CO-O-Alkyl,Alkyl usw.)

DOS   2 908 254   2. 3.79/24.10.80   Philagro
                                                 FR Prior. 2. 3.78

Keine Patentbeispiele mit O
statt Phenyl auch Heterocyclen.

US   3 988 350   29.10.71/26.10.76   Gaf Corp.
US   3 988 351   29.10.71/26.10.76   Gaf Corp.

Phenol-Lactam-Komplexe
z.B. N-Methylpyrrolidon x 2,4,5-Trichlorphenol
oder Polylactame z.B. Hexamethylen-bis(-2pyrrolidone)x Phenole

Unter anderem auch Herbizid-wirksam.

US   3 992 189   31.10.75/16.11.76   Du Pont

B = -C- oder =C
(Iso-indol-1-one)

A = (CH_2)_n oder CH_2-CH=CH-CH_2

US   4 046 775   11. 4.73/ 6. 9.77   Sterling Drug.

(H,Alkyl) NH-Ar (subst.)

US   4 129 729   3. 1.78/12.12.78   Monsanto

Q = H, Alkyl,Phenyl(subst)

JA   75 2 010 266   14. 7.75/26. 1.77   Nippon Soda

gegen Gräser

c)   5-Ringe 1 x S   Bd.5/602

US   4 049 421   9. 4.75/20. 9.77   Chevron Res.

u. a. auch herbizid
wirksam

BE   869 383   28. 7.77/29. 1.79   Philagro

SU   495 314   6. 8.74/21. 6.76   As. Sibe

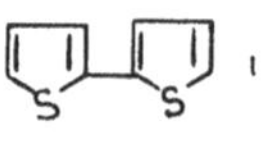 , 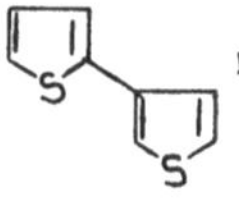 ,

u. a. auch Herbizide

d) **5-Ringe 1 x N + 1 x O bzw. S**

d1) **1 x N + 1 x O**

DOS   2 605 586   12. 2.76/26. 8.76   Stauffer Chem.
= BE   845 577   28. 8.75/28. 2.77   US Prior. 14.2. u. 3.11.75

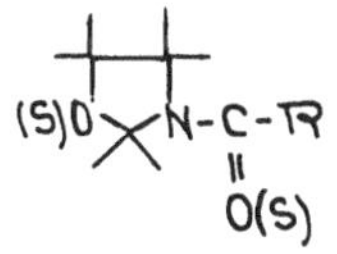

subst. durch H, Alkyl, Phenyl,
O-Alkyl, Alkylol

R = Halog.-Alkyl, Chlor-
    alkenyl., S-Alkyl

Gegen Herbizidschäden durch
Thiocarbamate oder $Cl-CH_2-CO-$
Anilide.

DOS   2 620 101   6. 5.76/18.11.76   Prod. Chim. Ugina
                                               Kuhlmann

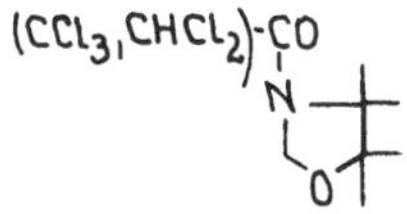

Antidot

DOS   2 637 580   20. 8.76/17. 3.77   Stauffer Chem.

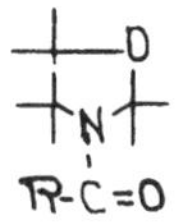

R = C_1-C_{10}Alkyl, Halog.-Alkyl,S-Alkyl

Antidots gegen Herbizid-Schäden
auch selbst herbizid wirksam.

DOS   2 820 655   11. 5.78/16.11.78   Monsanto
= BE   866 975   12. 5.77/13.11.78   Monsanto

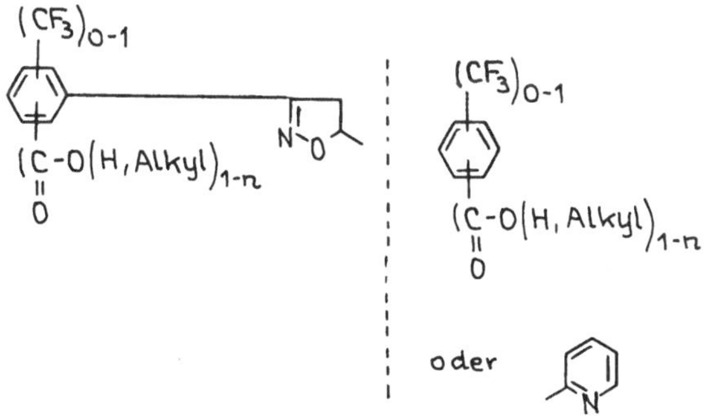

DOS   2 922 270   31. 5.79/ 6.12.79   Eszakmagyarurszagy Ung.

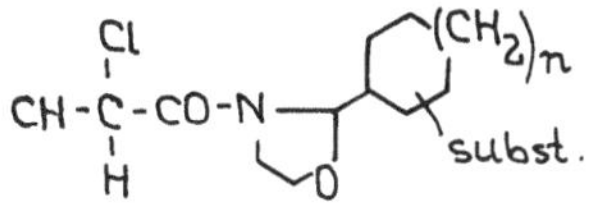

Antidot

EP   13 111   17.12.79/ 9. 7.80   Monsanto
                                                US Prior. 20.12.78

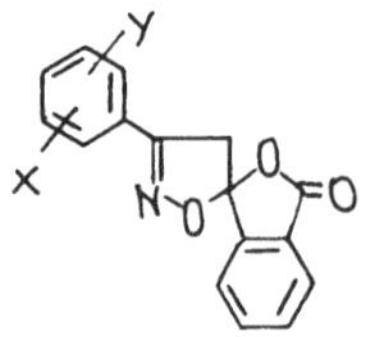

                                        X u. Y = H, Halog,
                                                Alkyl, O-Alkyl,
                                                O-Phenyl,
                                                Phenyl, CN

US   4 012 261   4. 3.74/15. 3.77   Tenneco Chem.

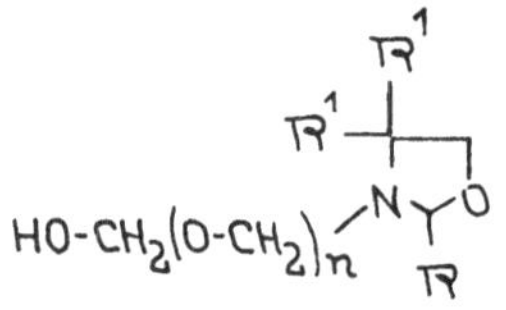

                                        R = H, Alkyl, subst. Phenyl,
                                            $(CH_2O)_m-CH_2OH$

                                        R^1= Alkyl, CH_2OH

Gegen oberfl. Wasserpflanzen.

US   4 022 607   9. 6.75/10. 5.77   Sandoz

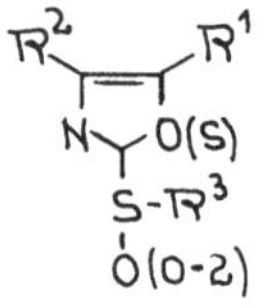

                                        R^1 u. R^2 = Alkyl, Phenyl(subst.)

                                        R^3 = Alkyl

US   4 038 284   28. 7.75/26. 7.77   Stauffer Chem.

neue Darstellungsverfahren

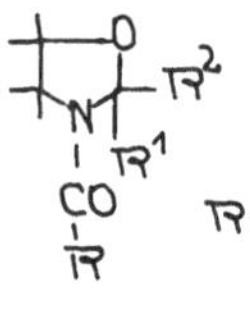

                                        R^1 u. R^2 = H, Alkyl, O-Alkyl,
                                                Alkyl-OH

                                        die anderen Substituenten:
                                        H, Alkyl, O-Alkyl, Alkyl-OH

R=Alkyl, Halog.-Alkyl, S-Alkyl

Siehe DOS 2 341 810
     US   3 959 304

Antidots u. z. T. Herbizide

352

US   4 062 670   31.12.75/13.12.77   Shell Oil

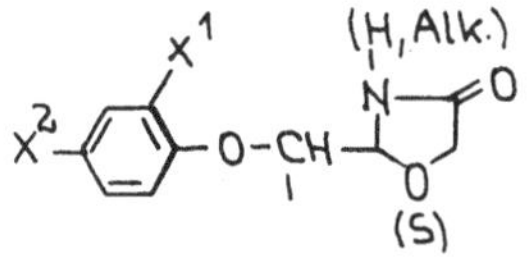

X^1 = H, Halog., CN,
    (O,S)-Alkyl, Alkyl
X^2 = H, Halog.,(O)-Alkyl

US   4 062 671   12.11.75/13.12.77   Amer. Cyanamid

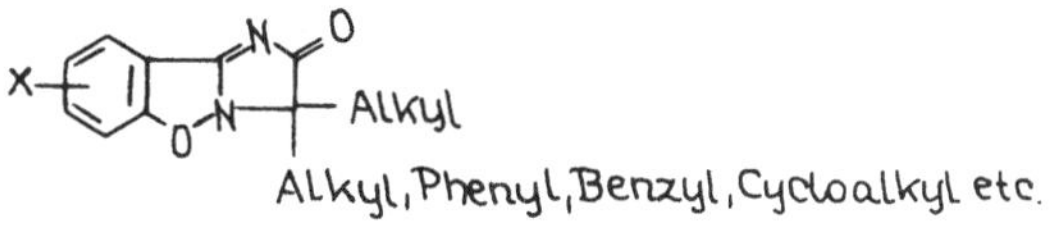

X = H, CH_3, NO_2,Cl,OCH_3
   usw.

US   4 064 384   21.10.75/20.12.77   Dow Chemical

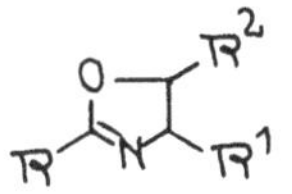

R = 1-Pyrrolidino, 1-Piperidino
    4-Morpholino, 1-Hexamethylenimino
R^1= H, Alkyl; R^2= H, Alkyl, Phenyl

US   4 065 463   11.12.75/27.12.77   Eli Lilly

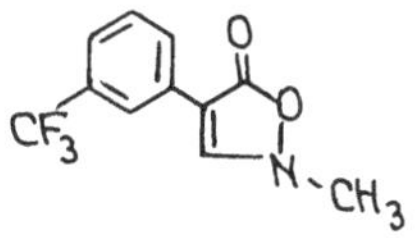

pre emergence herbizide

US   4 088 770   4. 5.71/ 9. 5.78   Eli Lilly

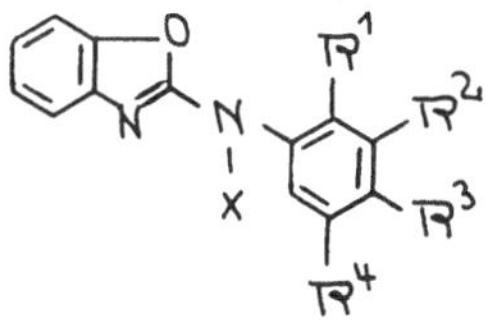

R^1 - R^4= Halog, CF_3,
    (O)-Alkyl,
    S-Alkyl, NO_2
    usw.
X = H,(O)-Alkyl, COOH,

   N<$^H_{Alkyl}$ usw.

u. a. auch Herbizid

US   4 139 366   12. 5.77/13. 2.79   Monsanto

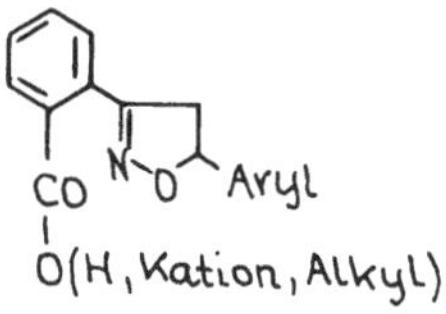

US   4 209 629   20.12.78/24. 6.80   Monsanto

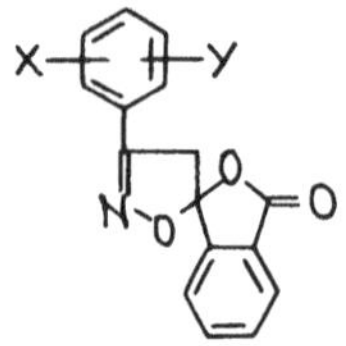

X u. Y = H, Halog., (O)-Alkyl,
          Halog.-Alkyl, CN,
          (O)-Phenyl

BE   841 320   7. 5.75/ 3.11.76   Ugine Kuhlmann

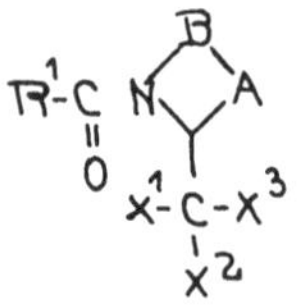

R^1 = Alkyl, Halog.-Alkyl,
       Alkenyl, Halog-Alkenyl,
       Aralkyl usw.

A = O,S,SO, SO_2, $\geq$ N- X^1 = Ce, Br, F

B = C_nH_2Alkyl   , Z = 2_{n+1} usw. X^2 u. X^3 = H, Ce, Br. F

Beispiel: Herbizid-Antidots

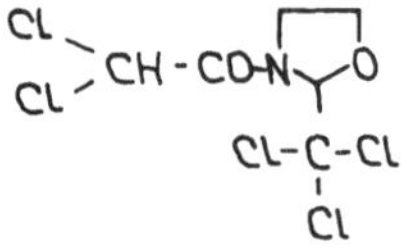

US   4 226 612   2. 7.79/ 7.10.80   Shell Oil Co.

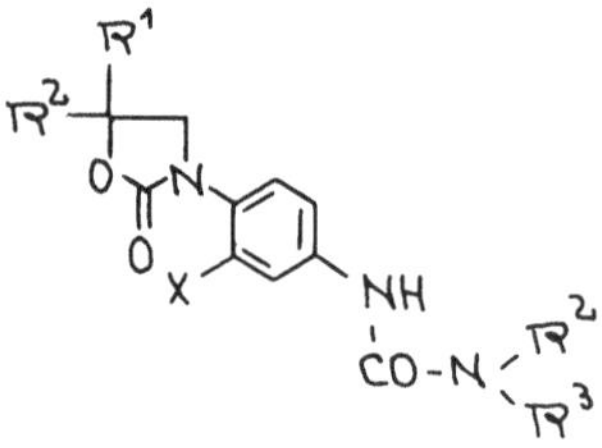

R u. R^1 = J, C_1-C_6-Alkyl
R^2 = C_1-C_6-Alkyl
R^3 = C_1-C_6-Alkyl,
       C_1-C_6-Alkoxy
X = Cl oder Br

354

CA   1 002 336   9. 8.71/28.12.76   Uniroyal

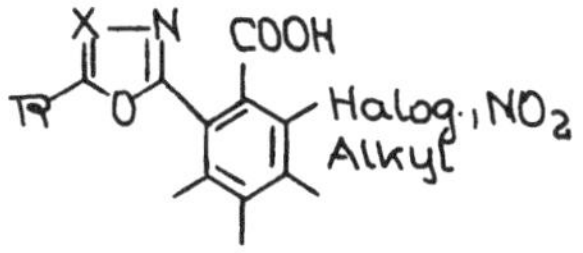

R = H, Alkyl, Heterocyc.
    Phenyl

$X = -\overset{|}{C}=$
(H, CH_3)

FR   2 300 084   10. 2.75/ 8.10.76   Stauffer Chem.

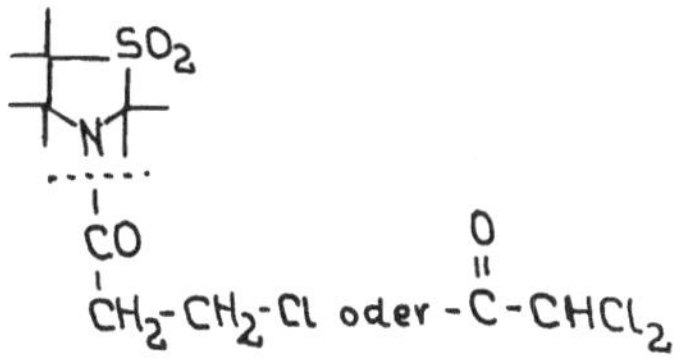

Antidots

JA   3 086 033   0. 0.77/29. 7.78   Shionogi

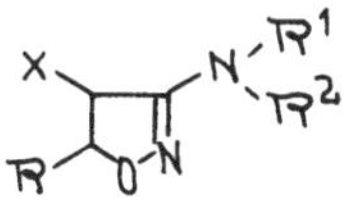

R = H, Alkyl, Cycloalkyl, Aryl
R^1= H, Alkyl; Alk(en,in)yl,
      Aryl, Aralkyl

R^2= Acyl, $-CO-N<$, $-CO-N\bigcirc$

X = H, Halogen, Alkyl

R u. X = Glieder (CH=CH) eines
         Ringes

JA   4 066 673   8.11.77/29. 5.79   Mitsui   Toatsu

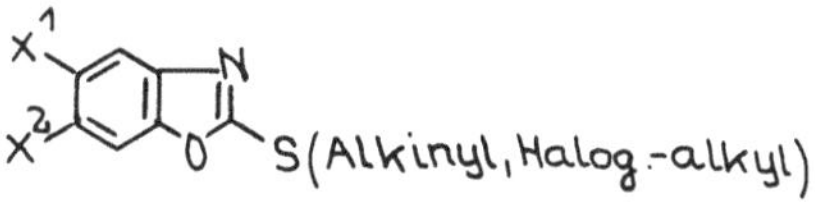

X^1 u. X^2u. Alkyl, Halog.,(CF_3)

Selektiv in Sojabohnen u. Erdnüssen

BE   855 602   14. 6.76/12.12.77   Eli Lilly

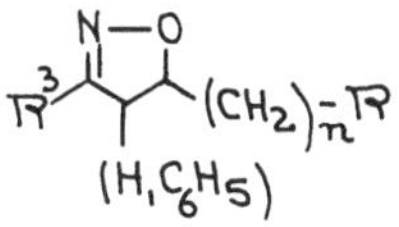

$R = N=C=S$, CN, $NHCS-N<^{R^1}_{R^2}$
R^3 = Phenyl, Pyridyl

Wasserherbizid

siehe auch unter d1; z.T. unter 1 x N + 1 x O mitbeansprucht

DOS   2 620 789   11. 5.76/25.11.76   Monsanto
= BE   841 692   US Prior. 12. 5.75

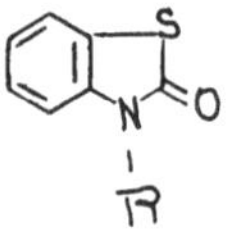

$R = (CH_2)_n CH, CH_2-COOH,$

$-CH(COOH)_2$ usw.

Erhöht Sacharosegehalt einiger
Pflanzen.

Siehe US 3 651 074 ähnliche Verb. als Herbizide

DOS   2 722 949   20. 5.77/ 1.12.77   Velsicol Chem.

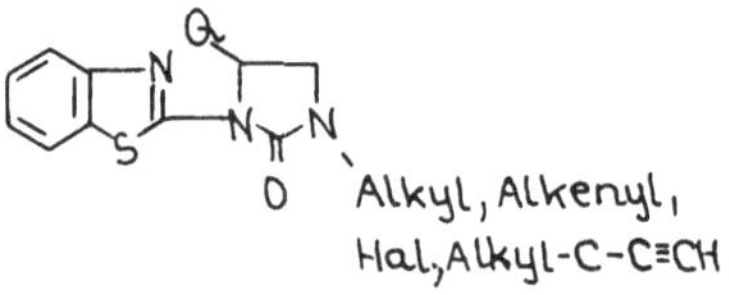

Q = OH, N$<$, O-CO-Alk(en,in)yl,
    Benzyl
X = (O)-Alkyl, Halogen

DOS   2 724 614   1. 6.76/15.12.77   Velsicol Chem.

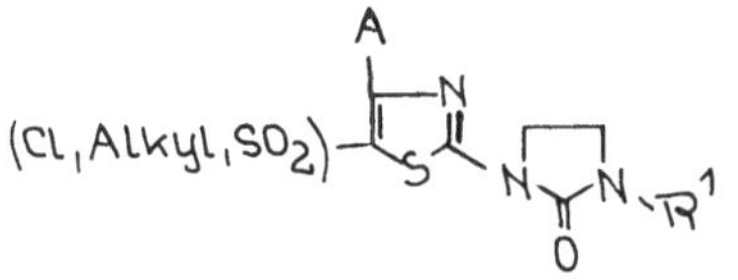

R^1= Alk(en)yl, C-C≡CH(Cl,Alkyl-
SO$_2$)-
A = OH, N$<$, O-CO-Alk(en,in)yl

DOS   3 007 556   28. 2.80/11. 9.80   Velsicol Chem.

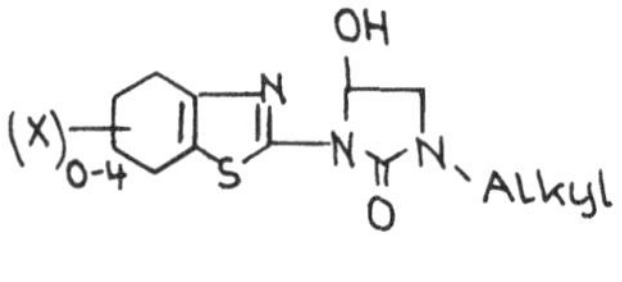

$X = (O,S)-Alkyl, N(Alkyl)_2,$
CH, Alk(en,in)yl

EP   1 349   19. 9.78/ 4. 4.79   Monsanto
                     (US 21. 9.77)

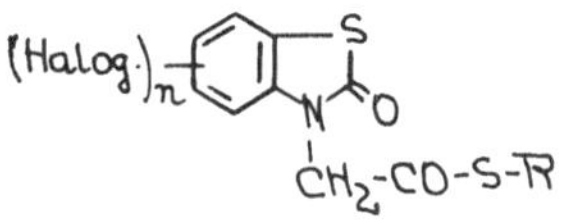

R = Alkyl, Benzyl, Chinolyl,
    subst.-Phenyl usw.

EP   4 129   31. 1.79/19. 9.79   ICI
                (GB   12.2.78 u. 17.2.78)

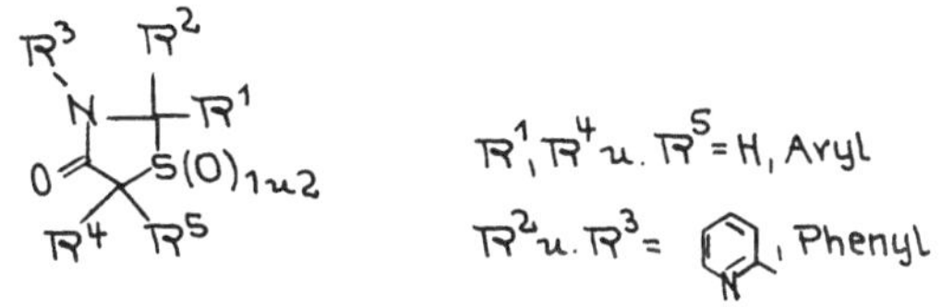

R^1, R^4 u. R^5 = H, Aryl

R^2 u. R^3 = Pyridyl, Phenyl

US   4 022 607   9. 6.75/10. 5.77   Sandoz

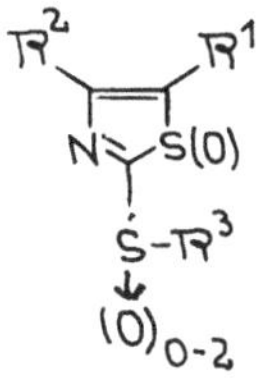

R^1 u. R^2 = Alkyl, Phenyl(subst.)

R^3 = Alkyl

US   4 045 446   17. 6.76/20. 8.77   Velsicol Chem.

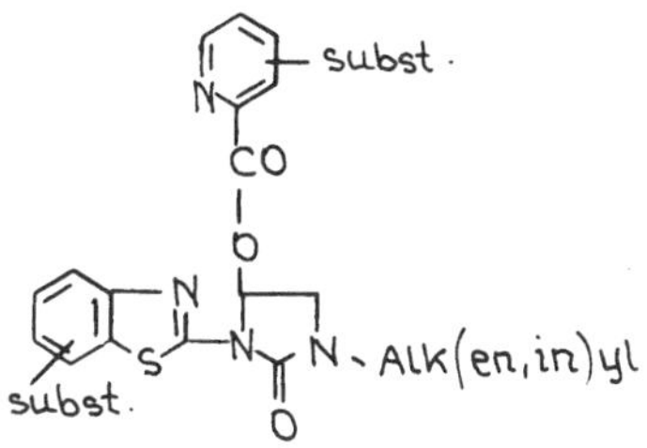

US   4 046 768   17. 6.76/ 6. 9.77   Velsicol Chem.

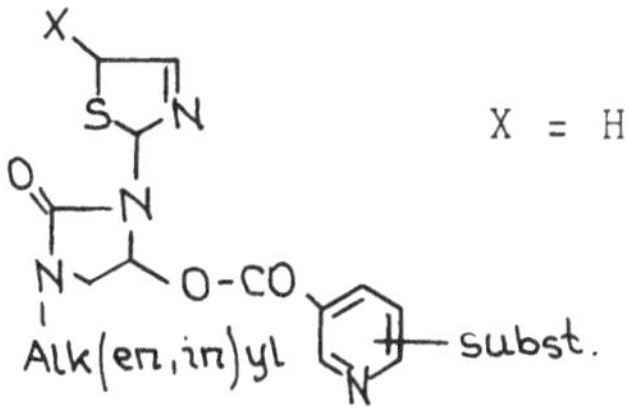

X = Halogen, $Alkyl-SO_2-$

US   4 032 322   18. 6.76/28. 6.77   FMC

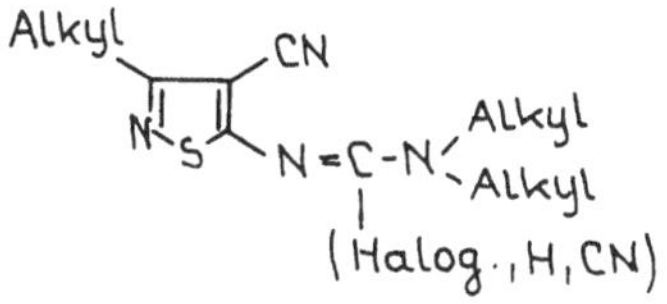

US   4 054 574   24. 5.76/18.10.77   Velsicol Chem.

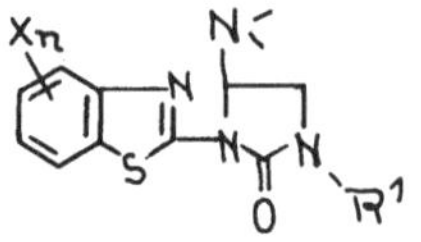

X = Halog.,CF$_3$,(O)-Alkyl;
    n=0-2
R^1= Alkyl, Alkenyl,
    Halog-Alkyl,
    $\overset{|}{\underset{|}{C}}$-C$\equiv$CH

US   4 086 240   1. 6.76/25. 4.78   Velsicol Chem.

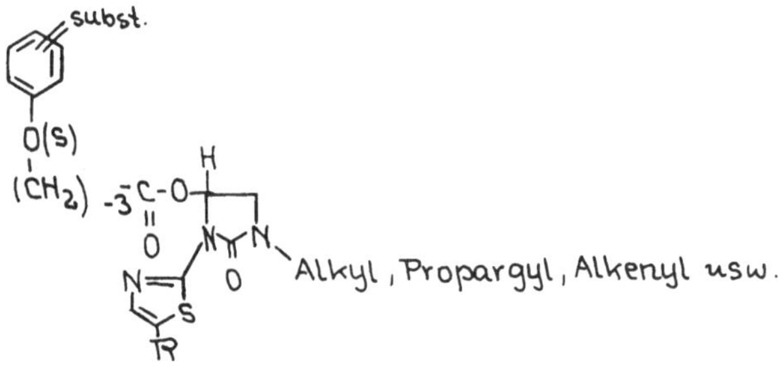

Alkyl, Propargyl, Alkenyl usw.

US   4 086 241   14. 4.76/25. 4.78   Velsicol Chem.

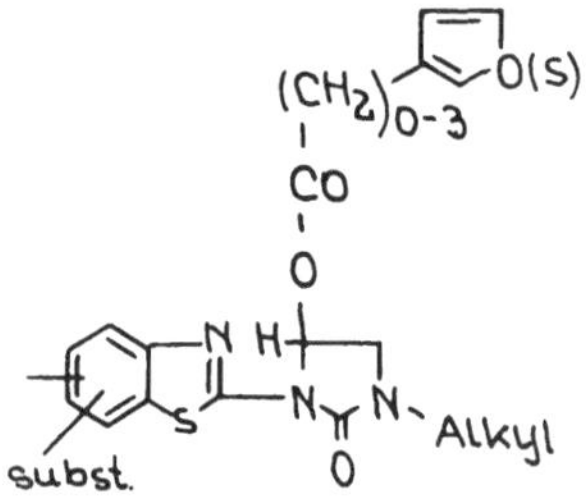

US   4 097 485   17. 6.76/28. 6.78   Velsicol Chem.

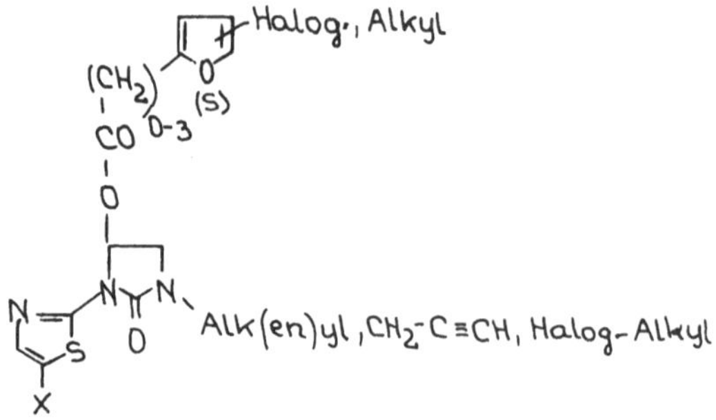

Alk(en)yl,CH$_2$-C$\equiv$CH, Halog-Alkyl

358

BE 880 849 25.12.79/24. 6.80 Monsanto
US Prior. 26.12.78

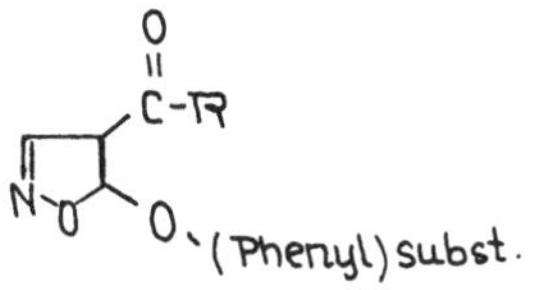

R = OH, O-Alkyl, O-Alkoxy-alkyl, N< , Cl

JA 54 145215 4. 5.78/13.11.79 Kuraray

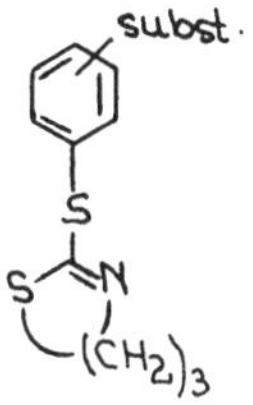

JA 54 163815 15. 6.78/26.12.79 Asahi Chem.

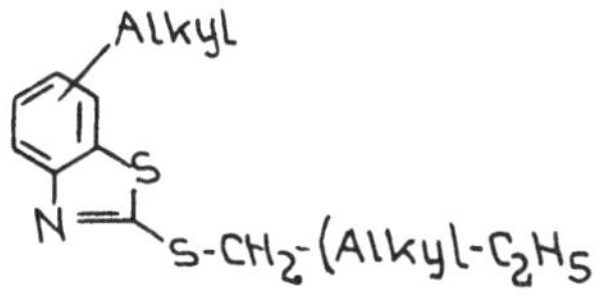

e) 2 x O oder 2 x S

Siehe auch unter "Cyclische Acetale von Di- und Tri-olen" 5.4.2.b Bd.5/605

EP 11 473 13.11.79/28. 5.80 Monsanto
US Prior. 15.11.78

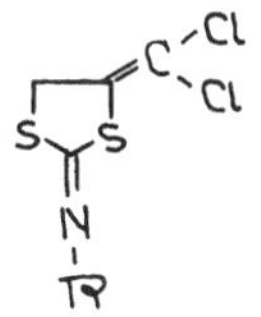

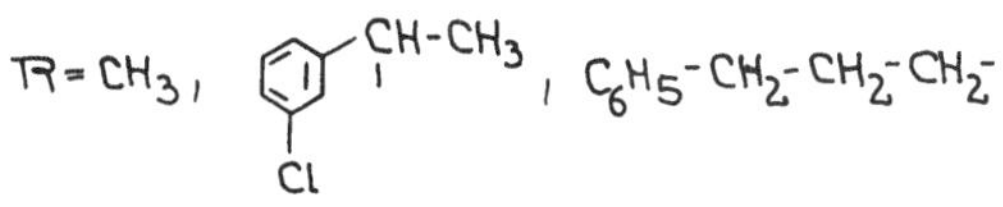

oder

Antidots für Thiocarbamate

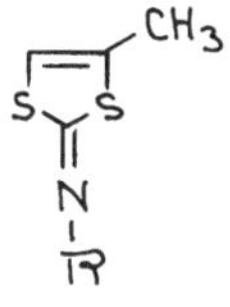

US   4 019 892   31.12.75/26. 4.77   Shell Oil

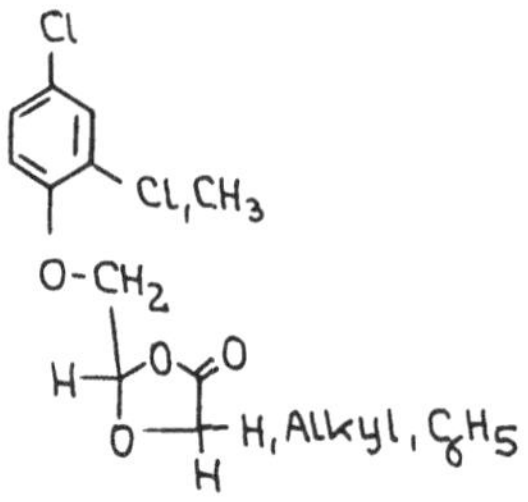

US   4 042 369   21. 6.74/16. 8.77   Shell Oil

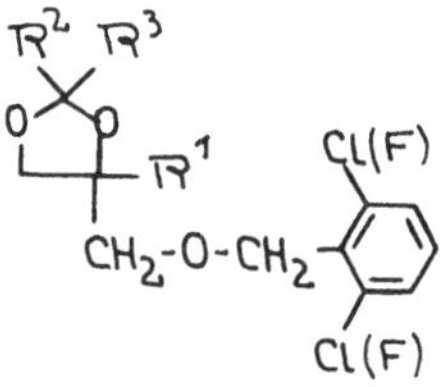

R^1 u. R^2 = CH_3, C_2H_5; R^3 = H, CH_3, C_2H_5

Spez. gegen Gräser in Weizen.

BE   846 802   30. 9.75/17. 1.77   Ansul

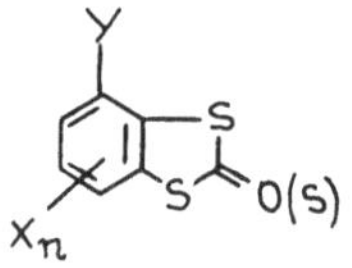

Y = H, CN, NO_2, CF_3, SO_2-Alkyl
X = Alk(en)yl, NO_2, CCl_3, CF_3,
    SO_2-CF_3, CO-N<
    N(Alkyl)$_2$,    N___O   usw.

DOS   2 655 843   9.12.76/23. 6.77   Ciba Geigy
CH Prior. 12. u. 24.12.75/15.11.76
s.a. DOS 2 655 843 Ciba-Geigy

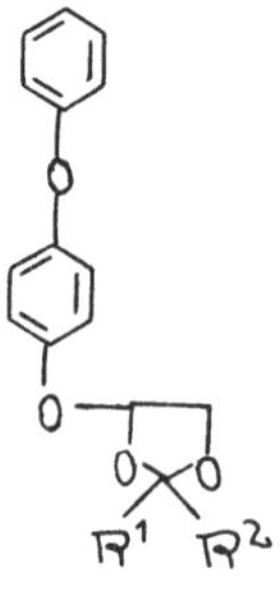

R^1 = H, Alkyl

R^2 = 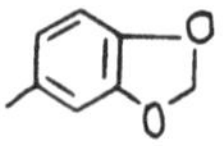   oder subst. Phenyl

360

f) 5-Ringe 1x N + 2 x S — Bd.5/606

EP 3 261 — 11. 2.78/ 8. 8.79 (US 30.12.77) — Chevron Res.

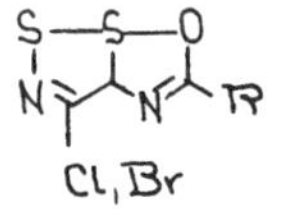

R = CH = CH_2, Alkyl, Alkyl-F, Cycloalkyl, Phenyl(subst.) CH_2-O-Phenyl usw.

g) 5-Ringverbindungen mit 2 x N — Bd.5/606

Pyrazole, Pyrazolone, Indazole, Imidazole, Benzimidazole, Imidazolone, Imidazolidinone, Imidazolidin-dione

g1) Pyrazole, Pyrazolone u. hydrierte Verbindungen (Pyrazine)

DOS 2 639 597 — 23. 9.75/24. 3.77 — Sankyo

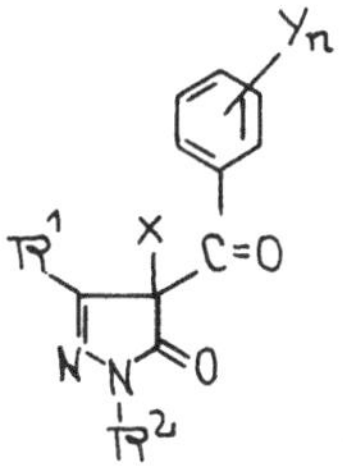

R^1 = H, Alkyl; R^2 = Alk(en)yl, X = Halog.; Y = Halog.,NO_2, (O)-Alkyl

n = 1-3

JA 51 138672 — 24. 5.75/30.11.76 — Sankyo
= DOS 2 627 223 — 18. 6.76/30.12.76 — Sankyo

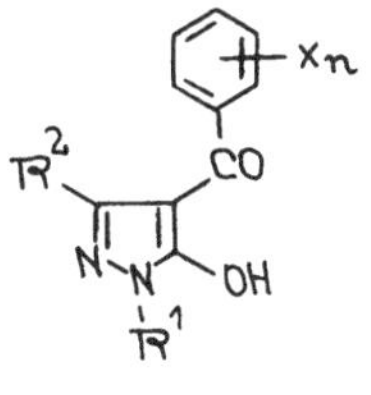

R^1 = Alk(en)yl
R^2 = Alkyl
X = Halogen

Darstellung u. Verw. als Herbizid.

GB 2 045 751 — 7. 3.79/ 5.11.80 — Ishihara Sangyo

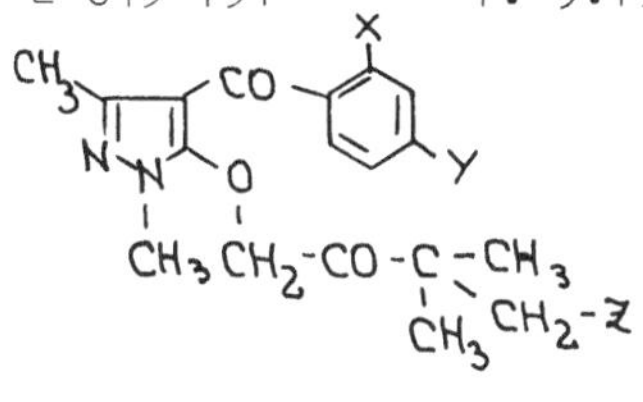

X u. Y = Cl, NO_2 od. CF_3
Z = H, Halog.

DOS 2 643 488 6.10.75/14. 4.77 American Cyanamid

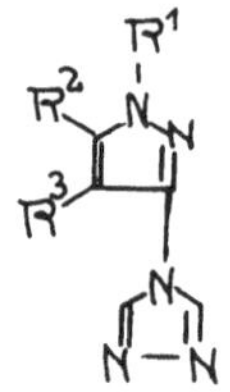

R^1 = H, C_1-C_3-Alkyl, Benzyl,

Acyl(C_2-C_4)

CO-N<

R^2 = H, CH_3, C_2H_5; R^3 = CH_3,

subst. Phenyl

DOS 2 646 628 15.10.76/21. 4.77 Du Pont
= BE 847 340 15.10.75 US Prior. 15.10.75; 12.12.75 u. 26. 8.76

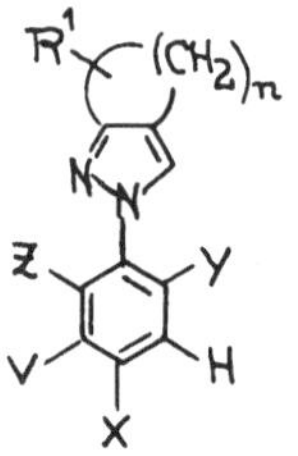

Cycloalkanpyrazole

R^1 = H, CH_3; Y = H, F, Cl;

Z = H, F

V = H, OCH_3, F, Cl

DOS 2 651 008 8.11.76/23. 6.77 Eli Lilly
= BE 849 092 US Prior. 11.12.76 u. 20. 9.76

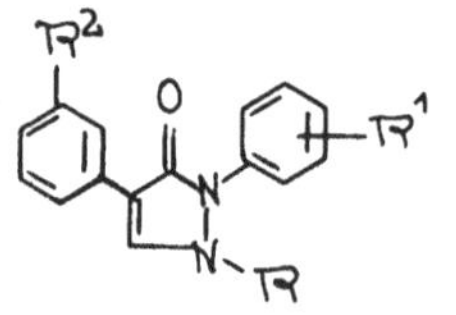

R = C_1-C_3-Alkyle

R^1 u. R^2 = H, Cl, Br, F, CH_3, CF_3

Darstellungsverfahren

siehe US 3 644 355, 3823 135

DOS 2 701 467 14. 1.77/28. 7.77 Du Pont
= BE 850 388 16. 1.76/14. 7.77 US Prior. 16.1.76; 6.2.76; 9.9.76; 24.9.76; 27.9.76

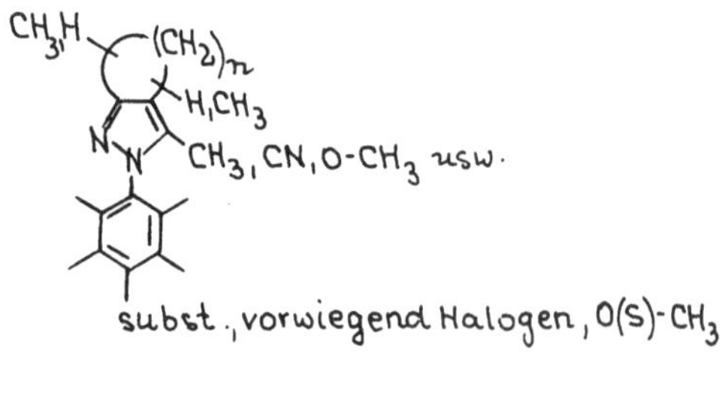
CH3,H
(CH2)n
H,CH3
CH3, CN, O-CH3 usw.
subst., vorwiegend Halogen, O(S)-CH3

108 Patentansprüche

DOS   2 747 531   22.10.77/26. 4.79   BASF

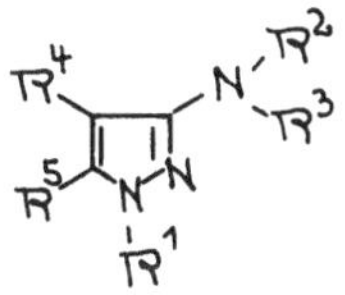

oder

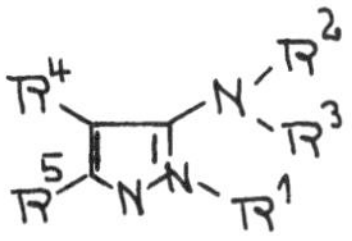

R^1 = H, Alkyl, Acyl, CN, Alkyl-SO_2, $>$N-SO_2, Alkoxycarbonyl usw.

R^2 = H, (O)-Alkyl, Acyl

R^3 = Alkyl, Alkoxy-carbonyl

R^4 = Halog.,CN usw.

R^5 = H, Alkyl, Halog., O-Alkyl, OH usw.

DOS   2 829 289   4. 7.78/24. 1.80   BASF

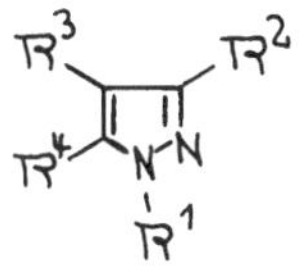

oder

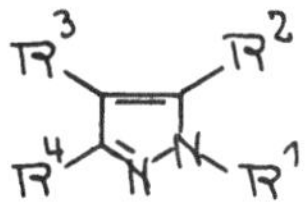

R^1 = H, Alkyl, CN, SO_2-Alkyl, ArSO_2 usw., CO(S)-Alkyl usw., CO-N$<$

R^2 = O(S)-Alkyl, Aryl usw.

R^3 = Halog., CN, NO_2. CO(S)-Alkyl, CO-N$<$, CO(S)-CH_2(Alkyl, Aryl u.

   Heterocycl.)

R^4 = H, Alkyl, Halog., O(S)-Alkyl usw.

EP   7 990   21. 6.79/20. 2.80   BASF
= DOS   2 819 289   DOS Prior. 4. 7.80

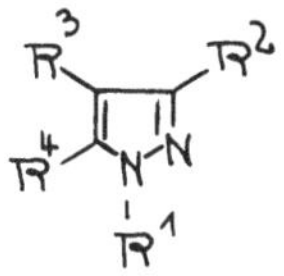

R^1 = H, CN, (subst.)Alkyl,
       SO_2HN-Alkyl, Alkyl-SO_2,

       ArSO_2, CO-O-Alkyl usw.

R^2 = O-(Alkyl, Aryl, Heteroc.),
       S-Alkyl usw.

R^3 = Halog., CN, NO_2,

       CO(S)-Alkyl, CO(S)-N$<$
       usw.

R^4 = H, O(S)-Alkyl, Phenyl usw.

US   3 979 409   20.10.75/ 7. 9.76   Amer. Cyanamid

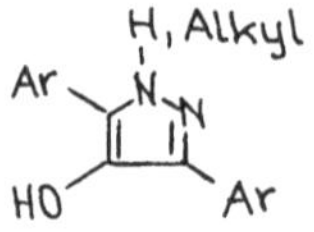

Darstellungsverfahren

US   4 008 249   27. 2.75/15. 2.77   BASF

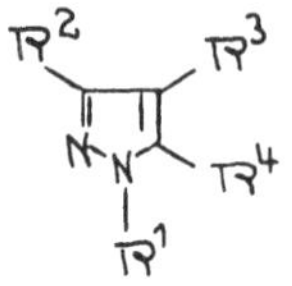

R^1 = H, Alkyl, subst. Phenyl,
    CO-N$<$ usw.
R^2 = subst. Phenyl
R^3 = H, Cl, Br, Alkyl
R^4 = S-Phenyl

US   4 021 227   22.11.74/ 3. 5.77   Amer. Cyanamid

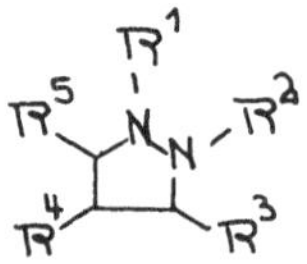

R^1 u. R^2 = C_1-C_4-Alkyl
R^3 u. R^5 = C_3-C_7-Cycloalkyl,
    Benzyl, subst.Phenyl

Gegen Wildhafer und Unkräuter.

US   4 042 372   2. 6.76/16. 8.77   Du Pont

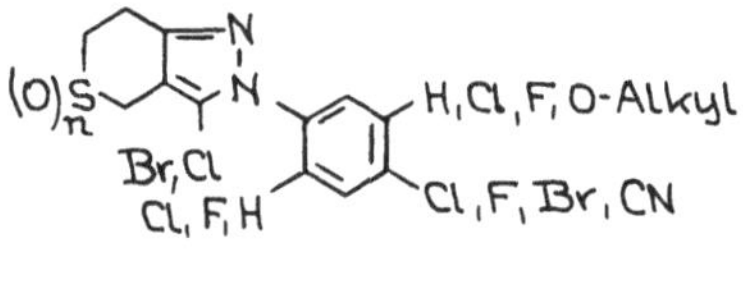

n = 0-2

Selektiv in Reis und Weizen.

US   4 066 776   15. 3.71/ 3. 1.78   Eli Lilly

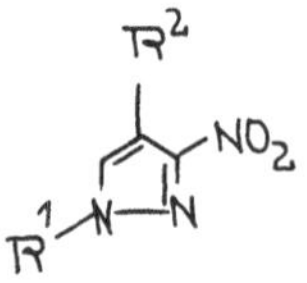

R^1 = H, Alkyl, Alk(en,in)yl, Cycloalkyl, O, (S)-Alkyl, CN,
    Nitrophenyl usw.

US   4 084 055   24. 3.77/11. 4.78   Du Pont

Indazol-Derivat

US   4 084 955   19. 2.70/18. 4.78   Upjohn

$X = Cl, Br$

Gegen Gräser, Ackerfuchsschwanz, Wildhafer.

US   4 087 612   22.11.74/ 2. 5.78   Amer. Cyanamid
     Teil aus 3 948 936

R = Cycloalkyl, Benzyl, Alkyl
R^1 = H, Alkyl, O-Alkyl

US   4 089 672   19. 2.70/16. 5.78   Upjohn

R = H, Alk(en)yl
R^1 u. R^2 = H, Alk(en,in)yl,
     Aralkyl, Aryl,
     Cycloalkyl.
R^1 u. R^2 = Glieder
     eines 3-7-Ringes

US   4 115 096   4.11.77/19. 9.78   FMC

US   4 124 373   3. 1.77/ 7.11.78   Du Pont

GB   2 002 375   12. 8.77/21. 2.79   Ishihara Sangyo

Gegen Gräser wirksam.

JA   511 06738   12. 3.75/21. 9.76   Sankyo KK

R^1 = H, Alkyl
R^2 = Alk(en)yl
X = Cl, NO_2, n=1-4

Gräserherbizid

JA   51 146464   7. 6.75/16.12.76   Sankyo

R^1 = Alkyl, CH_3O-CO-CH_2-
R^2 = Alkyl; X = Halog.,
NO_2

JA   52 083552   1. 1.76/12. 7.77   Mitsubishi

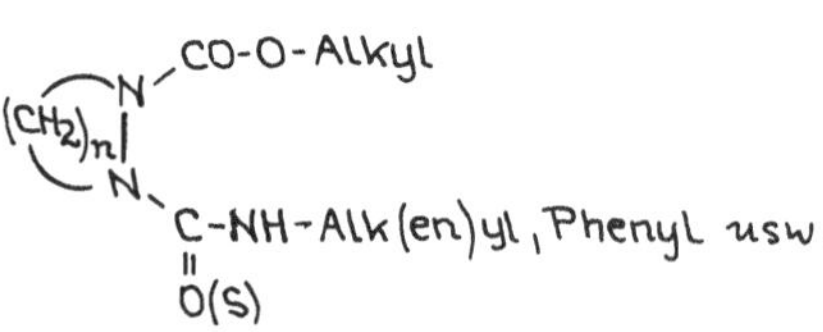

JA   52 151163   9. 6.76/15.12.77   Sankyo

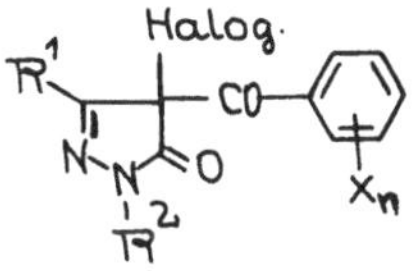

X = Halog., NO$_2$, Alkyl,O-Alkyl,

n=0-3

R^1 = H, Alkyl; R^2 = Alk(en)yl

JA   53 018566   30. 7.76/20. 2.78   Sankyo

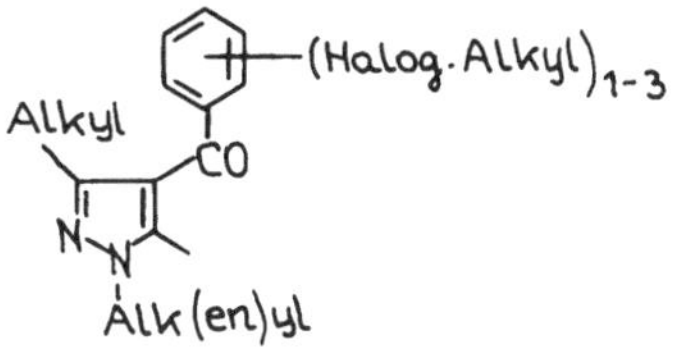

JA   54 009279   24. 6.77/24. 1.79   Ishihara Ind.

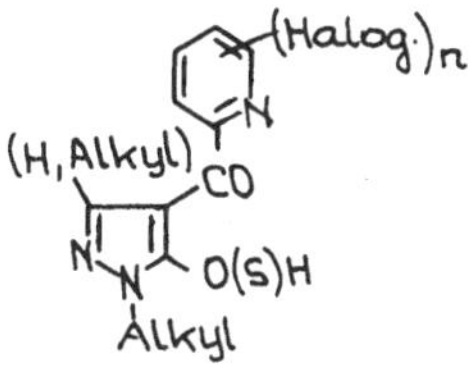

JA   55 033413   29. 8.78/ 8. 3.80   Ishihara Sangyo

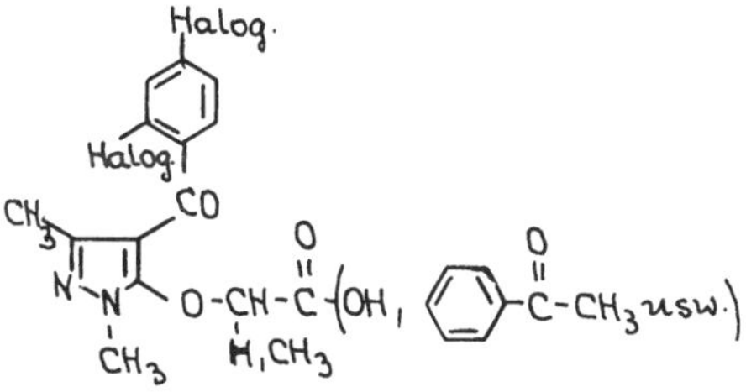

JA   55 033454   1. 9.78/ 8. 3.80   Sankyo

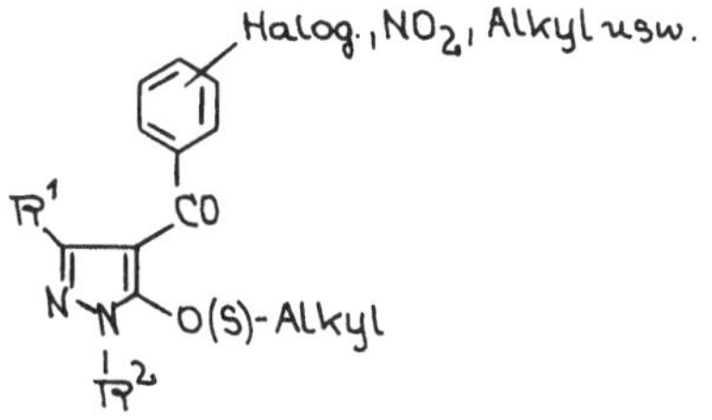

R^1 = H, Alkyl usw.
R^2 = H, Alk(en)yl,
        Phenyl

JA   55 035035   4. 9.78/11. 3.80   Sankyo
JA   55 035037/38/39 4. 9.78/11. 3.80   Sankyo

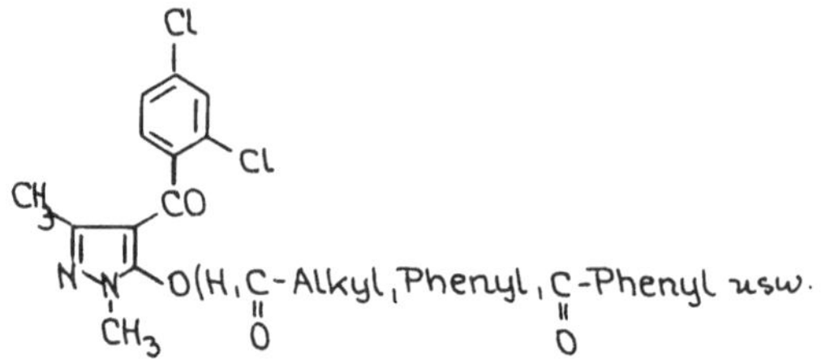

O(H,C-Alkyl,Phenyl,C-Phenyl usw.

JA   55 035036   4. 9.78/11. 3.80   Mitusi Toatso
JA   55 064573   8.11.78/15. 5.80   Ishihara Sangyo

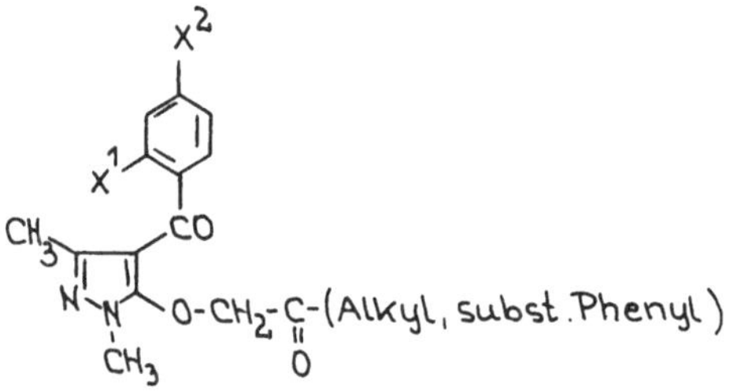

X^1 u. X^2 = 1XHalog.,
1XCF$_3$

O-CH$_2$-C-(Alkyl, subst. Phenyl)

JA   55 055105   20.10.78/22. 4.80   Sankyo

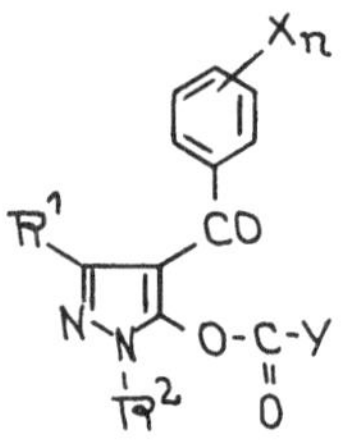

R^1 = Alkyl
R^2 = Alkyl, Alkenyl
X = Halog., NO$_2$,
   (O)-Alkyl; n=1-3
Y = Heterocycl. Ring

JA   55 073660   28.11.78/ 3. 6.80   Ishihara Sangyo

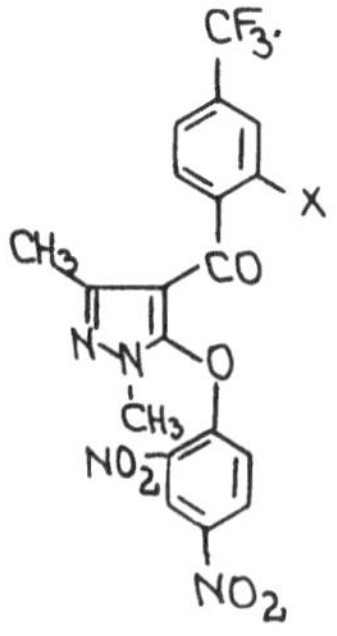

X = H, Halog.

Selektiv in Reis.

368

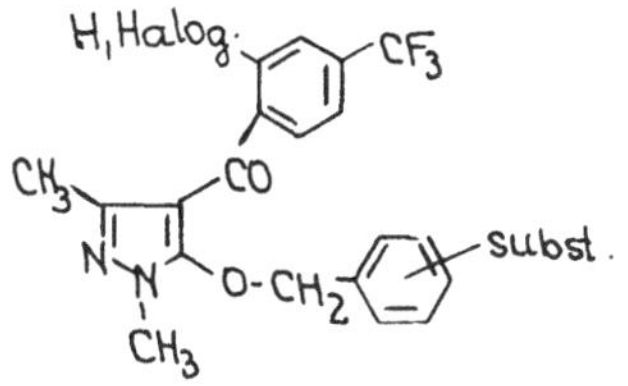

g2) <u>Imidazole</u>   Bd.5/608

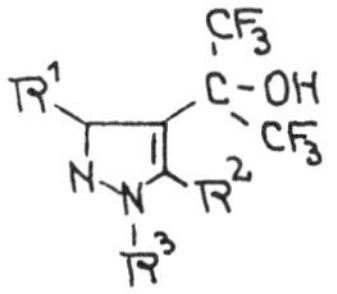

DOS   2 610 527   12. 3.76/29. 9.77   Bayer
= BE   852 313

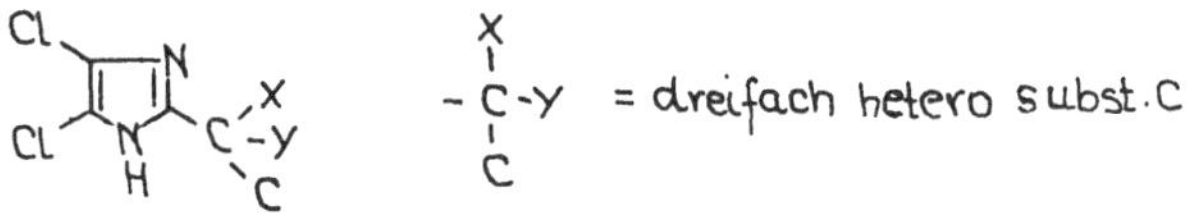

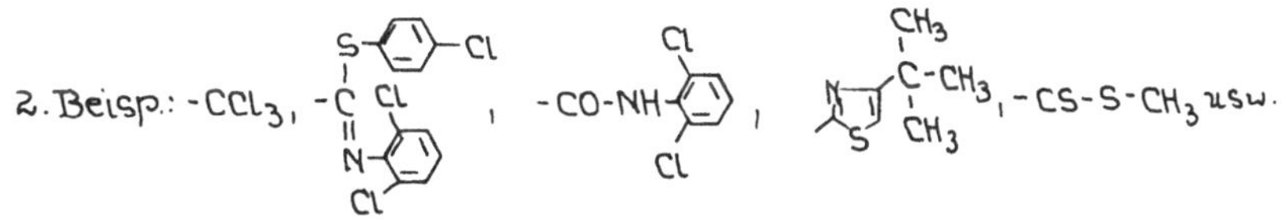

Sehr umfassendes Patent.

DOS   2 634 053   29. 7.76/23.2.78   Bayer

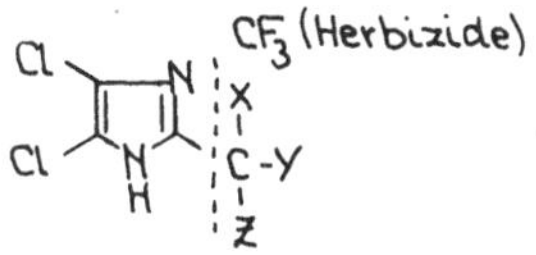

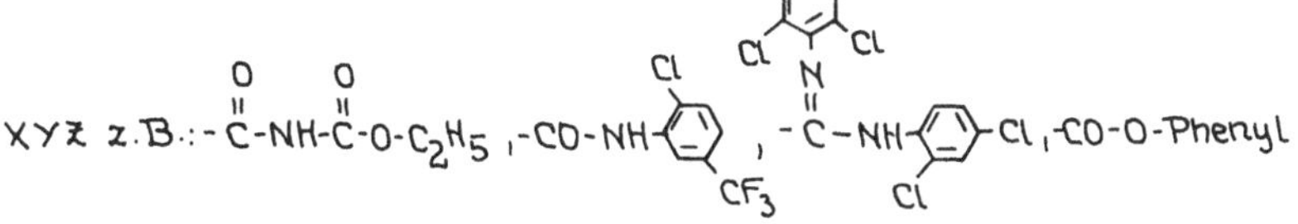

Insektizide, Fungizide, z. T. auch Herbizide.

DOS   2 646 142   13.10.76/20.4.78   Bayer

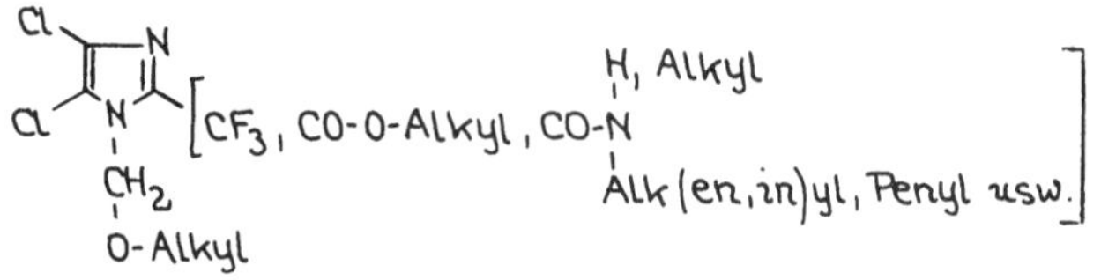

DOS   2 646 143   13.10.76/20. 4.78   Bayer

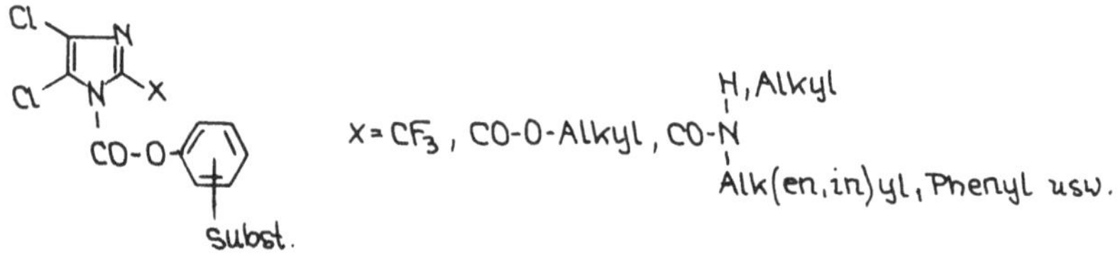

DOS   2 646 144   13.10.76/20. 4.78   Bayer

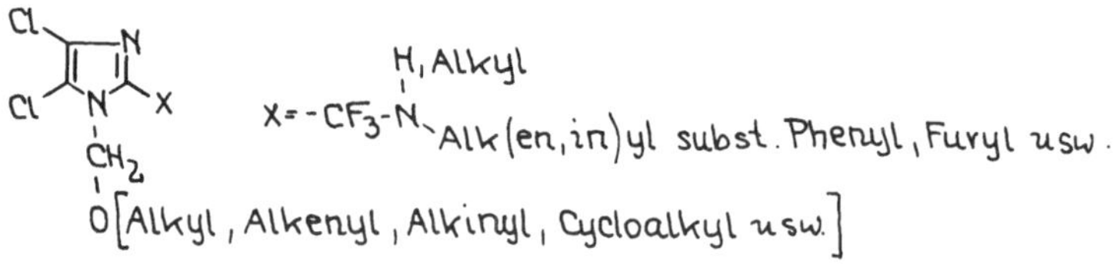

DOS   2 809 022   4. 3.77/ 7. 9.78   May und Baker

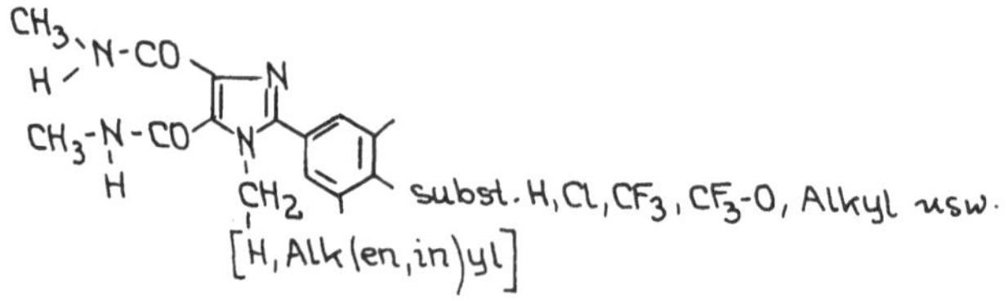

DOS   2 901 862   18. 1.79/31. 7.80   Bayer

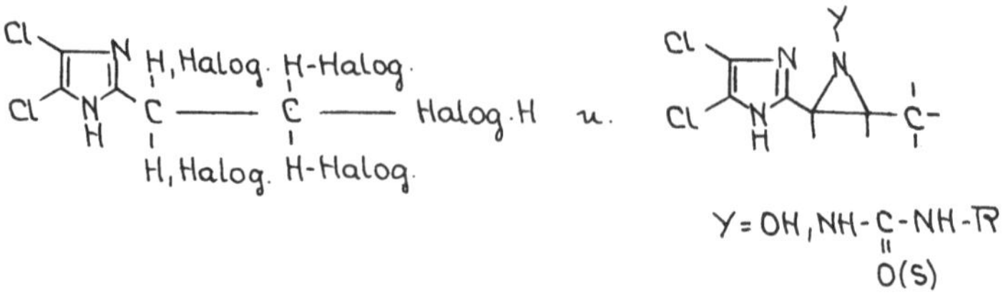

EP   373   6. 7.38/24. 1.79   Höchst
= DOS   2 732 531   19. 7.77

EP   9 298   27. 7.78/ 2. 4.80   ICI

R^1 u. R^2 = Alkyl, Phenyl-subst.
                 Cycloalkyl
Z^1 u. Z^2 = -CO- oder -CHOH-
Y = N- oder CN

2. Beisp.:

EP   13 905   10. 1.80/ 6. 8.80   Bayer
= DOS   2 901 862   18. 1.79

X^1 = H, Halog.
X^2, X^3 u. X^4 = H, Cl, Br
X^5 = H, Cl
X^1 u. X^3 = C-C Bindung
X^2 u. X^4 = C-C Bindung
X^1 u. X^2 = O oder HN-CO-NHR

US   4 220 466   3.11.75/ 2. 9.80   Gulf Oil

R^1 = CO-S(CH_3, C_2H_5)
R = H, Alkyl, O-C_2H_5,
        Phenyl usw.

371

BE   859 628   13.10.76/12. 4.78   Bayer

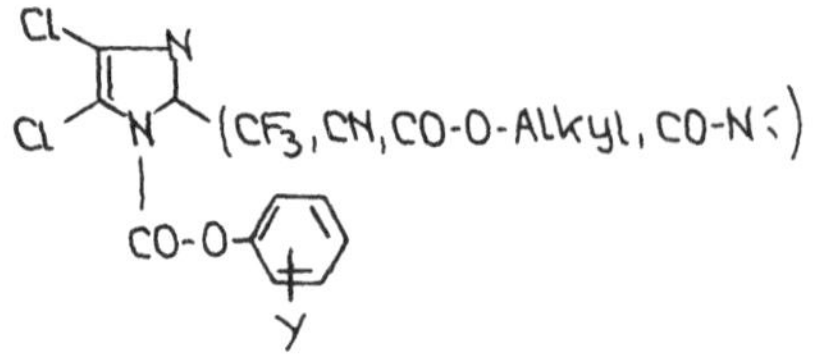

Y = Halog., C_1-C_6-Alkyl,
    O(S)-Alkyl, CF_3, NO_2,
    CN, COOAlkyl

BE   859 629   13.10.76/12. 4.78   Bayer

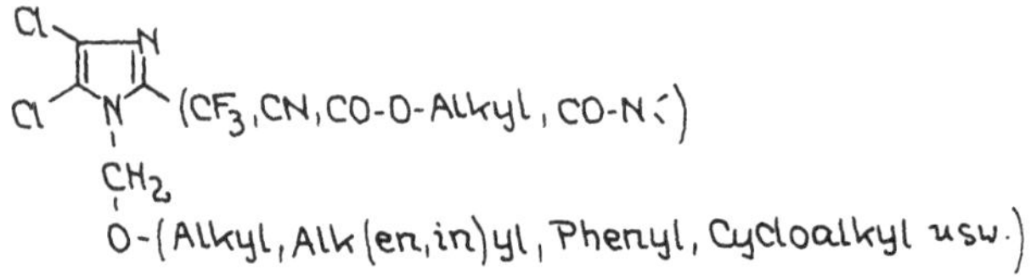

SU   505 637   24.12.74/21. 6.76   Museon Fine Ch. T.

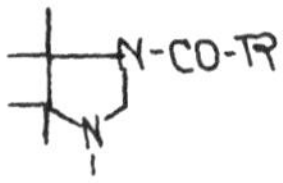

R = Ak(en)yl, Aryl, Arylkyl

u. a. auch Herbizide

SU   527 426   24. 5.75/20. 5.77   Kazan

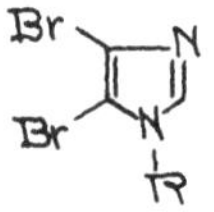

JA   53 065 879   24.11.76/12. 6.78   Nippon Soda

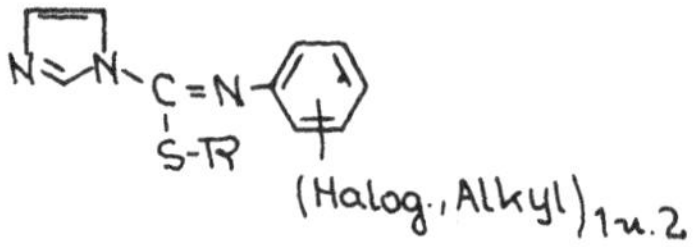

R = Alkyl

DOS   2 526 192   12. 6.75/30.12.76   Eli Lilly

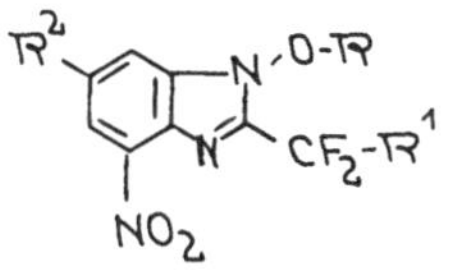

R^1 = H, F, CF_3 usw.

R^2 = CF_3, CF_2Cl, CHF_2

R^3 = Alkyl, Cycloalkyl,
        Benzyl, Phenäthyl,
        CO-OR, CO-N<

DOS   2 822 127   20. 5.78/21.12.78   US Borax
                                                   US Prior. 17. 6.77

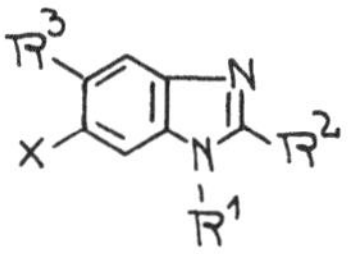

R^1 u. R^2 = Alkyl;

R^3 = Iso-Alkyl, Cycloalkyl

X = H, Halogen

Das Patent erfaßt aucht die N-Oxyde.

US   2 042 593   21.12.70/16. 8.77   Eli Lilly
US   4 000 145   21.12.70/28.12.76   Eli Lilly
Siehe US 3 932 428 u. 3 818 022

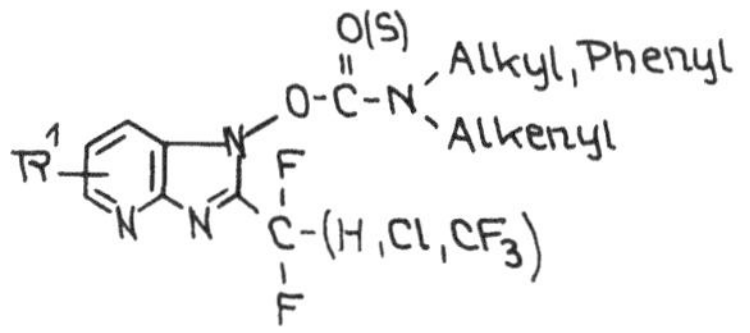

R^1 = Halog. NO_2

CF_3,CH_2-SO_2-Alkyl

US   4 000 145   21.12.70/28.12.76   Eli Lilly

Teil des US 3 932 428 u. 3 818 022

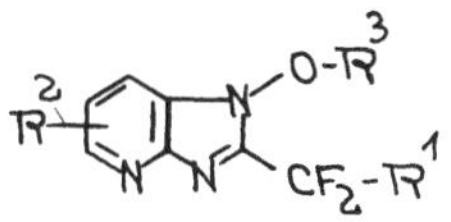

R^1 = H, Cl, F, CHF_3, CF_3 usw.

R^2 = Halog., NO_2, CF_3,

        Alkyl-SO_2 usw.

R^3 = CO(S)-N <Phenyl, Alkyl / Alk(en)yl

US   4 042 593   21.12.70/16. 8.77   Eli Lilly

$R = H, Cl, F, CHF_2, CF_3$ usw.

$R^1 = Halog., NO_2, CF_3, Alkyl-SO_2$
CHF_2

$R^2 = CO-O(S)-Alkyl, COO-(S)-$
Phenyl, SO_2-(Alkyl, Phenyl, Benzyl)

US   4 174 959   24. 4.78/20.11.79   US Borax

R^1 u. R^2 = H, Alkyl, N< usw.

R^3 = H, Alkyl, Phenyl
R^4 = Alkenyl
R^3 u. R^4 = Glieder eines Ringes

DDR   123 053   23.12.75/20.11.76   Grimmecke

oder

R^1 = Ar, ArCH = CH- oder heterocycl.-Ring, Ar(subst.)
R^2 = H, Alkyl, OAlkyl, Aryl, COOH, $CO-CH_3 NO_2 HN_2$ usw.
R^3 = H, $CO-CH_3$, $C_6H_5-CH_2-$

g4)   <u>Imidazolidin 2-one und 4-one</u>   Bd.5/614

Siehe auch unter „Überbrückte heterocyclisch-
aliphatische Harnstoffe", z.B. 5.5.3i

DOS   2 633 211   23. 7.76/27. 9.77   Sagami Chem. Res.

Neue Darstellung

Vorwiegend Wuchsregulatoren.

NO_2 siehe später

374

DOS   2 833 274   28. 7.78/22.2.79   Amer. Cyanamid

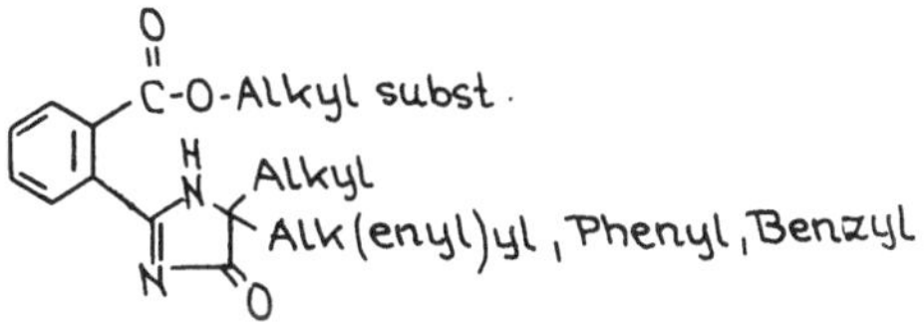

C-O-Alkyl subst.
Alkyl
Alk(enyl)yl, Phenyl, Benzyl

US   4 017 510   12.11.75/12. 4.77   Amer. Cyanamid

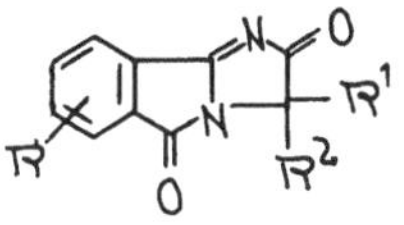

R   = H, CH_3, OCH_3, NO_2, Cl,

       SCH_3

R^1 = C_1-C_4-Alkyl; R^2=Alk(en)yl

       Phenyl, Benzyl

R^1 u. R^2 zus. Cycloalkylring

US   4 041 045   12.11.75/ 9. 8.77   Amer. Cyanamid

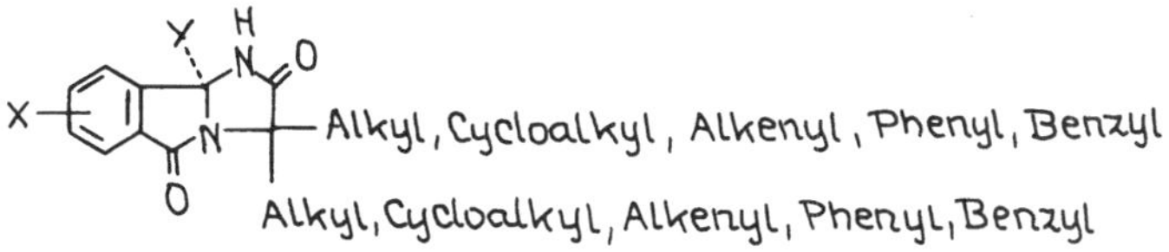

Alkyl, Cycloalkyl, Alkenyl, Phenyl, Benzyl
Alkyl, Cycloalkyl, Alkenyl, Phenyl, Benzyl

X = H, Cl, CH_3, NO_2O(S)-CH_3

Y = H, O(S)-Alkyl, N$\lessdot$

Herbizid- und Pflanzenwachstumsregulatoren.

US   4 122 275   8. 8.75/24.10.78   Amer. Cyanamid

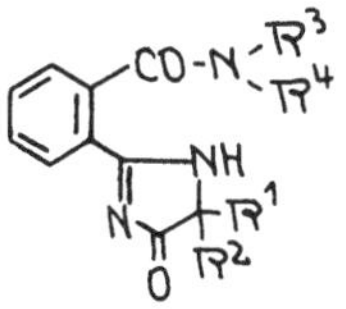

R^1 = Alkyl; R^2 = (Cyclo)-Alkyl
R^3 u. R^4 = H, Alkyl,
             Alk(en,in)yl, Benzyl

SZ   529 160   21. 2.75/14. 4.77   Faddeera Mi

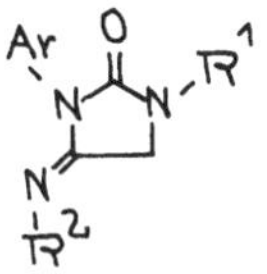

R^1 u. R^2 = Alkyl, subst. Phenyl

US 4 214 891 9. 8.77/29. 7.80 Du Pont

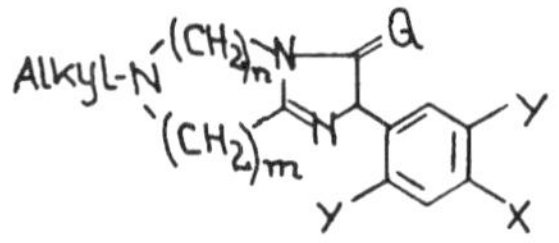

V = Halog., CH_3, O-Alkyl

X = Halog., CN, CH_3, NO_2, OCH_3

Y = H, Halog., CH_3

m + n = 0 = 4, Q = O,S

g6) <u>Imidazolidin-2,4-dione (Hydantoine)</u> Bd.5/615

DOS 2 604 989 9. 2.76/19. 8.76 Mitsubishi
JA Prior. 10.2; 30.9.1975

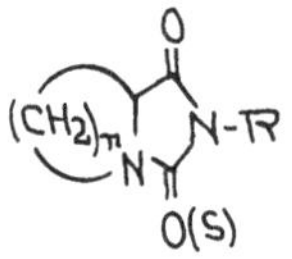

R = subst. Phenyl, z.B. 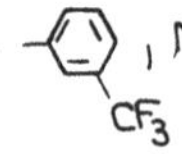, Naphthyl

n = 3 u. 4

Herbizide und z.T. Fungizide.

DDR 129 553 7. 2.77/25. 1.78 Werhan H.G.

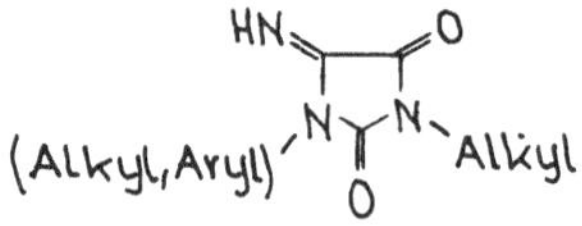

g7) <u>Imidazolidin-trione</u> Bd.5/617

DOS 2 916 647 25. 4.79/ 6.11.80 BASF

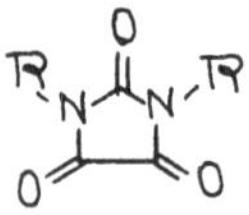

R^1 = (O)-Alkyl, Cycloalkyl, Alk(en,in)yl

R = 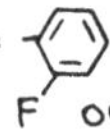 (subst.) durch CN, NO_2, Aryl, Aralkyl oder 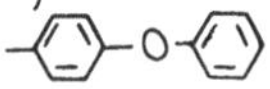

US 3 992 402 29. 3.72/16.11.76 Chevron Res.

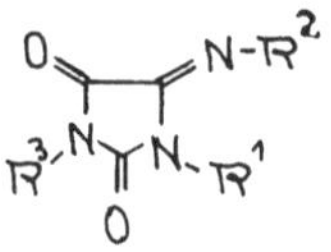

R^1 u. R^3 = Alkyl, Alkenyl, Naphthyl

R^2 = $CCl_2{=}CCl-$

Herbizide und Fungizide

h) <u>5-Ring-Verbindungen 2 x N + 1 x O</u>   Bd.5/618

h3) <u>1,2,4-Oxadiazolidine</u>   Bd.5/620

BE   873 447   12. 1.78/12. 7.79   Schering

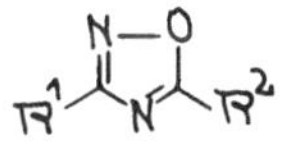

R^1 u. R^2 = (Cyclo)-Alkyl, Halog-alkyl, Phenyl (subst)

z.B. —NH-CO-(Alkyl, N<)

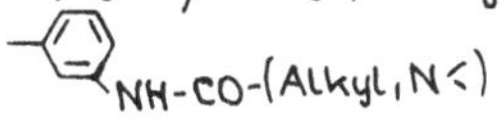

US   4 000 155   11.12.75/28.12.76   Eli Lilly

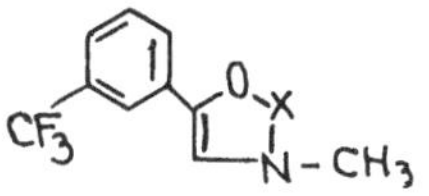

X = N-CH$_3$,

h4) <u>1,2,4-Oxadiazolidin-mon- und di-on(e)</u>   Bd.5/620

DOS   2 625 848   9. 6.76/16.12.76   Eszakmagyarorszagi.

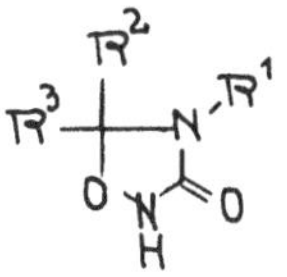

R^1 = Phenyl oder Naphthyl subst.
R^1 u. R^3 = Alkyl oder
              $(CH_2)_n$-Ring

US   4 088 472   13.12.65/ 9. 5.78   Monsanto

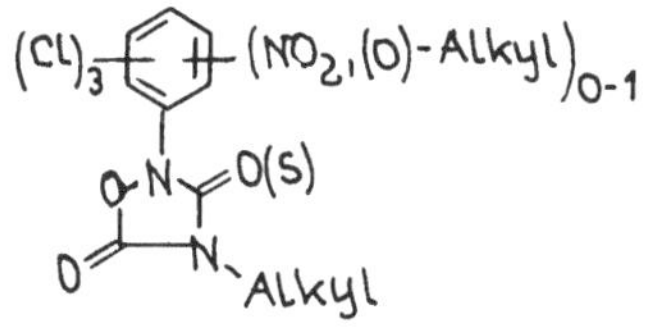

BE   842 740   9. 6.75/10.10.76   Eszakmagyar Vegymer

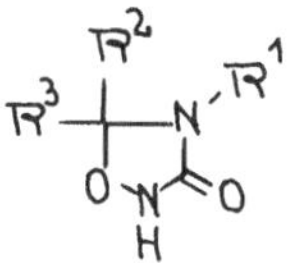

R^1 = Phenyl, Naphthyl subst.
R^2 u. R^3 = Alkyl, Aryl oder
              zusammen $(CH_2)_n$

DDR   121 518   30. 5.75/ 5. 8.76   Werchan H. G.

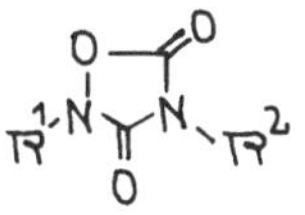

R^1 = Alkyl, Cycloalkyl, Aryl,
      Aralkyl-subst.
R^2 = Alkyl

i1)   1,3,4 Thiodiazolidine u. Thiodiazolidinone u. 1,2,3
       Thiodiazole

       Heterocyclisch-aliphatische Harnstoffe (5.5.3.i.5.3.)
       welche durch zweifache Kondensation im Harnstoffteil
       zu. sog. "überbrückten" Harnstoffen führen, wobei ein
       neuer heterocyclischer 5 oder 6 Ring entsteht, werden
       entsprechend dem Ausgangsheterocyclus registriert.

DOS   2 548 847   31.10.75/28.10.76   Velsicol Chem.
= US   4 063 924   29.10.76/20.12.77   Velsicol Chem.
u.US   4 097 486   29.10.76/28. 6.78   Velsicol Chem.

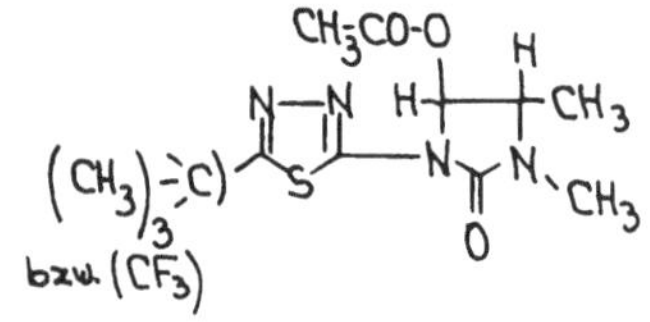

DOS   2 614 831   6. 4.76/20.10.77   Bayer
= BE   853 304

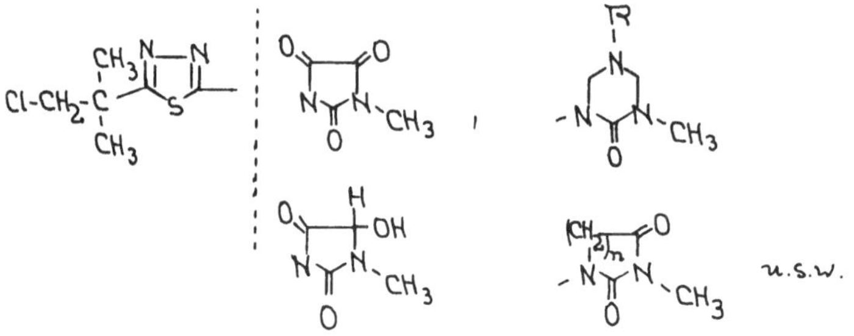

DOS   2 708 243   25. 2.77/22. 9.77   Velsicol Chem.
= BE   852 149   10. 3.76/ 1. 7.77   Velsicol Chem.

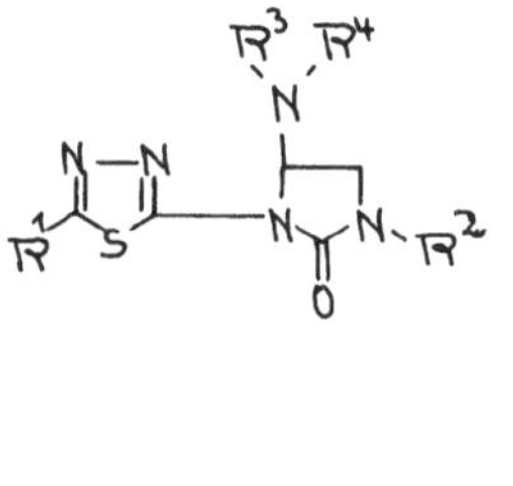

R^1 = Alk(en)yl, Cycloalkyl,
       O(S)-Alkyl, S(O)$_{1u.2}$-Alkyl

R^2 = Alk(en)yl, Alkyl-halogenid

       H
       |
       C-C≡CH
       |
       H, Alkyl

R^3 u. R^4 = Alk(en)yl, Cycloalkyl
               Phenyl, Benzyl usw.

378

DOS   2 717 742   21. 4.77/26.10.78   Gulf Oil

Siehe ähnlich US 3 522 267

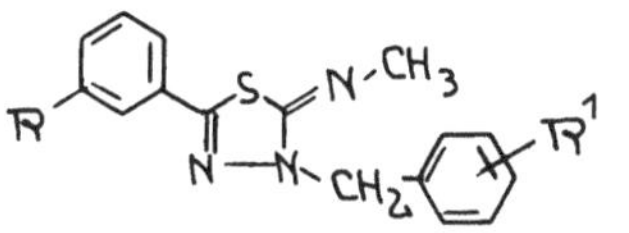

R = CH$_3$, Cl

R^1 = CH$_3$, CF$_3$, C$_2$H$_5$CH$_2$-O-, Cl,F

US   3 984 228   2. 2.72/ 5.10.76   Velsicol Chem.

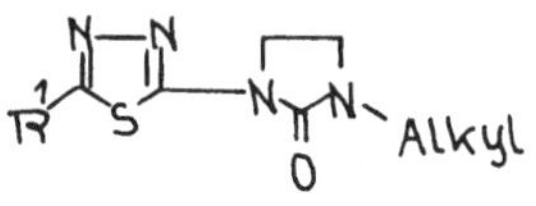

R^1 = CH$\langle{}^{CH_3}_{CH_3}$, C(CH$_3$)$_3$, CF$_3$

US   3 988 143   1.11.74/23.10.75   Velsicol

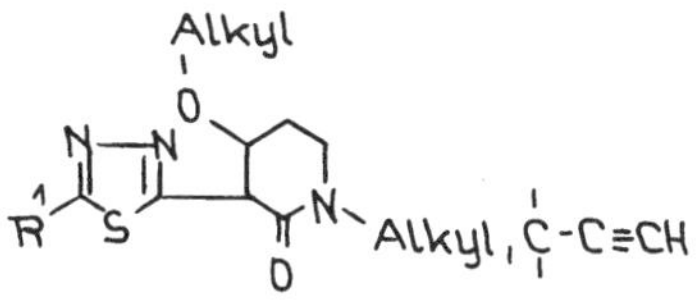

US   3 990 882   31.10.75/ 9.11.76   Velsicol

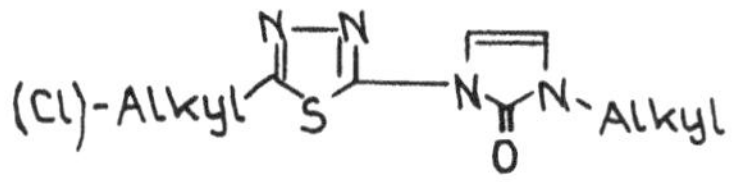

siehe 5.5.3.i.5.3.

US   4 007 031   28. 3.74/ 8. 2.77   Velsicol

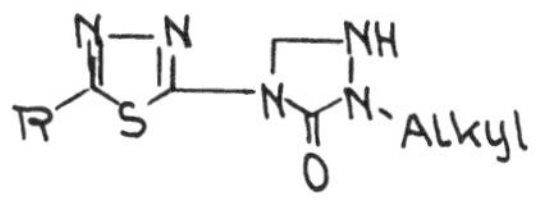

siehe 5.5.3.i.5.3.

US   4 008 068   8. 3.76/12. 2.77   Velsicol

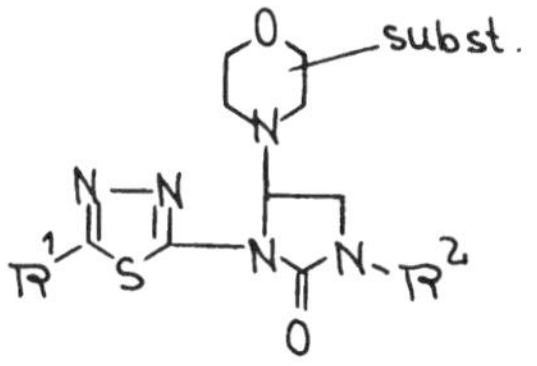

R^1 = Alk(en,in)yl, S(O)$_n$-Alkyl

n = 0-2, CF$_3$, Halog-Alkyl

R^2 = Alkyl, Halog-Alkyl,
        Propargyl usw.

US   4 012 223   15. 2.74/15. 3.77   Velsicol

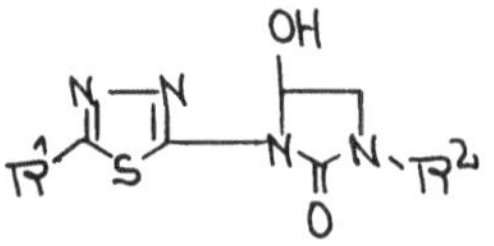

R^1 = C_3-C_7 Cyclo-Alkyl,

  ev. substit. durch
  (O)-Alkyl, CL, Br, OH
R^2 = nied. Alkyl

US   4 020 078   4. 9.74/26. 4.77   Gulf Oil

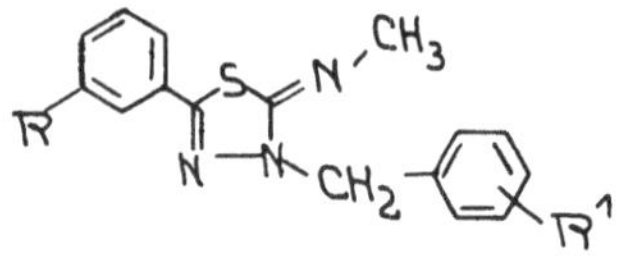

R u. R^1 = H, CH_3, Cl, F

US   4 023 957   26.12.73/17. 5.77   Velsicol

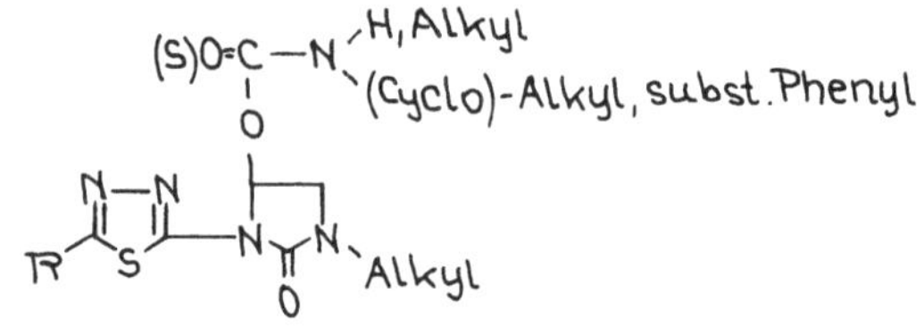

US   4 033 753   5. 4.76/ 5. 7.77   Velsicol

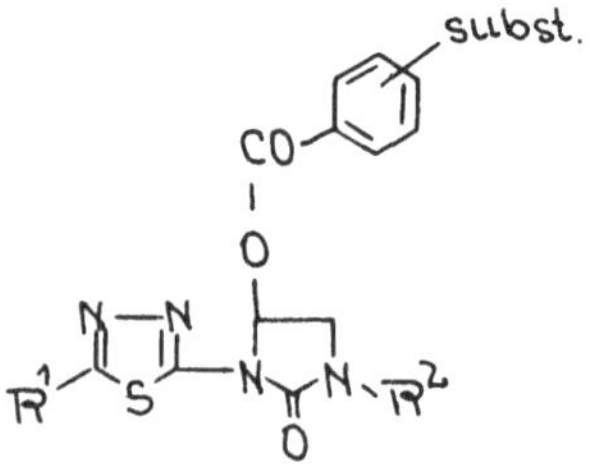

R^1 = (Cyclo)-Alkyl, O(S)Alkyl
  S-Alkyl
  (O)1 u. 2

R^2 = (Halog.)-Alkyl, Alkenyl
  C-C ≡ CH

US   4 036 848   15. 3.76/19. 7.77   Velsicol

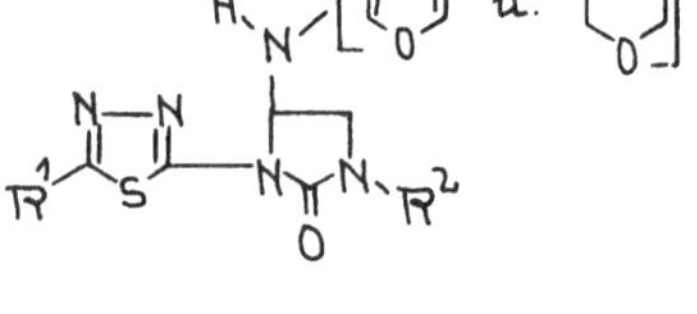

R^1 = Alk(en)yl, O-Alkyl
  Halog.-Alkyl, Alkyl-SO_2 usw.

R^2 = Alk(en)yl, Propargyl,
  Halog.-Alkyl

380

US   4 040 812   18. 6.75/ 9. 8.77   Velsicol

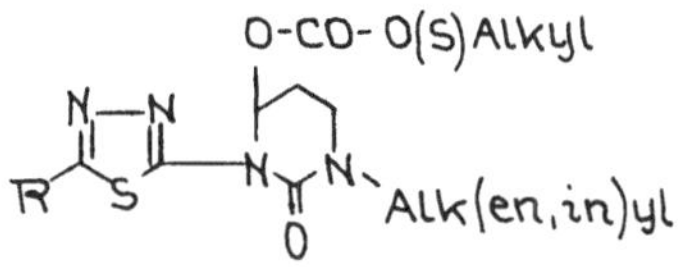

O-CO-O(S)Alkyl   siehe 5.5.3.i.5.3.

R—[1,3,4-thiadiazol]—N—N—Alk(en,in)yl

US   4 052 192   18. 3.76/ 4.10.77   Velsicol

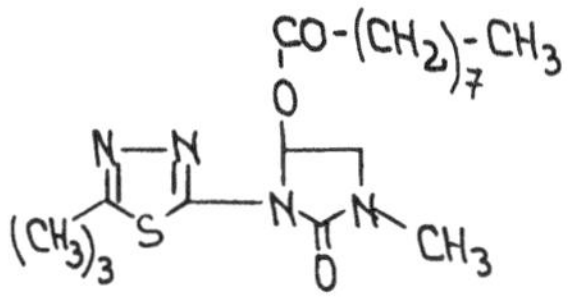

$CO-(CH_2)_7-CH_3$

$(CH_3)_3$—[thiadiazol]—N—N—CH_3

US   4 052 193   18. 3.76/ 4.10.77   Velsicol

wie zuvor: $O-\overset{O}{\underset{||}{C}}-(CH_2)_6-CH_3$

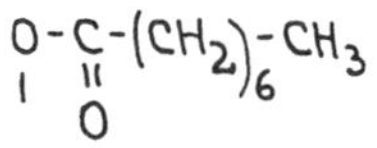

US   4 053 298   18. 9.74/11.10.77   Velsicol

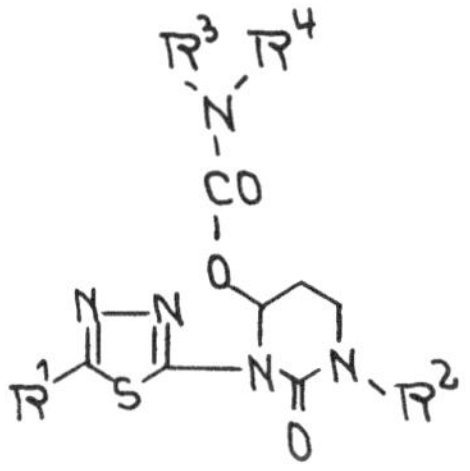

R^3 R^4
N
CO
O
R^1—[thiadiazol]—N—N—R^2

R^1 = Alk(en)yl, CF_3, O(S)-Alkyl
      Alkyl-SO_2-
R^2 = Alk(en)yl, Chloralkyl,
      $-C-C\equiv CH$
R^3 u. R^4 = H, Alkyl, Cycloalkyl
              subst. Phenyl

US   4 053 480   26. 4.76/11.10.76   Velsicol

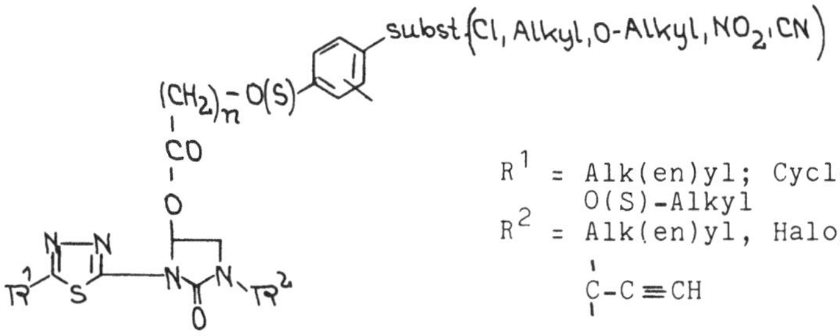

subst (Cl, Alkyl, O-Alkyl, NO_2, CN)
$(CH_2)_n-O(S)$—[phenyl]
CO
O
R^1—[thiadiazol]—N—N—R^2

R^1 = Alk(en)yl; Cyclo-Alkyl,
      O(S)-Alkyl
R^2 = Alk(en)yl, Halog-Alkyl,
      $C-C\equiv CH$

US   4 086 238   21. 6.76/25. 4.78   Velsicol

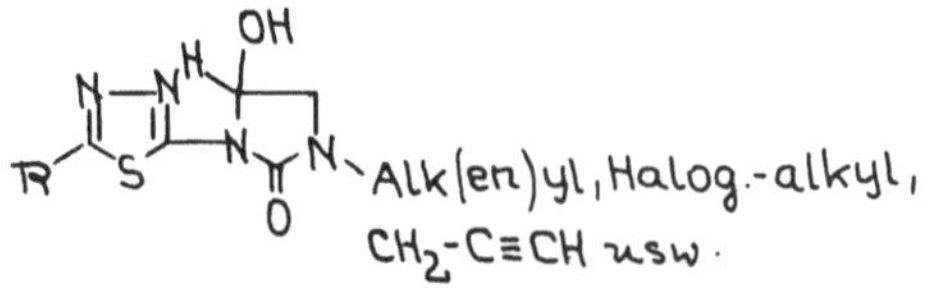

R = Alkyl, S-Alkyl, CH_2-C≡CH   usw.
         ↓
       $(O)_{0-2}$

US   4 111 949   25. 3.76/ 3. 9.78   Velsicol

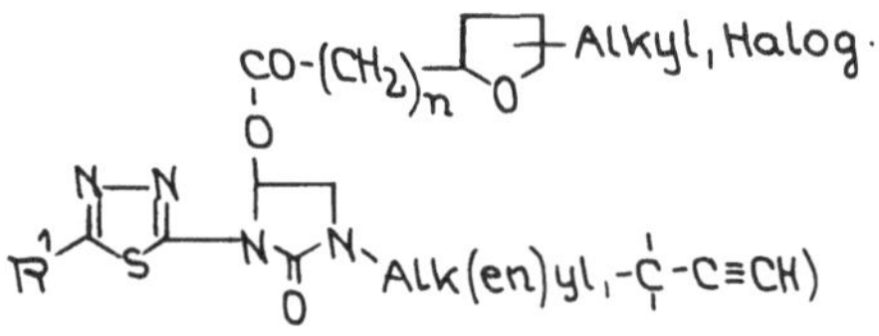

R^1 = Alk(en)yl, Halogenalkyl, Cycloalkyl, O(S)-Alkyl, SO_2-Alkyl

US   4 218 236   18. 6.79/19. 8.80   PPG

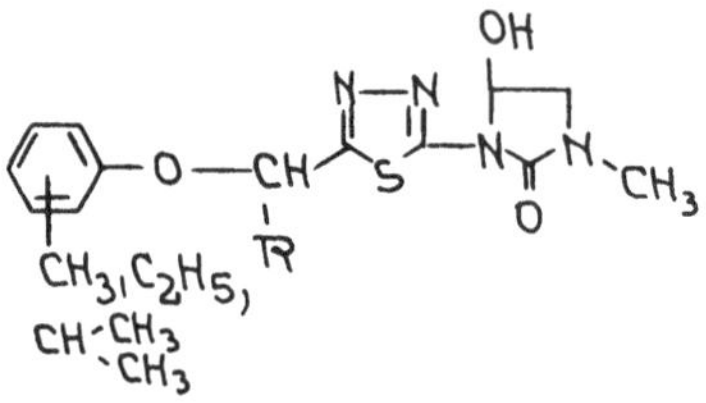

BE   850 394   16. 1.76/14. 7.77   Schering

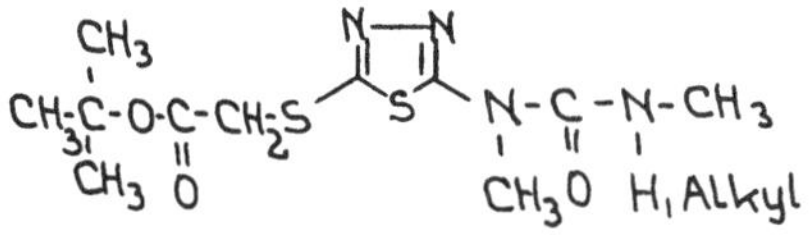

BE   850 395   16. 1.76/14. 7.77   Schering

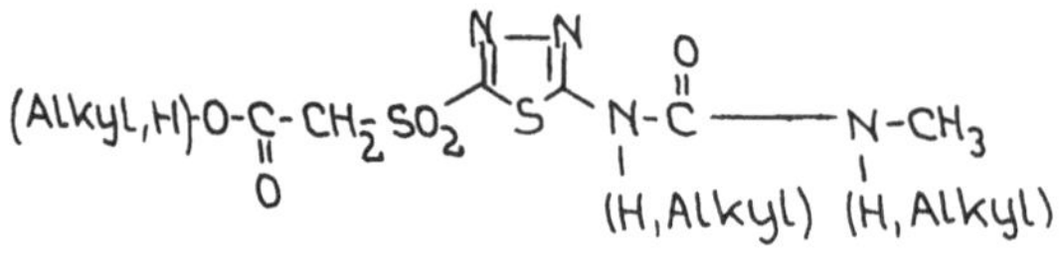

382

JA   52 076 432   18.12.75/27. 6.77   Nissan Chem.

CH$_2$ / S — Halogen

R–C–O(S)
‖
O

R = (O)-Alkyl

i2)   <u>1,2,4-Thiadiazole u. 1,2,5-Thiadiazole</u>   Bd.5/625

US   4 081 453   12.11.71/28. 3.78   Scott

Cl — N
    S — Cl

Zwischenprodukt für Herbizide.

SZ   530 029   28. 1.75/ 3. 1.77   Armn. Agr. Inst.

CH$_3$
HN–C–CH$_3$
O=S–N–NH
      Cl,H
      Cl,NO$_2$

DDR   134 184   28.11.77/14. 2.79   VEB Komb. Bitterf.

H,R^1
  N
R^2 — N
        N
        S
R^3

oder

R^4
SO$_2$-NH
Hal. — N
         N
         S
Hal.

R^1 = Acetyl, Phenoxyacetyl
R^2 = H, Halog.,SCN usw.
R^3 = H. Halog.,SCN usw.
R^4 = C–N$<$
        ‖
        O

US   4 228 289   24. 4.79/14.1080   Olin Corp.

R^1
    N   R^2 O
         N–C–S–R^3
N   S

R^1= CCl$_3$, CF$_3$, R^2= H, Alkyl
R^3= C$_1$-C$_4$-Alkyl

US   4 228 290   25. 1.79/14.10.80   Olin Corp·

R^1
    N
         O
         ‖
N–(CH$_2$)$_n$–C–O-Alkyl
N   S
    R^2

R^1= CCl$_3$, CF$_3$
R^2= H, Alkyl

i3)   1,2,4,-Thiadiazolidone   Bd.5/627

DDR   121 519   10. 7.75/ 5. 8.76   Werchan H. G.

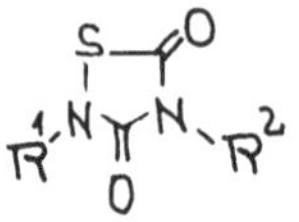

R^1 = subst. Aryl
R^2 = Alkyl

u..a. auch Herbizide

k)   3 x N oder 3 x S   Bd. 5/628

DOS   2 615 878   10. 4.76/20.10.77   BASF

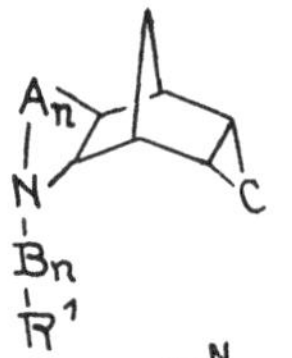

A = -N═N-
B = SO_2, CO(S), S
C = -N = N-, -N-N-
R^1 = H, Alk(en,in)yl, Phenyl
       usw.

z.B.

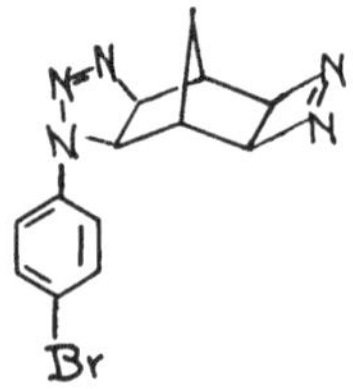

DOS   2 643 488   27. 9.76/14. 4.77   American Cyanamid

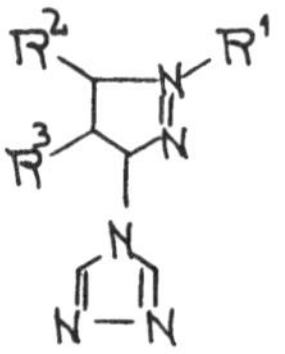

R^1 = H, Alkyl, Benzyl, Acyl,
       CO-N<
R^2 = H, CH_3, Phenyl
R^3 = CH_3, subst. Phenyl

DOS   2 650 831   6.11.76/11. 5.78   BASF

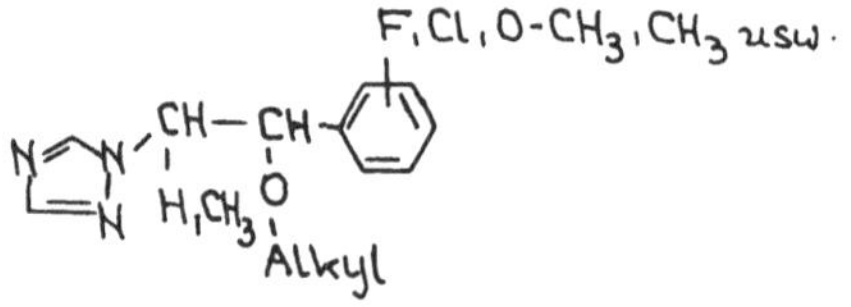

als Salz z.B. der HNO_3

Wuchshemm-mittel u. Herbizid.

DOS   2 713 871   29. 3.77/12.10.78   American Cyanamid

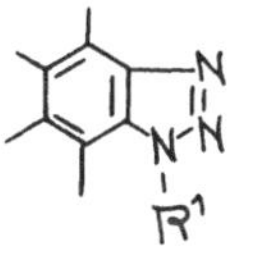

oder

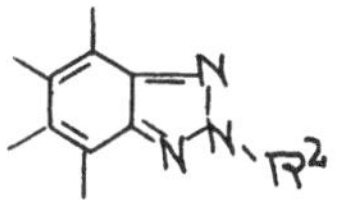

R^1 u. R^2 = sec. C_3-C_7-Alkyl, tert. C_4-C_7, Alkoxy, Benzyl,

Cycloalk(en)yl, $(CH_2)_{0-1}$

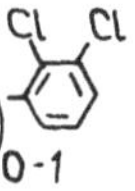

DOS   2 720 416   6. 5.77/ 1.12.77   Fisons
                                                   GB Prior. 21.5. u. 26.11.76

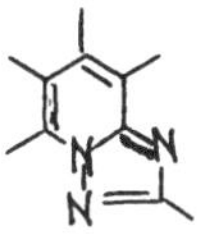

und

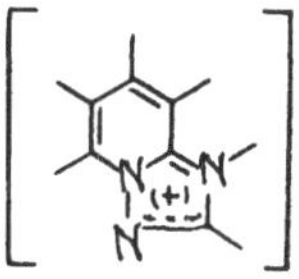

subst. durch H, Halog. Alkyl

DOS   2 738 640   26. 8.77/ 2. 3.78   ICI
= BE   858 113   GB Prior. 27.8.76

$\overset{X}{\underset{R^2}{C}}=CH-\overset{O}{\underset{\|}{C}}-R^1$

X = 1·(1,2,-4-Triazolylrest 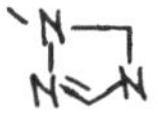
    1-Imidazolylrest
R^2 = Alkyl, Cycloalkyl, subst. Phenyl usw.
R^1 = (Cyclo)-Alkyl, Phenyl usw.

DOS   2 738 725   27. 8.77/ 8. 3.79   BASF

R^1-CH-CH$_2$-CH-R^2 R^1 u. R^2 = Alkyl, Cycloalkyl,
   OH   A   Furanyl, Thianyl,
                                                Pyridyl, Phenyl usw.

Az = 1,2,4 ~ 1,2,3-Triazolyl

z.B. CH$_3$-$\overset{CH_3}{\underset{CH_3OH}{C}}$-CH-CH$_2$-CH$-\langle$ $\rangle$-Cl

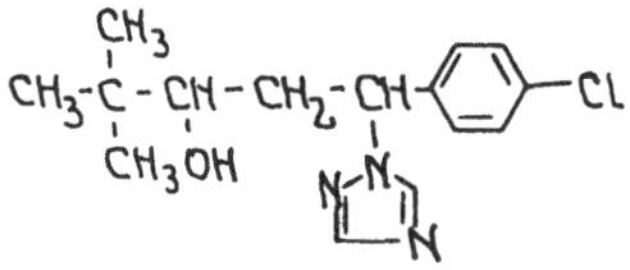

vorwiegend Wuchsregulatoren

385

DOS   2 801 429   13. 1.78/20. 7.78   Du Pont
                                       US Prior. 13.1.77 u. 17.11.77

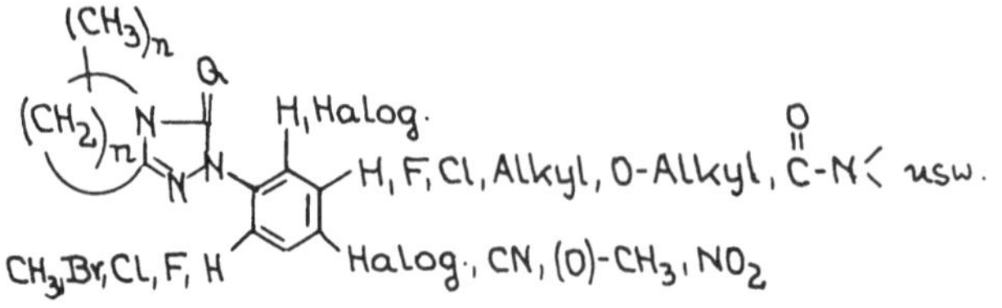

Q = O,S   Beispiel:

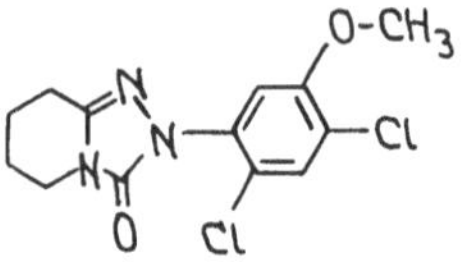

DOS   2 842 801   30. 9.78/10. 4.80   BASF

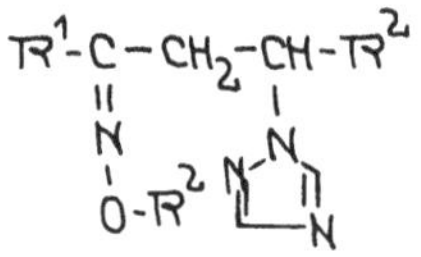

R^1 u. R^2 = Alkyl, Furanyl,
             Phenyl usw.
R^3 = H, Alkyl, Benzyl,
        CO-Alkyl usw.

vorwiegend Wuchsregulatoren

DOS   2 952 685   29.12.79/21. 8.80   Chevron Res.
                                       US Prior. 9.2.79

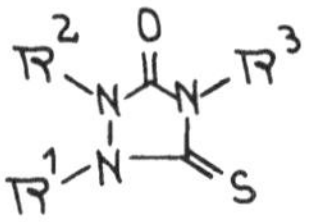

R^1 = Alkyl

R^2 = Alkyl, Phenyl (subst.)

R^3 = Phenyl durch Halogen
        substituiert, Alkyl

DOS   3 010 560   19. 3.80/ 2.10.80   Sumitomo Chem.
       JA Prior. 5.4.79; 10.9.79; 25.9.79 u. 30.1.80

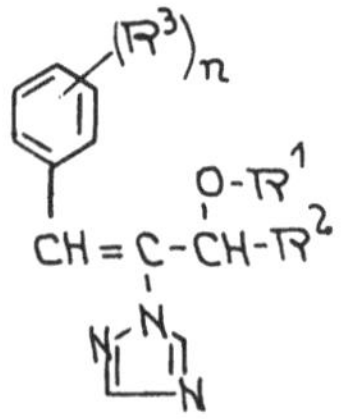

R^1 = H, C_1-C_4-Alkyl, Alkenyl,
        Propinyl
R^2 = Alkyl, Cyclopropyl
        1-Methyl-Cyclopropyl
R^3 = Halog., Alken (subst.),
        O-Alkyl, Phenoxyl, Phenyl,
        CN, NO_2

n = 0-3

u. a. auch Herbizide

386

EP   14   1. 6.78/20.12.78   Höchst
= DOS   2 725 148   3. 6.77

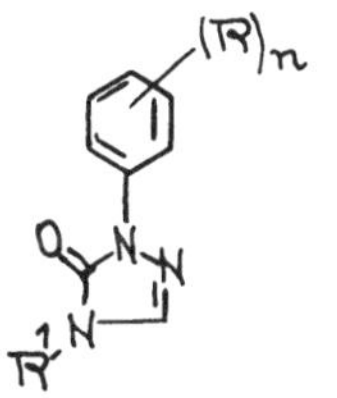

R = H, Halog., NO_2, CN, Alkyl,
(O) oder (S)-Alkyl usw.

R^1 = Alkyl, Alkoxy-alkyl,
N(Alkyl)$_2$

u. a. auch Herbizide

EP   3 884   13. 2.79/ 5. 9.79   ICI
GB Prior. 23.2.78

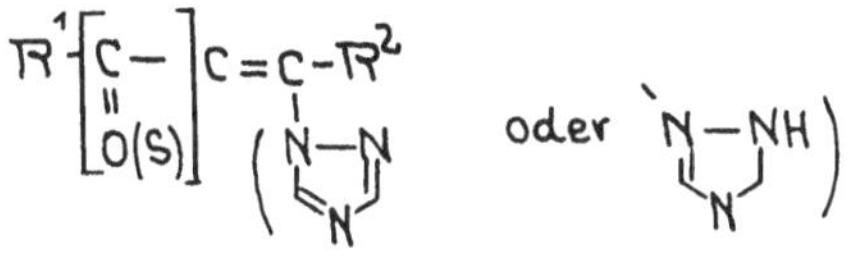

oder   statt C=O

auch CH-OH

R^1 = Alkyl, Cycloalkyl, subst. Phenyl

EP   9 228   27. 6.79/ 2. 9.80   ICI
GB Prior. 27.7.78

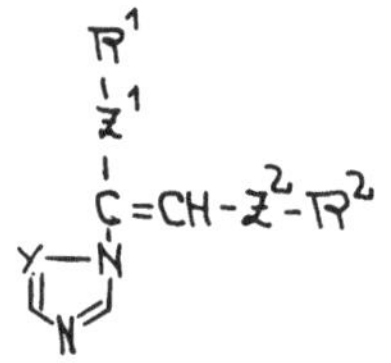

R^1 u. R^2 = (Cyclo)-Alkyl,
Phenyl
Y = -N = oder CH
Z^1 u. Z^2 = CO oder CHOH

EP   9 298   27. 7.78/ 2. 4.80   ICI

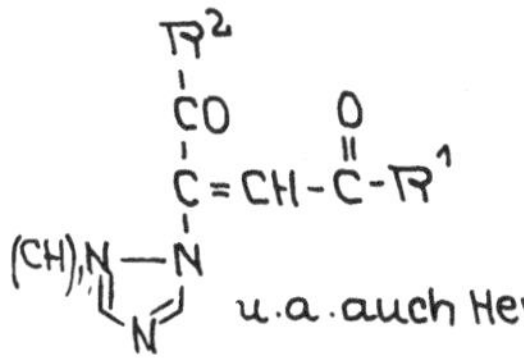

R^1 u. R^2 = Alkyl, Cycloalkyl,
Phenyl-subst.
statt C = O auch 1 x CHOH

u.a.auch Herbizid wirksam

US   4 002 636   11. 8.71/11. 1.77   Höchst

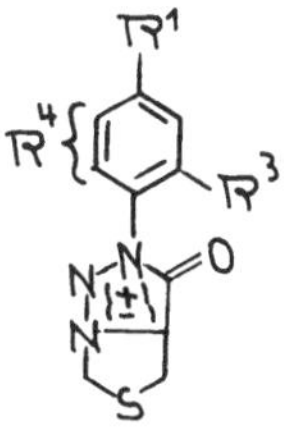

R^1 = Halog. R^3～ R^4 = H,Halog.

mesoionische Oxo-Thiazolo-(3,4-)
triazoliniumsalze

wirksam gegen Gräser

387

US   4 013 448   29.12.75/22. 3.77   Monsanto

R^1 u. R^2 = H, Alkyl

US   4 086 242   22. 7.74/25. 4.78   American Cyanamid

oder

R = (sec. u. tert.)Alkyl
    Benzyl, Alkyl-CO-O
O -Alkyl

US   4 139 364   9. 8.77/13. 2.79   Du Pont

BE   862 884   13. 1.77/13. 7.78   Du Pont

subst. durch Halogen, CN, NO_2, (O)-CH_3

BE   869 555   4. 8.77/ 1.12.78   Ansul

Y = H, CN, Alkyl-SO_2, NO_2, CF_3
X = Alk(en)yl, NO_2, CCl_3, CF_3,
    CF_3-SO_2, CN
    N< , N-$^H_{Aryl}$ usw.

u. a. auch Herbizid

BE   882 335   20. 3.79/16. 7.80   Sumitomo

CH=C-(CH-OH oder -C-)(Alkyl,Cyclopropyl usw.)

u. a. auch Herbizide

JA   52 113972   17. 3.76/24. 9.77   Mitsubishi

R^1 u. R^2 = Alkyl

JA   54 014971   5. 7. 77/ 9. 2.79   Nippon Carbide

JA   54 032 470   16. 8.77/ 9. 3.79   Sankyo

1)   2 x N + 1 x P, 1 x O + 1 x P usw.

US   3 980 618   6. 8.73/14. 9.76   Monsanto

Siehe auch US   3 904 654

Herbizid

US   4 219 510   6. 7.79/26. 8.80   Monsanto

a)   6-Ringe mit 1 x O

DOS   2 740 801   9. 9.77/16. 3.78   Inter. Developm.
= BE   858 646   Research
             GB Prior. 3.9.76 = BE 858 646   Centre

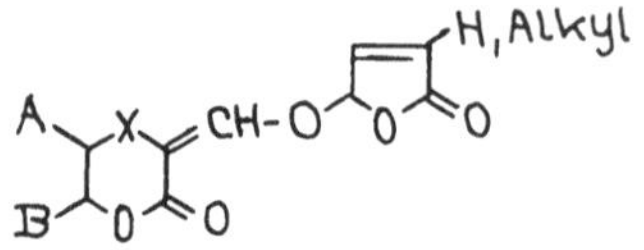

A u. B = H oder zusammen O, CH_2 Glieder eines Ringes
X = Einfachbindung   zum 5-Ring oder CH_2

DOS   2 910 283   15. 3.79/ 4.10.79   Philagro
= BE   874 842   15. 3.78/14. 9.79   Philagro

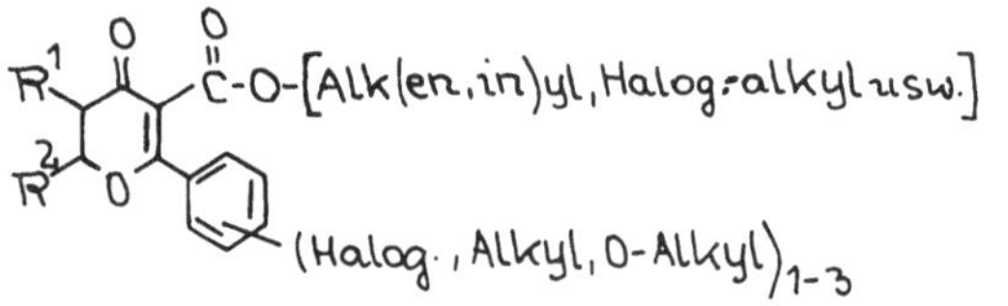

R^1 u. R^2 = H, Alkyl, CH_2-Glieder eines Ringes.

JA   50 129565   13. 3.74/13.10.75   Nippon Soda

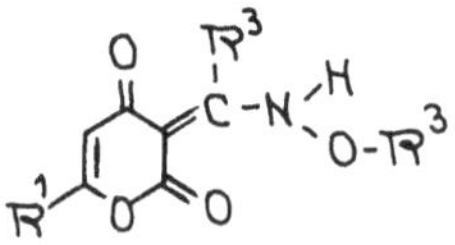

                                    R^1 = Alk(en)yl, Ph-CH = CH_2
                                    R^2 = Alkyl, Ph-CH_2, Ph, Ph-O-CH_2
                                    R^3 = H, Alk(en,in)yl, Ph-CH_2

JA   50 154 263   4. 6.74/12.12.75   Nippon Soda

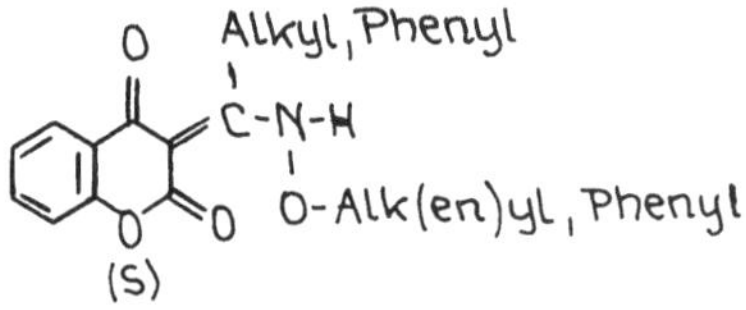

390

JA   52 041233   19. 7.72/30. 3.77   Nippon Soda

R^1 = Alk(en)yl

selektiv in Sojabohnen

JA   52 151175   17. 1.76/15.12.77   Nisshin Floor

X = $\underline{O}$, NH
R^1 u. R^2 = Alkyl; CH_2-Ar.

JA   53 034776   13. 3.74/31. 3.78   Nippon Soda

R = Alkyl, Phenyl

Spez. gegen Gräser wie Ackerfuchsschwanz.

JA   54 024875   22. 7.77/24. 2.79   Nihon Noyaku

JA   76 046814   19. 7.72/11.12.76   Nippon Soda

R = Alkyl, O-Alkyl, CH_2Cl
O-C_2H_5, CH_2-C_2H_5
R^1 = Alk(en)yl; X = H oder Br

b) <u>6-Ringe mit 1 x N, Pyridine und Pyridin-N-oxyde</u>

Pyridincarbonsäure Derivate siehe 5.5.5 e (heterocycl. Carbonsäuren)

b1) <u>Pyridine mit beliebigen Substituenten außer OH und NH_2</u>

DOS 2 602 186 21. 1.76/22. 7.76 Sumitomo

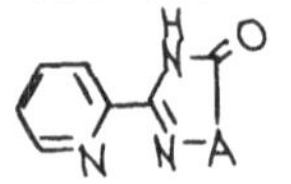

A = NH-CH- oder N= C-
R = H, Alkyl

Herbizide und Fungizide

z.B.

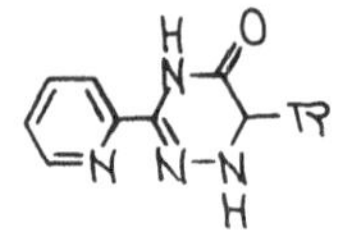

EP 8 145 8. 8.78/20. 2.80 Shell Int. Res.
GB Prior. 8.8.78

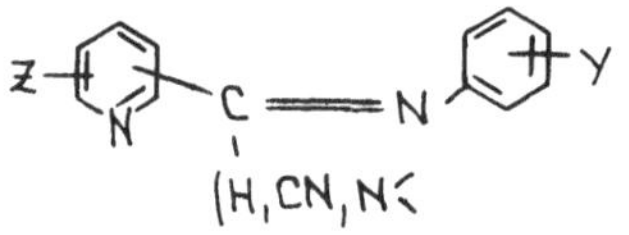

Y u. Z = Halog., Alkyl, O-Alkyl,
NO_2

u.a. auch Herbizide

EP 12 117 3.12.79/11. 6.80 Ciba Geigy
CH 5.12.78/23.10.79

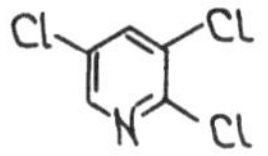

neues Darstellungsverfahren

US 3 987 050 22. 1.74/19.10.76 Dow Chem.

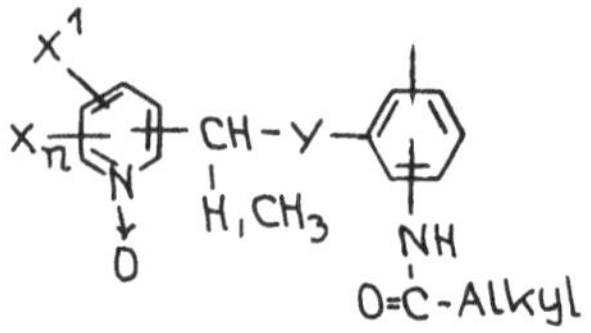

X^4 = CCl_3, NH_2, CF_3, CN(O,S)CH_3
n = 0 - 4
X^1 = Halog.
Y = O, S, SO_2

US 3 991 068 22. 1.74/ 9.11.76 Dow Chem.

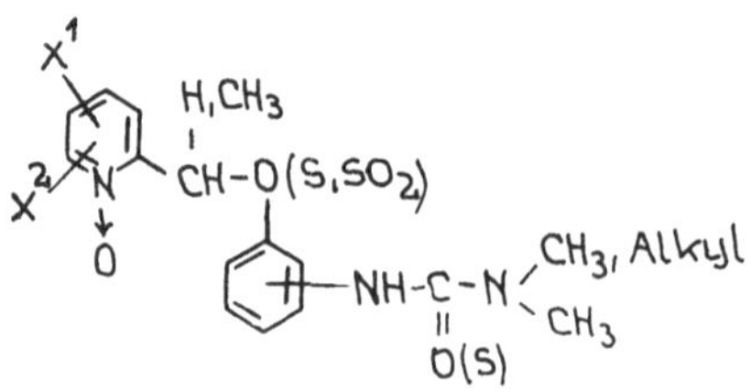

US   4 043 790   2. 4.76/23. 8.77   Eli Lilly

C-Alkyl, Cycloalkyl, Phenyl-Halog. subst.
Alkyl, Cycloalkyl, Phenyl-Halog. subst., Phenyl-O-CH$_3$
oder   HO-CH<Substituenten wie zuvor

} Wasser-Herbizid

US   4 043 791   2. 4.76/23. 8.77   Eli Lilly

C-Alkyl, Cycloalkyl, Phenyl
Trifluormethoxy-phenyl

US   4 085 278   5.11.73/18. 4.78   Dow Chem.

X = Alkyl, O(S)-Alkyl, Phenyl,
    NO$_2$, N< usw.
Z = NHSO$_2$N< 5 oder 6 Ring mit
                1 x N
T = OH, O-Alkyl

GB   2 015 524   6. 3.78/12. 9.78   ICI

R^1 u. R^2 = H, O-Alkyl, Pyridyl,
             Thionyl, Furyl
R^3 u. R^4 = H, Alkyl, Aryl,
             Glieder eines Ringes

u.a. auch Herbizide

JA   53 116378   16. 3.77/11.10.78   Sumitomo Chem.

C-(Cyclo)-alkyl
CH$_2$ Halog., NO$_2$, Alkyl

b2) <u>Pyridine mit OH oder SH am Ringsystem</u>

ihre Ester und Ether, siehe auch
Pyridyl-Phenylether 5.4.2.c. und heterocyclische
Oxyalkancarbonsäuren 5.7.3.f.2

DOS   2 609 204   5. 3.76/ 7.10.76   Uniroyal
                                                   US Prior. 17. 3.75

Struktur: Pyridin-N-oxid mit S–CH an Position 2, Substituenten: H, Alkyl, Phenyl, Cyclo-hexyl, 2·2-bichlor cyclopropyl, Propenyl, Benzyl, Naphthyl, Pyridyl, Benthiazolyl usw. (O)$_{1 u.2}$

Siehe auch US 3 107 994 mit Pyridin-N-oxid–S–CH<

Spezif. wirksam gegen Gräser.

DOS   2 637 886   23. 8.76/10. 3.77   Ciba Geigy
DOS   2 637 911   23. 8.76/10. 3.77   Ciba Geigy
                                                     CH Prior. 28. 8.75

Struktur: B–Pyridin–A–O-G-C-O(S)Alkyl   A = Halog., Alkyl
G = CH$_2$ B = Halog.

                                             Wuchsregulatoren

Ähnliche Verbindungen im US 3 761 486 u. 3 755 339.
als Herbizide. Darstellung: Acta. Chem. Scand. <u>23</u>,1791/ 1969

Ähnliche Verbindungen wie im vorangehenden Patent als Amide:
Wuchsregulatoren und Mittel gegen Herbizidschäden.

DOS   2 719 904   4. 5.77/ 9.11.78   Celamerk

Struktur: Pyridin mit O-CO-CH$_3$ und Cl.

                                             Darstellungsverfahren

US   4 025 333   17. 1.75/24. 5.77   Dow Chem.

                                             u.a. auch Herbizide

Struktur: (H$_2$N,C,H)O–Pyridin mit Br, Br, C-(Alkyl, N<)

394

US   4 066 438   25. 5.70/ 3. 1.78   Dow Chem.

$R = C_4H_9$, Alkyl subst. durch OH; $R^1 = H$, NH_2,
$N\begin{smallmatrix}H\\\diagdown Alkyl\end{smallmatrix}$, $C_6H_5-(CH_2)_{0-4}-NH_2$ usw.
$R^2 = H$, NH_2, $N\begin{smallmatrix}H\\\diagdown Alkyl\end{smallmatrix}$

CH   591 449   5. 8.74/17. 1.78   Phytopled.

BE   866 699   4. 5.77/ 3.11.78   Celamerk

$R = H$, Kation, CO-Alk(en)yl,
CO-Phenyl subst. usw.

b4)   <u>Pyridine mit anellierten Ringen</u>   Bd.5/638
Pyridino-imidazole siehe auch bei
5-Ring 2 x N   5.14.2 g3

US   4 042 593   21.12.70/16. 8.77   Eli Lilly

$R\ = H$, Cl, F, CHF_2, CF_3 usw.
$R^1 = $ Halog., NO_2, CF_3, CHF_2
       usw. Alkyl-SO_2
$R^2 = $ CO-O(S)-Alkyl, CO-O(S)-
       Phenyl, SO_2-Alkyl, Phenyl,
       Benzyl

US   4 054 573   21.12.70/18.10.77   Eli Lilly

X = Halogen, n = 2 oder 3
R = Alk(en)yl, Benzyl,
    Phenethyl usw.

US   4 082 534   21.12.70/ 4. 4.78   Eli Lilly

$R = Hal.$, NO_2, CF_3, SO_2-Alkyl usw.

US   4 087 432   20. 3.70/ 2. 5.78   Eli Lilly

395

BE   854 423   15. 5.76/ 9.11.77   Fisons
                                                  GB Prior. 15. 5.76

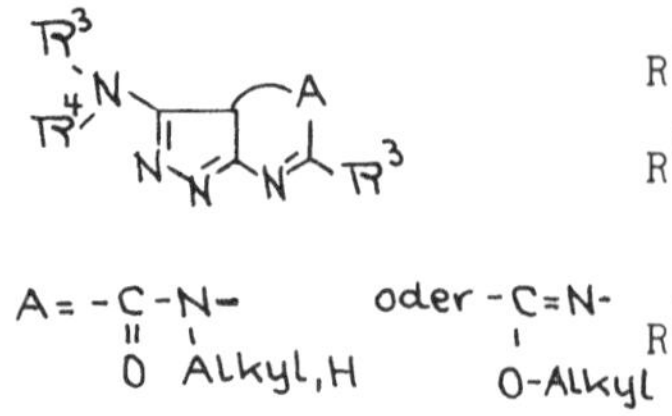

R^1 = H, (Cyclo)-Alkyl, Phenyl, Hetero-
        cyclyl
R^2 = H, Halogen, OH, Alk(en,in)yl,
        O-Alkyl, $S:(O)_{0-2}$ -Alkyl, Aryl
        Aralkyl, CF$_3$

A = -C-N-   oder -C=N-   R^3 u. R^4 = H, Alkyl, Aryl, Aralkyl,
      ‖ | | Acyl, Alkylsulphonyl,
      O Alkyl,H   O-Alkyl   Carbamoyl, N< , Heterocyclic
                                           etc.

Selective pre emergence Herbizide.

BE   854 689   21. 5.76/16.11.77   Fisons
                                                  GB Prior. 26.11.76

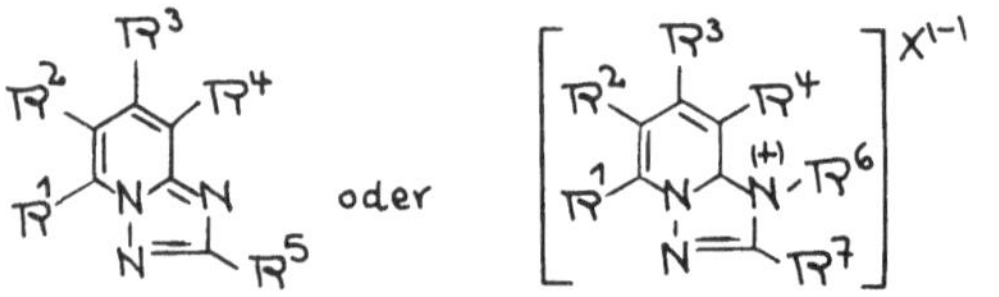

oder

R^1 - R^4 = H, Halogen, Alkyl, Aryl, Aralkyl, NO$_2$, CN

R^5 = Alkyl,

R^7 = Alkyl oder R^5, R^6 = Alkyl in einem Heteroring mit quartärem N

X = Anion

<u>b5)   1 x N teilweise oder vollständig hydriert</u>   Bd.5/639

EP   6 540   28.·6.78/ 9. 1.80   Bayer
= DOS   2 828 293

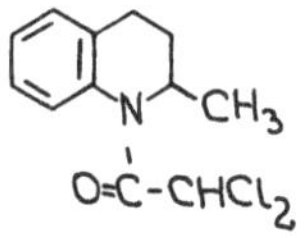

Safener gegen Herbizidschäden
durch Acetanilide C$_6$H$_5$-N$\langle$ $^{CH_2-R}_{CO-CH_2-Hal.}$ R = Heterocycl.
                                                                          5-Ring mit
                                                                          2 x N oder 3 x N

EP   6 542   28. 6.78/ 9. 1.80   Bayer

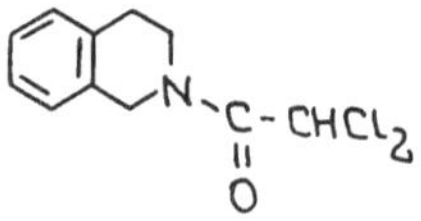

Safener gegen Herbizidschäden

durch Acetanilide C$_6$H$_5$-N$\langle$ $^{CH_2-R}_{CO-CH_2-Hal.}$

R = Heterocycl. 5-Ring mit 2 x N oder 3 x N

396

DOS   2 628 992   26. 6.76/20. 1.77   Eli Lilly
                                      US Prior. 3. 7.75 u. 20.5. 76

                                oder

R = CH$_3$, C$_2$H$_5$
R^1 = H, O(S)-C$_2$H$_5$, (O)-Alkyl
R^2 = F, Br, CF$_3$
R^3 = Alk(en,in)yl; R^4 = H, (O,S)-Phenyl, (O,S)-Alkyl
R^5 = Alkyl, Halog. CF$_3$; subst. Phenyl

DOS   2 828 331   28. 6.78/10. 1.80   Bayer

                                                     Safener gegen Herbizid-
                                                     schäden.

EP   10 799   15.10.79/14. 5.80   Shell Inter. Res.
                                GB Prior. 27.10.78 u. 14. 5.79

US   4 038 065   11.10.74/26. 7.77   Rohm u. Haas

                                        R = Alkyl; R^1 u. R^2 = CH$_3$ oder
                                             C$_2$H$_5$
                                        R^3 = CO-(OH , O-Alkyl, NH$_2$)
                                        Wuchsstoffe mit herbiziden
                                        Eigenschafften in höherer Dosis

EP   2 859   13.12.78/11. 7.79   Shell
                                                GB Prior. 22.12.77

                           R^1 = H, Alk(en,in)yl, Aralkyl, CO(S)-N< usw.
                           R^2 = -CO(S)-N< usw.

US   4 127 581   8. 9.74/28.11.78   Eli Lilly

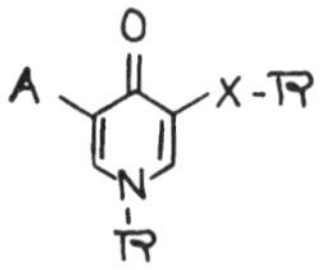

Breitspektr. Herbizide.

$A = $ Phenyl, Halog., Alkyl usw.

$R = $ Alkyl, Alk(en,in)yl, N$<$,

$X = O,S$

BE   843 481   3. 7.75/28.12.76   Eli Lilly

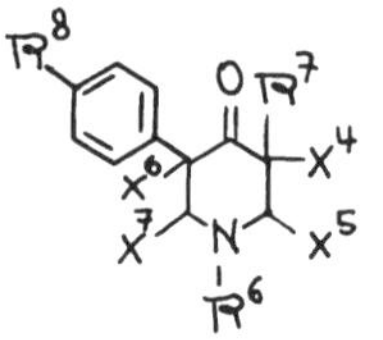

$R^6 = C_{1-3}$Alkyl, C_2-C_3-Alkenyl, Propargyl

$R^7 = H$, C_6-H_5-O, C_6-H_5-S-, O(S)-Alkyl, Alkyl
        Phenyl subst. durch Halog.;CF_3,(O)-Alkyl

$R^8 = $ Halog.,CF_3, (O)-Alkyl

X^4 u. $X^5 = $ H oder C-C Bindung

X^6 u. $X^7 = $ H oder C-C Bindung

BE   868 393   27. 6.77/27.12.78   Eli Lilly

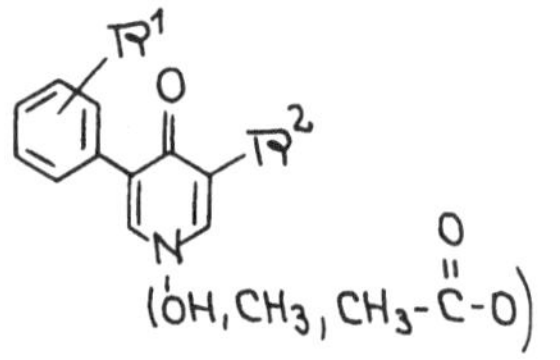

$R^1 = $ Halog., Alkyl, NO_2, O(S)-Alkyl
        usw.

$R^2 = $ CH, OH, SO-Alkyl usw.

DDR   126 221   12. 5.76/ 6. 7.77   Metschiner

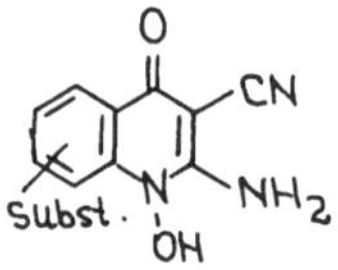

                                                 u.a. auch herbizid
                                                 wirksam

ZA   7 505 425   28. 8.74/22. 9.76   Eli Lilly

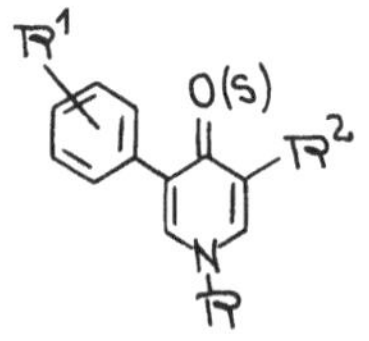

R = Alkyl subst. durch CN, Halog., COOR;
    Alkenyl, Alkinyl, N$<$

$R^1 = $ Halog., Cycloalkenyl

$R^2 = $ H, CN, Halog., Alk(enyl) usw.

Selektiv in Baumwolle

JA   50 154272   4. 6.74/12.12.75   Nippon Soda

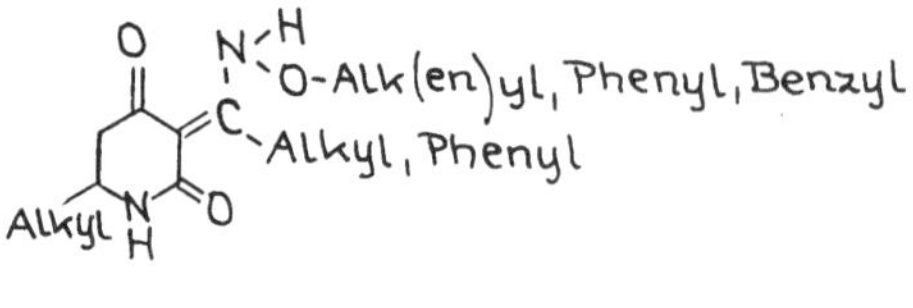

JA   3 52 151175   17. 1.76/15.12.77   Nippon Soda

R^1 u. R^2 = Alkyl, Benzyl

JA   54 070283   14.11.77/ 5. 6.79   Kumiai

H,(0)-Alkyl, Halogen

Gegen Gräser und Unkräuter

c)   <u>6-Ringe, 2 x 0; 2 x S; 1 x 0 + 1 x S; 1 x N + 1 x 0;</u>
      <u>1 x N + 1 x S u. 2 x N</u>

c1)   <u>2 x 0 u. 2 x S</u>   Bd.5/640

      Siehe auch bei cyclischen Acetalen von Diolen und Triolen 5.5.2

DOS   2 739 067   31. 8.76/ 2. 3.78   FMC

        R^1 = Phenyl, Furyl, Thienyl, Pyridyl usw.
              subst.
        R^2 = H, Alkyl, Phenyl; R^3 = H
        R^2 u. R^3 = $(CH_2)_n$; R^4 = Alkyl,Halog.-Alkyl
                  Cyanalkyl usw.
        R^5 = H, Alkyl;
        R^6 = H, Alkyl, Halog-Alkyl, Cyanalkyl

US   4 007 206   3. 8.72/ 8. 2.77   Hoffmann La Roche

        R = Alkyl, Alkenyl, Alkinyl
        R^1 - R^4 = Alk(en)yl

US   4 035 178   31. 8.76/12. 7.77   FMC

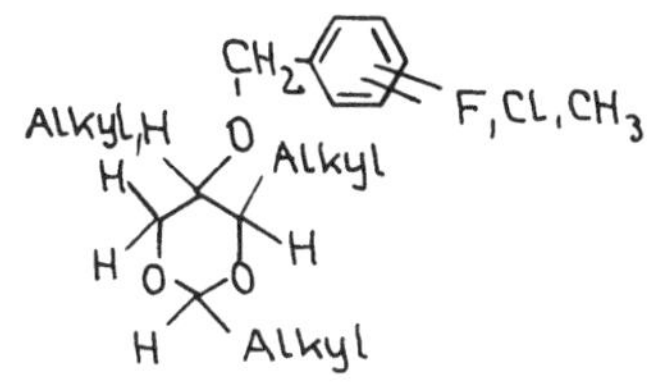

US   4 077 982   19. 1.68/ 7. 3.78   FMC

R^1 = H, Alk(en,in)yl,Cycloalkyl,
       Phenyl usw.
R^2 = H, Alkyl
R^3 = H, Cyanalkyl, Halog.-alkyl

Gegen Gräser.

US   4 193 787   31. 8.78/18. 3.80   Stauffer Chem.

R u. R^1 = Alkyl, O-Alkyl usw.

US   4 207 088   31. 8.76/10. 6.80   FMC

R^2 = H, Alkyl, Phenyl
R^{2a} = H
R^2 u. R^{2a} = Polymethylen
                   eines Ringes
R^4 = Alkyl, subst. Alkyl
R^5 = H, Alkyl
R^6 = H, (CN, Cl)-alkyl

JA   53 044578   30. 9.76/21. 4.78   Nihon Noyaku

R^1 = H, CH_3, CO-OAlkyl
R^2 = H, CH_3
R^3 = H, Alkyl, CH_3-CO-O
R^4 = H, CH_3; R^2u.R^3 = Glieder
                                 eines
                                 Ringes

c3)   <u>1 x N + 1 x O; 1x N + 1 x S</u>   Bd.5/641

DOS   2 710 571   11. 3.77/13.10.77   US Borax
= US   4 049 422   US Prior. 29. 3.76

X = Halog.-alkyl, SO_2-Alkyl,
    SO-Alkyl, Halogen,Alkyl,CF_3
R^1 = H, Halog., Alkyl
R^2 = H, Halog., (O)-Alkyl,
       CN, NO_2, N<
R^3 = H, Alkyl

400

DOS   2 914 915   12. 4.79/3010.80   BASF

R^1 = H, Halog., NO_2, Alkyl,
CO-Alkyl, CO-N , $S(O)_{0-2}$-
Alkyl usw.

R^2 = cycloaliph. Rest, Pyrimi-
dinrest, subst. Phenyl,
Triazolrest, Furan usw. Ring

z.B.

$(NH-\overset{O}{\overset{||}{C}}-S-CH_3, NH-\overset{O}{\underset{||}{\overset{||}{C}}}-N\big\langle$ usw.$)$

DOS   2 915 406   17. 4.79/25.10.79   Sumitomo
         JA Prior.17. 4.78 = GB 201 874

Halog.

US   4 020 061   2. 4.75/26. 4.77   Shell Oil

u. a. auch Herbizid

$(CH_2)_n$

S-N-C-Alk(en,in)yl
NO_2C O
O=C-O-Alkyl

US   4 029 657   14. 6.76/14. 6.77   Dow Chem.

H, Alk(en)yl, Cycloalkyl

u. a. auch Herbizid

Cl
Cl

US   4 030 906   27. 9.74/21. 6.77   Chevron

Siehe auch US 3 917 592

O=C-CXYY

Alkyl, Cl

R^1
R^2
R^4 R^3

X = Cl, Br u. F; Y = H, Cl,
Br u. F

R^1 - R^4 = H u. Alkyl

US   4 080 499   20.12.76/21. 3.78   Dow Chem.

Alk(en)yl usw.

Herbizide und Fungizide.

US   4 094 663   12. 9.77/13. 6.78   US Borax

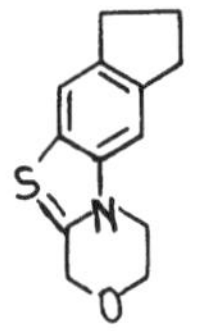

US   4 123 251   2. 6.77/31.10.78   US Borax

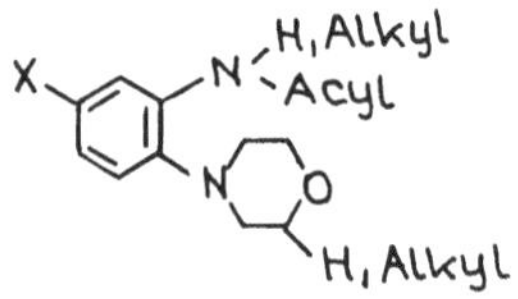

US   4 133 672   7. 2.77/ 9. 1.79   US Borax

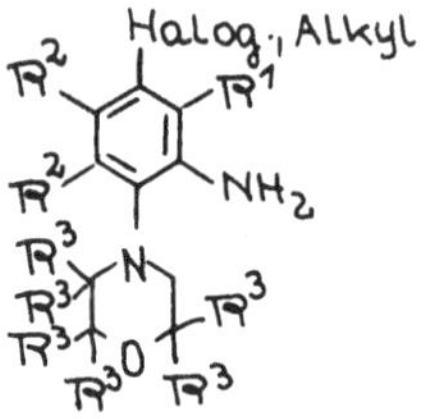

R^1 = H, Halogen

R^2 = H, Halogen, Alkyl, O-Alkyl, NH_2

R^3 = H

US   4 164 407   7.10.77/14. 8.79   Sandoz

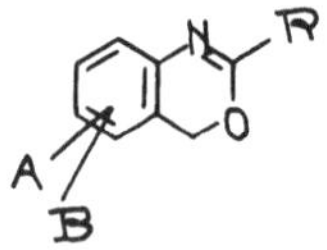

A u. B = H, Cl

$R = N\big\langle{\text{Alk(en,in)yl} \atop \text{Alk(en,in)yl}}$, $N\big\langle{\text{H} \atop \text{C(CH}_3)_3}$

US   4 179 276   1. 2.77/18.12.79   Du Pont

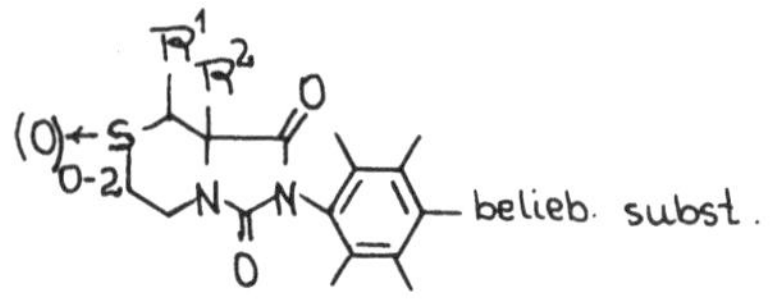

R^1 u. R^2 = H, Cl, R^1 u. R^2 = Doppelbindung

US   4 214 889   7.10.77/29. 7.80   Sandoz

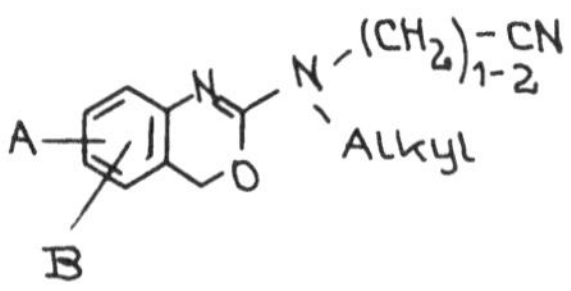

BE   874 361   22. 2.79/18. 6.79   Nitro Kema

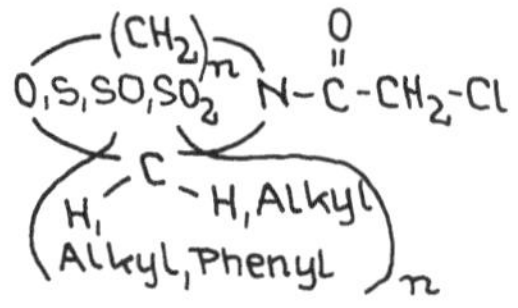

Antidot für Thiolcarbamate, Triazine, Chloracetanilide u. Phenoxyessigsäuren

SU   586 171   17. 8.76/ 1. 2.78   Bogomolova

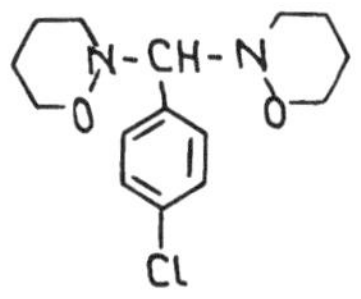

d)   6-Ring-Verbindungen mit 2 x N

d1)   1,2-Pyridazine   Bd.5/643

DOS   2 706 701   17.12.77/24. 8.78   Consortium f. elektrochem·
                                                 Ind.

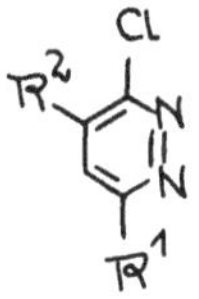

                                          R^1 = H, Alkyl-subst.
                                          R^2 = Cl, Br

EP   9 616   28. 8.79/16. 4.80   Chemie Linz

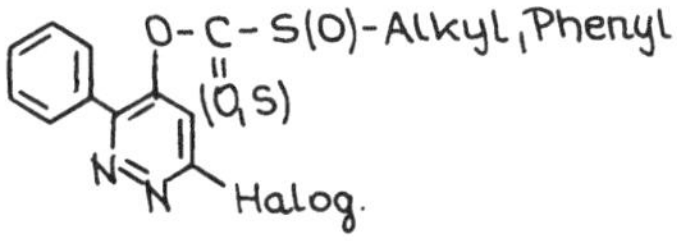

US   4 083 713   14. 3.75/11. 4.78   Rohm u. Haas

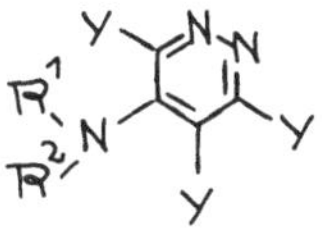

                                          Y = F, Cl, R^1 u. R^2 = H, Cl,
                                                          $ClCH_2$-CO

                                          sel. gegen Gräser auch Acker-
                                          fuchsschwanz u. Wild-Hafer usw.

JA   53 040785   27. 9.76/13. 4.78   Mitsubishi

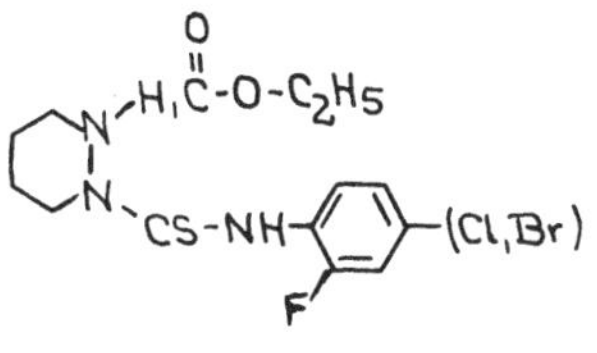

d2)   1,2-Pyridazinone mit Substituenten am N   Bd.5/647

DOS   2 526 643   14. 6.75/30.12.76   BASF

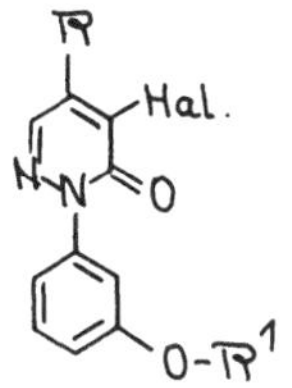

                                          R^1 = F-Alkyl

                                          R = NH_2, N$\langle$H, Alkyl, N(Alkyl)$_2$

                                          N$\langle$H, Alkyl, CO-CH_2Cl usw.

DOS   2 628 990   28. 6.76/20. 1.77   Eli Lilly
                                                    US Prior. 3. 7.75

R^1 u. R^2 = H, CF_3, CH_3Halog.

DOS   2 706 700   17. 2.77/24. 8.78   Consortium f. elektro-
                                                    chem. Ind.

R^1 = H, Alkyl, Cyclo(pentyl,-Hexyl)
R^2 = H. Alkyl, Cl, Br
R^3 = H, Alkyl, Cycloalkyl (substit.)

Darstellung in Verwendung als Herbizid.

DOS   2 742 034   19. 9.77/29. 3.79   BASF

$$A = -N=N-, -N=\overset{\uparrow O}{N}-, -N-\overset{O}{\overset{\|}{C}}-, -N-N-$$

R^3 u. R^4 zusammen = $(O, -C-, S_3, -O-N=C\prec)-N=\overset{\uparrow O}{N}-C-$ usw.
R^1 u. R^2 = H, Alk(en,in)yl, Aryl, Heterocycl. usw.

DOS   2 808 193   25. 2.78/ 6. 9.79   BASF
= EP   3 805

R u. R^1 = O-Alkyl
R^2 = CF_2H, F, Cl, O(S)-CH_2Cl,
        O(S)-CH_2-CH_2-Cl
R^3 = H oder O(S)-Halog.-Alkyl

DOS   2 925 110   21. 6.79/10. 1.80   Oxon Italia
                                                    IT Prior. 22. 6.77

Darstellung der reinen Verbindung.

EP   5 516   19. 5.78/28.11.79   BASF
= DOS   2 821 809   19. 5.78   BASF

X = Halog.
R^1 = Alkyl
R^2 = H, Alkyl, Halog.-Alkyl
R^3 = Phenyl, Cyclohexyl

US   4 013 658   3. 7.75/22. 3.77   Eli Lilly

R = H, Alkyl
R^1 u. R^2 = CF_3, H, F, Cl, CH_3, -N$\prec$
-N$\bigcirc$O, -N$\square$

Neues Darstellungsverfahren.

404

US   4 033 751   20. 4.76/ 5. 7.77   Du Pont

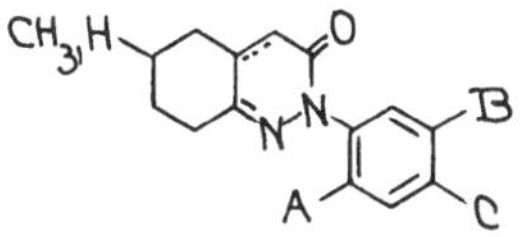

A = Cl, F, CF_3, CH_3
B = H, CH_3
C = Halogen

BE   842 926   14. 6.75/14.12.76   BASF

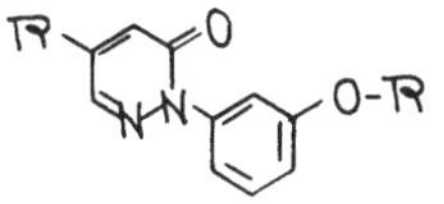

R^1 = niedr. Fluoralkyl
R = NH_2, $N(Alkyl)_2$, O-Alkyl,
     $NHCO-CH_2Cl$ usw.

DDR   124 918   27. 2.76/23. 3.77   Unverricht AG

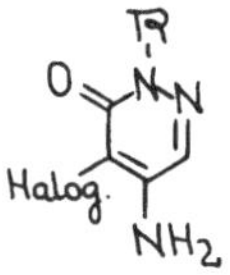

R = H, subst. Alkyl oder Aryl

Darstellungsverfahren

DDR   131 172   30.12.76/ 7. 6.78   Chinoin Gyogyszer

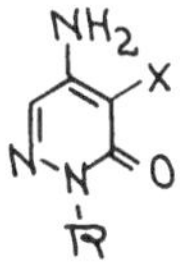

R = Phenyl, Heterocycl.
X = Halogen

JA   52 083552   1. 1.76/12. 7.77   Mitsubishi

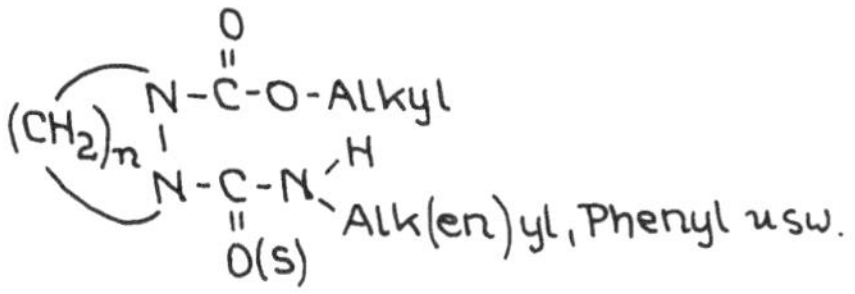

JA   55 028901   16. 8.78/29. 2.80   Consortium f. Elektro-
                                                    chem.

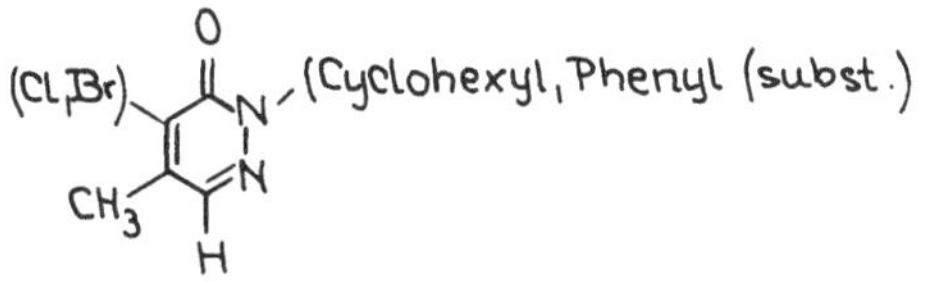

JA   55 053205   18. 4.80/17.10.80   Sumitomo

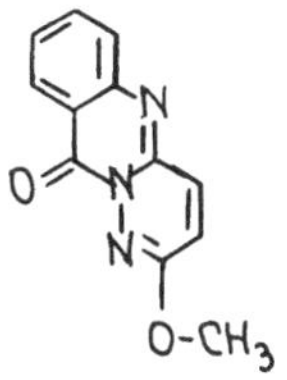

Pyrimidine, Pyrimidinone, Pyrimidin-di-one
(Uracile) u. Chinazoline

e1)   Pyrimidine

DOS   2 630 140   5. 7.76/21. 1.77   Ciba Geigy
                                                    SZ Prior. 7.7.75

R^1 u. R^3 = C_1-C_9-Alkylen substit.,
        Alk(en,in)yl, Cyclopropyl
R^2 u. R^4 = H, C_1-C_4-Alkyl
X = Halog., C_1-C_4O(S)Alkyl
Y = H, Halog.

DOS   2 734 827   2. 8.77/ 9. 2.78   ICI
                                                    GB Prior. 2.8.76

R = H, Phenyl(subst.), Alk(en)yl(subst.),
    Aralkyl, usw.
R^1 = Alkyl; R^2 = H, Alkyl; R^3 = Alk(enyl)
       (subst.)

DOS   2 738 653   26. 8.77/ 2. 3.78   FMC
                                                    US Prior. 27.8.76/30.6.77

R^2 = Halog., $N{<}^H_H$, Alkyl, Cycloalkyl usw!
R^4 = Halog. O-Alkyl
R^6 = H, $N{<}^H_H$, Alkyl, Cycloalkyl usw.

EP   681   21. 7.78/ 7. 2.79   Kuhlmann
                     (FR 8. 7.77 u. 5. 2.78)

X^1, X^2, X^3 = Cl, Br, $N{<}^H_{Alkyl}$, $N(Alkyl)_2$

EP   9 419   26. 9.79/ 2. 4.80   Du Pont
                     (US 27. 9.78 u. 7. 8.79)   Siehe auch 5.5.3k4

$CH_3(O)$ — Z — Alkyl, subst. Alkyl, CH_2Cl usw.

N-(H,CH_3)
C=O(S)
N-H,CH_3
SO_2-R

R = Thienyl oder (subst.)-subst.
Z = CH

EP   9 298   27. 6.79/ 2. 4.80   ICI
                                                   Siehe auch 5.14.5

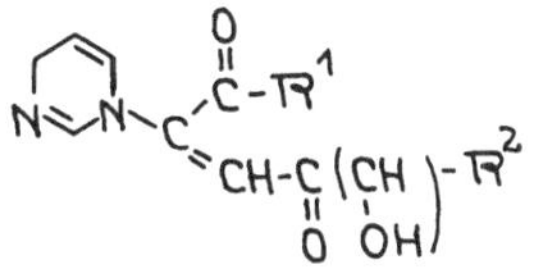

US   4 002 628   18. 6.73/11. 1.77   Eli Lilly

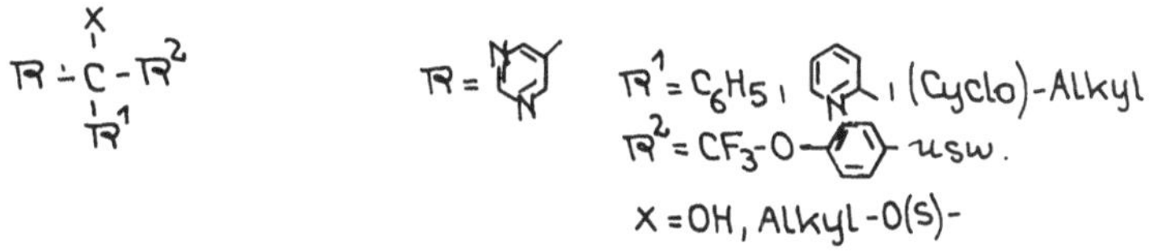

$$R \doteq \overset{X}{\underset{R^1}{C}} - R^2$$

$R = $ (Pyridyl) $R^1 = C_6H_5$, (Naphthyl), (Cyclo)-Alkyl

$R^2 = CF_3-O-$(Phenyl)$-$ usw.

$X = OH,\ Alkyl-O(S)-$

US   4 118 217   6. 1.77/ 3.10.78   Farmland Ind.

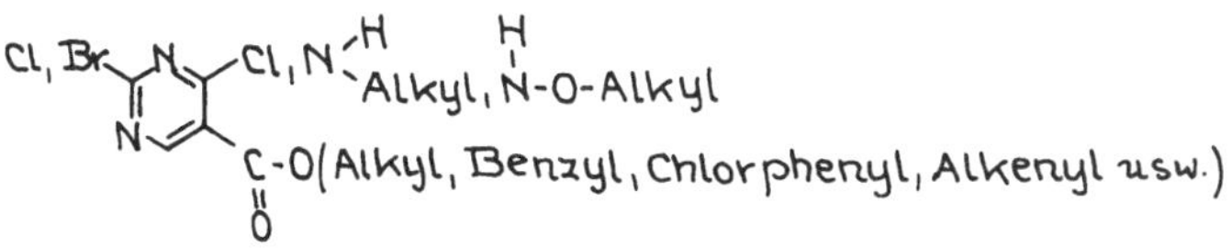

Cl, Br-(Pyrimidin)-Cl, N-H-Alkyl, N-O-Alkyl

C-O(Alkyl, Benzyl, Chlorphenyl, Alkenyl usw.)

BE   843 820   7. 7.75/ 6. 1.77   Ciba Geigy

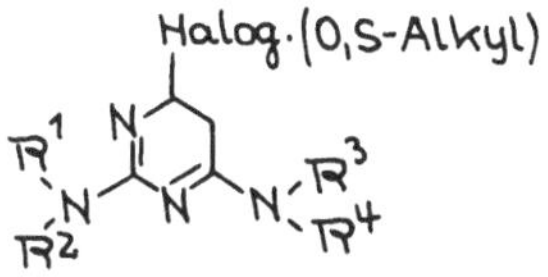

Halog.(O,S-Alkyl)

R^1 u. R^3 = Alkyl subst. durch CN, O-Alkyl,
                 S-Alkyl, Cyclopropyl, Phenyl

R^2 u. R^4 = H, Alkyl

(R^1 oder R^3 1 x Cyclopropyl)

BE   865 034   23. 3.77/18. 9.78   ICI

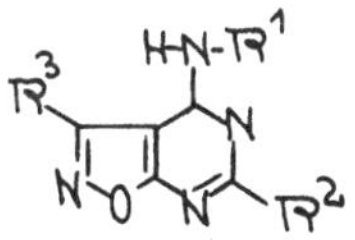

R^1 = C_2-Alkyl, Alkoxy-Alkyl, Aralkyl

R^2 = H, Alkyl

R^3 = Alkyl, Alkyl-S-, Alkyl-SO_2-, Alkyl-O-,
         CN, N$<$

u. a. auch herbizid wirksam

GB   2 041 359   10. 1.79/10. 9.80   ICI

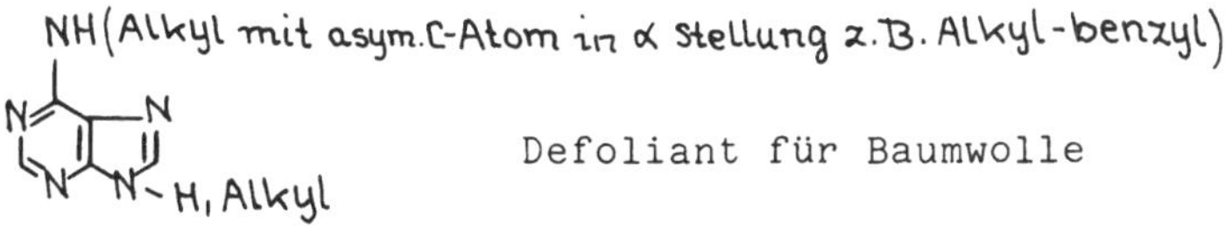

NH(Alkyl mit asym. C-Atom in α Stellung z. B. Alkyl-benzyl)

Defoliant für Baumwolle

JA   54 027580   2. 8.77/ 1. 3.79   Mitsui Toatsu

Alk(en)yl, Cl, S-CH$_3$, NH$_2$

JA   54 117468   3. 3.78/12. 9.79   Sankyo KK

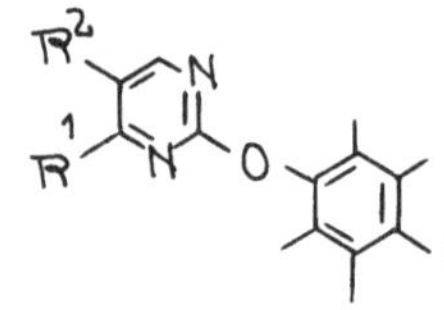

R^1 = H, CH_3; R^2 = Halog.

belieb. subst. z.B. Alkyl, CF_3, C-Alkyl, CH_2-CO-O-Alk. usw.

JA   55 066505   15.11.78/25. 5.80   Nissan Chem.

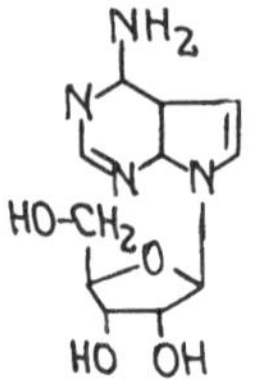

Antibioticum "Tubercidin" als Herbicid
J. Antibiot. 10A201 (1957)

e2)   <u>Pyrimidin-one</u>   Bd.5/656

DOS   2 654 095   29.11.76/23. 6.77   Eli Lilly
                      US 11.12.75

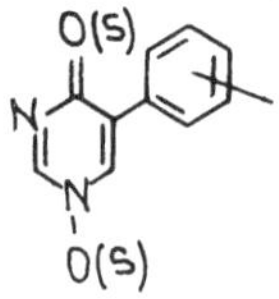

Cl, Br oder CF_3 in m-Stellung

DOS   2 720 442   6. 5.77/ 8.12.77   Fisons

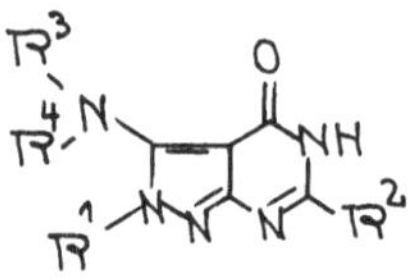

R^1 = H, Alkyl(ev. substituiert)
       Cycloalkyl, Phenyl
R^2 = H, OH, Halogen, Alkyl, Alk(en,
       in,)yl, Cycloalkyl, Aryl,
       Heterocycl. usw.
R^3 u. R^4 = H, Alkyl, Alkyl-SO_2,
       Alkyl-O-CO usw.

DOS   2 734 827   2. 8.76/ 9. 2.78   ICI

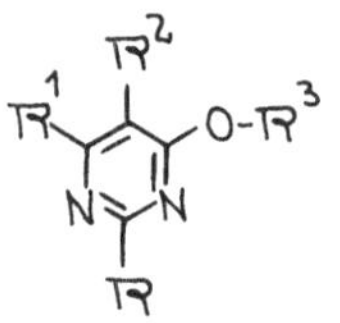

R = H, (subst.)-Phenyl, Alkyl, Alkenyl,
    Phenyl, S-Alkyl, Furyl, Pyridyl, N<
R^1 = Alkyl; R^2 = H, Alkyl,
R^3 = (Cyclo)-Alkyl, Phenyl, Pyridyl usw.

US   4 012 388   11.12.75/15. 3.77   Eli Lilly

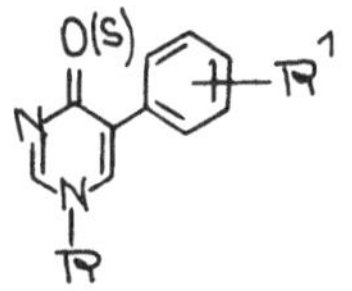

R = C_1–C_3–Alkyl
R^1 = Cl, Br, CF_3

JA   52 156 858   22. 6.76/27.12.77   Takeda Chem.

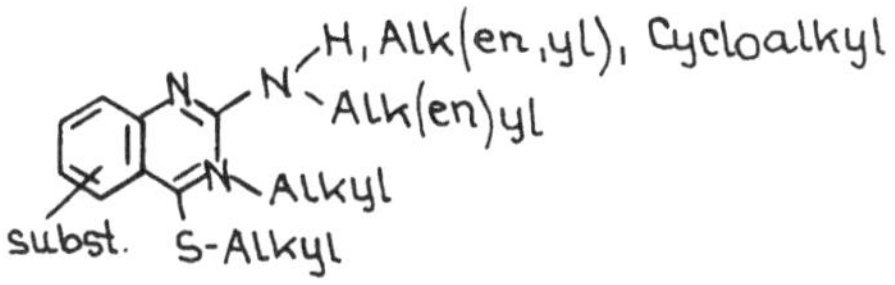

JA   54 005035   13. 6.77/16. 1.79   Nissan

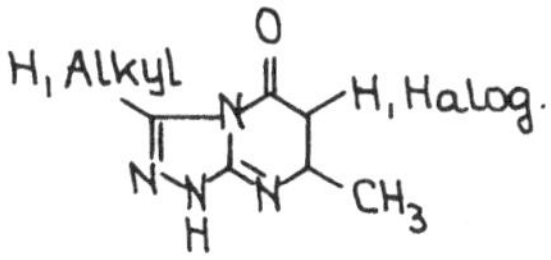

gegen Unkräuter in Reis

JA   55 053 205   17.10.78/18. 4.80 Sumitomo

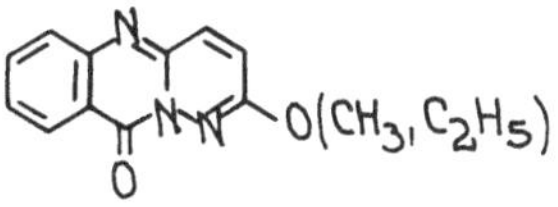

e3)   Dioxo-pyrimidine(Uracile u. Dihydrouracile)   Bd.5/659

DOS   1 967 031   25. 6.69/12. 5.77   Amchem
US Prior. 26. 8.68

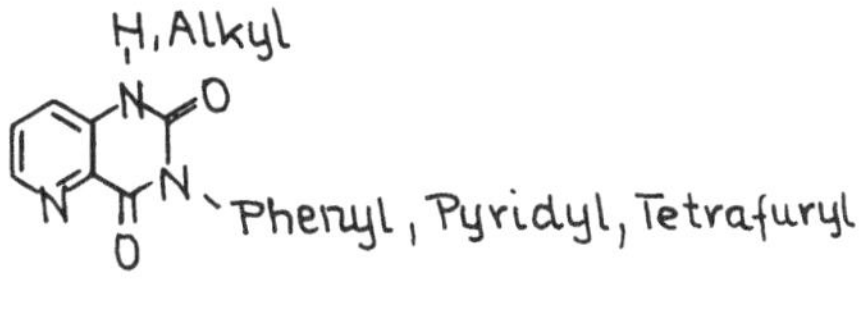

DOS   2 801 008   11. 1.78/12. 7.79   Bayer

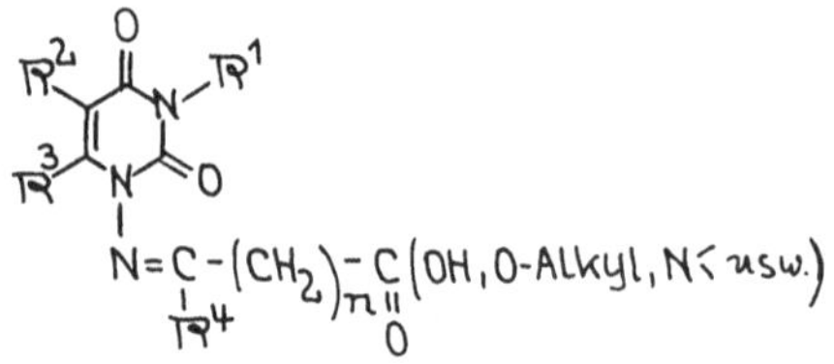

R^1 = Alk(en,in)yl, O-(S)-Alkyl
Aryl, Cycloalkyl,Ar-Alkyl
R^2 = H, Halog., Alkyl
R^3 = Alkyl,
R^2 u. R^3 = mehrgl.CH_2-Brücke,
ev durch Heteroatom
unterbrochen; R^4 = H,
Alkyl, Phenyl

BE   845 387   21. 8.75/21. 2.77   Roussel Uclaf

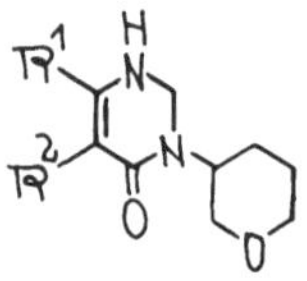

R = C_3H_7; R^2 = Cl oder R^1 = CH_2 und
R^2 = CH$<^{CH_3}_{CH_3}$

selektiv in Getreide; pre emergence Anw.

FR   2 293 432   5.12.74/ 6. 8.76   Roussel Uclaf

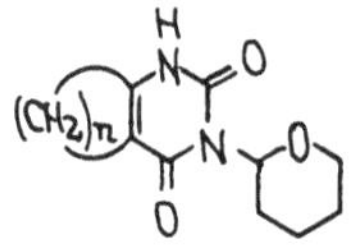

n = 4 oder 5

Selektiv in Getreide spez. Hafer,
Weizen und Gerste

FR   2 335 511   18.12.75/19. 8.77   Roussel Uclaf

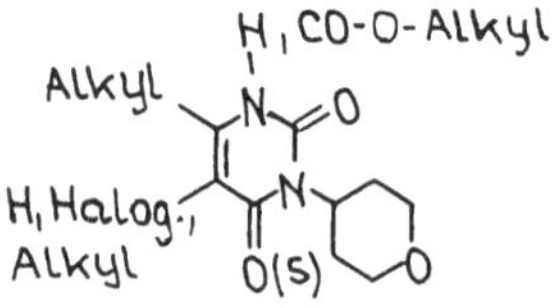

Tetrohydropyranyl-uracile

FR   2 371 445   18.12.75/21. 7.78   Roussel Uclaf

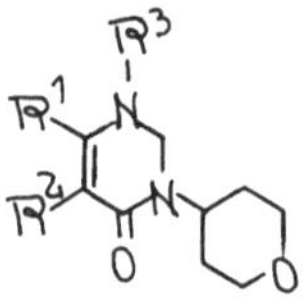

R^1 = Propyl, iso-Propyl, Ethyl
R^2 = Cl
R^3 = H, CO-OC_2H_5

Selektiv in Rüben und Getreide.

JA   54 073122   17.11.77/12. 6.77   Katayama Kasaku

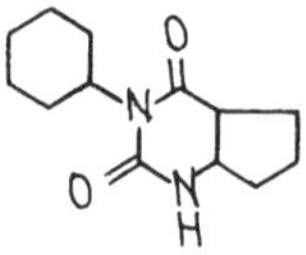

JA   55 108 858   14. 2.79/21. 8.80   Kyowa Hakko

A u. B verschieden,1(S)subst. Phenyl
anderes, Alkyl, Cycloalkyl, Aryl

f)   <u>1,4-Pyrazine</u>   Bd.5/665

DOS   2 800 010   2. 1.78/13. 7.78   ICI
                                              GB Prior. 6. 1.77

DOS   2 854 603   18.12.78/ 5. 7.79   Kyowa Gas Chem.
= NL   7 812 456   JA Prior.22.12.77/20. 1.78;
= BE   872 944   26. 7.78;28. 9.78

A = H, Alkyl, Phenyl, Benzyl,
    O-(S)-Alk(en,in)yl,O-Aryl,
    Heterocycl., N<
B = Halog., Phenyl, Benzyl. O-(S)-
    Alk(en,in)yl, O-Aryl, Heterocycl.
    N<

US   4 194 899   17.12.75/23. 3.80   Eli Lilly

u.a. auch Herbizide

GB   1 574 429   28. 4.76/10. 9.80   Shell Int. Res.

X = -N =u. R = H, O-Alkyl, O-Phenyl oder
X = -NH-,R = = O

JA	52 090628	23. 1.76/30. 7.77	Nissan

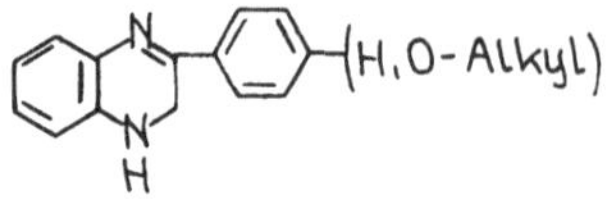

(H,O-Alkyl)

JA	52 090629	23. 1.76/30. 7.77	Nissan

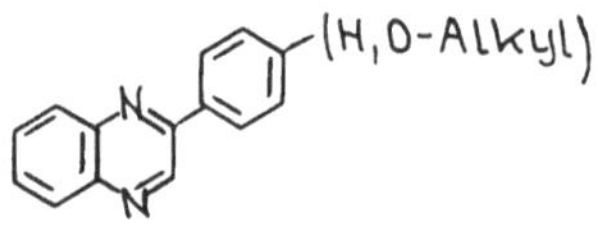

(H,O-Alkyl)

JA	52 156927	21. 6.76/27.12.77	Nissan

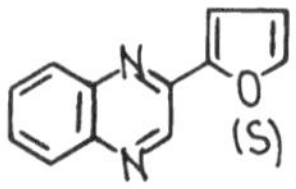

(S)

Neue Herbizidklasse,Kontakt u. Bodenherbizid

JA	54 106479	7. 2.78/21. 8.79	Kyowa Gas Chem.
JA	54 106480	7.2.78/21. 8.79	Kyowa Gas Chem.

H, Alkyl, Benzyl, Aryl, bzw. Cl

Alkyl

H, Alkyl, NH-CH-CO-(O-Alkyl, N (CH$_2$)$_n$)

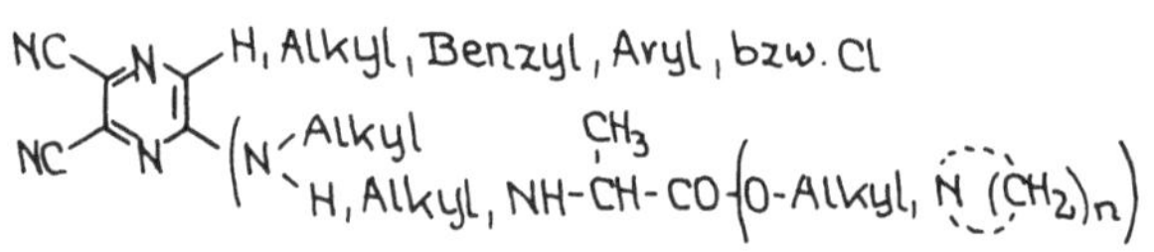

JA	55 017341	25. 7.78/ 6. 2.80	Meiji Seika

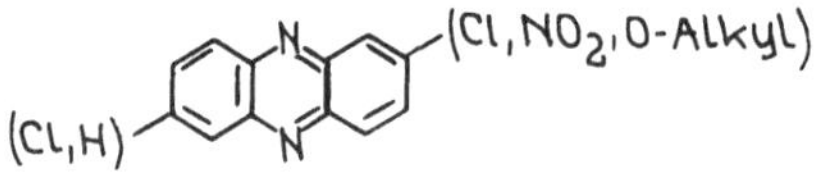

(Cl,NO$_2$,O-Alkyl)

(Cl,H)

JA	55 028908	21. 8.78/29. 9.80	Kyowa Gas Chem.

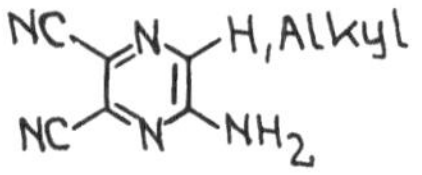

H,Alkyl

NH$_2$

JA	55 045647	29. 9.78/31. 3.80	Kyowa Gas Chem.

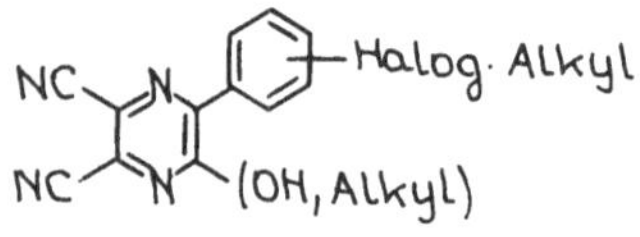

Halog. Alkyl

(OH,Alkyl)

selektiv im Wassereis

JA	55 079307	11.12.78/14. 6.80	Kyowa Gas Chem.

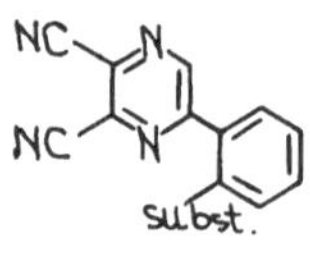

subst.

JA   55 079378   11.12.78/14. 6.80   Kyowa Gas Chem.

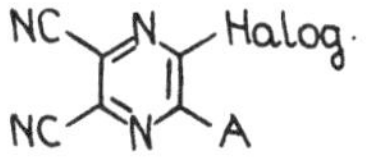

A = O(S)-Alkyl, O(S)Phenyl oder Benzyl,
$$-N\begin{array}{c}\text{Alkyl}\\\text{Alkyl}\end{array}$$

5.14.4   6-Ring-Verbindungen mit 3 x N   Bd.5/666

   a) Symmetrische Triazine

DOS   2 505 703 u.4   12. 2.75/26. 8.76   Degussa

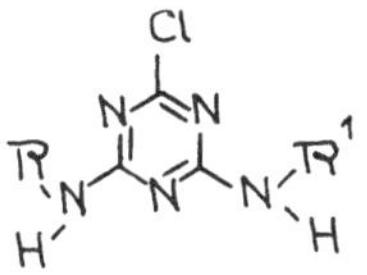

                                               spezielles Darstellungsverfahren

DOS   2 614 762   6. 4.76/30. 6.77   Oxon Sa.

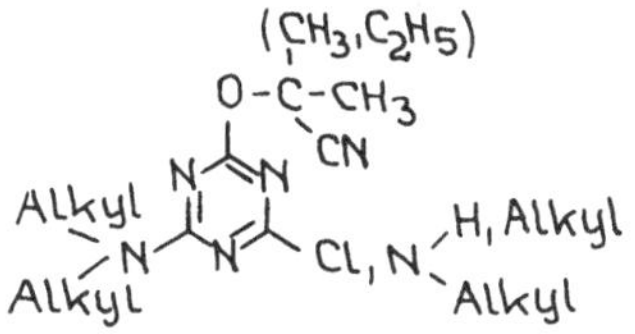

DOS   2 638 069   24. 8.76/15.12.77   Richter Gedern
                                               HU Prior. 1.6.76

Herbizide wie symmetrische Triazine + K_2SO_4 oder $NaHSO_3$, $KHSO_4$

DOS   2 657 944   21.12.76/22. 6.78   Armin Agr. Inst.

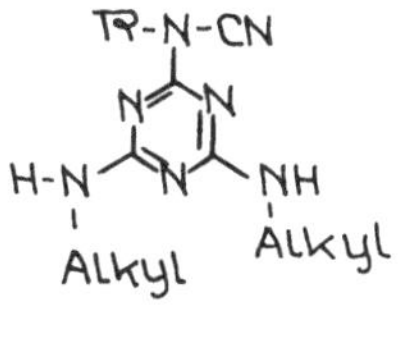

                                               R = Alk(en,in)yl

DOS   2 815 689   11. 4.78/19.10.78   Ciba Geigy
= BE   865 935   14. 4.77/13.10.78   CH Prior. 14. 4.77

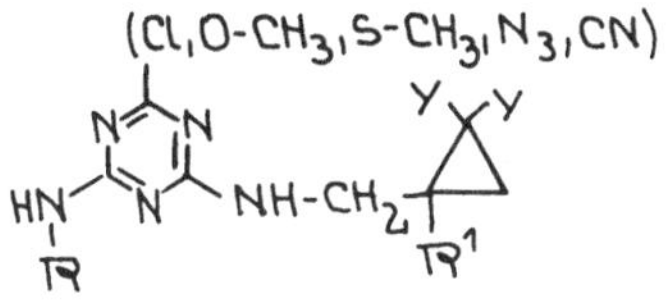

                                               R = H, C_1-C_6-Alkyl, Alkyl subst.
                                                  durch CN,OH,O-Alkyl,Cyclo-
                                                  Propyl oder CH_2-Cyclopropyl
                                               R^1 = H,CH_3; Y = Halogen

EP   3 749   22. 1.79/ 5. 9.79
                                          CH Prior. 3.2.78 Ciba Geigy

S-CH$_3$

                                          Anwendung als Meerwasseralgizide

HN—[Triazin]—NH—CH—CH(CH$_3$)(CH$_3$)
 |
 (Cyclopropyl)   CH$_3$

                    auch. ähnl. Verbindungen

EP   5 986   29. 5.79/12.12.79   Du Pont
                                                        US Prior. 30. 5.78

X—Y—X (Triazin)
NH—C=N—SO$_2$-(Thienyl) Ar
     |
     S-R

X= CH$_3$, O-CH$_3$, CH$_2$-O-CH$_3$, O-C$_2$H$_5$
Y = N, CH

EP   9 419   26. 9.79/ 2. 4.80   Du Pont

CH$_3$(O)—[Triazin]—Alkyl, subst. Alkyl usw.
                      H, CH$_3$
CH$_3$, H-N-C-N-SO$_2$-(2-Thienyl, Ar. subst.)
            ||
            O(S)

US   2 522 290   20. 5.75/ 2.12.76   Degussa
= BE   842 009   20. 5.75/19.11.76   Degussa

N$_3$
R-HN—[Triazin]—NH-R

US   4 127 405   7. 4.76/28.11.78   Du Pont

(H, Cl, S-CH$_3$, CF$_3$, CH$_2$-O-CH$_3$ usw.)
R^1-SO$_2$-NH-C-NH—[Triazin]
              ||
              O(S)   (O-CH$_3$, CH$_3$)

R^1= (Tetrahydrofuranyl)-subst., (Tetrahydrothienyl)-subst., (Aryl)-subst.

BE   845 808   5. 9.75/31.12.76   Oxon Sa.

                    CH$_3$, C$_2$H$_5$
            O-C-CN
                    CH$_3$
Alk, H
      N—[Triazin]
Alk   Cl, N(Alkyl)$_2$, S-Alkyl

                                          Herbizide u. Wuchsregulatoren

414

BE   862 101   22.12.76/21. 6.78   Ciba Geigy

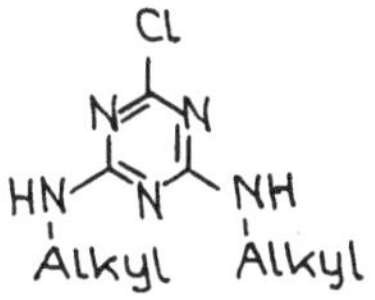

                                                    verbessertes Darstellungs-
                                                    verfahren

SU   430 641   11. 1.71/23.11.77   Melnikova

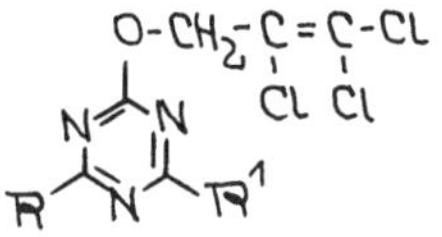

                                    R u. R^1 = Phenyl, -O-CH$_3$, -S-CH$_3$,
                                    CCl$_2$ = CCl-CH$_2$-O, N $<$ H, Alkyl, Phenyl
                                                         H, Alkyl, Phenyl

SU   503 573   3. 6.74/25. 3.76   Plant Chem.

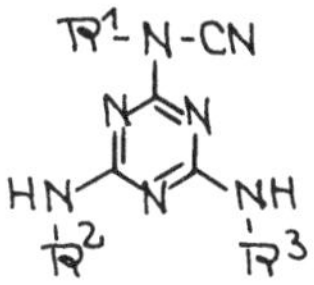

                                    R^1 = Alk(en)yl
                                    R^2 u. R^3 = Alkyl

CH   601 999   29. 5.74/14. 7.78   Ciba Geigy

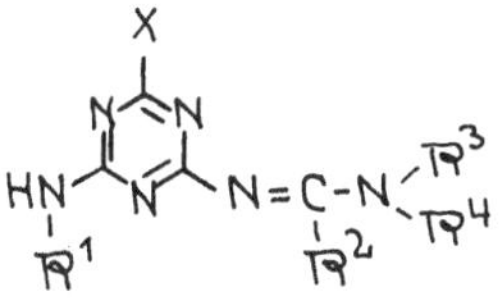

                                    R^1 = (Cyclo)-Alkyl; X = Halog.,O(S)-Alkyl
                                    R^2 = H, Alkyl
                                    R^3 u. R^4 = Alk(en,in)yl oder einen
                                                Polymethylenring bildend;
                                                Ring mit O, S oder N

CH   605 853   31. 1.75/13.10.78   Sandoz

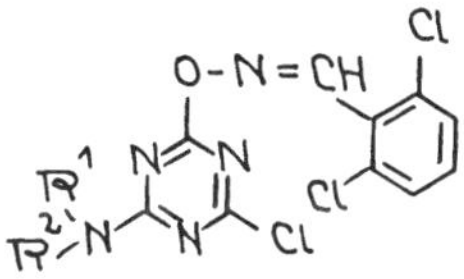

                                    R^1 = (CH$_2$)$_n$ N$<$ O(S)Alkyl;

                                    R^2=(CH$_2$)$_n$ N$<$ O(S)Alkyl oder N$<$

JA   53 028186   30. 8.76/16. 3.78   Kyowak

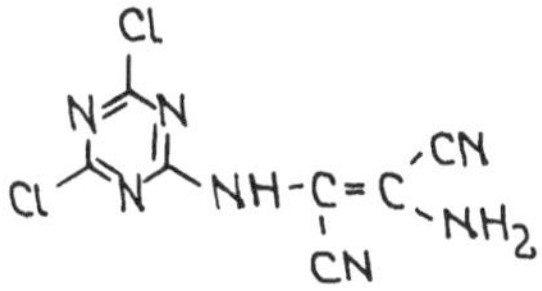

                                    herbizid und fungizid wirksam

JA   54 014987   1. 7.77/ 3. 2.79   Kyowak

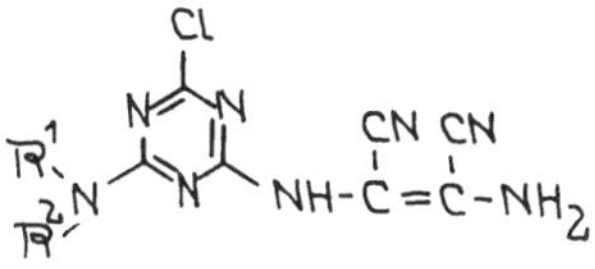

                                    R^1 = H, Aralkyl; R^2 = H,
                                          Aralkyl, Allyl

b) <u>Teil- oder voll-hydrierte 1,3,5-Triazine</u>

<u>sowie 1,3,5-Triazinone</u>

DOS 2 254 200 6.11.72/16. 5.74 Bayer

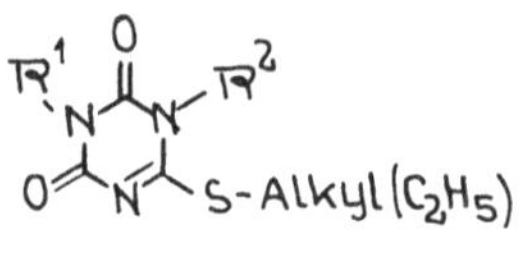

R^1 = Alkyl, Cycloalkyl, subst. Aryl
O(S)Aryl, Neopentyl
R^2 = H, Alkyl, Cycloalkyl, Aralkyl, OH,
O-Alkyl, N<, NH_2

breites z.T. auch selektives Wirkungsspektrum z.B. auch gegen
Flughafer

DOS 2 603 180 28. 1.76/19. 8.76 ICI
GB Prior. 5. 2.75

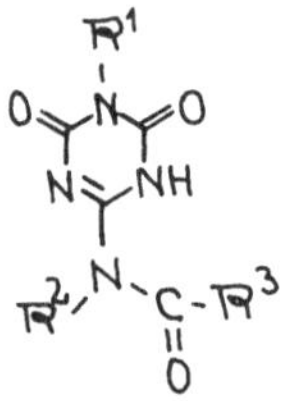

R^1 = C_1-C_5-Alkyl mit sec. oder tert.C am N
R^2 = Alkyl, verzweigt
R^3 = CH_3, C_2H_5 u. Δ

Selektiv gegen Gräser in Erdnüssen, Sojabohnen, Luzernen.

DOS 2 638 519 26. 8.76/ 3. 3.77 Du Pont
= US 4 060 404 26. 8.77/29.11.77

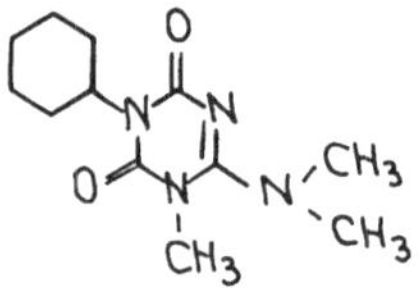

statt Cyclohexyl auch Alkenyl
selektiv in immergrünen Kulturen

DOS 2 645 558 8.10.76/21. 4.77 Du Pont
= BE 847 117

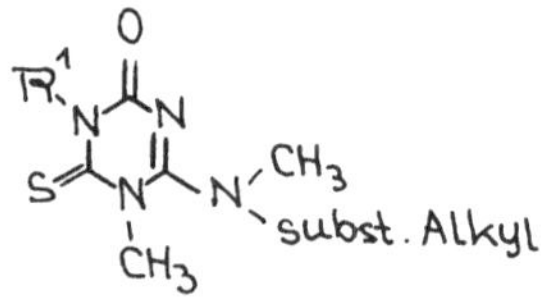

R^1 = C_3-C_6-Alkyl, subst. Cycloalkyl
Cyclohexenyl, subst. Phenyl

DOS 2 720 792 12. 5.76/24.11.77 ICI
GB 12. 5. 76

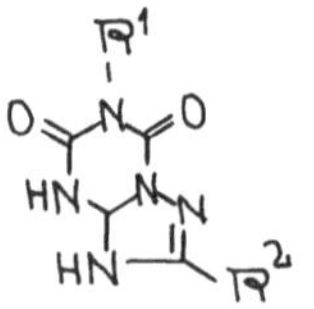

R^1 = (Cyclo)-Alkyl
R^2 = H, (Cyclo)-Alkyl, S-Alkyl, N<

selektiv in Mais, Weizen, Gerste

DOS   2 751 437   17.11.77/24. 5.78   Eli Lilly
                                               US Prior. 19.11.76

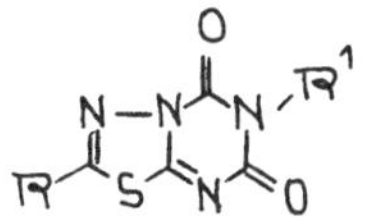

R = C_1-C_4-Alkyl, SO_2-$N(CH_3)_2$

CF$_3$, CClF$_2$, SO$_2$-N⌒O

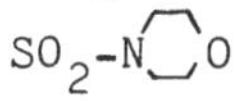

R^1 = H; C_1-C_4-Alkyl, Allyl, Phenyl(subst.)

DOS   2 801 029   11. 1.78/12. 7.79   Bayer

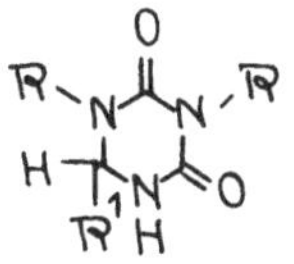

R = Alkyl, Cycloalkyl, Phenyl,C_2-H_5-SO_2-,
                              usw.
R^1 = H, Aryl(Cyclo)-Alkyl

EP   4 171   5. 3.79/19. 9.79   ICI
                   (GB 10. 3.78)

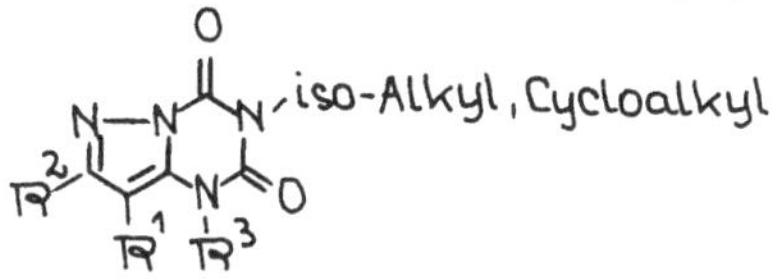

R^1 = Halog., CN, NO_2 CO-N$<$
        -N-CO-Alkyl
        |
R^2 = H, O(S)-Alkyl,-N-Alkyl;
                       |
                       H

US   3 907 795   24. 5.74/23. 9.75   Du Pont
(irrtümlich an anderer Stelle im Bd.5 eingeordnet)

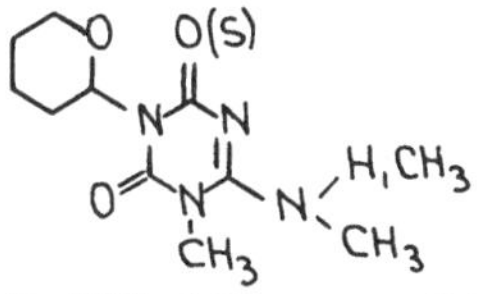

US   3 983 116   24. 5.72/28. 9.76   Du Pont

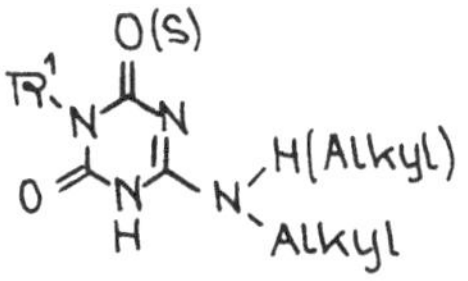

R^1 = (Cyclo)-Alkyl subst., Alk(en,in)al

US   4 042 372   19.11.76/16. 8.77   Eli Lilly

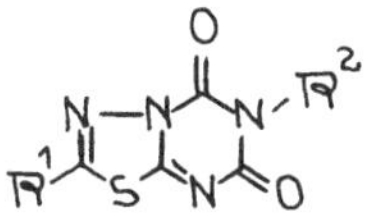

R^1 = CH_3; ClF_2C, CHF_2, CF_3, C_2F_5

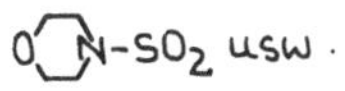

O⌒N-SO_2 usw.

R^2 = H, CH_2, C_2H_5, CH_2-CH = CH_2,C_6H_5
        (subst.)

u.a. auch herbizid wirksam

US   4 152 516   9. 2.78/ 1. 5.79   PPG

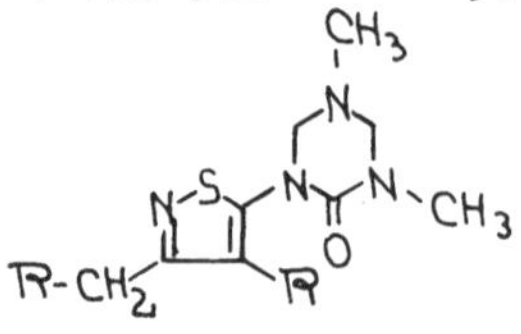

US   4 178 448   5. 5.75/11.12.79   Du Pont

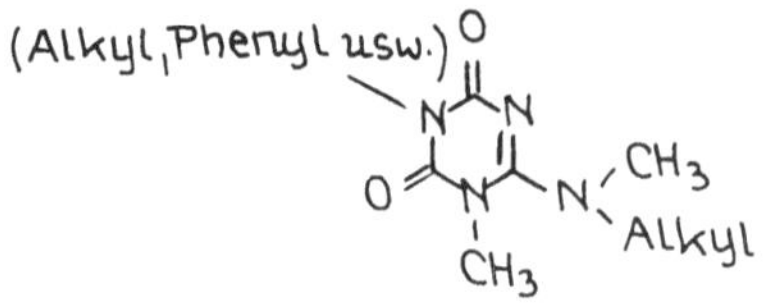

Neues Verfahren

SU   600 142   18.12.75/ 7. 6.78   Armin Agric. Inst.

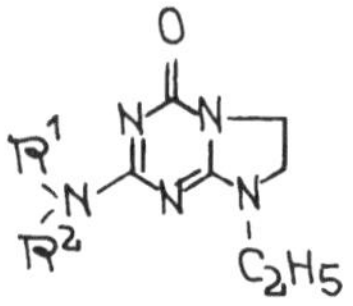

R^1 u. R^2 = H, CH_3, C_2H_5, iso-C_3H_7

DDR   121 517   10. 6.75/ 5. 8.76   Werchan H.G.

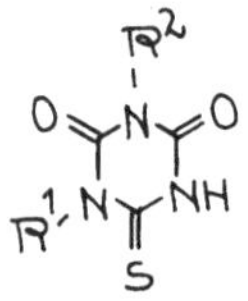

R^1 = Alkyl, subst. Aryl
R^2 = Alkyl

5.14.5   <u>1,2,4-Triazine u. Triazin-5-one sowie dione</u>   Bd.5/683

<u>(Azauracile) und Dihydroverbindungen</u>

DOS   2 517 654   22. 4.75/ 4.11.76   Bayer
= BE   840 947   22. 4.75/21.10.76   Bayer

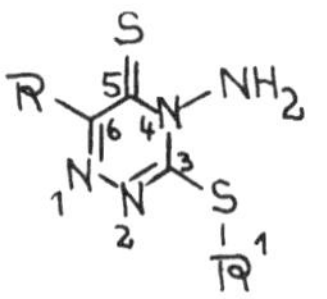

R = (Cyclo)Alkyl, Alkenyl, Phenyl
     (subst.)
R^1 = H, Alkyl(C_1-C_9)

DOS   2 527 490   20. 6.75/30.12.76   BASF

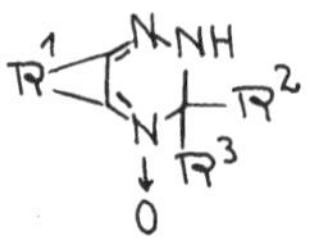

R^1 = aliph. Rest
R^2 u. R^3 = Alkyl, Aryl, Cycloalkyl

418

DOS   2 528 302   25. 6.75/13. 1.77   Bayer

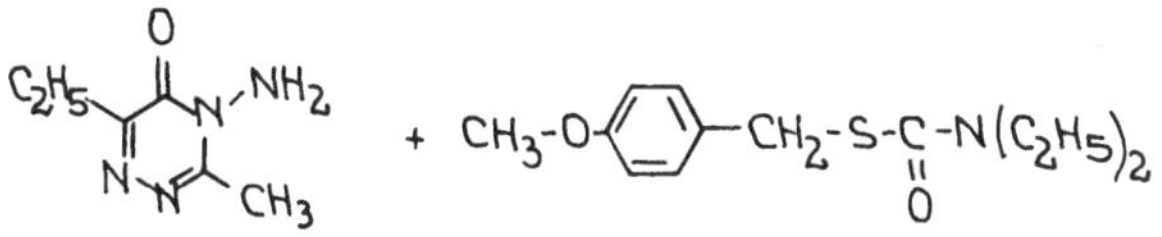

$+\ CH_3-O-\langle\rangle-CH_2-S-\underset{O}{\overset{O}{\underset{||}{C}}}-N(C_2H_5)_2$

Gegen einige spez. Gräser wie Wildhafer, Digitaria sanguinalis u.
Unkräuter wie Galium aparine.

DOS   2 537 290   21. 8.75/ 3. 3.77   Bayer

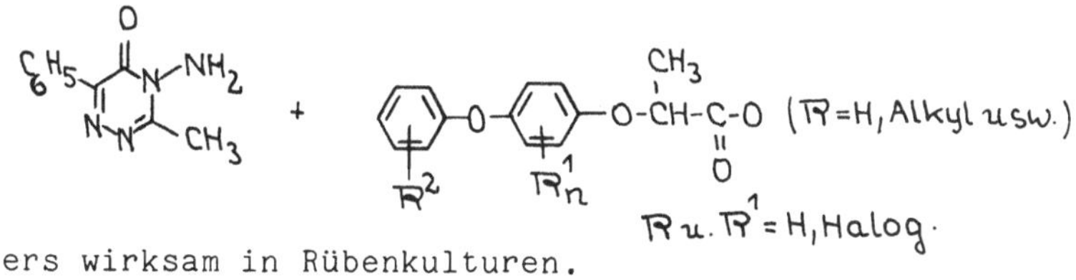

$+\quad(R = H, Alkyl\ usw.)$

$R\ u.\ R^1 = H, Halog.$

Besonders wirksam in Rübenkulturen.

DOS   2 540 958   13. 9.75/17. 3.77   Bayer
= BE   846 128   13. 9.75/14. 3.77   Bayer

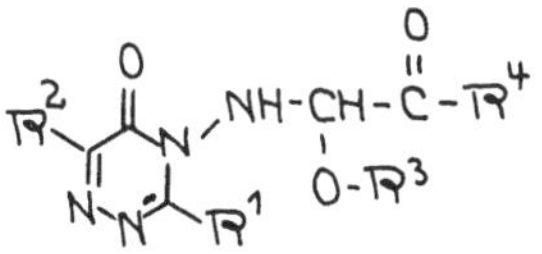

R^1 = (O,S)-Alkyl, N
R^2 = (Cyclo)-Alkyl, subst. Phenyl
R^3 = H, Alk(en,in)yl
R^4 = OH, O-Alkyl, S-Alkyl, N$<$

DOS   2 602 186   21. 1.76/22. 7.76   Sumitomo

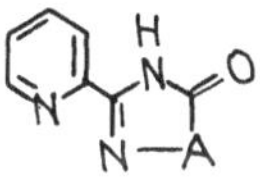

A = -NH-CRH- oder -N $=$ CR
R = H, Alkyl

Herbizide und Fungizide

z.B.

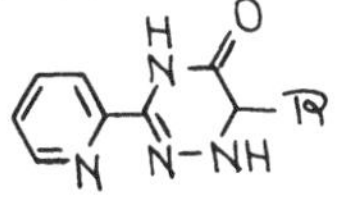

DOS   2 613 434   30. 3.76/13.10.77   Bayer

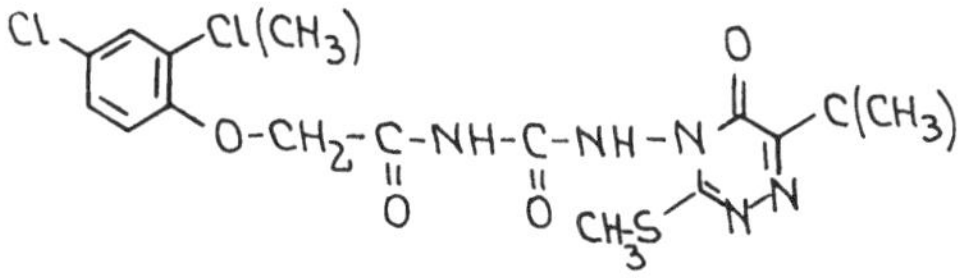

DOS   2 620 370   8. 5.76/17.11.77   Bayer

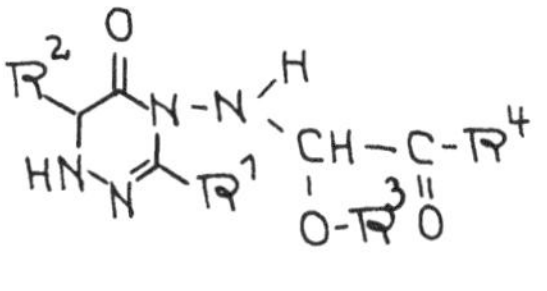

R^1 = (O,S)Alkyl, N$<$
R^2 = (Cyclo)Alkyl, Phenyl(subst.)
R^3 = H, Alk(en,in)yl
R^4 = OH, O-Alkyl, N$<$

DOS   2 647 460   21.10.76/27. 4.78   Bayer

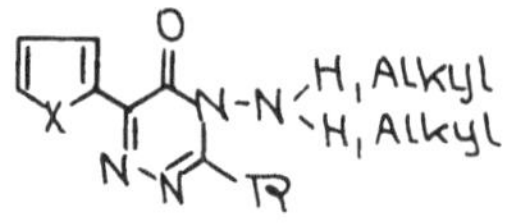

X = O, S, NH

R = H, Alkyl, Alkenyl usw.

DOS   2 729 761   1. 7.77/ 2. 2.78   Mobay USA u. Bayer
                                            US Prior. 26. 7.76

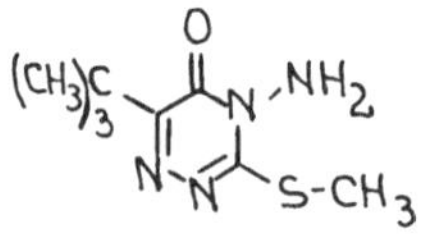

Spezielles Methylierungsverfahren
bei der Darstellung.

DOS   2 732 797   20. 7.77/ 8. 2.79   Ciba Geigy
= BE   869 138   20. 7.77/19. 1.79   Ciba Geigy

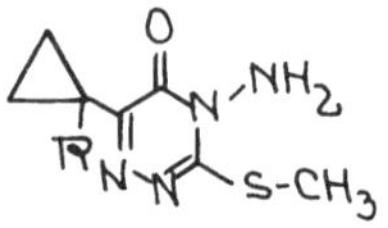

R = iso(n)-Propyl, CH_3, C_2H_5

DOS   2 733 180   22. 7.77/25. 1.79   Degussa
= BE   869 139   22. 7.77/19. 1.79   Degussa

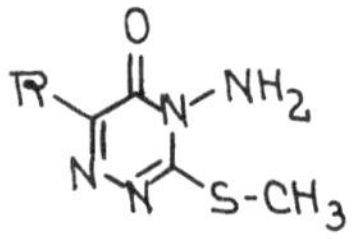

R = $(CH_3)_3$C oder

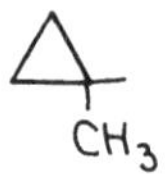

Darstellungsverfahren

DOS   2 804 435   2. 2.78/17. 8.78   American Cyanamid
= BE   863 567   3. 2.77/ 2. 8.78   US Prior. 3. 2.;18.10;
                                            7.11.77

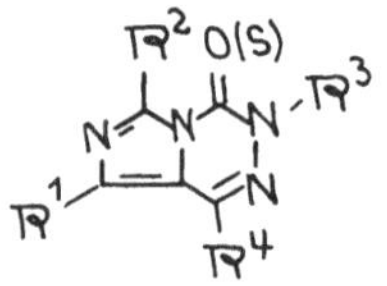

R^1 = H, Alkyl, Halogen-Alkyl

R^2 = H, Alkyl. Cycloalkyl,$-CH_2-O-CH_3$,
$C_6H_5-CH_2$, Phenyl(subst.) usw.
NH_2, N$\begin{smallmatrix}Alkyl\\H, Alkyl\end{smallmatrix}$

R^3 = H, Alkyl, Alkenyl, Alkinyl

Arzneimittel und Herbizide. R^4 = H, Alkyl

DOS   2 856 750   29.12.78/17. 7.80   Ciba Geigy u. Degussa

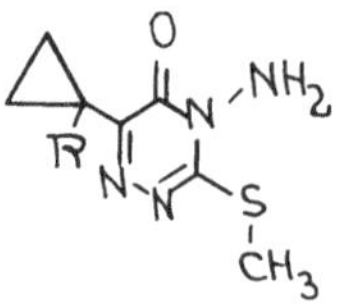

R = iso, n-Propyl oder C_2H_5

siehe DOS 2 732 797

420

DOS   2 908 963   7. 3.79/18. 9.80   Bayer

DOS   2 908 964   7. 3.79/18. 9. 80   Bayer

DOS   3 016 304   28. 4.80/ 6.11.80   Sumitomo

R^1=H, CH_2-OH, CH_2-N$\begin{smallmatrix}H, Alkyl, CH_2-C\equiv CH, Phenyl\ usw.\\ H, Alkyl, CH_2-C\equiv CH, Phenyl\ usw.\end{smallmatrix}$

CH_2-N (O,N) Heterocyclus, hydriert

Spezifisch wirksam gegen Wildhafer in Getreide.

EP   9 298   27. 6.79/ 2. 4.80   ICI
                                                 GB Prior. 27. 7.78

R^1 u. R^2 = Alkyl, Cycloalkyl
              Phenyl

statt 1x =N⁻ und =CH-

EP   15 452   22. 2.80/17. 9.80   Bayer
= DOS   2 908 963, 2 908 964 u. 2 938 384

R^1 = sec. oder tert. Butyl,
       Cyclohexyl

US   4 002 624   26. 1.76/11. 1.77   Dow Chem.

R^1 = H, C_1-C_3-Alkyl

oder

R^2 u. R^3 = C_6H_5-CH_2, p-Cl-C_6H_4-CH_2-,
              $(C_6H_5)_2CH$-

Allgemeine Wirkung, auch Herbizide.

US   4 002 625   26. 1.76/11. 1.77   Dow Chem.

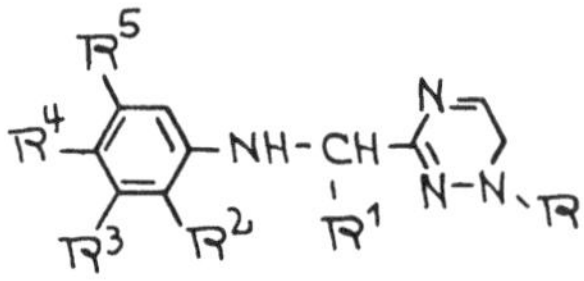

R u. R^1 = H, Alkyl
R^2, R^3, R^4 u. R^5 = H, CH_3,
              CCl_3, CF_3, Halog.

US   4 080 192   14. 4.77/21. 3.78   Du Pont

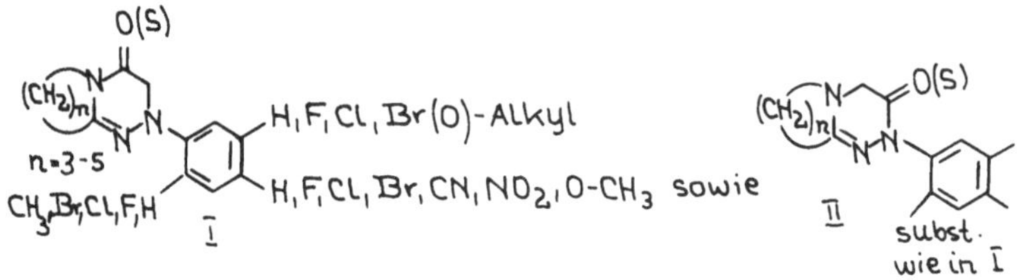

US   4 107 308   18.10.77/15. 8.78   Amer. Cyanamid

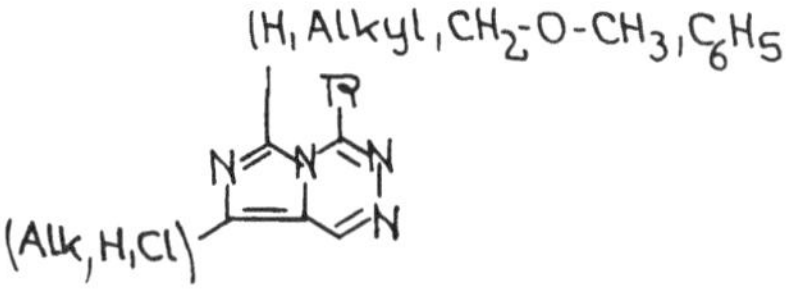

R = $S-CH_2-CH = CH_2$, $N\!\!<^H_{Alkyl}$,
S-Alkyl, Pyrazolyl-amino
usw.

US   4 126 444   22.11.77/21.11.78   Amer. Cyanamid

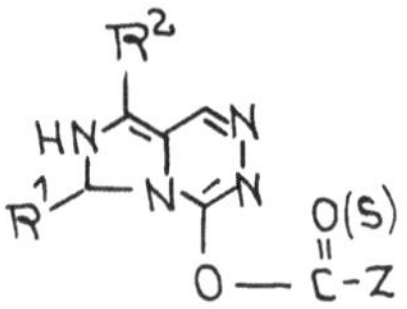

R^1 = (Cyclo)-Alkyl, Phenyl
R^2 = H, Alkyl, Halogen
Z = Alkyl, $C_6H_5-CH_2-$

US   4 168 964   18.10.77/25. 9.79   Amer. Cyanamid

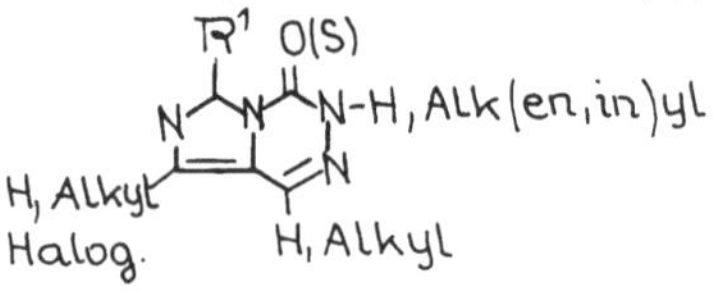

R^1 = H, Alkyl, Cycloalkyl,
       Phenyl usw.

BE   845 594   29. 8.73/28. 2.77   Ciba Geigy

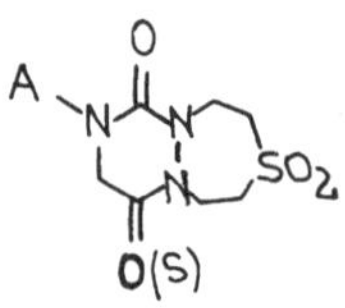

A = Phenyl in p.Stellung subst. durch
    Cl, (O)-Alkyl, CF_3, $SO_2-N<$, CN,
    CO-N$<$, SO_2-Alkyl

Wirksam gegen Galium-Arten.

BE   859 936   21.10.76/20. 4.78   Bayer

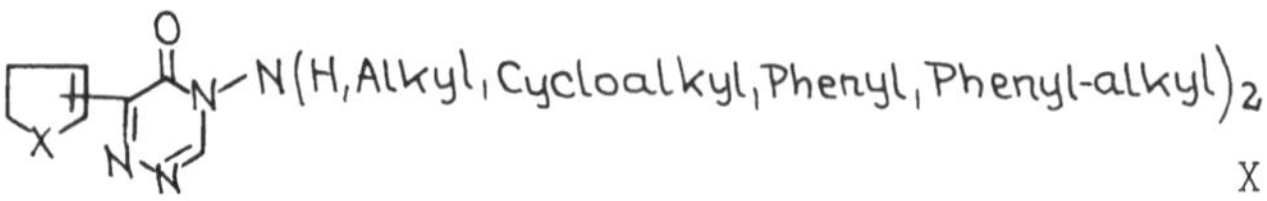

X = O, S, NH

422

5.14.6   <u>6-Ring-Verbindungen mit drei verschiedenen oder</u>

          <u>mehr Heteroatomen</u>

     a)   <u>1 x P + 2 x O</u>

JA   54 032491   17. 8.77/ 9. 3.79   Sumitomo

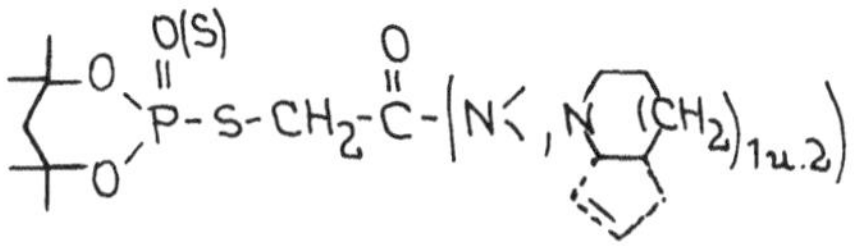

$$\text{[Dioxaphospholan]}O(S)=P-S-CH_2-C(=O)-\left(N<,\ N\ (CH_2)_{1\ u.2}\right)$$

     b)   <u>2 x N + 1 x O oder S, 2 x N + 1 x P</u>

     b1) <u>2 x N + 1 x O</u>   Bd.5/687

DOS   2 640 464   8. 9.76/14. 4.77   Philagro
= BE   846 083   FR Prior. 11. 9.75

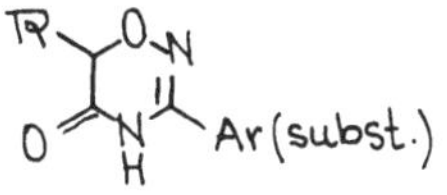

$$R\text{-[1,2,4-Oxadiazin-5-on]}-Ar(subst.)$$

                                    1,2,4-Oxadiazin-5-on Derivate
                                    R = H, Alkyl, Phenyl

DOS   2 754 238   6.12.77/15. 6.78   Philagro
                                                   FR Prior. 6.12.76

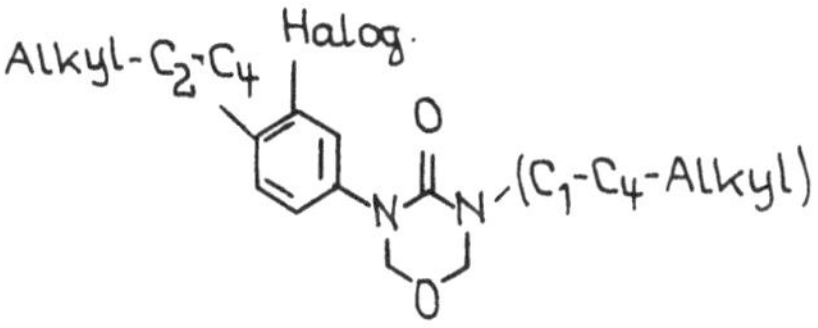

$$\text{Alkyl-}C_2\text{-}C_4,\ \text{Halog.-[Phenyl]-N-C(=O)-N-}(C_1\text{-}C_4\text{-Alkyl})\ \text{[oxadiazin]}$$

BE   861 546   6.12.76/ 6. 6.78   Philagro

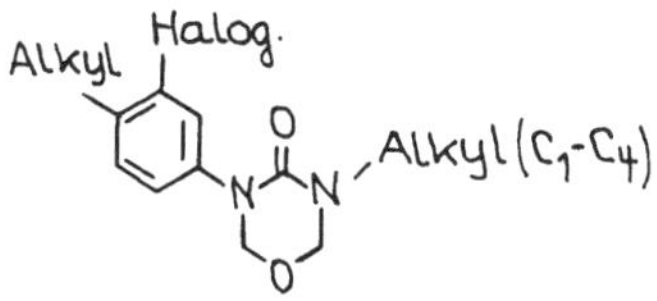

$$\text{Alkyl, Halog.-[Phenyl]-N-C(=O)-N-Alkyl}(C_1\text{-}C_4)\ \text{[oxadiazin]}$$

DOS   2 553 209   17.11.75/16. 6.77   BASF
= BE   848 694

R^1 u. R^2 = H, (Halog.)-Alkyl,Alk(en,
                in)yl, Cycloalkyl,
                O(S)-Alkyl

$>$N-CO, CO-O-Alkyl usw.
X = Halog.;NO_2, Alkyl, Cycloalkyl,SCN,
     COOR, CF_3, m = 0 - 4;Y = O oder S

DOS   2 619 090   3. 5.76/17.11.77   Celamerk
= BE   854 184

X = Cl, O(S)-Alkyl, SO_n-Alkyl, Cycloalkyl,
     Phenoxy(subst.), CO-O-Alkyl

DOS   2 631 413   13. 7.76/ 3. 2.77   Siegfried AG
                                               CH Prior. 18. 7. 75

DOS   2 656 290   11.12.76/15. 6.78   BASF

R^1 u. R^2 = Alkyl, Alk(en,in)yl, Halog-
                Alkyl, Cycloalkyl, Halog-
                Alk(en,in)yl, Dialkyl-Keton,
                Azido-Alkyl, Cyan-Alkyl,
                CO-N$<$, Aryl(subst.)
R^1 = HO-Alkyl, Alkyl-O-
                          Alkyl $>$N-PS-, Alkyl-
                                       S
Alkyl-CS-S-, Alkyl-S-   , CF_3,SO_2-N$<$
                     $(O)_{0-2}$
      usw.

DOS   2 710 382   10. 3.77/14. 9.78   BASF

R^1, R^2, R^3 = H, Halog., (O-)Alkyl,
                 SO_2-Alkyl, SO_2-N ,CF_3usw.
R^4 = H, Alkyl

Darstellungsverfahren

US   3 989 507   5.11.73/ 2.11.76   Dow Chem.

$X =$ Alkyl, Halog-Alkyl, Cycloalkyl, SO_2Alkyl, CCl_3, CF_3, N<, NO_2,
    SO_2N< usw.

R^1 u. $R^2 =$ H, Alkyl, Benzyl, $SCCl_3$, CO-N<, NH_2, CH_2-CH_2-N<,

    CH_2-CH_2-N⊐   usw.

US   4 051 130   17. 9.73/27. 9.77   Dow Chem.

$X =$ Alk(en,in)yl, Halog-Alkyl, Benzyl,
    Phenyl, Cycloalkyl usw; SCN,
    CO-O-Alkyl, SO_2N<, CF_3

$R =$ Alk(en,in)yl, Halog-Alkyl, Benzyl,
    N<, S-Alkyl, $(CH_2)_n$-CN

$R^1 =$ H, CO-N<, SO_2-Alkyl usw.

Selektiv in Baumwolle.

US   4 139 700   27. 7.77/13. 2.79   Monsanto

R, R^1 u. $R^2 =$ H, Alk(en,in)yl, O-Alkyl,
                Cycloalkyl usw.
A = Alkylen, Phenylen, Heterocycl.
Neue Darstellungsmethode.
u.a. auch herbizid wirksam

US   4 155 746   17. 9.73/22. 5.79   Dow Chem.

R u. $R^1 =$ H, Alkyl, Halog-Alkyl,
    Cycloalkyl, Phenyl, Benzyl,
    Alk(en,in)yl, $SCCl_3$, CO-N<,
    SO_2-Aryl, CO-O-Alkyl

$$-\underset{\underset{CH_3}{|}}{\overset{\overset{CH_3}{|}}{C}}-CN, \quad -\overset{|}{\underset{|}{C}}-C\equiv CH \quad usw.$$

(Halog, NO_2)

US   4 182 623   27. 7.77/ 8. 1.80   Monsanto

A = Glieder eines Heteroringes bis zu
    7 Ringglied. nur S bzw. S
                    $(\overset{\downarrow}{O})_{1-2}$

subst.O

enthaltend

$R = R^1 =$ H, Alk(en,in)yl, Ar usw.

US   4 208 514   23. 2.76/17. 6.80   Dow Chem.
                 (US 23. 2.76)

R = Halog.-Alk(en)yl, Cyan-Alkyl,Alkyl-
        thio-Alkyl usw.
R^1 = CO-O-Alkyl, CO-S Alkyl, CO-N< ,
        SO_2-N< usw.

BE   861 594   11.12.76/ 7. 6.78   BASF

R^1 = Alk(en,in)yl, Halog-Alkyl, Alkoxy-
        alkyl, Alkylmercaptoalkyl,
        -CO-Alkyl, Dialkyl-carbamoylalkyl
        usw.
R^2 = CN; SCN; N< ; S-O Alkyl, -SO_2N<
        usw.

JA   52 105188   27. 2.76/ 3. 9.77   Ube Ind. u. Sankyo

JA   52 105189   27. 2.76/ 3. 9.77   Sankyo

JA   54 154780   24. 5.78/ 6.12.79   Nihon Noyaku

JA   51 118853   7. 4.75/19.10.76   Kumiai

Siehe entspr. BASF Patent.

<u>c)   3 x N + 1 x S</u>   Bd.5/690

US   4 007 175   24. 5.71/ 8. 2.77   Chevron

                    R = Alkyl; R^1 = (O)-Alkyl; R^2 = Phenyl
                                                      (subst.)

426

5.14.7   <u>Höhergliedrige Hetero-Ringsysteme</u>   Bd.5/692

(> 6 Glieder)

DOS   2 638 543   26. 8.76/ 3. 3.77   Ciba Geigy

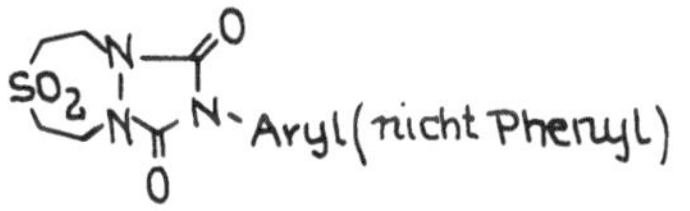

SO$_2$-ring-N-C(=O)-N-C(=O) / N-Aryl (nicht Phenyl)

5.15 Phosphor enthaltende Verbindungen

5.15.1   <u>Phosphite</u>   Bd.5/692

US   4 140 514   24. 9.75/20. 2.79   Chevron

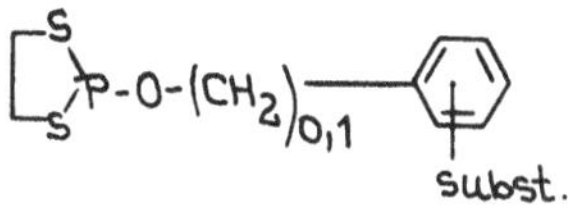

$\left[\begin{array}{c} S \\ S \end{array} \right\rangle P-O-(CH_2)_{0,1} - \text{Phenyl} \quad \text{subst.}$

SU   488 575   25. 4.74/16. 1.76   Kazan Kirov

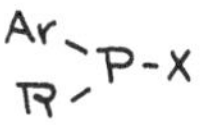

$\begin{array}{c} Ar \\ R \end{array} \hspace{-4pt} \diagdown P-X$

A = Phenyl oder Naphthyl (subst.)

R = C_2-C_5-O-Alkyl; p-subst. Phenyl

X = C_2-C_5-O-Alkyl, O-Phenyl, $N(CH_3)_2$

5.15.2   <u>Phosphorsäure–ester u. Thio-ester</u>   Bd.5/593

DOS   2 638 804   28. 8.75/ 3. 3.77   Sumitomo

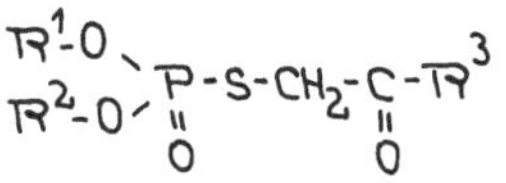

$\begin{array}{c} R^1O \\ R^2-O \end{array} \hspace{-4pt} \diagdown \underset{O}{P} - S - CH_2 - \underset{O}{C} - R^3$

R^1 u. R^2 = subst. Phenyl, $-CH-(CH_2)_n-$Halog
H, CH_3

R^3 = Piperidino und andere subst. hydrierte Heterocyclen

DOS   2 928 855   31. 8.78/13. 3.80   Gaf Corp.

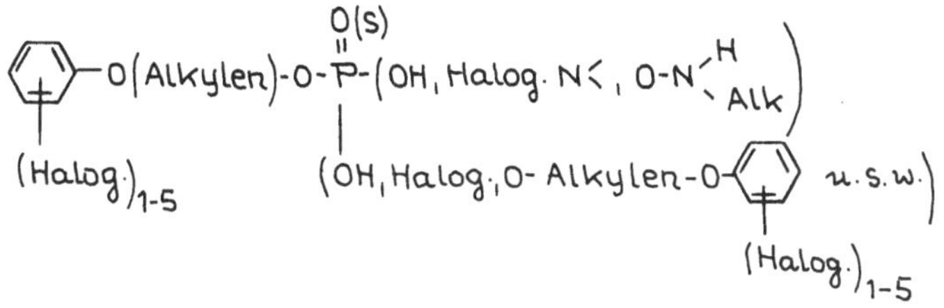

Phenyl-O(Alkylen)-O-$\overset{O(S)}{\underset{}{P}}$-(OH, Halog. N<, O-N<H Alk)

(Halog.)$_{1-5}$ (OH, Halog., O-Alkylen-O-Phenyl u.s.w.)

(Halog.)$_{1-5}$

US   4 036 628   13.11.72/19. 7.77   Stauffer

Alkyl O,Cl-Alkylen-O $-\overset{O}{\underset{}{P}}$ $\left(\text{O-CH}_2\text{-CH}_2\text{Cl(Br)}\right)_2$ Antidot gegen Herbizidschäden durch Thiocarbamate.

BE   847 538   22.10.75/22. 4.77   Bayer

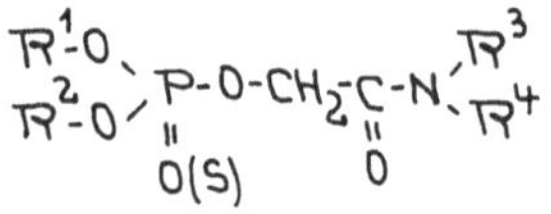

$R^1 = C_1-C_6$-Alkyl, Cyclohexyl,Phenyl
R^2 u. $R^3 = C_1-C_6$-Alkyl
R^4 = Cyclohexyl, Phenyl

BE   847 539   22.10.75/22. 4.77   Bayer

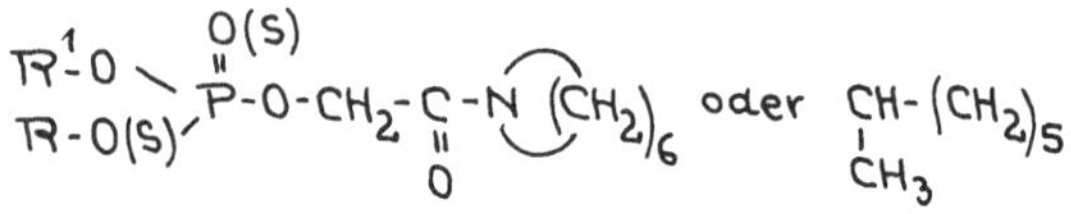

BE   848 254   14.11.75/12. 5.77   Ciba Geigy

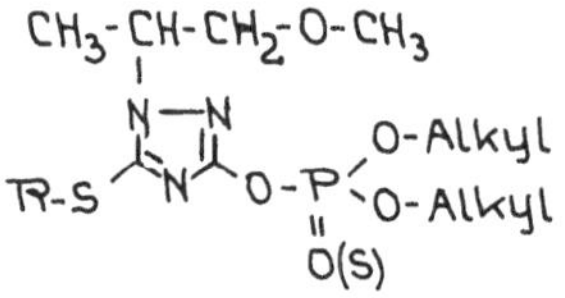

R = Alkyl

u.a. auch Herbizide

CH   602 774   1.12.75/31. 7.78   Ciba Geigy

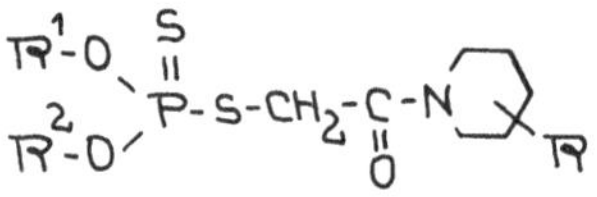

$R = CH_3$, C_2H_5
R^1 u. R^2 = Alk(en)yl,
Alkyl-O$(CH_2)_n$-

SU   702 026   4.11.76/ 5.12.79   Kazan,Lenin-Univers.

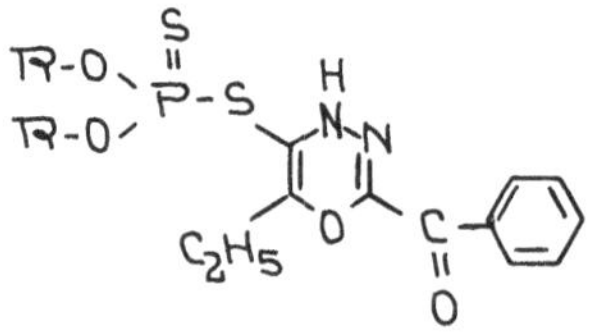

DDR   122 471   24.10.75/12.10.76   Kochmann

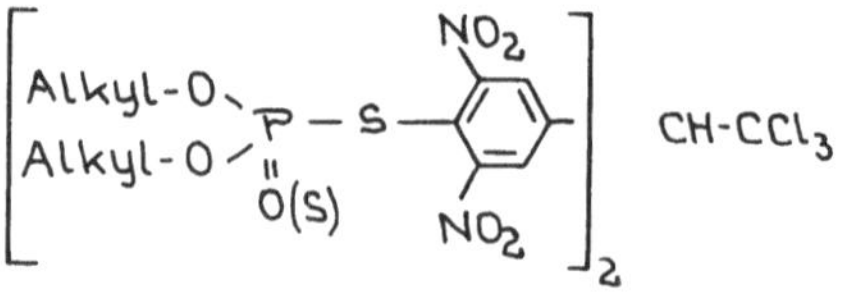

428

| DDR | 137 359 | 11. 5.78/11. 7.79 | Akad. Wissensch. DDR |

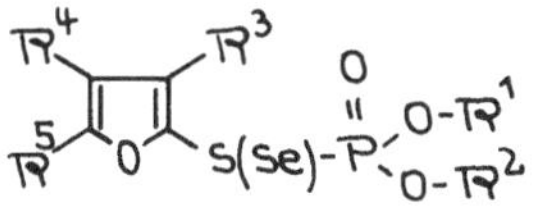

R^1 u. R^2 = Alkyl

R^3 u. R^4 u. R^5 = H, Halog., Alkyl

u.a. auch herbizid wirksam

| GB | 2 028 823 | 31. 8.78/12. 3.80 | Gaf Corp. |

| JA | 52 059140 | 10.11.75/16. 5.77 | Sumitomo |
| JA | 52 059141 | 10.11.75/16. 5.77 | Sumitomo |

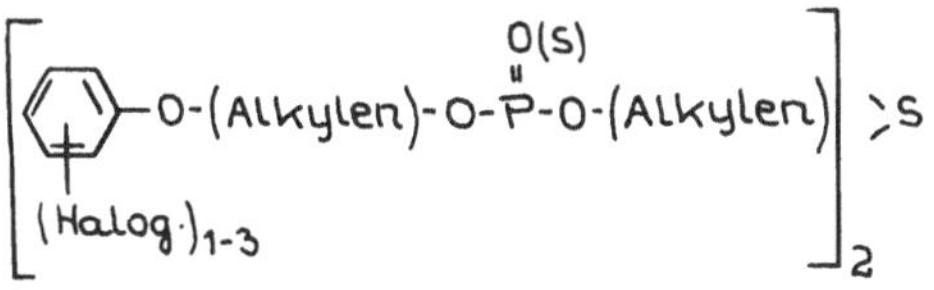

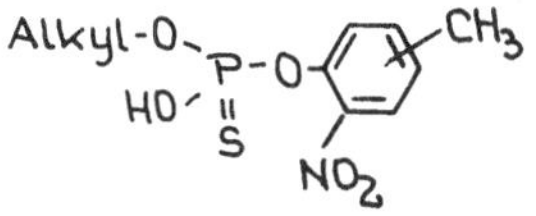

Spaltung in die optischen Antipoden mittels Brucin oder Strychnin

5.15.3 Phosphorsäure-ester-amide

a) Monoamide   Bd.5/696

| DOS | 2 167 007 | 24. 9.71/19. 6.77 | Sumitomo |

JA Prior. 25. 9.70

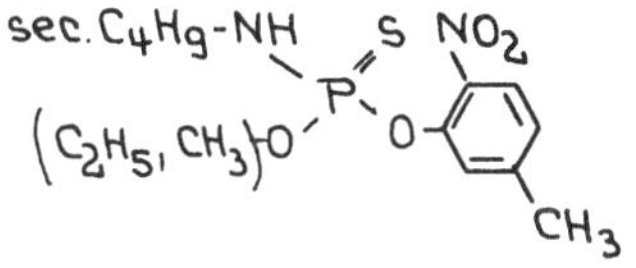

selektiv in Reis

| DOS | 2 604 224 | 4. 2.76/11. 8.77 | Höchst |
| = BE | 851 126 | | |

| US | 4 043 794 | 21. 5.76/23. 8.77 | Du Pont |

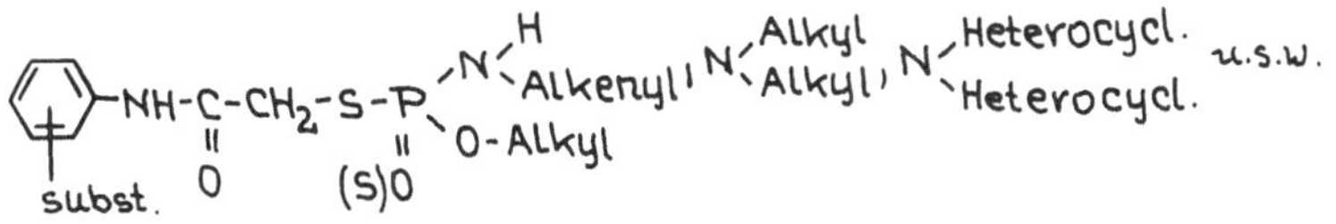

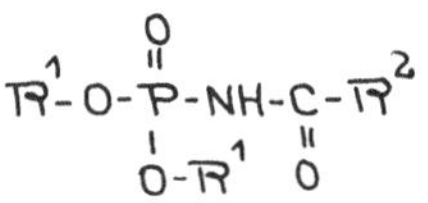

R^1 = CH_3, C_2H_5; R^2 = CH_3, C_2H_5 subst.

durch Cl, OCH_3 oder $O-C_2H_5$

US   4 126 442   23. 3.78/21.11.78   US Borax

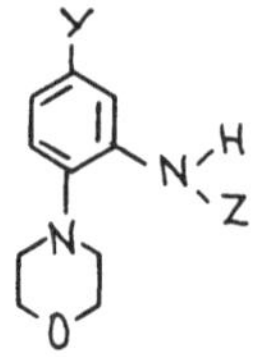

R^1 = Alk(en)yl, Benzyl, Phenethyl
       Phenyl

R^2 = H oder = R^1

$Y = C_3-C_6$-Alkyl(iso) $Z = P(O-R^1)_2$, $(S)O=P(O-R^2)_2$

JA   51 144729   5. 6.75/13.12.76   Ishihara Sangyo

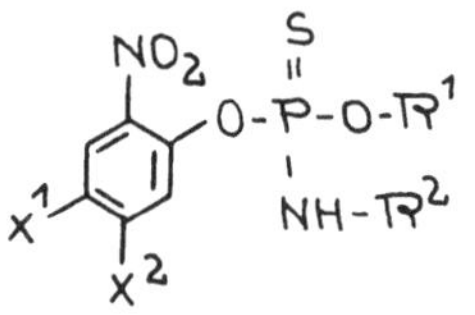

X^1 = H, CF_3, $S-CH_3$ $\downarrow$ $(O)_{0-2}$

X^2 = H, $S-CH_3$ $\downarrow$ $(O)_{0-2}$

JA   52 057324   7.11.75/11. 5.77   Sumitomo

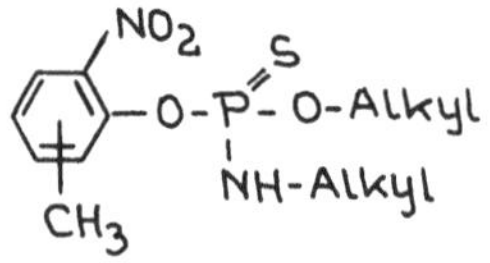

JA   55 102592   6. 4.71/ 5. 8.80   Hodogaya Chem.

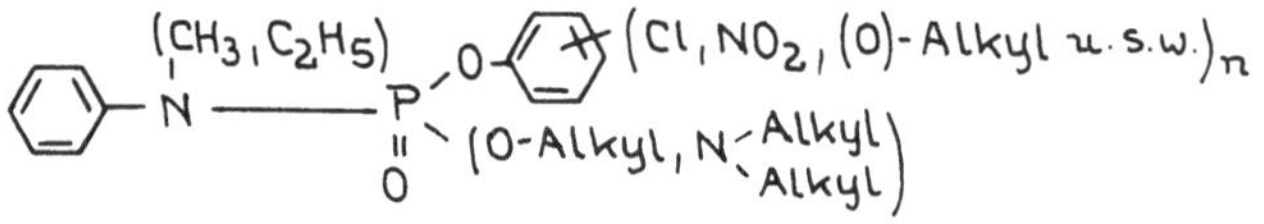

5.15.4   Phosphonsäure-ester und Amide   Bd.5/698

   a)   Ester u. z.T. Amide

DOS   2 604 224   4. 2.76/11. 8.77   Höchst

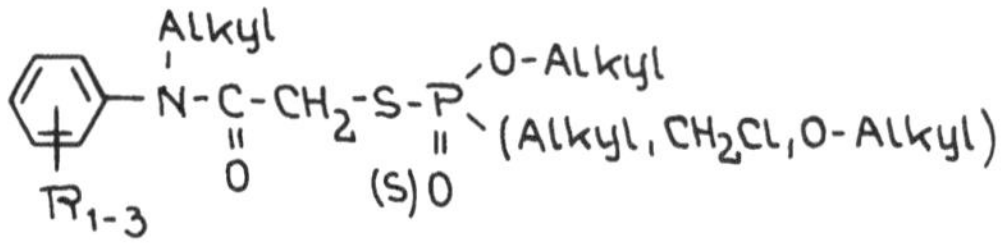

R = Alkyl, O-Alkyl, Halog.-Alkyl, NO_2, SO_2-CH_3, SO_2-NH_2, COOH, CN

spez. z.B. gegen Ackerfuchsschwanz

430

DOS   2 638 706   27. 8.76/ 3. 3.77   Du Pont
                                                 US Prior. 29. 8.75

$$\text{Alkyl-O-}\overset{\overset{O}{\|}}{\underset{\underset{O(H, NH_4, Na, K\ usw.)}{|}}{P}}\text{-}\overset{\overset{O}{\|}}{C}\text{-NH}_2$$

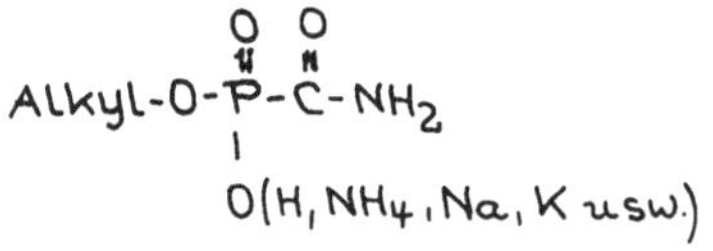

DOS   2 638 754   27. 8.76/ 3. 3.77   Du Pont
                                                 US Prior. 29. 8.75

$$R^1\text{-O-}\overset{\overset{O}{\|}}{\underset{\underset{O\text{-}R^2}{|}}{P}}\text{-}\overset{\overset{O}{\|}}{C}\text{-NH}_2$$

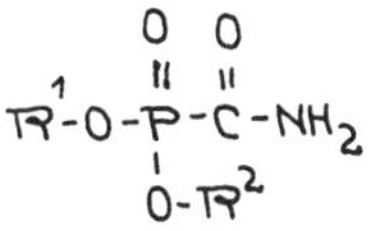

R^1 u. R^2 = C_1-C_5-Alkyle, Allyl

bes. R^1 u. R^2 = CH_3

spez. wirksam gegen Convolvulus spp. (Bindweed)

DOS   2 703 363   21. 1.77/10. 8.78   Plants Chem.

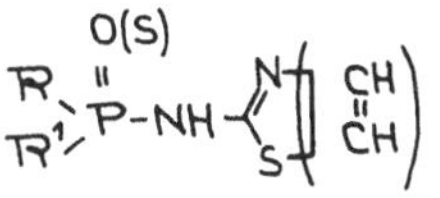

R = (Halog.)-Alkyl, Alkyl-O-Alkylen

R^1 = O-Alkyl, O-C_2-H_5(subst.),

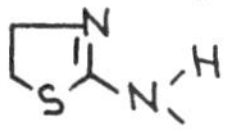

DOS   2 717 440   20. 4.77/ 1.12.77   Höchst
= BE   854 753   CH Prior. 17. 5.76

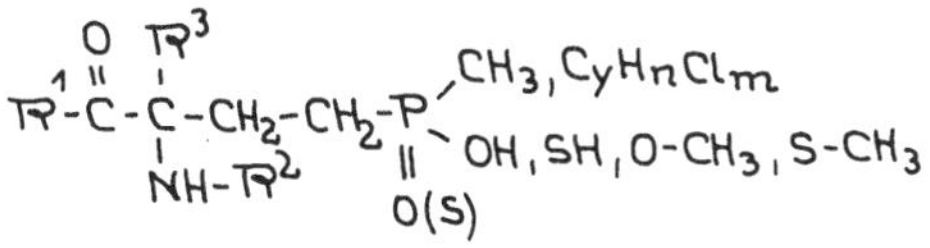

R^1 = OH, SH, O-CH_3, S-CH_3; R^2 = H, Acyl; R^3 = H, Alkyl

Total-Herbizide

DOS   2 719 721   3. 5.77/16.11.78   Bayer

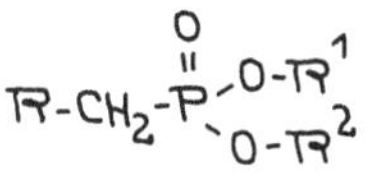

R = NH_2, NH_3^+/Anion, R^1 u. R^2 = H, Alkyl

Entblätterungsmittel für Baumwolle

DOS   2 812 659   22. 3.78/25.10.79   Vsesojuznyi Inst.
                                                 fiziologii

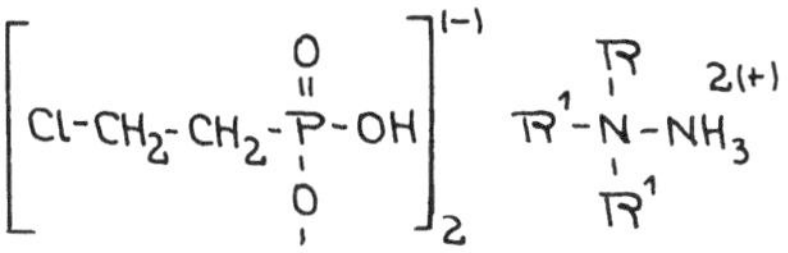

DOS   2 848 869   10.11.78/22. 5.80   Ciba Geigy

$$CH_3-NH-CH_2-P\overset{O}{\underset{\parallel}{<}}\overset{OH}{_{OH}}$$

DOS   2 914 294   9. 4.79/25.10.79   Nitrokemia
                                                    HU Prior. 12. 4.78

$$HO-\overset{O}{\underset{\parallel}{P}}\overset{H,Alkyl}{\underset{OH}{\underset{|}{C}}} \underline{\qquad} N\overset{H,Alkyl,OH}{_{H,Alkyl,OH}}$$

EP   1 018   25. 8.78/ 7. 3.79   Du Pont

                US Prior. 26. 8.77/14. 6.78/23. 6.78

$$\overset{R}{\underset{R^1}{>}}\overset{Cl}{\underset{\parallel}{P}}-\underset{|}{CH}-\overset{\parallel}{C}-O(S)-Alkyl, Cycloalkyl$$

R u. R^1 = O-Alkyl, Alkyl
          O-Phenyl (subst.)

$$R = \text{Phenyl}-N\overset{(CH_2)}{\underset{(CH_2)}{<}}N- \; ; \quad R^1 = -N\bigcirc O \; usw.$$

EP   1.331   25. 8.78/ 4. 4.79   Pont de Nemours
                        (US 14. 6.78 u. 23. 6.78)

subst.

$$\text{Phenyl}-\overset{O}{\underset{O}{P}}-\overset{Cl}{\underset{Cl}{C}}-\overset{O}{C}-O-Alkyl (iso)$$

Alkyl

EP   2 031   19.11.77/30. 5.79   Ciba Geigy
= GB   2 006 215   12.10.77/25. 5.79   Ciba Geigy

$$HO-\overset{O}{\underset{\parallel}{P}}-\underset{H (H,Alkyl)}{CH}-NH_2-CH_3$$

                                    Wuchsregulator und Herbizid

EP   7 210   10. 7.78/23. 1.80   Monsanto

$$CH_3-\overset{}{\underset{O}{C}}-N\overset{CH_2-C-O(Alkyl, Phenyl)}{_{CH_2-P-[O(Alkyl, Benzyl, Naphthyl subst.]_2}}$$

EP   9 348   3. 9.79/ 2. 4.80   Fisons
                        (GB 22. 9.78)

$$R^1-\overset{}{\underset{O}{C}}-\overset{}{\underset{O(S)}{P}}\overset{O-CH_3, H, Alkyl, Aryl usw.}{_{(O-Kat., N< usw.)}}$$

                                    auch herbizid wirksam

EP   14 684   31. 1.80/20. 8.80   Ciba Geigy
                                                     CH Prior. 6. 2.79

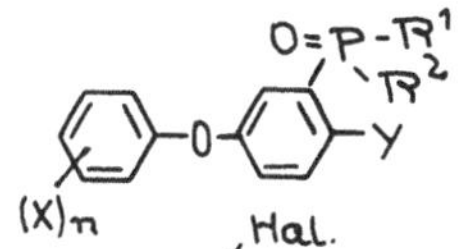

R^1 u. R^2 = OH, O(S)-Alkyl, N<

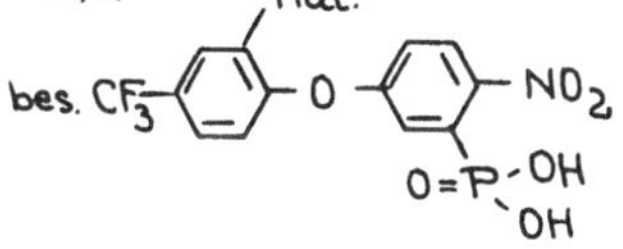

Y = NO_2,OH,CN,NH_2,NH-CO-Alkyl

X = CF_3, NO_2CN usw.

US   4 043 793   21. 5.76/23. 8.77   Du Pont

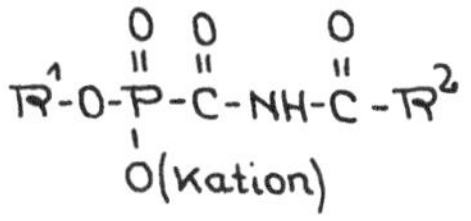

R^1 = CH_3, C_2H_5
R^2 = CH_3, C_2H_5 subst. durch Cl,
          OCH_3, OC_2H_5

spez. wuchsverhindernd

US   4 163 850   19. 2.71/ 7. 8.79   Petrolite Corp.

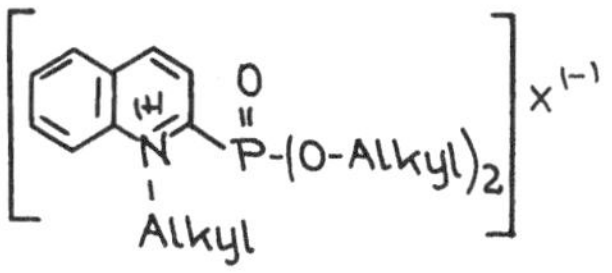

u. Stellungsisomere

DDR   123 423   9. 2.76/20.12.76   Gotschel

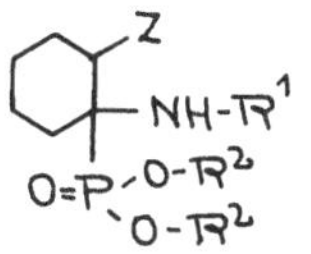

Desiccant in Kartoffeln

Z = H, CH_3, Halog.
R^1 u. R^2 = Alkyl

DDR   136 501   11. 5.78/11. 7.79   Akad. Wissensch. DDR

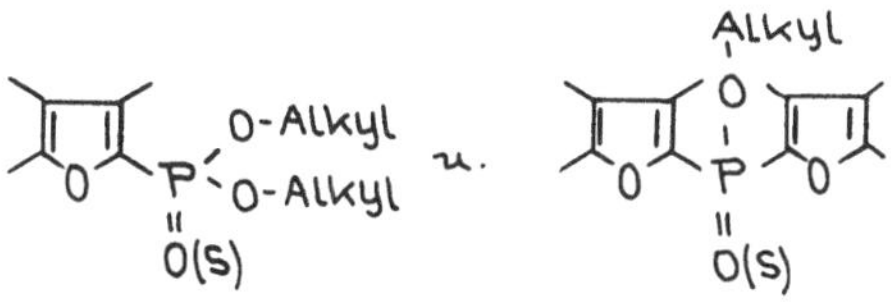

subst. durch Alkyl oder NO_2

GB   1 508 772   5. 4.74/26. 4.78   Shell Ind.

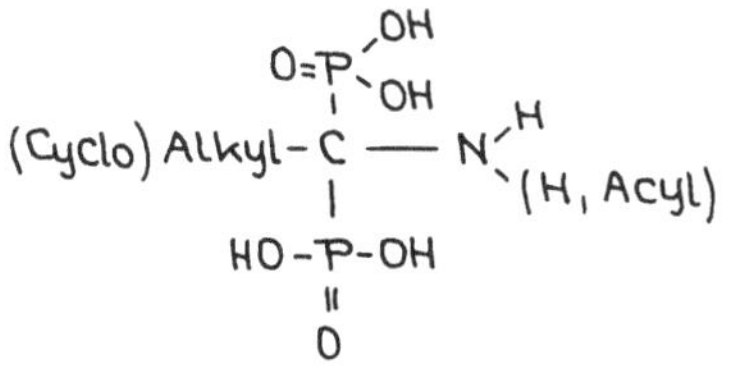

SU   509 583   30. 9.74/ 4. 6.76   Melnikov NN

$Q = -CH_2-CH_2-,\ -CH = CH-$

R^1 = Alkyl; R^2 = Aryl

SU   570 954   21. 1.76/ 1.11.77   As. Uzb. Vegation

R = Alkyl, Aryl
R^1 = H, Alkyl, Aryl
R^2 = Alkyl, Aryl

JA   53 006431   25.11.76/13. 6.78   Nissan Chem.

JA   53 059674   25.10.76/29. 5.78   Nissan Chem.

JA   54 144383   27. 4.78/10.11.79   Nissan Chem.

JA   54 147925   11. 5.78/19.11.79   Nissan Chem.

JA   55 089210   28.12.78/ 5. 7.80   Nissan Chem.

starkes Herbizid

JA   55 098193   22. 1.79/25. 7.80   Nissan Chem.

$$(HO)_2P(=O)-C(X)(H, Halog. \cdot H, Alkyl)-\text{Pyridyl}, \quad P(OH)_2(=O)$$

X = O, S, CN_2

JA   55 098194   22. 1.79/25. 7.80   Nissan Chem.

$$(HO)_2P(=O)-CH-P(=O)(OH)_2$$

mit N-Ring: X^1 u. X^2 = H, Alkyl oder zusammen

$$\geq C = C \leq$$

$$Y = (CH_2)_n \text{ oder } CH_2-CH = CH-CH_2$$

b)   <u>Verbindungen der Formel:</u>

$$(HOOC-CH_2)_n-N(H)_x-\left(CH_2-\underset{O}{P}\!\!<^{OH}_{OH}\right)_{n^1} \quad \text{oder} \quad \left(CH_2-\underset{O}{P}\!\!<^{Alkyl}_{Alkyl}\right)_{n^1}$$

$n + n^1 = 3$ u. 2, X = 0 u. 1

DOS   2 366 060   30. 5.73/ 6.10.77   Monsanto
                                                          US Prior. 31.12.72 u. 16. 2.73

$$(HO)_2\underset{O}{P}-CH_2-NH-CH_2-COOH$$

Neues Darstellungsverfahren.

DOS   2 622 837   21. 5.76/ 2.12.76   Monsanto
                                                          US Prior. 25. 5.75

$$C_nF_{2n+1}-\underset{O}{C}-N(CH_2-COOH)-CH_2-\underset{OH}{\overset{O}{P}}-OH$$

DOS   2 641 318   14. 9.76/ 7. 7.77   Monsanto
= US   3 991 095 u. 4 035 177

$$HO-\underset{O}{C}-CH_2-N(CH_2-\underset{O}{P}(OH)_2)(C(=O)-S-R)$$

R = Alkyl, Phenyl, Benzyl subst.

DOS   2 659 172   28.12.76/ 7. 7.77   Monsanto
= BE   849 907   US Prior. 29.12.75

$$(R-O)(R^1-O)-CH_2-NH-CH_2-\underset{O}{C}-O(H, Alkyl)$$

R = Phenyl, Benzyl, Naphthyl
     ev. subst.
R^1 = H oder R

| DOS | 2 700 017 | 3. 1.77/14. 7.77 | Monsanto |
| = BE | 850 059 | | US Prior. 5. 1.76 |

$$\begin{array}{c} R^1\text{-O} \\ R^1\text{-O} \end{array} \!\!\!\! \overset{\displaystyle O}{\underset{\displaystyle \|}{P}}\!\!-CH_2-N\overset{H}{\underset{\displaystyle CH_2\text{-}COOR}{\diagdown}}$$

R = Alkyl, R^1 = Alkyl

neues Darstellungsverfahren

| DOS | 2 719 583 | 2. 5.77/24.11.77 | Monsanto |
| = BE | 854 167 | | US Prior. 3. 5.76 |

$$Z\text{-}O\text{-}CO\text{-}R\text{-}\overset{\displaystyle}{\underset{\displaystyle}{C}}\text{=}O$$
$$Y\text{-}O\text{-}\overset{}{\underset{\displaystyle O}{\overset{\|}{C}}}\text{-}CH_2\text{-}N\text{-}CH_2\text{-}\overset{}{\underset{\displaystyle O}{\overset{\|}{P}}}\overset{OH}{\underset{OH}{\diagup}}$$

Y = H, Alkalimetall, Alkyl

R = Vinylen, Alkylen, Phenylen, Pyridyliden usw.

Z = H, Alkalimetall

| DOS | 2 751 630 | 13.12.76/20. 6.77 | Monsanto |

$$\left[(\text{Phenyl oder Naphthyl-O})_2 - \overset{\displaystyle O}{\underset{\displaystyle \|}{P}}-CH_2-N-CH_2-CN \right]_2$$
$$\underset{\displaystyle CH_2}{|}$$

| DOS | 2 751 631 | 18.11.77/15. 6.78 | Monsanto |
| | | | US Prior. 13.12.76/ 3. 1.77 |

$$(\text{subst. Aryl-O})_{2-b} - \overset{\displaystyle O(S)}{\underset{\displaystyle |}{\overset{\|}{P}}}-CH_2-N\overset{H}{\underset{CH_2\text{-}CN}{\diagdown}} \quad (\text{Salz einer starken Säure})$$
$$\underset{(OH)_b \qquad b=0}{}$$

| DOS | 2 810 080 | 8. 3.78/21. 9.78 | Monsanto |
| = US | 4 084 953 | 9. 3.77/18. 4.78 | Monsanto |

$$HO\text{-}\overset{}{\underset{\displaystyle O}{\overset{\|}{C}}}\text{-}CH_2\text{-}\overset{}{\underset{\displaystyle OH}{N}}\text{-}CH_2\text{-}\overset{}{\underset{\displaystyle O}{\overset{\|}{P}}}\overset{OH}{\underset{OH}{\diagup}}$$

| DOS | 2 813 581 | 1. 4.77/ 5.10.78 | Ciba Geigy |

$$\left(R^1\text{-}O\text{-}\overset{}{\underset{\displaystyle O}{\overset{\|}{C}}}\text{-}CH_2\text{-}\overset{}{\underset{\displaystyle R}{N}}\text{-}CH_2\right)_2 \overset{\displaystyle O}{\underset{\displaystyle \|}{P}}\text{-}O(\text{Alkyl}, CH_2\text{-}CH_2\text{-}CN)$$

R = H, Alkyl, Benzyl, Trityl; R^1 = OH; O-Alkyl N< , NH-N<

| DOS | 2 828 915 | 30. 6.78/ 1. 2.79 | Monsanto |

$$\overset{H(CH_3)}{\underset{O}{\bigvee}}\;N\diagdown_{CH_2\text{-}\overset{\displaystyle O}{\overset{\|}{P}}\overset{OH}{\underset{O\text{-Alkalimet.}}{\diagup}}}$$
$$(O)_{n=0\text{-}1}$$

DOS   2 829 174   3. 7.78/22. 2.79   Monsanto

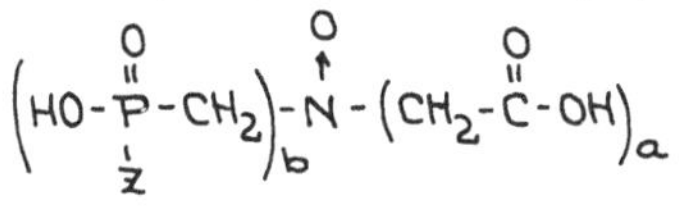

$$\left(HO-\overset{O}{\underset{Z}{\overset{\parallel}{P}}}-CH_2\right)_b-\overset{\uparrow}{\underset{}{N}}-\left(CH_2-\overset{O}{\overset{\parallel}{C}}-OH\right)_a \qquad a+b=3$$

Z = H, Alkyl, Phenyl, $-CH_2-\overset{\bullet}{\underset{O}{N}}-(CH_2CO-OH)_2$

Erhöhung des Sacharose-Gehaltes von Zuckerpflanzen.

DOS   2 831 578   18. 7.78/ 1. 2.79   Nissan Chem.
                    JA Prior. 20. 7, 27. 7.77; 11. 4.78

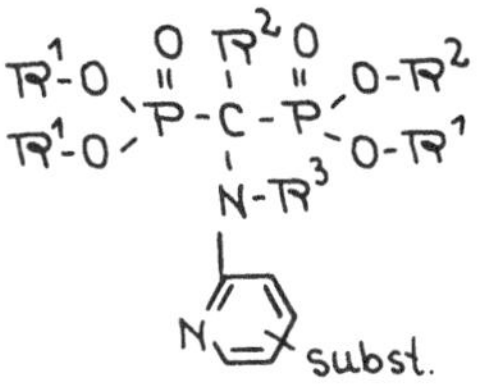

$R^1 - R^3$ = H, Alkyl

DOS   2 900 024   2. 1.79/12. 7.79   Monsanto
                                        US Prior. 3. 1.78

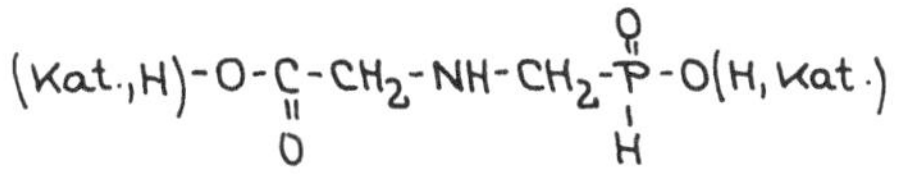

$$(Kat.,H)-O-\overset{}{\underset{O}{\overset{\parallel}{C}}}-CH_2-NH-CH_2-\overset{O}{\underset{H}{\overset{\parallel}{P}}}-O(H,Kat.)$$

Darstellungsverfahren

DOS   2 900 025   2. 1.79/12. 7.79   Monsanto

Saure Salze des Glyphosate, siehe voranstehendes Patent.

DOS   2 942 898   24.10.79/ 8. 5.80   Biolog. Chem. Activit.
                    CH Prior. 24.10.78

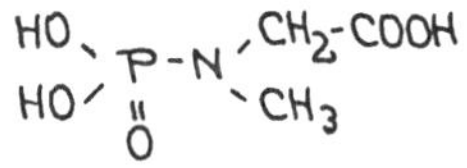

$$\overset{HO}{\underset{HO}{\diagdown}}\overset{}{\underset{\underset{O}{\parallel}}{P}}-N\overset{CH_2-COOH}{\underset{CH_3}{\diagup}}$$

DOS   2 944 598   15.11.79/22. 5.80   Sandoz
                                          GB Prior. 20.11.78

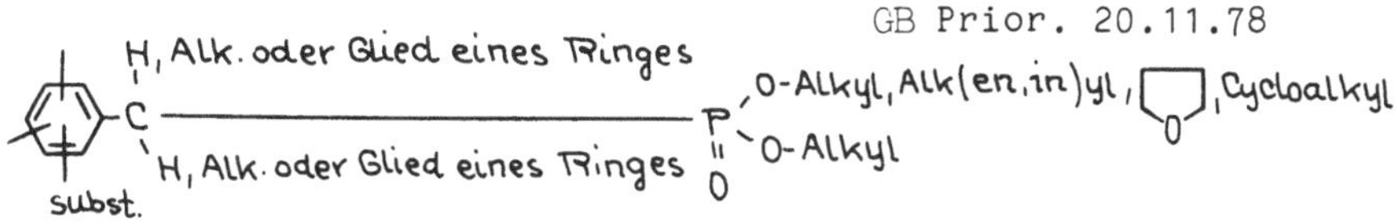

EP   767   2. 8.78/21. 2.79   Ciba Geigy
                 (CH 11. 8.77)

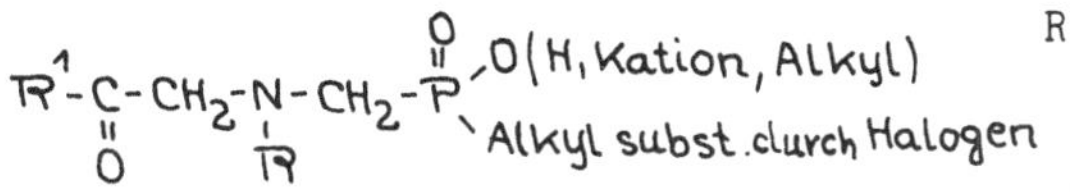

R = H, Alkyl, Benzyl,
    $CH(C_2H_5)_2, C(C_2H_5)_3,$
    CH_2-COOH
R^1 = OH, O-Alkyl, NH_2

EP   5 321   4. 4.79/14.11.79   Monsanto
                (US   6. 4.78)

$$CF_3\text{-}\underset{\underset{O}{\|}}{C}\text{-}\underset{\underset{CH_2\text{-}\underset{\underset{O}{\|}}{P}Cl_2}{|}}{N}\text{-}CH_2\text{-}CO\text{-}(Cl,O\text{-}Alkyl)$$

EP   7 210   6. 7.79/23. 1.80   Monsanto
                                                US Prior. 10. 7.78

$$CF_3\text{-}C\text{-}N\big\langle\overset{CH_2\text{-}C\text{-}O(Alkyl,Phenyl)}{CH_2\text{-}P\text{-}O\text{-}R}$$

R = Allyl, Naphthyl, Benzyl subst. d. CN, NO$_2$, CF$_3$ usw.

EP   7 211   19. 3. /23. 1.80   Monsanto
                (US   6. 7.79/   US Prior. 10. 7.78

$$CF_3\text{-}C\text{-}N\big\langle\overset{CH_2\text{-}C\text{-}O\text{-}Alkyl}{CH_2\text{-}P\text{-}R^1}$$

R^1 u. R^2 = O(S)-Alkyl, O(S)-Phenyl,
$N\big\langle^H_{Alkyl}$ N(Alkyl)$_2$ N$\bigcirc$O

R^1 aber nicht = R^2

EP   7 684   25. 5.79/ 6. 2.80   Monsanto
                                                US Prior. 30. 5.78

R = Phenyl, Tolyl, Phenyl durch CF$_3$
    subst.

EP   8 158   6. 7.79/20. 2.80   Monsanto
                                                US Prior. 10. 7.78
Siehe US 4 195 983   26.12.78/ 1. 4.80   Monsanto

$$CF_3\text{-}C\text{-}N\big\langle\overset{CH_2\text{-}CO\text{-}O\text{-}(Alkyl,Alkylclorid)}{CH_2\text{-}P\text{-}S\text{-}R}$$

R = Alkyl, Phenyl, Alkenyl, Benzyl usw.

EP   8 852   19. 3.78/19. 3.80   Monsanto
                (US 10. 7.78)

$$Alkyl\text{-}O\text{-}C\text{-}CH_2\text{-}N\text{-}CH_2\text{-}P\big(N\text{-}Alk(en,in)yl, Cycloalkyl, N\big\langle\big)$$
$$\underset{\underset{O}{\|}}{C}\text{-}CF_3 \quad (Alk(en,in)yl)$$

EP   10 066   1.10.79/16. 4.80   Ciba Geigy
                          (GB 5.10.78)

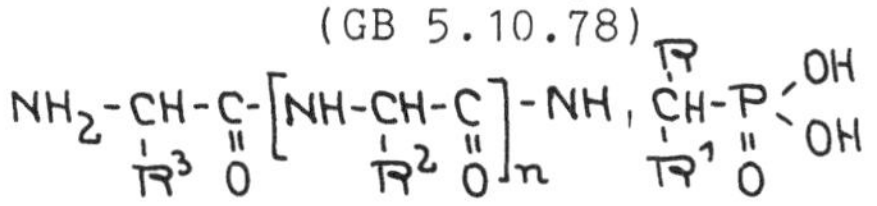

$$NH_2-CH-C-[NH-CH-C]_n-NH, CH-P{<}^{OH}_{OH} \qquad n = 0-3$$

R u. R^1 = H, Alkyl, Alk(en,in)yl, Cycloalkyl, Aryl, Heterocycl.
            Glieder eines Ringsystems usw.
R^2 u. R^3 = H, Alkyl, Cycloalkyl, Heterocycl. usw.

EP   10 067   1.10.79/16. 4.80   Ciba Geigy
                          (GB 5.10.78)

$$R^7-NH-CH-C-[NH, CH-C]_n-NH-C-P{<}^{R^6}_{OH}$$

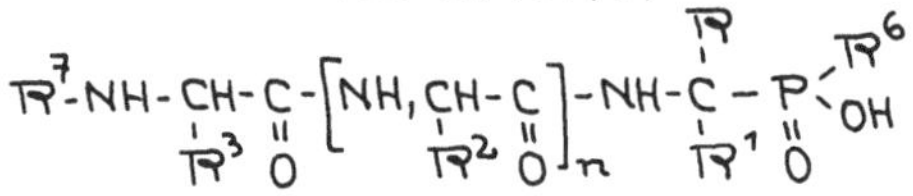

R u. R^1 siehe zuvor, R^2 u. R^3 siehe zuvor,
R^6 = OH, CH_3, Phenyl; R^7 = H, CO-O Alkyl, CO-O-CH_2-

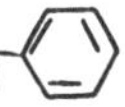

US   3 979 200   4. 9.73/ 7. 9.76   Monsanto

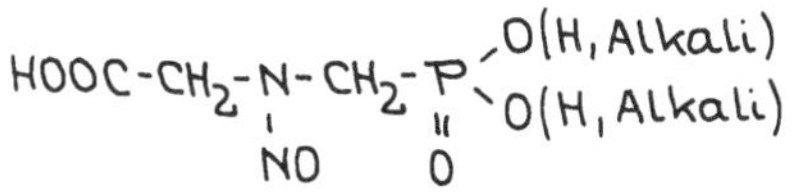

$$HOOC-CH_2-N-CH_2-P{<}^{O(H,Alkali)}_{O(H,Alkali)}$$

US   3 989 727   6. 8.73/ 2.11.76   Monsanto

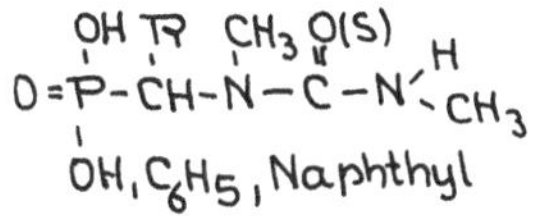

$$O=P-CH-N-C-N{<}^{H}_{CH_3}$$

$$R = C_1-C_5-Alkyl, Alkenyl, -CH_2-S-C_2H_5, \quad , C_6H_5$$

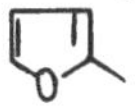

US   4 031 170   6. 8.73/21. 6.77   Monsanto
siehe auch 3 954 860

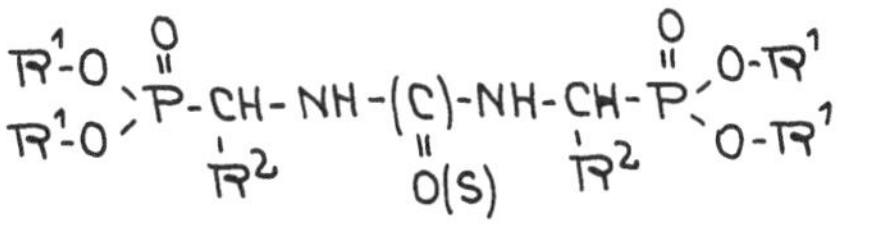

$$R^1-O{>}P-CH-NH-(C)-NH-CH-P{<}^{O-R^1}_{O-R^1}$$

R^1 = Aryl subst. durch
        Alkyl
R^2 = Alk(en)yl, Aryl
        subst.

u.a. auch Herbizide

US   4 035 176   23. 5.75/12. 7.77   Monsanto

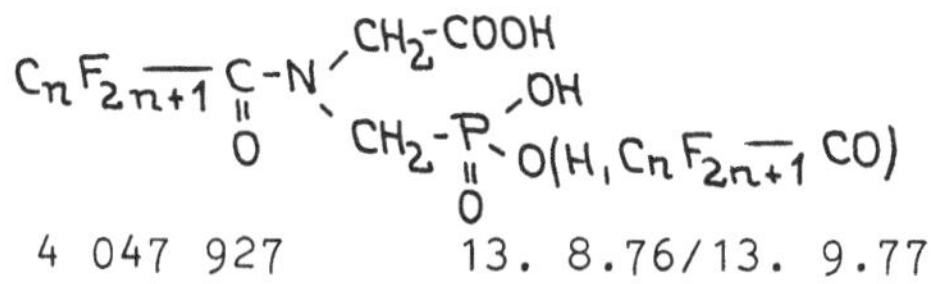

$$C_nF_{2n+1}C-N{<}^{CH_2-COOH}_{CH_2-P{<}^{OH}_{O(H,C_nF_{2n+1}CO)}}$$

n = 1 - 4

US   4 047 927   13. 8.76/13. 9.77   Monsanto

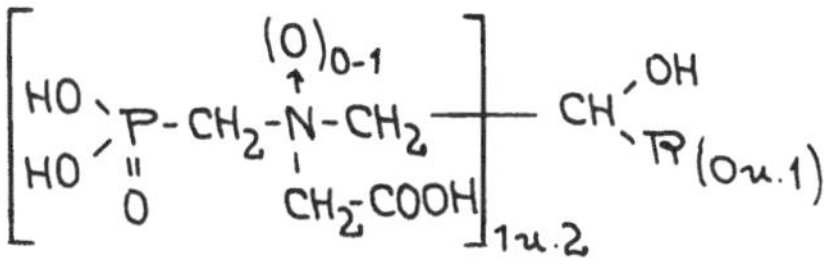

$$\left[HO{>}P-CH_2-N-CH_2\right]_{1u.2}-CH{<}^{OH}_{R(0u.1)}$$

US   4 089 672   19. 2.70/16. 5.78   Upjohn

$$\left(HOOC\text{-}CH_2\right)_n\text{-}N(H)_x^{-}\left(CH_2\text{-}P\!\!<^{OH}_{OH}\right)_{n^1}$$

$n + n^1 = 3$
$X = 0 - 1$

US   4 104 050   20. 6.77/ 1. 8.78   Monsanto

$$R\text{-}O\!\!>\!\!\overset{O}{\underset{}{P}}\text{-}CH_2\text{-}NH\text{-}CH_2\text{-}\overset{NH}{\underset{}{C}}\text{-}O\text{-}R^1 \cdot 2HCl$$

$R = \langle \text{Ring} \rangle$; $R^1 =$ Phenyl, (Cyclo)-Alkyl
Alkyl

US   4 106 932   3. 3.71/15. 8.78   Monsanto

$$\left[Alkyl\text{-}O\text{-}\overset{}{\underset{O}{C}}\text{-}CH_2\text{-}NH\text{-}CH_2\text{-}\overset{O}{\underset{OH}{P}}\right]_2 Ox(H_2O)_{0\text{ u.}1}$$

US   4 119 430   13. 8.76/10.10.78   Monsanto

$$CH_3\text{-}O\!\!>\!\!\overset{(O)_{0\,oder\,1}}{\underset{O}{P}}\text{-}CH_2\text{-}N\text{-}CH_2\text{-}COO(H, Alkalimetall)$$
$$HO\nearrow$$
$$CH_2\!\!-\!\!CH(CH_2\text{-}OH)\text{-}OH_{1\,oder\,2}$$
$$\text{1 oder 2}$$

US   4 130 412   11.12.72/19.12.78   Monsanto

$$Alkyl\text{-}O\text{-}\overset{O}{\underset{O}{C}}\text{-}CH_2\text{-}\overset{O}{\underset{(CH_2)_{1\text{-}10}\text{-}COOH}{N}}\text{-}CH_2\text{-}P\!\!<^{OH}_{OH}$$

US   4 131 448   11.12.72/26.12.78   Monsanto

$$R\text{-}O\text{-}\overset{O}{\underset{O}{C}}\text{-}CH_2\text{-}\overset{O}{\underset{R^3}{N}}\text{-}CH_2\text{-}P\!\!<^{O\text{-}R^1}_{O\text{-}R^2}$$

R^1 u. $R^2 = H$, Alkali, Alkyl
$R^3 =$ Alkyl, Allyl, Cycloalkyl, Phenyl-Alkyl usw.
$R = R^1$

US   4 175 946   10. 7.78/27.11.79   Monsanto

$$O=\overset{R^1}{\underset{R^1}{P}}\text{-}CH_2\text{-}\overset{}{\underset{\overset{C\text{-}CF_3}{O}}{N}}\text{-}CH_2\text{-}\overset{O}{C}\text{-}O\text{-}Alkyl$$

$R^1 =$ S-Alkyl, S-Alkenyl
S-CH_2-Phenyl, S-Phenyl
subst.

| US | 4 180 394 | 10. 7.78/25.12.79 | Monsanto |

$$CF_3-\underset{\underset{O}{\|}}{C}-\underset{\underset{CH_2-CO-O-Alkyl}{|}}{N}-CH_2 \longrightarrow \underset{\underset{O}{\|}}{P} \diagdown^{O(S)Alkyl}_{(Bis)Alk(en,in)ylamino}$$

| US | 4 211 547 | 22.12.78/ 8. 7.80 | Monsanto |

$$\underset{\underset{O}{\|}}{\overset{HO}{\underset{Kat.O}{\diagdown}}}P-CH_2-N\diagup^{CH_2-CN}_{CH_2-CN}$$

| US | 4 211 732 | 26.12.78/ 8. 7.80 | Monsanto |

$$(Cl)-Alkyl-O-\underset{\underset{O}{\|}}{C}-CH_2-N\diagup^{CH_2-\overset{\overset{S}{\|}}{P}\diagup^{R^1}_{R^1}}_{CO-O-CH_2-C_6H_5}$$

$R^1 = O(S)-Alkyl, O,$
  $S(-Phenyl)$ subst.

| US | 4 217 312 | 26. 8.77/12. 8.80 | Du Pont |

$$(Alkyl,Phenyl)-\underset{\underset{O}{\|}}{P}\diagup^{CCl_2-CO-O(Alk(en)yl}_{(O-Alkyl, N\langle)}$$

| BE | 860 971 | 13.12.76/20. 6.78 | Monsanto |

$$\left[\underset{R-O}{\overset{R-O}{\diagdown}}\underset{\underset{O}{\|}}{P}-CH_2-\underset{\underset{CH_2-CN}{|}}{N}\longrightarrow\right]_2 CH_2$$

R = Phenyl, Naphthyl subst.

| BE | 879 076 | 29. 9.78/28. 3.80 | Monsanto |

$$(Alkyl, Alkyl-O-(CH_2)_x)-O-\underset{\underset{O}{\|}}{C}-CH_2-\underset{\underset{O=C-S-(Alkyl,Alk(en,in)yl,Benzyl)}{|}}{N}-CH_2-\underset{\underset{O}{\|}}{P}\diagup^{CH_3}_{CH_3}$$
$$(O)_{0-2}$$

| BE | 879 774 | 3.11.78/30. 4.80 | Monsanto |

$$(Alkyl, Alkoxyalkyl, Cl-Alkyl)O-\underset{\underset{O}{\|}}{C}-CH_2-\underset{\underset{\underset{O-CH_2-C_6H_5}{|}}{C=O}}{N}-CH_2-\underset{\underset{O}{\|}}{P}\diagup^{Cl,OH,O-Kat.}_{Cl,OH,O-Kat.}$$

| BE | 879 776 | 3.11.78/30. 4.80 | Monsanto |

$$(Alkyl, Cl-Alkyl)O-\underset{\underset{O}{\|}}{C}-CH_2-NH-CH_2-\overset{\overset{O}{\|}}{P}(S-Alk(en)yl, Phenyl)_2$$

BE   879 777   3.11.78/30. 4.80   Monsanto

$$(\text{Alkyl}, \text{Cl-Alkyl}, \text{Alkoxyalkyl})\text{O-CO-CH}_2\text{-}\overset{\overset{O}{\|}}{P}\left(\begin{matrix} \text{O-Alkyl}, \text{S-Alkyl}, \text{S-Alkenyl,} \\ \text{Phenyl usw.} \end{matrix}\right)_2$$
$$C_6H_5\text{-}CH_2\text{O-}\overset{|}{C}O$$

BE   880 847   26.12.78/24. 6.80   Monsanto
                                                        US Prior. 26.12.78

$$R\text{-O-CH}_2\text{-NH-CH}_2\text{-}\overset{\overset{R^1}{\diagup}}{\underset{\overset{\|}{S}}{P}}\diagdown R^1$$

R = Alkyl, CCl_3, Alkoxyalkyl
R^1 = S-Alkyl, S-Phenyl,
        O-Phenyl subst.

DDR   141 929   20. 2.79/28. 5.80   Issleb

$$\text{HOOC-CH}_2\text{-NH-CH}_2\text{-}\overset{\overset{O}{\|}}{P}\overset{\diagup \text{O(H,Alkyl,Phenyl)}}{\diagdown \text{O(H,Alkyl,Phenyl)}}$$

Darstellungsverfahren

RD   177 051   20.12.78/10. 1.79   Anonym

$$R^-_{(a)}N\left[CH_2\text{-}\underset{\overset{\|}{O(S)}}{P}\text{-(O-Y)}_2\right]_b$$

a = 1 dann b = 2
Y = 2 dann b = 1
u. entspr. N-O_{xy}Cl

R = Halog.-Alkyl,
    CN-Alkyl, $CH(COOH)_2$
    CH_2-CO-OH

JA   52 072823   16.12.75/17. 6.77   Nissan Chem.

$$\overset{\text{Phenyl-O}}{\underset{\text{Phenyl-O}}{\diagdown}}\underset{\overset{\|}{O}}{P}\text{-CH}_2\text{-NH-CH}_2\text{-COOR}$$

JA   52 095640   5. 2.76/11. 8.77   Nissan Chem.

$$\overset{\text{Phenyl-O}}{\underset{\text{Phenyl-O}}{\diagdown}}\underset{\overset{\|}{O}}{P}\text{-CH}_2\text{-NH-CH}_2\text{-CN}$$

JA   52 108955   11. 3.76/12. 9.77   Nissan Chem.

$$\overset{\text{Phenyl-O}}{\underset{\text{Phenyl-O}}{\diagdown}}\underset{\overset{\|}{O}}{P}\text{-CH}_2\text{-NH-CH}_2\text{-CO-O-Alkyl}$$

JA   52 142047   22. 5.76/26.11.77   Nissan Chem.

$$C_6H_5-O-P(OH)(=O)-CH_2-NH-CH_2-CN$$

JA   53 056620   4.11.76/23. 5.78   Nissan Chem.

$$R-O,\ R-O-P(=O)-CH_2-NH-CH_2-CN$$

Grasherbizide

$$R = H,\ Alkyl,\ Phenyl$$

d)   <u>Phosphinsäurederivate</u>

    siehe z.T. unter Phosphonderivaten mit erfasst

DOS   2 805 074   10. 2.77/17. 8.77   Ciba Geigy

$$NH_2-CH_2,\ NH_2-CH_2-P(=O)(OH)$$

Wuchsregulator und Herbizid.

DOS   2 813 581   29. 3.78/ 5.10.78   Ciba Geigy
                                                      CH Prior. 1 4. 77

$$HOOC-A-N(R)-CH_2,\ HOOC-A-N(R)-CH_2-P(=O)(OH)$$

A = Alkenyl, zusammen mit R Glieder
    eines gesättigten Ringes
R = Alkyl, Benzyl, Di- Tri-phenyl-
    methyl

DOS   2 848 224   7.11.78/10. 5.79   Meiji Seika
siehe auch US 3 832 394   JA Prior. 8.11.77

$$CH_3,\ HO-P(=O)-CH_2-CH_2-CH(NH_2)-CO-NH-CH(CH_3)-CO-NH-CH(CH_3)-COOH$$

Herbizid aus dem Antibiotikum SF-1293

DOS   2 856 260   27.12.78/ 5. 7.79   Meiji Seika
= GB   2 011 416   JA Prior. 28.12.77; 29.12.77; 30. 3.78

$$(Me^{(+)},H)O-\overset{O}{\overset{\|}{P}}(CH_3)-CH_2-CH_2-CH(NH_2)-CO-O(H, Na\ usw)$$

$$(HN(Alkyl)_3)^{(+)}$$

DOS   2 915 085   12. 4.79/25.10.79   Meiji Seika
                                    JA Prior. 15. 4.78; 4. 7.78

$$\left[CH_3-\overset{\overset{O}{\|}}{\underset{\underset{O^-}{|}}{P}}-CH_2-CH_2-\overset{}{\underset{\underset{(+)NH_3}{|}}{CH}}-CO-\overset{\overset{CH_3}{|}}{CH}-CO-NH-\overset{\overset{CH_3}{|}}{CH}-COO \right]^{(-)} (CH_3)_3\overset{(+)}{N}-CH_2-CH_2-OH$$

$$u. \left[CH_3-\overset{\overset{O}{\|}}{\underset{\underset{O^-}{|}}{P}}-CH_2-CH_2-\overset{}{\underset{\underset{(+)NH_3}{|}}{CH}}-COO \right]^{(-)} (CH_3)_3\overset{(+)}{N}-CH_2-CH_2-OH$$

DOS   2 939 269   28. 9.79/30. 4.80   Meiji Seika
Opt. Trennung von DL-2-Amino-4-Methylphosphino-Buttersäure

$$CH_3-\overset{\overset{O}{\|}}{\underset{\underset{OH}{|}}{P}}-CH_2-CH_2-\overset{}{\underset{\underset{NH_2}{|}}{CH}}-COOH$$

EP   1 018   26. 8.77/ 7. 3.79   Du Pont

$$R-\overset{\overset{O}{\|}}{\underset{\underset{\overset{|}{R^1} Cl}{|}}{P}}-CH-CO-O(S)-CH\overset{\diagup Alkyl}{\diagdown Alkyl}$$

R = Alkyl, Cycloalkoxy, Alkoxy,
    Phenoxy, N< usw.
R^1 = Alkoxy, Alkyl

EP   1 331   26. 8.77/ 4. 4.79   Du Pont

$$(Thienyl, Aralkyl, Alkyl)\,\overset{\overset{O}{\|}}{\underset{\underset{(subst.)Phenyl}{|}}{P}} \!-\!\!-\!\!-\! \overset{\overset{H,Cl}{|}}{\underset{\underset{Cl}{|}}{C}} \!-\!\!-\!\!-\! CO-O(S)Alkyl$$

EP   9 022   18.12.78/19. 3.80   Meiji Seika
                   (JA 19.12.77)

$$CH_3-\overset{\overset{O}{\|}}{\underset{\underset{OH}{|}}{P}}-CH_2-CH_2-\overset{}{\underset{\underset{NH_2}{|}}{CH}}-COOH$$

Fungizid und Herbizid

EP   11 245   8.11.79/28. 5.80   Höchst
= DOS   2 849 003   (DB 11.11.78)

$$\text{Alkyl, Aryl, Aralkyl, Cycloalkyl} \diagdown$$
$$\qquad\qquad\qquad\qquad\qquad \overset{\overset{O(S)}{\|}}{P}-CH_2-CH\overset{\diagup \overset{O(H, Acyl)}{\underset{\underset{H, Alkyl, Phenyl u.s.w}{|}}{C-CN}}}{\diagdown}$$
$$\text{Alkyl, Aryl, Aralkyl, Cycloalkyl-O} \diagup \qquad \underset{H, Alkyl, Benzyl, Phenyl u.s.w.}{|}$$

444

US   4 130 411   5.11.76/19.12.78   M. u. T. Chemicals

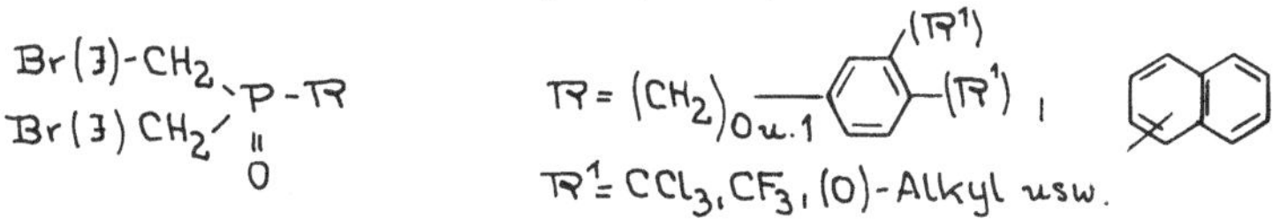

$Br(3)-CH_2$, $Br(3)CH_2$ $-P-TR$, $\|$ O

$TR = (CH_2)_{0\,u.1}$ —⟨ ⟩— (TR^1), (TR^1) ;

$TR^1 = CCl_3, CF_3, (O)-Alkyl$ usw.

US   4 225 521   23. 6.78/30. 9.80   Du Pont

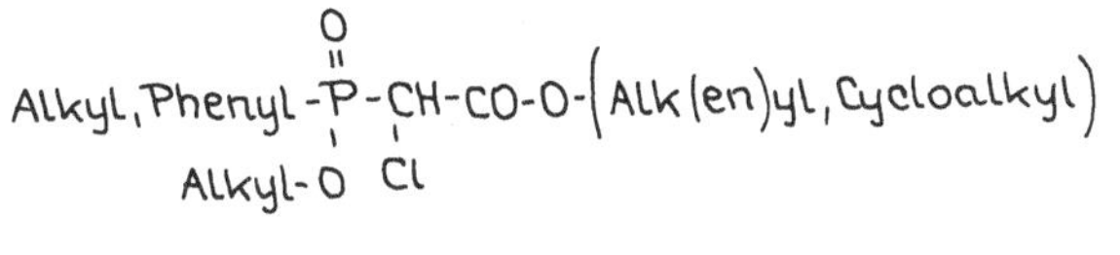

$$Alkyl, Phenyl \overset{O}{\underset{Alkyl-O}{-\overset{\|}{P}-}} \underset{Cl}{CH}-CO-O-\left(Alk(en)yl, Cycloalkyl \right)$$

JA   55 062096   1.11.78/10. 5.80   Nissan

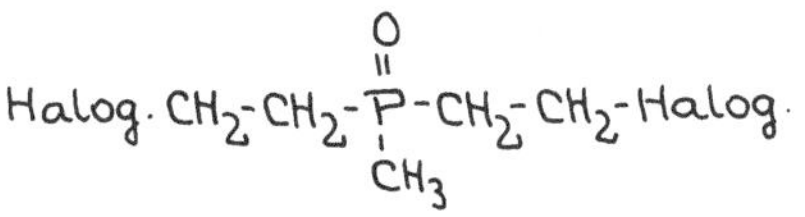

$$Halog.\ CH_2\text{-}CH_2\overset{O}{\underset{CH_3}{-\overset{\|}{P}-}}CH_2\text{-}CH_2\text{-}Halog.$$

JA   55 064595   7.11.78/15, 5.80   Nissan

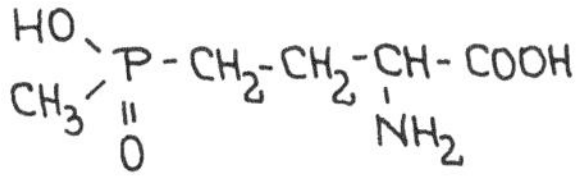

$$\underset{CH_3}{\overset{HO}{\diagdown}}\underset{\|}{P}-CH_2\text{-}CH_2\text{-}\underset{NH_2}{CH}-COOH$$

5.15.5 <u>Phosphoniumverbindungen</u>   Bd.5/711

DOS   2 520 315   7. 5.75/18.11.76   Bayer

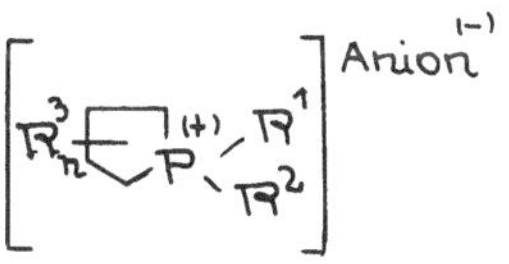

$$\left[R^3_n \boxminus \overset{(+)}{P} \diagup^{R^1}_{\diagdown R^2} \right] Anion^{(-)}$$

R^1 = (Cyclo)Alk(en,in)yl, Aralkyl
R^2 = (Cyclo)Alk(en,in)yl, Aralkyl
R^3 = Halogen, Alkyl

Vorwiegend zur Wuchsdämpfung.

US   4 173 462   16. 2.77/ 6.11.79   Gaf. Corp.

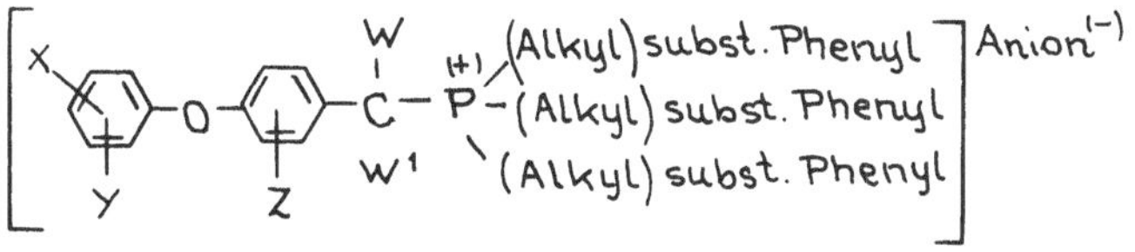

$$\left[X-⟨ ⟩-O-⟨ ⟩-\overset{W}{\underset{W^1}{C}}-\overset{(+)}{P}\diagup \begin{matrix}(Alkyl)subst.Phenyl\\(Alkyl)subst.Phenyl\\(Alkyl)subst.Phenyl\end{matrix} \right] Anion^{(-)}$$

Wuchsregulatoren u. Herbizide

US   4 173 463   31.10.78/ 6.11.79   Shell Oil

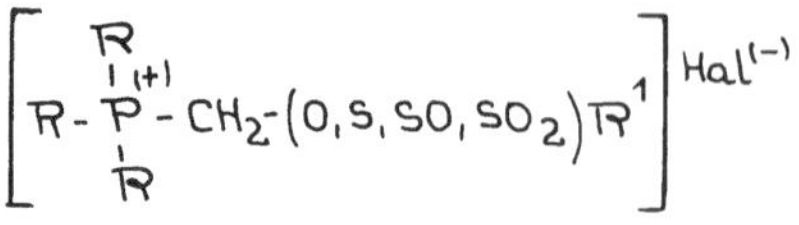

$$\left[\begin{array}{c} R \\ | \\ R-\overset{(+)}{P}-CH_2-(O,S,SO,SO_2)R^1 \\ | \\ R \end{array} \right] Hal^{(-)}$$

post emerg. Herbizid

R^1 = C_1-C_4-Alkyl, C_3-C_5-Alkenyl
Phenyl (subst.)

R = Alkyl, Cycloalkyl,
Phenyl (subst.)

BE   869 943   2. 9.77/26. 2.79   Agence Nat.

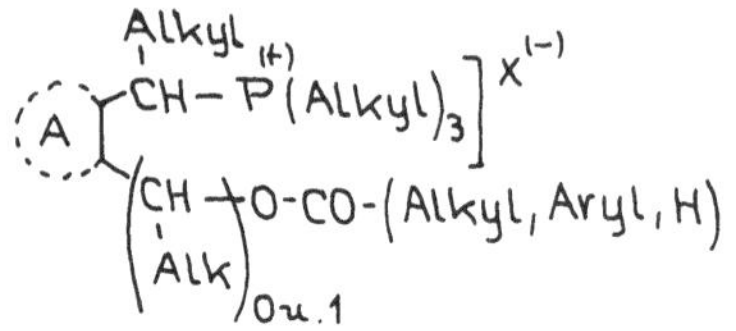

$$\left[A \begin{array}{l} CH-\overset{(+)}{P}(Alkyl)_3 \\ \\ \left(CH-O-CO-(Alkyl, Aryl, H)\right) \\ | \\ Alk \end{array} \right]_{0 u.1} X^{(-)}$$

u.a. auch herbizid
wirksam

SU   562 251   11. 3.76/17. 8.77   AS UZb Veg.

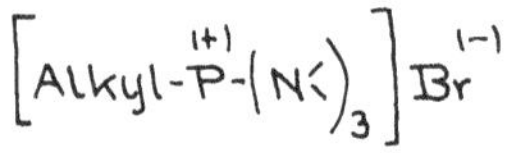

Defoliant für Baumwolle.

$$\left[Alkyl-\overset{(+)}{P}-(N{<})_3 \right] Br^{(-)}$$

5.17.2   <u>Sn-Verbindungen</u>   Bd.5/723

EP   10 143   28. 8.79/30. 4.80   Ciba Geigy
                    (CH   31. 8.78)

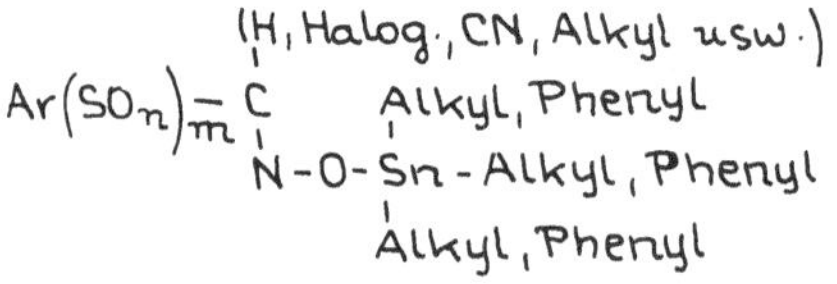

$$Ar(SO_n)_{\overline{m}} \,\, C \begin{array}{l} (H, Halog., CN, Alkyl usw.) \\ Alkyl, Phenyl \\ N-O-Sn-Alkyl, Phenyl \\ | \\ Alkyl, Phenyl \end{array}$$

n = 0 - 2
m = 0 - 1

Safener für Herbizide.

US   3 988 145   10. 2.70/26.10.76   Procter u.Gamble

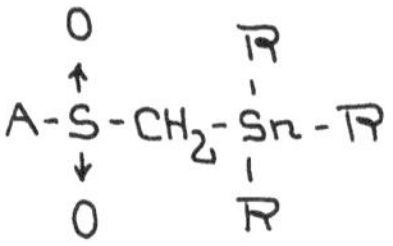

$$\begin{array}{c} O \qquad\quad R \\ \uparrow \qquad\quad | \\ A-S-CH_2-Sn-R \\ \downarrow \qquad\quad | \\ O \qquad\quad R \end{array}$$

A = C_1-C_{19}-Alkyl, Aryl, Ar-O-CH_2-
$N(Alkyl)_2$

R = C_1-C_{19}-Alkyl, Aryl

446

US   4 080 333   5. 5.77/21. 3.78   Chevron Res.

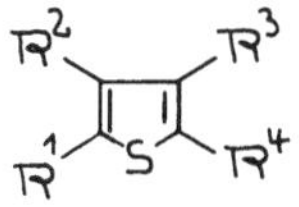

R^1 = H,F,Cl,Br,Alkyl,NO_2
R^2 = H,F,Cl,Br,Alkyl,NO_2
R^3 = H,F,Cl,Br,Alkyl,NO_2
R^4 = H,F,Cl,Br,Alkyl,NO_2

1 x SO_2N — Alkyl, Phenyl

Sn — Alkyl, Alkyl, Alkyl oder Phenyl

CA   999 155   1.12.71/ 2.11.76   Albricht u. Wilson

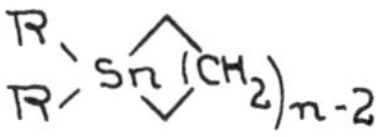

R = Alkyl, Aryl, Alk-Aryl usw.

u.a. auch Herbizide

DDR   142 887   11. 4.79/16. 7.80   Tzschach

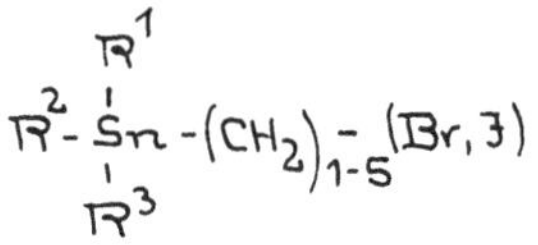

R^1, R^2 u. R^3 = Alkyl, Cycloalkyl, Aryl

Sachverzeichnis

The Handbook of Environmental Chemistry

Editor: O. Hutzinger

This handbook is the first advanced level compendium of environmental chemistry to appear to date. It covers the chemistry and physical behavior of compounds in the environment. Under the editorship of Prof. O. Hutzinger, director of the Laboratory of Environmental and Toxicological Chemistry at the University of Amsterdam, 37 international specialists have contributed of the first three volumes.

For a rapid publication of the material each volume will be divided into two parts. Part A of the first three volumes are now available, Part B will follow in 1981. Each volume contains a subject index.

The Handbook of Environmental Chemistry is a critical and complete outline of our present knowledge in this field and will prove invaluable to environmental scientists, biologists, chemists (biochemists, agricultural and analytical chemists), medical scientists, occupational and environmental hygienists, research geologists, and meteorologists, and industry and administrative bodies.

Springer-Verlag
Berlin
Heidelberg
New York

Volume 1 (in 2 parts)
Part A

The Natural Environment and the Biogeochmical Cycles

With contributions by numerous experts
1980. 54 figures. XV, 258 pages
ISBN 3-540-09688-4

Contents:
The Atmosphere. – The Hydrosphere. – Chemical Oceanography. – Chemical Aspects of Soil. – The Oxygen Cycle. – The Sulfur Cycle. – The Phosphorus Cycle. – Metal Cycles and Biological Methylation. – Natural Organohalogen Compounds. – Subject Index.

Volume 2 (in 2 parts)
Part A

Reactions and Processes

With contributions by numerous experts
1980. 66 figures, 27 tables. XVIII, 307 pages
ISBN 3-540-09689-2

Contents:
Transport and Transformation of Chemicals: A Perspective. – Transport Processes in Air. – Solubility, Partition Coefficients, Volatility, and Evaporation Rates. – Adsorption Processes in Soil. – Sedimentation Processes in the Sea. – Chemical and Photo Oxidation. – Atmospheric Photochemistry. – Photochemistry at Surfaces and Interphases. – Microbial Metabolism. – Plant Uptake, Transport and Metabolism. – Metabolism and Distribution by Aquatic Animals. – Laboratory Microecosystems. – Reaction Types in the Environment. – Subject Index.

Volume 3 (in 2 parts)
Part A

Anthropogenic Compounds

With contributions by numerous experts
1980. 61 figures, 73 tables. XIII, 274 pages
ISBN 3-540-09690-6

Contents:
Mercury. – Cadmium. – Polycyclic Aromatic and Heteroaromatic Hydrocarbons. – Fluorocarbons. – Chlorinated Paraffins. – Chloroaromatic Compounds Containing Oxygen. – Organic Dyes and Pigments. – Inorganic Pigments. – Radioactive Substances. – Subject Index.